ATLAS OF THE SKIES

Published by TAJ Books 2003
27 Ferndown Gardens
Cobham
Surrey
KT11 2BH
UK
www.tajbooks.com

Reprinted 2004
Reprinted 2005

Scientific consultant
Prof. Fabrizio Mazzucconi,
ARCE Astrophysical Observatory, Florence

Graphical design and page layout
Enrico Albisetti

Tabulation and computer graphics
Enrico Albisetti
Bernardo Mannnucci

Stellar maps
Bernardo Mannucci
Based on stellar map models created using
the software program Planetario

Acknowledgements
We thank the following for their valuable collaboration:
The Florence Astronomy Circulation,
who have made the stellar maps available,
consultable also at the web site www.cd.astro.org

ISBN 1-84406-011-x

Printed in China.
2 3 4 5 06 05 04

ATLAS OF THE SKIES

JOURNEYING BETWEEN
THE STARS AND PLANETS
IN THE DISCOVERY OF THE UNIVERSE

TJ

Legends of the Figures on Pages 8-9: astrolabe attributed to Egnazio Danti, cosmographer to Cosimo I dè Medici. The characteristic instrument of the la test generation of astronomical instruments preceding telescopic observation, this astrolabe has a diameter of 84 cm and is made of brass and has "a single drum for the latitude of 45°24' corresponding to the city of Padova". It was also used by Galileo.
CENTRE: the Hubble Space Telescope, one of the modern instruments which has contributed towards revolutionising the means of observing the Universe.
26-27: hurricane Fran approaches the coast of Florida, captured by satellite. The image is computer enhanced.
CENTRE: the Saturn 5 which will launch the Apollo 11 mission towards its lunar rendezvous takes off from the launch pad.
28: the horn of Africa seen from a satellite in orbit.
42: the footprint of Neil Armstrong, the first man on the Moon.
62-63: this image, taken in the visible range by the second wide field planetary camera (WFPC2) of the Hubble Space Telescope, shows a fragment of the Dorado 30 nebula. Through infrared analyses it is observed that in this area of space numerous new stars are forming.
CENTRE: a late seventeenth century Austrian celestial sphere with the patterns of the constellations.
64: map of the constellations in a seventeenth century Czechoslovakian print.
76: the main stars in the constellation of Orion.
120: the movement of the celestial vault on the shoulders of Gemini, one of the most advanced telescopes of the Mauna Kea observatory.
136-137: a panorama of the Sun during the development of an immense protuberance captured by SOHO. The hottest areas are almost white, the coolest have darker red colours.
CENTRE: the Solar and Heliospheric Observatory (SOHO) which has furnished unthinkable raw data on the structure and the solar phenomena.
138: a beautiful image of a ring-shaped protuberance captured by SOHO.
154: photomontage with the planets of the solar system. The Earth excluded. The furthest are the terrestrial planets, the closest are these outlying.
186-187: images of the galactic nebula *NGC3603* captured by Hubble, show the blue supergiant denominated Sher25, a cluster dominated by the "young" and hot Wolf-Rayet stars, and by O type stars, which produce ionising radiation and stellar winds which interact with the surrounding nebular materials, creating, amongst others, the gigantic gaseous pinnacles. Lower, to the right, the more obscure signals are Bok globules, probably the initial stages in the formation of new stars.
CENTRE: collision between the galaxies *NGC6872* and IC4970 observed by Hubble.
188: the Egg nebula (*CRL2688*), 3,000 light years away from us, seen by Hubble in the infrared and reproduced in false colour: two beams of radiation emerge from the star by way of inference of its evolutive course, hidden by the cloud of powder and gas which are expanding at a velocity of 20 km/s.
208: the spiral galaxy NGC1232, in the Eridanus constellation, is found at 20° below the celestial equator, at approx. 100 million light years from earth. It is captured by the European Southern Observatory (ESO) Very Large Telescope.

ICONOGRAPHIC REFERENCES

pp. 23, 121, 195: Nigel Sharp, NOAO/NSO/Kitt Peak FTS/AURA/NSF; p. 123: NOAO/AURA/NSF; p. 125: NOAO/AURA/NSF; p. 141: NOAO/AURA/NSFM; p. 145: T.Rimmele, M.Hanna/NOAO/AURA/NSF; p. 196: NOAO/AURA/NSF; p. 210:C. Howk (JHU), B. Savage (U. Wisconsin), N.A. Sharp (NOAO)/WIYN/NOAO/NSF; pp. 10, 13, 25, 29, 32b, 33, 34b, 34, 36,37,39,40,51-51,52,69,70,72,73,74,75,77, 111, 112, 113, 114, 115, 116, 117, 118, 118-119, 119, 120, 124, 149, 189, 198: Corbis/Grazia Neri and Science Photo Library/Grazia Neri; pp. 8-9, 10a, 11, 12, 14, 15, 16, 17, 18, 19, 20, 21, 22, 23a, 24, 42, 53, 54, 63, 64,70,71,72,73,75,80, 81, 82b, 83, 84, 118c,148bda,155bd, 166bas, 182,183a, 183bs, 189bsb; 192b, 198d, 208d; 208b: Giunti Iconographic Archive; pp. 27, 28, 34c, 42, 43, 44, 45,46, 47,48, 49, 501, 53b, 55, 56, 57, 58, 59, 60, 61, 701a, 76, 126, 127, 128, 129, 130, 131, 132a, 133, 134as, 134b, 1461, 154, 155a, 155c, 155bs, 158, 159, 160, 161, 162, 163, 164, 165, 166, 167, 168, 170bs, 1711, 173,175 b, 177 b, 178, 179as, 179b, 180, 181, 202c: NASA/NSSDC; pp. 26-27: Nasa/Goddard Space Flight Centre; p. 38: Nasa/Goddard Space Flight Centre; p. 28:NOAA; pp. 711a, 1321, 135b: ESA; p. 134: ESA/NASA/JPL/Caltech; pp. 121,136-137, 133, 135a, 136, 137, 138, 139, 141, 142, 143, 144, 145, 146, 147, 148, 149, 152, 153: Solar & Heliospheric Observatory (SOHO). SOHO is a project of international cooperation between ESA and NASA; pp. 120, 122, 123a,123c: ESO; pp. l87, 221cd: ESO (VLT ANTU/UT1+FORS1), p. 200a: ESO (VLT ANTU/UT1+ISAAC); p. 204: ESO (VLT KUEYEN/UT2+FORS2); p. 207b: ESO (VLT KUEYEN+FORS2+FIERA); p. 208as: ESO (VLT ANTU/UT1+FORS1), p. 211a: ESO (VLT ANTU+FORS1), p. 224ad: ESO (VLT KUEYEN/UT2+FORS2); 131b: Edward A. Guinness, Washington University in St. Louis; 165b: Mary A. Dale-Bannister, Washington University in St. Louis; pp. 62-63: NASA, John Trauger (Jet Propulsion Laboratory) and James Westphal (California Institute of Technology); p. 64: Don Figer (Space Telescope Science Institute) and NASA; p. 701b: Steve Lee (Univ. Colorado), Todd Clancy (Space Science Inst., Boulder, CO), Phil James (Univ. Toledo), and NASA; p. 711a: Erich Karkoschka (University of Arizona), and NASA; p. 82: NASA/K.L.Luhman (Harvard-Smithsonian Centre for Astrophysics, Cambridge, Mass.)/G.Schneider, E. Young, G. Rieke, A. Cotera, H. Chen, M. Rieke, R. Thompson (Steward Observatory, University of Arizona, Tucson, Ariz.); pp. 9, 126,127,128,129,130, 131, 133,134b, 154, 155, 158-159, 160-161, 162-163, 164, 165, 166, 167, 168, 171 c, 173, 175 b, 177 b, 178, 179as, 179b, 180as, 180ad, 180b, 217ad: NASA; pp. 162-163, 160c: NASA/USGS; p. 166a: GSFC/NASA; p. 164b: Steve Lee (University of Colorado), Jim Bell (Cornell University), Mike Wolff (Space Science Institute) and NASA, p. 1641: Jim Bell (Cornell University), Justin Maki (JPL); and Mike Wolff (Space Sciences Institute) and NASA; p. 164s: Jim Bell (Cornell University), Justin Maki (JPL), and Mike Wolff (Space Sciences Institute) and NASA; p. 166as: Phil James (Univ. Toledo), Todd Clancy (Space Science Inst., Boulder, CO), Steve Lee (Univ. Colorado), and NASA: p. 169: Ben Zellner (Georgia Southern University), Peter Thorn as (Cornell University) and NASA; 170bd: Reta Beebe (New Mexico State University) and NASA; 171 b: NASA/ESA, John Clarke (University of Michigan); 174a:J. Trauger (JPL) and NASA; p. 174bs: Erich Karkoschka (University of Arizona), and NASA; p. 174bd: Erich Karkoschka (University of Arizona Lunar & Planetary Lab) and NASA; p. 175c: NASA and The Hubble Heritage Team (STScI/AURA). Acknowledgment: R.G. French (Wellesley College), J. Cuzzi (NASA/Ames), L. Dones (SwRI), and J. llssauer (NASA/Ames); p. 177as: Heidi Hammel (Massachusetts Institute of Technology) and NASA; p. 177ad: Erich Karkoschka (University of Arizona) and NASA; p. 179ad: Lawrence Sromovsky (University of Wisconsin-Madison), NASA; p. 181: Dr. R. Albrecht, ESA/ESO Space Telescope European Coordinating Facility; NASA; p. 183bd: NASA/Harold Weaver (The John Hopkins University)/the HSY Comet LINEAR Investigation Team/the Unversity of Hawaii; p. 184: anRodger Thompson, Marcia Rieke and Glenn Schneider (University of Arizona) and NASA; pp. 186-187: Wolfgang Brandner (JPL/IPAC), Eva K. Grebel (Univ. Washington), You-Hua Chu (Univ. Illinois Urbana-Champaign), and NASA; p. 188a: R. Sahai and J. Trauger (JPL), the WFPC2 Science Team and NASA; p. 188b: Yves Grosdidier (Universitie de Montreal and Observatoire de Strasbourg), Anthony Moffat (Universitie de Montreal), Gilles Joncas (Universite Laval), Agnes Acker (Observatoire de Strasbourg), and NASA; p. 189a: C. Burrows and J. Krist (ST Scl) and NASA; p. 189bda: CfA and NASA; p. 190: C.A. Grady (National Optical Astronomy Observatories, NASA Goddard Space Flight Centre), B. Woodgate (NASA Goddard Space Flight Centre), F. Bruhweiler and A. Boggess (Catholic University of America), P. Plait and D. Lindler (ACC, Inc., Goddard Space Flight Centre), M. Clampin (Space Telescope Science Institute), and NASA; p. 191a: Don F. Figer (UCLA) and NASA; p. 191bs: C.A. Grady (National Optical Astronomy Observatories, NASA Goddard Space Flight Centre), B. Woodgate (NASA Goddard Space Flight Centre), F. Bruhweiler and A. Boggess (Catholic University of America), P. Plait and D. Lindler (ACC, Inc., Goddard Space Flight Centre), M. Clampin (Space Telescope Science Institute), and NASA; p. 191bd: The Hubble Research Team, led by Paul Kalas (Space Telescope Science Institute, Baltimore, Md.), consists of John Larwood (Queen Mary and Westfield College, London, United Kingdom), Bradford Smith (University of Hawaii, Institute for Astronomy, Honolulu, Hawaii), and Alfred Schultz (Space Telescope Science Institute); p. 192c: Don Figer (STScI) and NASA; p. 198s: NASA/C:R: O'Dell and S.K. Wong (Rice University); 198d:NASA/K.L.Luhman (Harvard-Smithsonian Centre for Astrophysics, Cambridge, Mass.)/G. Schneider, E. Young, G. Rieke, A. Cotera, H. Chen, M. Rieke, R. Thompson (Steward Observatory, University of Arizona, Tucson, Ariz.); p. 199: A. Caulet (ST-ECF, ESA) and NASA; 200 b: STScI and NASA; p. 201: Chris Burrows (STScI), the WFPC2 Science Team and NASA; 202as: NASA, The Hubble Heritage Team (STScI/AURA); p. 202ac: Hubble Heritage Team (AURA/STScI/NASA); p. 202ad: Matt Bobrowsky (Orbital Sciences Corporation), K. Sahu (STScI) and NASA; p. 202c: Harvey Richer (University of British Columbia, Vancouver, Canada) and NASA; p. 203ad: Bruce Balick (University of Washington), Vincent Icke (Leiden University, The Netherlands), Garrelt Mellema (Stockholm University), and NASA; p. 203bs: NASA/the Hubble Heritage Team (AURA/STScI); p. 203bd:T. Nakajima and S. Kulkarni (Caltech), S. Durrance and D. Golimowski (JHU), NASA; p. 204as: NASA, Peter Challis and Robert Kirshner (Harvard-Smithsonian Centre for Astrophysics), Peter Garnavich (University of Notre Dame), and the SINS Collaboration; p. 204ac: Roeland P. van der Marel (STScI), Frank C. van den Bosch (University of Washington), and NASA; p. 205a:Gary Bower, Richard Green (NOAO), the STIS Instrument Definition Team, and NASA; p. 205ba: Dave Finley (National Radio Astronomy Observator), Bill Junor (University of New Mexico), Space Telescope Science Institute and NASA; p. 205bd, NASA and John Biretta (STScI/JHU); p. 205bc, National Radio Astronomy Observatory/Associated Universities, Inc.; p. 206c: Mike Shara, Bob Williams, and David Zurek (STScI); Roberto Gilmozzi (European Southern Observatory); Dina Prialnik (Tel Aviv University); and NASA; p. 207as: C. Barbieri (Univ. of Padua), and NASA/ESA; p. 207ad: D. Golimowski (Johns Hopkins University), and NASA; p. 207c: NASA and The Hubble Heritage Team (STScI/AURA); p. 209: The Hubble Space Telescope Key Project Team, and NASA; p. 211b: Jeff Hester (Arizona State University)/NASA; p. 212bd: NASA and The Hubble Heritage Team (STScI/AURA); p. 213a: NASA and Jeff Hester (Arizona State University); p. 213b: STScI and NASA; p. 215: NASA, Brian D. Moore, Jeff Hester, Paul Scowen (Arizona State University), Reginald Dufour (Rice University); p. 216: NASA, The Hubble Heritage Team (AURA/STScI); p. 217as: NASA, John Trauger (Jet Propulsion Laboratory) and James Westphal (California Institute of Technology); p. 217b: NASA, Donald Walter (South Carolina State University), Paul Scowen and Brian Moore (Arizona State University); p. 218a: STScI and NASA; p. 218c: Michael Rich, Kenneth Mighell, and James D. Neill (Columbia University), and Wendy Freedman (Carnegie Observatories) and NASA; p. 218b: NASA, ESA, and Martino Romaniello (European Southern Observatory, Germany). Acknowledgments: The image processing for this image was done by Martino Romaniello, Richard Hook, Bob Fosbury and the Hubble European Space Agency Information Centre; p. 219s: Hubble Heritage Team (AURA/STScI/NASA); p. 219d: R. Saffer (Villanova University), D. Zurek (STScI) and NASA; p. 220, Ground-based image: Allen Sandage (Carnegie Observatories), John Bedke (STScI). WFPC2 image: NASA and John Trauger (JPL). NICMOSimage: NASA, ESA, and C. Marcella Carollo (Columbia University); 221ad: Hubble Heritage Team (AURA/STScI/NASA); p. 221cs: NASA, Jayanne English (University of Manitoba), Sally Hunsberger (Pennsylvania State University), Zolt Levay (Space Telescope Science Institute), Sarah Gallagher (Pennsylvania State University), and Jane Charlton (Pennsylvania State University) Science Credit: Sarah Gallagher (Pennsylvania State University), Jane Charlton (Pennsylvania State University), Sally Hunsberger (Pennsylvania State University), Dennis Zaritsky (University of Arizona), and Bradley Whitmore (Space Telescope Science Institute); p. 221b: G. Fritz Benedict, Andrew Howell, Inger Jorgensen, David Chapell (University of Texas), Jeffery Kenney (Yale University), and Beverly J. Smith (CASA, University of Colorado), and NASA; p. 222a: Hubble Heritage Team (AURA/STScI/NASA); p. 222b: Brad Whitmore (ST Scl) and NASA; p. 223a: Allan Sandage (The Observatories of the Carnegie Institution of Washington) and John Bedke (Computer Sciences Corporation and the Space Telescope Science Institute)/NASA/ESA and Reynier Peletier (University of Nottingham, UK); p. 223b: NASA and The Hubble Heritage Team (STScI/AURA); p. 224as. NASA, William C. Keel (University of Alabama, Tuscaloosa); p. 224bs: NASA, Andrew S. Wilson (University of Maryland); Patrick L. Shopbell (Caltech); Chris Simpson (Subaru Telescope); Thaisa Storchi-Bergmann and F. K. B. Barbosa (UFRGS, Brazil); and Martin J. Ward (University of Leicester, U.K.); p. 224bd: John Hutchings (Dominion Astrophysical Observatory), Bruce Woodgate (GSFC/NASA), Mary Beth Kaiser (Johns Hopkins University), Steven Kraemer (Catholic University of America), and the STISTeam. and NASA; p. 225a: Christopher D. Impey (University of Arizona), p. 225b: Karl Gebhardt (University of Michigan), Tod Lauer (NOAO), and NASA; p. 226: R. Williams and the HDF Team (STScI) and NASA; p. 227: NASA, A. Fruchter and the ERO Team, STScI, ST-ECF.

TABLE OF CONTENTS

OBSERVING THE HEAVENS

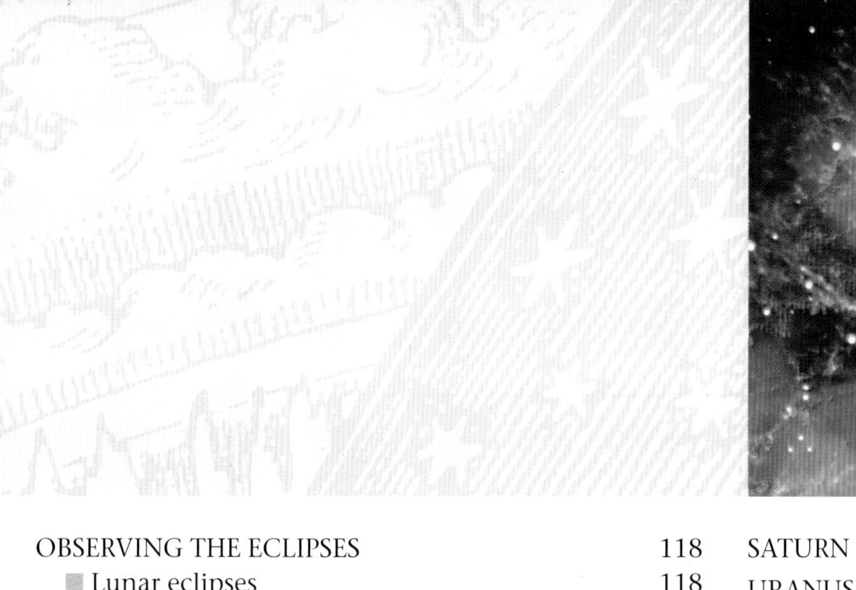

THE SUN AND ITS PLANETS

STARS, GALAXIES AND BEYOND

AVGVSTVS
CAPRICORNVS
AQVARIVS
ARIES
AQVILA·VEL·VVLTVR·VOLANS·CAVDA·DELFINI
CAVDE·CAPRIC·PRECEDENS
SAGITT
RICTVS·PEGASEI
CA·DRACONIS
LIRA
CAPVT·HERCVLIS
TIBIA·AQVARII
HVMERVS·PEGASEI
CRVX·PEGASEI
ANDROMEDA·INASPICE·TRIANGVLI·P·S·PLEAD
CAVDA·CETI
CETI·NARES

A hundred thousand years of astronomy

A LITTLE HISTORY

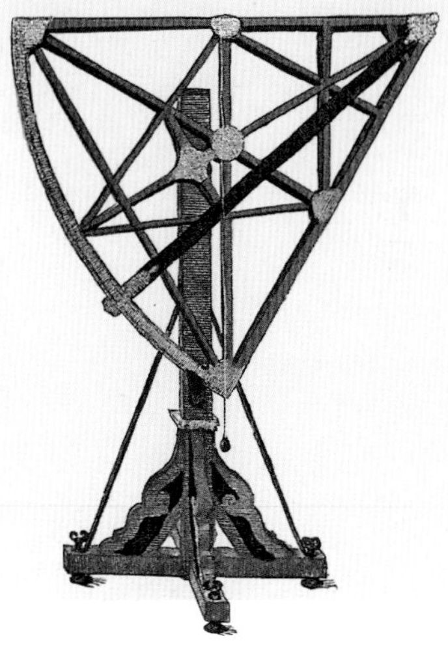

Astronomy, i.e. the study of the heavenly bodies, of their movements, of the phenomena associated with them, is certainly the oldest science: one can say it was born with man, intimately bound to his nature of being a thinker, to his need to measure time, to "organise" the things he understood (or believed he understood), to his need to find direction, to orientate himself in his travels, to organise his agricultural work, more in general, to command nature and the seasons and to plan the future.

It is certainly not by accident that even in the oldest archaeological remains one finds surprising astronomical correspondences.

Stonehenge, the most famous Palaeolithic monument, has been constructed on the basis of precise astronomical understanding; but it is not the only example; an astronomical role was certainly performed from the arrangement of cromlech and Breton monoliths, of English triliths, of Irish stones and barrows, from the medicine wheel of the North American Indians, from the Casa Rinconada of the Anasazi Indians, as is the astronomical-religious importance certain of the Mayan sites of Uaxactun, Copan and Caracol, of the Inca towers of Cuzco or of Machu Picchu and the exquisitely scientific function of the out and out ancient astronomical observatories of the Indians, Arabs, Chinese...

The more archeo-astronomical studies proceed, the more numerous the proof of the astronomical understanding of the peoples of the past becomes, and the further back in time the date must be moved with which, with relative certainty, one can establish the sure participation of man in the principal celestial events. The last clue linking to the study of the sky, the cave paintings of Lascaux: whether this discovery is valid or not, it is however undoubted that the contemplation of the night sky must have provoked marvel, awe, questions since primordial times: what is the nature of the heavenly bodies? Why do they move, and how? Do they have an effect on each other? And above all: do they influence the Earth and the destiny of its inhabitants? Can we foresee these effects, and read the future in the movements of the planets? To similar questions, every culture, in every era, has given their own answers, and frequently they have been answers bound to complex cosmological myths.

Stonehenge, Approx. 2800 B.C.
Whilst the most spectacular part, which is made up of 30 pillars supporting stone lintels so as to form a circle with a diameter of almost 30 metres, 5 metres from the ground is certainly later, the original megalithic edifice is formed by an embankment around 100 metres in diameter, inside of which are excavated 56 holes, over 1 metre deep. Outside, a menhir which exceeded 6 metres in height, indicated to these watching from the centre, the exact direction in which the sun rose on the summer solstice. Next, and nearby the entrance to the building, holes served for the insertion of wooden poles, which in the same way assisted with the location of important standstills in the complicated trajectory of the Moon.

FROM THE FIRST ASTRONOMERS TO ARISTOTLE

The first astronomers in history, writing on clay tablets, have been the Sumerians. But they were certainly not the first to note that some illuminated points of the celestial vault moved with the passing of time whilst others remained still. Today, in the era of science fantasy and special effects, the distinction that they made between "fixed stars" and "wandering stars" (in Greek they say "planets") may seem a banality, but six, seven, eight thousand years ago this discovery must have represented a very significant event for knowledge. To distinguish with the naked eye a planet from a star and recognise it every time that, over the distance of time, it is shown in the night sky, is not as simple as it seems. Try to do it: without knowing anything of astronomy, with the naked eye, try to observe a sky crammed full of stars, like today we see them only if we are on some lost mountain or in the middle of the sea, far away from our dazzling cities. And in the midst of that immensity of shimmering spots, try to distinguish Mars from Jupiter or from Saturn. Lets suppose you succeed. Now try, night after night, to find again that same small moving light, and follow it through its course, to retrace it every time that it reappears after a long absence... In the best case you would need a lot of time and a lot of patience before succeeding in orienting yourself, but it is likely that you would quickly resign yourself to failure. Despite the obvious difficulty, all peoples, even the most ancient, understood well the movements of the stars, so regular that it comes spontaneously to talk of "celestial mechanics" when one begins to use mathematics for describing it. If the Sumerians were the first to measure exactly the planetary movements and to predict lunar eclipses[52] organising a perfect calendar, the best at using imagination to arrive at theoretical explanations, which did not depend solely on mythology were –as usual – the Greeks. In the 6th century B.C., following millennia in which the action of some Deity was sufficient to explain everything, it is here that one seeks a rational logic to the natural order which binds the phenomena: the natural philosophers are the first to assert mans possibility for understanding and describing nature by means of intelligence. A truly innovative idea.

The first gather together at Mileto: Thales, Anassimandro, Anassimene make astronomical observations with the sundial, draw nautical charts, propose hypotheses more or less adherent to the observed facts which relate to the structure of the Earth, the nature of the planets and the stars, the laws followed by the stars in their movements. At Mileto science, as understanding the rational interpretation of observations, takes its first steps. Naturally the majority of humanity continues to believe in gods and spirits. Nothing new: still today, in the era of super-technology, these that give faith to images of saints and horoscopes are many! That notwithstanding, even if these new philosophical attitudes to the world will involve, for centuries, only an elite of thinkers, the rational investigation of nature will not stop. In the 6th century B.C. the Pythagoras school is formed. In a climate of secret sects, Pythagoras and other philosophers believe that the world is held by two opposing principals:

Lascaux, 100,000 years ago
In a cave, illuminated only on the day of the winter solstice, men, only just having become "sapiens", have painted marvellous animals. According to recent interpretations, they would reproduce the constellations visible in the sky at that time. A leap towards the stars, in a cave, dark all year round.

Astronomical sextant
Above, near the title, an image of the Brahe sextant, used to measure the height, in degrees with respect to the horizon, of the celestial bodies.

The Maya in the 4th - 8th centuries A.D
Interested above all in the recurrence of astral phenomena, favoured by the equatorial position of their territory, the Maya developed a zenithal astronomy working out a complex calendar which attributed enormous importance to Venus. In the manuscript reproduced here, are annotated the movements of Venus over a period of 104 years: below, the two dying figures represent the Sun in eclipse[52] (left) and Venus in the invisible phase[180]; the figures in the centre are celestial divinities which kill the two stars.

the finite (i.e. the good, the cosmos, order) and the infinite, the unlimited (i.e. evil, chaos, disorder). Their mathematical studies have a magical and symbolic value: Pythagoras discovers numerical integer ratios behind every formal and musical harmony, and as with music finds the harmony in numbers, thus astronomy becomes the harmony of geometrical shapes. The Earth, therefore, must be spherical, the movements of the stars must follow the laws of perfect numbers, must be circular, perfect shapes. This is a vision, which will influence the way in which we look at the sky, and its phenomena for centuries. Even Aristotle (384-322 B.C.), the "polymath" of antiquity, considered in medieval times, to be the maximum reference point of knowledge, makes this idea of celestial perfection his own. Or better, going further: finds an "explanation" as to why things must be thus. The Earth, place of "the low" where land and water meet (two of the 5 elements which make up the world), cannot be found other than at the centre of the universe; air and fire remain "above", their "natural places". The ether, the fifth element, unknown to us, forms the celestial bodies which, by nature, move in circles, transported by a system of 55 homocentric spheres made of a special crystal, incorruptible and eternal. Around the Earth, immobile, revolve the Moon, Mercury, Venus, the Sun, Mars, Jupiter, Saturn and the last spheres of the fixed stars, kept in motion by love of the "divine immobile motor": it is exactly this sphere which establishes the rhythm of the day and the nightand transmits motion (uniform and circular) to all the

THE FIRST ASTRONOMICAL CALCULATIONS

1. Knowing that, if the Moon is found in the first or the last quarter[61] the angle between the Earth-Moon conjunction line and the Moon-Sun conjunction line is 90°, by measuring in these days, the angle between the Earth-Moon conjunction line and the Earth-Sun conjunction line, one obtains the ratio between the sides of the Sun-Earth-Moon triangle. Aristarch measured an angle of 87° and concluded that the distance of the Sun from the Earth was equal to 19 times the Earth-Moon distance. Today, we know that this angle, in reality, is 89°45', only 2°45' less with respect to that of the real value, but 2 degrees and 45 minutes which moves the Sun much further away: instead of 19 times, it is 389 times further away than the Moon.
2. Convinced of the spherical nature of the Earth, Eratostene

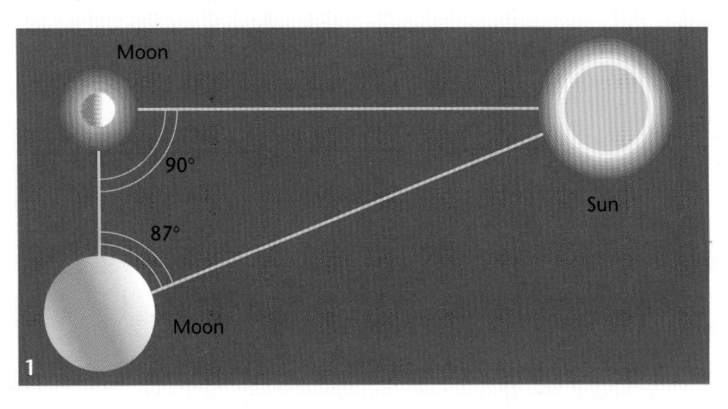

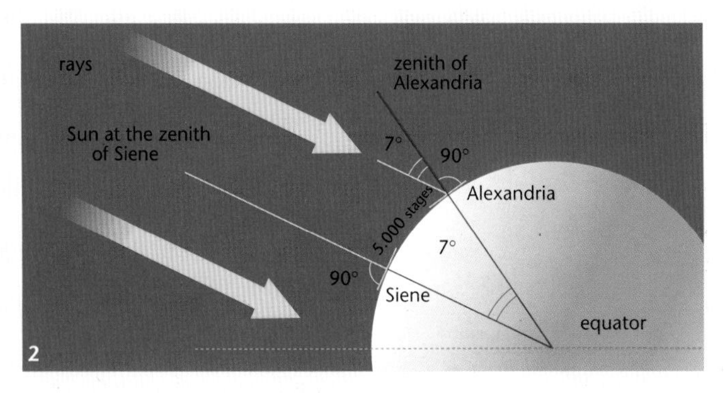

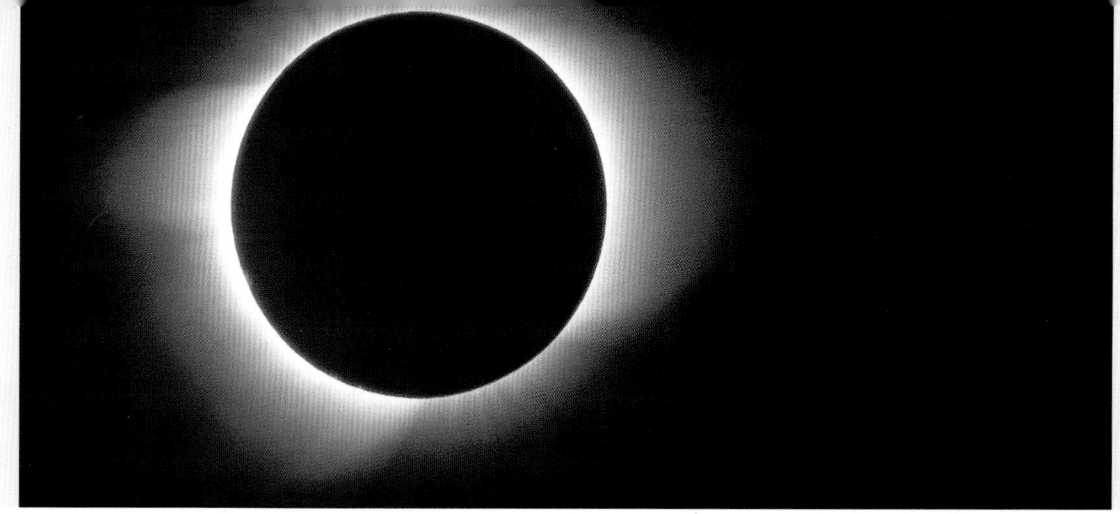

Eclipses of the sun
The visual angle occupied by the Sun and that occupied by the Moon differ by a very small amount. Hipparch's error was that of basing his work on the results of Aristarch: in reality, in fact, the Sun is much further away and - therefore –also much larger than the Moon.

system of spheres. Little by little as one descends from the exterior towards the Earth, the motion diminishes and, below the sphere of the Moon, the movements are rectilinear. Here the continuous remixing of the four fundamental elements gives rise to all the substances we know. It is a description which harmonises mysticism and physics, celestial mechanics and fantasy: it is liked by many, for a long time.

MORE GREEKS

The prestige and the fame which Aristotle went about acquiring even in other spheres (from philosophy to politics, from economy to physics, from metaphysics to natural sciences) contributed to the success of this geocentric idea of the universe. It is in fact without doubt, that already in the 4th century B.C. it was known that, to explain the movements of the stars, at least two types of geocentric systems and one heliocentric system are considered likewise practically valid. In order to have that which is required by governors, agriculturalists or navigators, it is sufficient to be able to "foresee" the celestial phenomena, and to make that point on astral configurations, thus refinding the planets throughout their routes. The hypothesis for the reasons on that which one observes remain philosophical investigations, without the need for concrete proof. Thus, many scholars have their say on the universe, on it's structure and on it's mechanisms: frequently fantasy and nothing more, sometimes lucky intuition.
But there are also these who choose to measure. Aristarch of Samo (310-230 B.C.) is the first true astronomer in history. Not only are his convictions logical and – with hindsight -correct, but he is the first to use mathematical instruments in the investigation of the cosmos. He is convinced that the Earth rotates around the Sun along a circular orbit, and that the Sun is immobile at the centre of the sphere of stars, it also immobile. From the fact that no one succeeds in observing stellar parallax effects, he deduces that the stars must be found at enormous distances from the Earth. And tries to measure the enormity of this space by establishing the Earth-Sun distance as a function of that for the Earth-Moon: it is based on the measurement of

observed how, at the solstice, the height of the Sun at noon appeared differently in the sky over Alexandria in Egypt to that at Siene (now Asswan), two cities which are located almost on the same meridian. Using the sundial, one could establish that the difference in inclination of 7.2° corresponded to a distance of around 5,000 stages. By making a simple proportion, one thus obtains the length of the terrestrial meridian, corresponding to 360°: a length of 250,159 stages, equal to 39,400 km, a measurement exceptionally close to the real value.

Sole
Earth
Moon
r_L
r_T
r_S
distance Moon-Earth
distance Earth-Sun
3

3. Knowing that the Moon takes in the order of an hour to traverse a tract of the sky equal to its own diameter (about 0.5°), if one measures the time in which, in the phase of totality during a total eclipse, it takes to cross the shadow of the Earth, one deduces the lunar radius. Since the distance at which an object must be located in order to occupy a visual angle of approx. 0.5° is around 120 times it's size, the earth-Moon distance is approx 1/4 of the terrestrial diameter x 120 = 30 terrestrial diameters (or 60 terrestrial radii). As had happened to Aristarch, also Hipparch miscalculated the measurement of the angles: that of the lunar parallax by only 3 minutes (53' against the real 57'). That brings an increase of 7% to the Earth-Moon distance. Instead, in the case of the Sun, assuming the Earth-Moon distance found by Aristarch to be good and, considering that in the eclipse both the Moon and the Sun occupy the same visual angle.

Ptolemy
In this manuscript from the 16th century, Ptolemy establishes the position of the Moon with respect to the starry vault. At his shoulder, the Goddess of Astronomy guides him. There are still 100 years before the beginning of the Copernican revolution.

Ptolemy at work
Here Ptolemy is represented in a formella created by Andrea Pisano for Giotto's bell tower of the Florence cathedral (1337-1348). Also in this case the validity of the conclusions of the scientist and the acceptance of his work are underlined by the secondary figure: Jesus, at the centre of a theory of angels, faces the celestial sphere made evident by the symbols which, in the background, represent the signs of the zodiac. Ptolemy, from his studio open to the heavens, observes and works, blessed by God.

Ptolemaic universe
At the centre of the universe, the Earth is found in the sublunar world where the four elements form all that exists. Follow the sphere of the planets made from the ether: the Moon, Mercury, Venus, the Sun, Mars, Jupiter and Saturn. Finally the sphere of the fixed stars, where one reads the names of the zodiacal constellations. The reference to Aristotelian concepts is evident.

angles and on the simple calculations of geometry. Concluding that the Moon is located at 30 terrestrial diameters from our planet and that the Sun is 19 times further away (570 terrestrial diameters). We know that they are all values more or less mistaken due to the slight inaccuracy of the measurements "by eye", but this discrepancy takes nothing away from the conceptual and philosophical importance of this approach to the problem: it is the first time in history that someone tries to increase their understanding of the universe experimentally, using logic, the mathematical and geometrical laws which they know, making observations and measurements. A decisively modern approach to a complex mathematical problem. Eratostene of Cirene (276-194 B.C.) also works in the same way, who, with a simple but genial mathematical calculation, finds the real dimensions of our planet: the terrestrial meridian corresponds to 250,159 stages, approx. 39,400 km (a value surprisingly close to the average value known today of 40,009km).

Also Hipparch (188-125 B.C.) performs careful and intelligent observations: compiling a catalogue of 1,080 stellar positions and, comparing his observations with these carried out 154 years previously by Timocaris, discovers the precession of the equinoxes 40 and quantifies this very slow slide of the path of the Sun with respect to the equator, in approx. 47" per year (a value, even this, very close to that calculated today: 50.1"). The Earth, hence, is immense, and the Sun must be even more so. Space, then, assumes incalculable dimensions: only a few manage to accept this revolutionary assertion. Perhaps exactly for this reason, after Hipparch nothing more happens for over 300 years: it is simpler to accept as valid the thesis of the great Aristotle, and not seek anything beyond.

FROM PTOLEMY TO COPERNICUS

With the passing of time, however, with the various observations which go on accumulating, even the ideas of Aristotle begin to creak. The planets observed from the Earth have inexplicable movements with respect to the celestial sphere: they slow down, reverse their routes, stop, turn to move in the "correct" direction, plotting, sometimes, true and proper rings. The perfect Aristotelian model is revised. Thought Claudio Ptolemy (ca. 100 - ca. 170 A.D.): the Earth is spherical, fixed in the centre of the universe, the sky, also spherical, rotates around a fixed axis moved by an external sphere devoid of stars, all as Aristotle said. To explain the precession of the equinoxes and the "strange" movements of the planets it is sufficient to add other spheres or, as Apollonius indicated in approx. 200 years B.C., add new circles of rotation: deferential eccentrics, epicycles, epicycles of epicycles. The space around the Earth is filled with gears. Also, to Ptolemy, there is concern that the geocentric model of the universe corresponds to a physical reality: he defines his complicated system as a "useful mathematical instrument" to calculate the planetary positions. It is curious that this is the same definition that will be used to spread the opposing hypothesis without

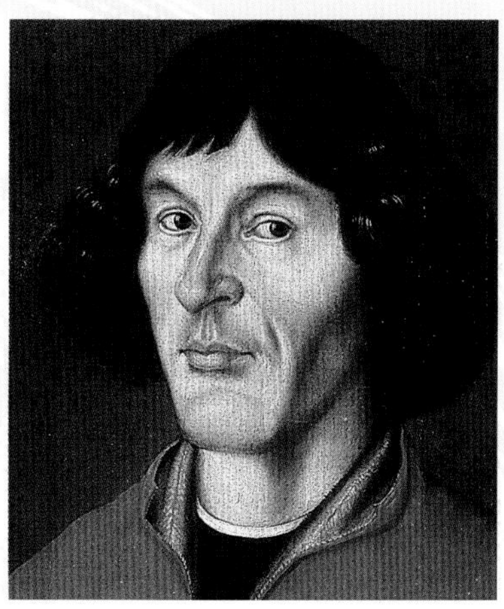

Representation of Copernicus
Calm, tranquil and respected man of the church, Copernicus has an obvious aversion to facing the judgement of the learned, to risking also only true tranquillity in defence of his ideas. Some, comparing him to Giordano Bruno who, inspired by the Copernican works, did not hesitate to end up on the stake for affirming his convictions on the infinity of the universe and of the inhabited words, have branded him a coward. But the discretion of Copernicus is due, very probably, to his liking for the Pythagorean philosophy, according to which, discoveries are shared with a tight group of the elite. Why talk to who does not want to hear or is not capable of understanding? To what end risk suffering due to incompetents? In short, even Copernicus had his reasons and his choices, in a certain sense, it is the same which Galileo will do when, in the end, he finds himself bound to silence if he wants to continue working.

provoking the criticisms of ecclesiastics and traditionalists. It is also strange that Ptolemy prefers to perfect the model of Aristotle making it even more complex rather than adopting the simple and innovative model of Aristarch: if he only had need of a mathematical instrument, that of Aristarch was certainly easier to use. He would have changed history: after Ptolemy, in fact, the recollection of the heliocentric hypothesis is also lost and, although his is "only a mathematical hypothesis", for over 1,300 years everyone thinks that our planet really is fixed, at the centre of a universe moved by very complicated circles. Instead he writes Mathematikè synthaxis (mathematical synthesis), renamed by the Arabs al-Magisti, perhaps from the Greek e meghistè: "the greatest", born in medieval times as Almagesto. It is a monumental work where Ptolemy reorganises all the astronomy of the past: it is thanks to his immense work that we know a large part of what was produced in the preceding centuries. Summarising and perfecting the ideas of Apollonius and Hipparch, integrating their calculations with the results of his own research, he elaborates a theoretical system which adapts well to the observations. "His" universe is pushed by 40 wheels which move in unison: like an immense mechanical clock which accumulates, over time, only very small differences, it is sufficient to update the situation from time to time, to continue to keep it working. Only a great mathematician could have constructed a work so vast and complex, and it is certainly for this that it has lasted with time and that, over the centuries, the geocentric system has always been known as the "Ptolemaic system". After Ptolemy, in fact, to have a different idea of the universe is almost impossible: The Almagesto is so complex that simplifying it would mean having the wrong results. Furthermore the Ptolemaic hypothesis is very much liked by the Christians, who are becoming ever more powerful: it is logical that the planet created by God expressly for man be found at the centre of the universe. From a mathematical instrument, Ptolemy's hypothesis becomes dogma and to have different ideas becomes dangerous. One must wait until another intellect of the capacity of Ptolemy overturns the perspective and destroys cycles, epicycles and deferents simplifying the panorama. We must wait until a great astronomer gathers an enormous mass of very accurate data, and that a great mathematician, with a mind uncluttered by prejudices, elaborates them, finding the unbiased proof of the validity of the new hypotheses. One must wait until another astronomer of sufficiently brave genius, asserts this new idea to the attention of the scientific world, challenging the ecclesiastical authority, revolutionising the way of looking at nature. One must wait over a thousand years, while Copernicus, Brahe, Kepler and above all Galileo, do not revolutionise astronomy.

THE ROCK IN THE POOL: COPERNICUS

Finally man adventures beyond the pillars of Hercules and discovers that the Earth, considered

The Copernican universe
In this print one sees the Sun at the centre; in the sublunar globe the orbits of Mercury and Venus; the Earth at the limit of the sublunar globe and, for the first time, the Moon which rotates around it. Followed by the orbits of Mars, of Jupiter and Saturn. As with Ptolemy, the sphere of the fixed stars, recognisable by the zodiacal constellations, surrounds and encloses the universe.

PORTRAITS OF TYCHO BRAHE AND JOHANNES KEPLER
In the style of the court painters of the era, these two portraits show Tycho Brahe (left) and Johannes Kepler (right) at the time of their work as court mathematicians.

THE TYCHO OBSERVATORY
External view of the Stjoerneiborg observatory realised by Brahe. The instruments and domes are the only visible elements, all the rest are underground.

ARMILLARE SPHERE
Drawing of an Armillare sphere feature from the Codice Settala preserved in the Biblioteca Ambrosiana, Milan. The sphere shows the universe and the celestial movements according to the Ptolemaic concept of the solar system.

flat despite Ptolemy, is in reality a boundless ball, all to be discovered. Finally the use of the printing press spreads: around the mid-fifteenth century, the doors are opened to the discovery of the world and to never seen skies, as with the circulation of ideas. The times are ready for radical changes. The first doctrine to be affected by the novelties is astronomy: for those who travel under foreign skies the Ptolemaic model is restrictive, and to "make the point" geographically, tables are required on the motions of the planets more precise than these available. Even the calendar is revised: that in use is still the calendar of Julius Caesar! Society has need of something new and the attempts to save the Ptolemaic system by adding new deferents and epicycles transfer the universe into a tangle of rotating circles. It is to this point that Nicholas Copernicus (the Anglicisation of the Polish name of Nikolas Koppernigk, 1473-1543) launches his message of renewal. He refuses that which he has learned, denies that philosophers, scientists, theologians have, over the centuries, recounted the truth, denies that that which appears as self evident to everyone – i.e. that it is the Sun that rises, moves across the sky and sets; corresponds to the true nature of things. Removing the sons of God from the centre of the universe, in a period in which one could end up being burned at the stake for less, has the courage to declare that the planet of men is n o more than one of the many which rotate around the Sun.

But his ideal is that of the Pythagorean school: it is better to communicate ones own ideas in a whisper and to only a few of the initiated. Thus is his work, purely theoretical, pursued in silence: Copernicus makes few direct observations, he trusts the data of the ancients of which he reads the originals and collects criticisms and doubts on the Ptolemaic system. As he writes in De revolutionibus orbium coelestium (on the revolution of the celestial spheres), it is exactly the variety of opinions, uncertainties and incongruencies which he finds, that convince him that in the Ptolemaic theory there is something that does not fit. As with Ptolemy, however, also his construction is exquisitely mathematical and his thought essentially Aristotelian. It's true, the Sun is at the centre and around it rotate the planets, but all the rest remains as before: the perfectly circular orbits, the "natural motion" of the Earth not subjected to forces; Earth, Sun and universe, spherical because "this shape is the most perfect of all, a total entirety attributing itself to the divine bodies"... But there is something radically new: Copernicus, against all evidence, believes that the movement of the Earth is real and that the astronomical geometry describes the true workings of the celestial machine.

To work out his heliocentric system takes around 25 years, during which time, love for secrecy and the fear of the "bites of the slanderers" prevail. At sixty three years of age he has still published nothing, but the word about his work has spread. In 1539, Retico, a young Lutheran professor of the University of Wittenberg, studies the manuscript of De revolutionibus and extracts from Copernicus the authorisation to write a summary. Published in 1540, it has an immediate and great success. Hailed as the new Ptolemy, finally Copernicus broadcasts his work. And in 1542: will die before seeing the effects. It will remain for the preface – not his – where it is declared that the theory described is only an opinion amongst many, it will be for the optimal relations which Copernicus has always had with the ecclesiastical authority, the fact that the book is not banned by the church until 1616. There is a reaction, obviously, but it remains within the academic circle. For things to change for real, more time must pass.

KEPLER AND BRAHE: THE MATHEMATICAL PROOFS

The German, Johannes Kepler (1571-1630) studies the mathematics and astronomy of ancient texts, writes in Latin and makes astrological forecasts of meteorology and agriculture for which he becomes famous. Religious, mystic, he sees in astrology an essential element for interpreting the mysterious bond between man and the cosmos: he is

convinced that in every phenomenon one can find a purpose, a higher order, geometrical harmony. A convinced Copernican, already in Mistero cosmografico he explains how some observations which Ptolemy did not succeed in clarifying find a simple solution in the system proposed by Copernicus. But the questions he poses are inspired by his research into harmony, and the explanations that he gives re-enter into a complex picture where astrology, symbolism, religion and the need for geometrical and mathematical perfection play essential roles. Copernican spheres, vertices, faces and sides, perfectly interposed in planetary orbits, the same orbits and their mathematical relationships... Kepler places everything into relationships, everything is connected by a single harmony: in the making the world, God had certainly followed mathematical and geometrical laws, and the Copernican theory adapts itself to this plan. The opportunity of finding the numerical proofs for this idea arrives when Tycho Brahe makes available his own immense archive of observations. The Dane, Tycho Brahe (1546-1061) is obsessed by precision and, making more exact observations, personally builds new instruments. His fame exceeds the borders of his country, and the King of Denmark offers him the island of Hveen to build a true and proper observatory. Uraniborg, the first observatory in Europe, is futuristic: towers, domes, pendulums, solar quadrants, celestial globes, a mural quadrant of over 4 and a half metres in diameter, a bronze celestial globe of 1.6 m in diameter and, in the basement, the workshops to construct the instruments, the alchemy laboratory, printshop, paper mill and the prison to punish the menials... It is not enough: he also builds a second observatory, Stjoernerborg, where everything is underground except the dome.

Brahe passes his life accumulating data, measurements, observations; uses new methods of tracking and contains the errors to between l'-2', an exceptional result if one thinks that up to now no one has ever made errors of less than 8'-10'. For years, day after day, he notes every celestial phenomenon: the positions of the stars, Sun and planets, the distance and motion of the comets, observes the explosion of a nova[206] in 1572 which takes a little over a year to disappear. And he understands that the hypotheses of Ptolemy are not enough to explain what he sees.

Brahe is not a Copernican: the Earth "serious and lazy" cannot move, it would be against all physical and religious evidence. But he is not Ptolemaic: also the stars are not immovable, and the comets have an orbit not "exactly circular but a little elongated, such as the shape commonly called oval" which breaks the crystal sphere. Thus he proposes his idea of the universe which puts in agreement the phenomena and the writings, and which everyone likes: physicists, philosophers, roman and protestant theologians. It is of little importance if his universe is even more complicated than that of Ptolemy: to ensure that the positions of the planets correspond to the previous. It is at this point that Kepler and Brahe meet. Following a brief troubled period of collaboration, Brahe dies, leaving to Kepler his position as Imperial mathematician and all the collected data: the best ever in existence. Working on them, Kepler discovers that it is

Brahe at work
This ancient print shows Tycho in his Uraniborg observatory, of wehich one recognises the semi-underground structure. Outside, celestial globes, quadrants and various instruments are used by assistants and other scholars whilst Tycho, in the centre, indicates the direction to the observatory to the right, which measures the height of a star aiming directly from the ocular of the great mural quadrant. Below to the left, behind the dog, one can make out a part of the alchemy laboratory.

KEPLER'S THREE LAWS
1. Kepler's first law: the planets follow elliptical orbits around the Sun which occupies one of the foci.
2. Kepler's second law: the Sun-planet vector ray sweeps across equal areas in equal time (i.e. areas proportional to the considered times). The velocity of the planets therefore increases with their approaching the Sun and diminishes when they move away.
3. Kepler's third law: the squares of the periods of revolution of the planets remain proportional to the cubes of the larger semiaxes of their orbits. This allows one to express all the distances of the planets from the Sun on the basis of just one of them
allowing the construction

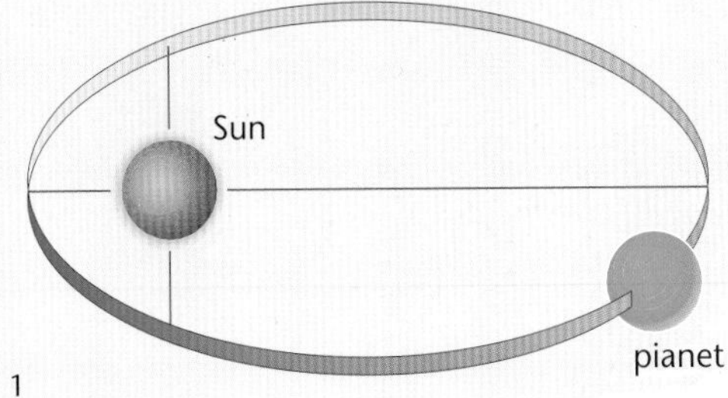

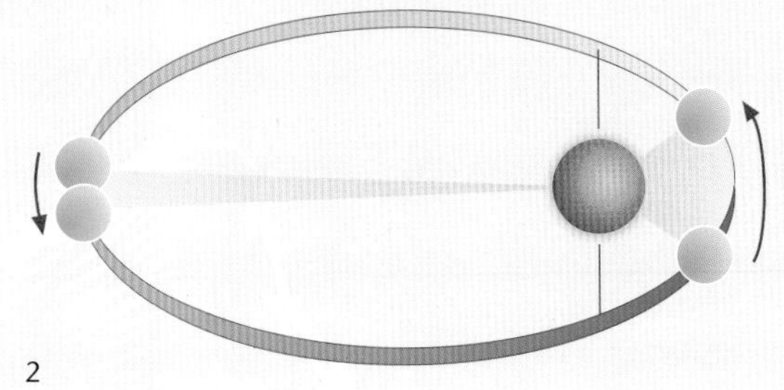

SIDEREVS
NVNCIVS
MAGNA, LONGEQVE ADMIRABILIA
Spectacula pandens, suspiciendaque proponens
vnicuique, præsertim verò
PHILOSOPHIS, atq; ASTRONOMIS, quæ à
GALILEO GALILEO
PATRITIO FLORENTINO
Patauini Gymnasij Publico Mathematico
PERSPICILLI
QVATVOR PLANETIS
MEDICEA SIDERA
NVNCVPANDOS DECREVIT.
VENETIIS, Apud Thomam Baglionum. M DC X.
Superiorum Permissu, & Priuilegio.

TITLE PAGE OF THE SIDERUS NUNCIUS
The 24 pages of this "messenger of the stars" are rich in revolutionary news: studies on the movements of Venus, on the conformation of the Moon and the rings of Saturn, on Sun spots.

mathematically impossible that the Sun remains at the centre of the solar system, and he is convinced that from it must propagate a mysterious force which acts on all the planets. He also finds that, strangely, the planets are moved through their orbits with variable velocity: The hypothesis that the orbits are ellipses proceeds, but the repulsion for the "imperfect shapes" is such that he prefers to keep it as a "substitute hypothesis" and to recalculate it all from the beginning starting from the data relating to the Earth. But he discovers that even the Earth rotates at non constant velocity around the Sun. The truth is in the counting.
It is from his examination of the numbers, comes the intuition: the velocity varies because the distance from the Sun varies, and with that varies the forces to which they are subjected. It is "Kepler's second law". But the orbit of Mars still presents problems: Kepler repeats the observations, measurements, calculations. This time starting from raw data to decide which type of orbit best fits the observations. It is the decisive turn: he notices that every problem disappears if one only considers the elliptical shaped orbit, with the Sun in one of the foci. It is "Keplers first law", which "puts in place" all the observations. He will discover almost by chance the "third law": whilst preparing a universal summary which harmonises science, religion, astrology, art, philosophy, geometry and music, he notices the ratios which relate between them the squares and cubes of the planetary distances. Antique astronomy is outdone. It has taken 5 years of work, repeating the calculations 70 times but, for the first time in history, a proposed model is not just a theoretical hypothesis: it is a representation of the real universe. Thus, as had already happened to the work of Copernicus, also that of Kepler is put in the Index.

GALILEO, PALADIN OF THE REVOLUTION

Copernicus, Brahe, Kepler, with their revolutionary innovations barely scratch the popular Ptolemaic tradition, both because they write in Latin to the exclusive advantage of colleagues, and because they limit the exposure of their hypotheses without the pretence of imposing them onto contemporaries. The doubts are many: although the new model is sustained by solid data, if it was really the earth that moved, all that which is found on its surface would have to fly away! It is one thing, however, to make models, and another to claim that things are really thus. But behold Galileo Galilei (1564-1642), shrewd spirit and nonconformist, proud, ironic, controversial, literary man and musician, lover of debate, great worker and perfect artisan, builder and inventor of new instruments and experiments, comes to cast the foundations of modern physics and, conceiving scientific method, opens the doors to the modern era. Initially he is interested in magnets, thermometers,motion and mechanics deducing the laws and asserting that bodies all tend to descend through the effect of gravity. He believes that the movements of the planets are "natural", uniform and circular, challenging Kepler, who has sent him his Mistero cosmogmfico, the blind faith in the data of Brahe: who has built instruments and done experiments, knows well how the exactness of the measurements are disputable! He is convinced that the truth can be understood only through "ideal" experiments,

COPERNICAN SYSTEM
This small model of the Copernican system designed by Galileo and published in his Dialogo sopra i massimi sistemi (1632) shows the Medicei satellites around Jupiter and the Moon rotating around the Earth. Discovered on the 7th of January 1610, the satellites of Jupiter constitute for Galileo irrefutable proof of the exactness of the thesis of Copernicus.

of a scale model of the solar system.
These empirical laws find an explanation in the laws of gravitation elaborated by Newton which, vice versa, implicate the three Kepler laws.

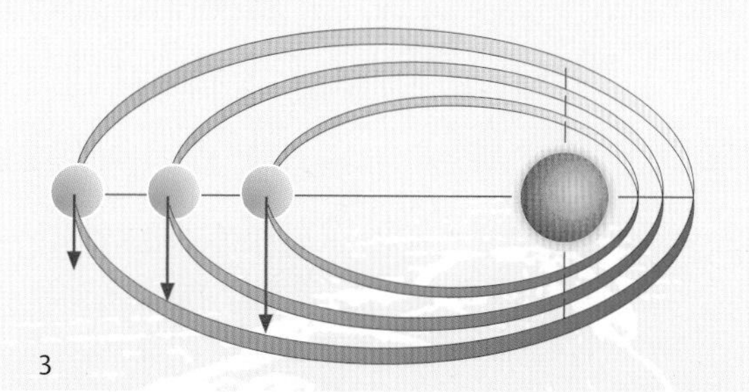

Portrait of Galileo
This famous portrait of Galileo housed in the Uffizi has been painted by J. Susterman.

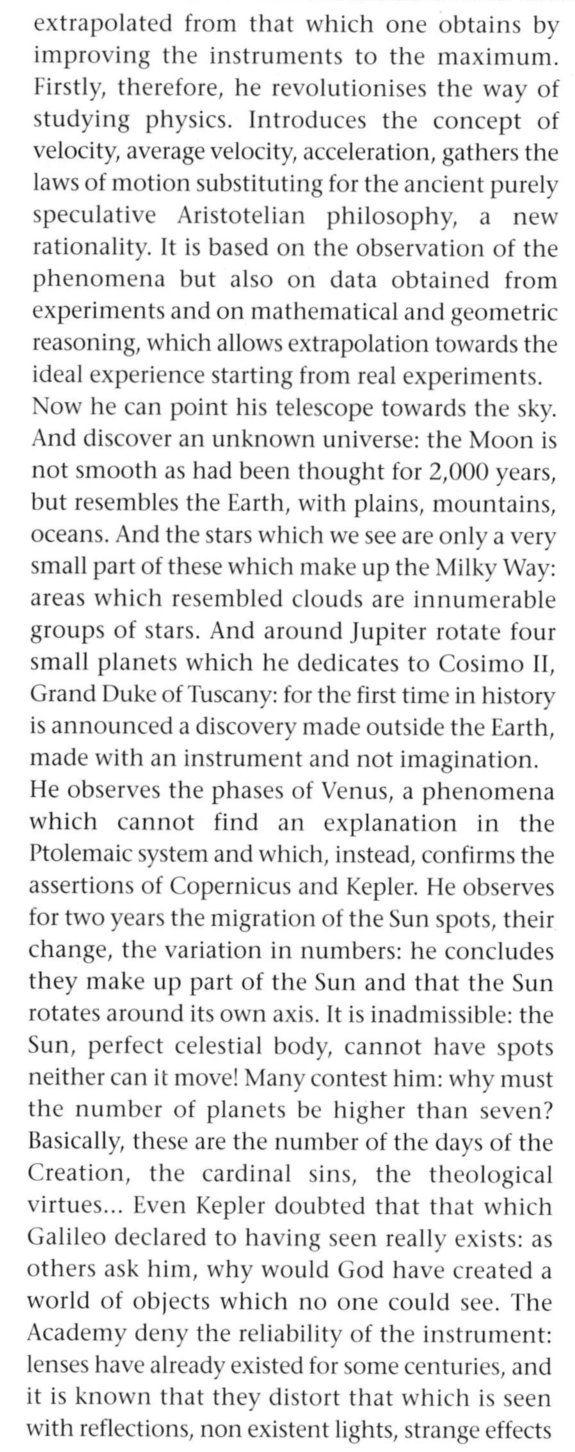

extrapolated from that which one obtains by improving the instruments to the maximum. Firstly, therefore, he revolutionises the way of studying physics. Introduces the concept of velocity, average velocity, acceleration, gathers the laws of motion substituting for the ancient purely speculative Aristotelian philosophy, a new rationality. It is based on the observation of the phenomena but also on data obtained from experiments and on mathematical and geometric reasoning, which allows extrapolation towards the ideal experience starting from real experiments.
Now he can point his telescope towards the sky. And discover an unknown universe: the Moon is not smooth as had been thought for 2,000 years, but resembles the Earth, with plains, mountains, oceans. And the stars which we see are only a very small part of these which make up the Milky Way: areas which resembled clouds are innumerable groups of stars. And around Jupiter rotate four small planets which he dedicates to Cosimo II, Grand Duke of Tuscany: for the first time in history is announced a discovery made outside the Earth, made with an instrument and not imagination.
He observes the phases of Venus, a phenomena which cannot find an explanation in the Ptolemaic system and which, instead, confirms the assertions of Copernicus and Kepler. He observes for two years the migration of the Sun spots, their change, the variation in numbers: he concludes they make up part of the Sun and that the Sun rotates around its own axis. It is inadmissible: the Sun, perfect celestial body, cannot have spots neither can it move! Many contest him: why must the number of planets be higher than seven? Basically, these are the number of the days of the Creation, the cardinal sins, the theological virtues... Even Kepler doubted that that which Galileo declared to having seen really exists: as others ask him, why would God have created a world of objects which no one could see. The Academy deny the reliability of the instrument: lenses have already existed for some centuries, and it is known that they distort that which is seen with reflections, non existent lights, strange effects and optical illusions... who could believe that which they see through an entire series of these objects? But Galileo is convinced of being right: he builds dozens of telescopes and gives them to friends, scholars and Princes throughout Europe. Kepler can observe the same things which Galileo has seen: he becomes enthusiastic and in a few months publishes Diottrica, a treatise on the geometrical theory of lenses which explains the functioning of the Galilean telescope and the principles of the objective lens. It is the spring of 1611: following millennia of darkness, the two geniuses have turned the spotlights on space. The telescopes become, in every effect, an extension of the view. But the attitude of Galileo is mistaken: certain of that which he has seen and of his own conclusions, he purports to know more than Aristotle and to all these following in the preceding 2000 years, sustaining that his scientific method is the only type of valid research. Not only: in his opinion, the discords with the Scriptures are only due to errors in interpretation because that which the scientific discoveries show is the work of God, and God cannot contradict himself. It is decidedly an act of war in the opinion of traditionalists - in the overall majority – and above all of the Church. The academic world and ecclesiastical powers, discovering the destructive powers of these affirmations, try to make him stay silent: they prohibit him from teaching and order him to sustain Copernican theory. And Galileo remains silent for a while. But it does not last:
In 1623 he dedicates to his friend and admirer Maffeo Barberini, who became Pope with the name of Urbano VIII, Il Saggiatore, a work, the first in vernacular, which became a milestone of modern science. Here it states that nature is studied with humility, astuteness and imagination, observing and posing questions, always heedful to distinguish reality and appearance, objectivity and subjectivity. Mathematics, geometry, rational thinking are the only means which we have to extrapolate from the imperfect reality, the ideal laws which regulate the created. It is the new

New instruments
Two telescopes used by Galileo for his revolutionary observations of the sky. In the centre, framed, the lens made by Galileo in person.

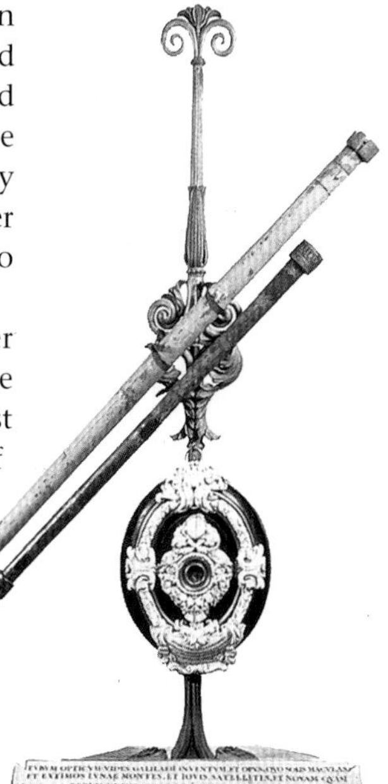

Blind Galileo
Galileo, in the last days of his life, by now blind and helped only by his most attached disciples, lives in the villa "il Gioiello", near to Arcetri, a few kilometres from Florence, assisted by his daughter Sister Mary Celeste. His work and his image have been redeemed by the Church only in 1992, 350 years after his death.

philosophy of understanding.
A short time after follows "Dialogue over the two maximum systems of the tolemaic and copernic" world wherein, refinding courage, Galileo makes two very grave errors. First of all he asserts that the tides are due to the rotation of the Earth: a topic which he was expressly forbidden to deal with. But that which is worse, derides the position of the Pope, which had been very clear: God almighty can make whatever He wants happen, and the phenomena can be verified in a thousand different ways; the observation of natural events, therefore, cannot lead to the understanding of truth. And if so the Pope says that...
In clear vernacular, Simplicio, who incarnates the obtuse Aristotelian mentality and worthy curatrix of every scorn, declares that, although the hypothesis of the rotation of the Earth to explain the tides seems to be the best, it is refused in favour of "a very solid doctrine, which already gives very able and eminently learned people, and to which it is quietened, strength". At the same time, Salviati, who gives voice to the Galilean conviction, repeats that man can reach an understanding of created things equal to that of God: "of these few understandings of the human intellect, I believe that "knowledge equals divine" in the objective certainty, since it arrives at understanding the necessity, above which there cannot be greater certainty". Exactly the opposite of that stated by the Pope. The many that Galileo had openly ridiculed and insulted in the past take an easy revenge: the book is an attack on the Church, to its authority over science, to its infallibility, and what is more is in Italian, and gives to anyone who knows how to read, access to these subversive and diabolical ideas! The condemnation must be exemplary. Galileo risks the stake, where Giordano Bruno ended up a short time ago. And, fortunately, is scared. Or perhaps he understands that against the crazy there is no way to win, or perhaps he notices not being able to rely on the Grand Ducal protectors. Perhaps he thinks that, however, that which is done is done: has other ideas which he wants to finish, for which it is worthwhile bowing his head. He submits, then, shows humility, regrets. Asks understanding for his "decrepit age". That notwithstanding he is judged "vehemently suspected of heresy", obliged to publicly disavow: "curse and detest the aforementioned errors and heresies". He is condemned – as we say today –to house arrest. His work is placed in the Index, together with that of Copernicus and Kepler. Defeated, in 1637 he also goes blind. But continues to work, to find supporting elements of his new method, to deny Aristotelian physics based on the imagination. Despite poor health, the work which he carries out in his final home is his utmost contribution to physics. Discorsi e dimostrazioni matematiche intorno a due nuove scienze attinenti alla meccanica ed i movimenti locali (Discourses and mathematical demonstrations around two new sciences pertinent to mechanics and local movements) it is his penultimate work where definitions, concepts, theorems, their

DIALOGO
DI
GALILEO GALILEI LINCEO
MATEMATICO SOPRAORDINARIO
DELLO STVDIO DI PISA.
E Filosofo, e Matematico primario del
SERENISSIMO
GR.DVCA DI TOSCANA.
Doue ne i congressi di quattro giornate si discorre
sopra i due
MASSIMI SISTEMI DEL MONDO
TOLEMAICO, E COPERNICANO;
CON PRI VILEGI.
IN FIORENZA, Per Gio:Batista Landini MDCXXXII.
CON LICENZA DE' SVPERIORI.

Title page of Dialogo
This work published in 1632 costs Galileo condemnation by the inquisition and lifelong house arrest.

PORTRAIT OF ISAAC NEWTON
Generous, profoundly against any type of violence and hypocrisy, Newton has a rare intellectual honesty. He says of himself: "I don't know how the world will judge me; but I to myself seem like a child playing on the sea shore, having fun collecting now a more polished stone, now a sea shell more brilliant then usual, whilst the unbounded ocean of truth extends unexplored in front of him", and affirms, alluding to Cartesian, creator of analytical geometry, to Kepler, of whom he uses the laws on planetary motion, to Galileo, of whom he uses the laws of dynamics: "If I could see further than others it's because I found myself on the shoulders of giants". As have so many others...

demonstrations and their corollaries form the coherent body of the new physics, formulating all the problems which will be faced and resolved, in the following decades, by his disciples and by scholars, until Newton. It is the 8th of January 1642: Galileo Galilei dies. The organisation of a solemn burial in Santa Croce is blocked by Rome "to not horrify the good" and not "offend the reputation" of the Holy Inquisition. The works of this blaspheming heretic remain forbidden until 1757. But luckily this prohibition, like all senseless prohibitions, is repeatedly infringed, and the work of Galileo quickly becomes a breeding ground for new, fertile ideas.

NEWTON AND GRAVITATION

In the same year in which Galileo exhales his last breath, in England Isaac Newton is born (1642-1727): it will be up to him to gather all the signs of predecessors and contemporaries to plot a new universe, that which we know. In an England bloodied by civil war, he studies in Trinity College, Cambridge reading everything and elaborating the method of the infinite series, the first step towards infinitesimal calculus. But the bubonic plague spreads: The University is closed and Newton returns home where, in two years, he invents "fluxional calculus" (differential and integral), begins his experiments on the colours of light developing a corpuscolar theory opposed to that undulatory of Huygens and Hooke, and invents the reflective telescope.
He is convinced that all movements have something in common, and that if the nature of the celestial bodies is analogous to these on Earth, as maintained by Galileo, all the celestial bodies must have a "gravity", like the Earth. Kepler had thought that it was perhaps a magnetic force that bound the planets to the Sun: perhaps, instead, it was gravity. Newton is not the only one to be attracted to this idea: Boulliau suggests that gravitation is proportional to the mass considered, and inversely proportional to the square of their distances (1645); Hooke proposes the hypothesis that the planets are subjected to a reciprocal attraction which causes their movement (1674) and that the attraction between the Sun and the planets is inversely proportional to the square of their distances (1679). No one, however, has ideas so clear and mathematically defined as Newton: in 1687 he publishes Philosophiae Naturalis Principia Mathematica (Mathematical principals of natural philosophy) which introduces theoretical physics to science, organises mechanics in a definitive manner and, in particular, defines once and for all the laws of gravitation. It will become one of the fundamental books in the history of man.
Newton ascertains the inevitable existence of this mysterious "action at a distance", unacceptable from the philosophical point of view. Two bodies are attracted with a force proportional to the product of their masses and inversely proportional to the square of their distance: and that is all. The empirical laws of Kepler are a logical consequence of this law. Even, this law

NEWTONS REFLECTIVE TELESCOPE
A concave parabolic mirror placed inside the tube makes the rays emitted by the observed object converge in a single beam. An additional planar mirror, inclined at 45° with respect to the axis of the instrument, makes it so that the trajectory of the beam is deviated by 90°, and that the image forms on the side of the tube, where it may be easily observed.

Ever more powerful instruments
The centuries following Newton see the growing importance of instruments for the realisation of ever more accurate observations. In this print is depicted a reflector telescope made in Ireland for William Parson Lord of Rosse (1800-1867): a mirror with a diameter of 1.80 m popularly known as "the Parsonstown Leviathan " which allows them to discover the spiral structures of some galaxies.
Below: two prints from the era reproduce a moment in the construction of the instrument.

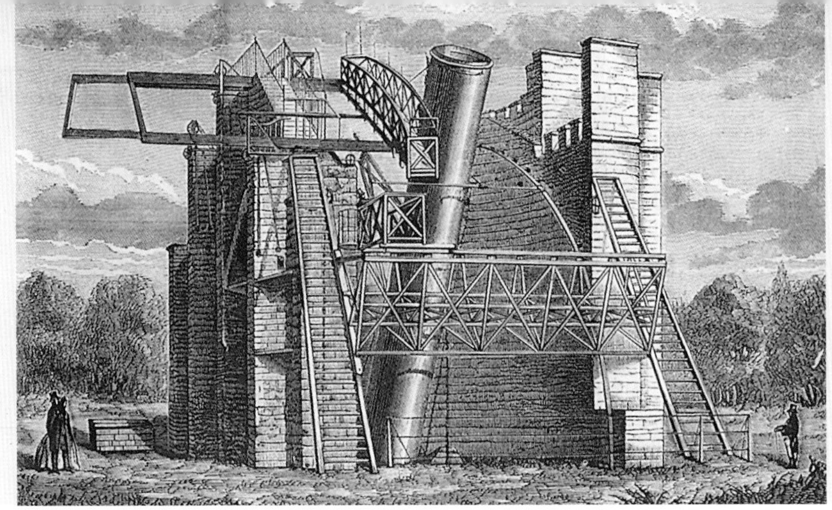

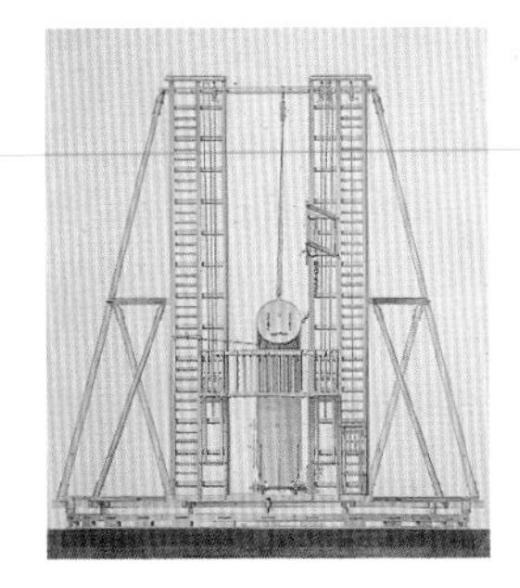

perfects Kepler's third law allowing one to evaluate the influence of the mass of the single planets which escaped Kepler, in as much as being negligible with respect to that of the Sun. To deny the Copernican idea had become by now very difficult. This law, in fact, with its simplicity, gave an answer to many astronomical problems: the form and the velocity of the orbit of planets and comets around the Sun and of the satellites around the planets, the precession of the equinoxes, the shape of the Earth, the movements of the objects on the Earth, the tides...
This the perception was consolidated; that every phenomenon was regulated by a few fundamental natural laws which can be identified though observation and experimentation, and translated into simple mathematical form. Exactly as Galileo had said. In the premise to the third book of the Principia, Newton sets out the four "rules of philosophy" which describe this new attitude:

1) "To the natural things must not be granted reasons more numerous than these which are true and sufficient to explain the phenomena;
2) "thus, while it can be made, the same reasons are attributed to natural effects of the same type";
3) "The quality of bodies which cannot be increased and reduced, is that which belongs to all bodies on which it is possible to establish experiments, it must be retained in the quality of all bodies";
4) "In experimental philosophy, the propositions obtained by the induction of phenomena must, despite hypotheses to the contrary, be considered true or rigorously or as much as possible whilst other phenomenon do not intervene, through which or are made more precise or are subdued with exception".

In the following centuries all mechanics will be the development of Newtonian theory. Not even the theory of relativity succeeds in invalidating it: our world, the solar system, the physics of our Galaxy are still Newtonian. That notwithstanding, there are many who refute the ideas of Newton, and not only for personal competition: Leibniz, Kant and Goethe are decisive and competent adversaries, and Hegel even underlines "the impropriety and the incorrectness of the observations and the experiments , and not least the blandness of them, even, as Goethe has shown, their bad faith. And equally the bad quality of the reasoning's, of the inferences and of the demonstrations, made through these impure empirical data". It is not to be wondered at if Newton, tired of controversy, spite and pettiness, after the controversy following the publication of the theory on the formation of colours forgoes publishing Lessons in optics occupying himself until 1684 above all in the study of theology and alchemy. Despite everything, however, when he dies, luxurious honours are bestowed: only 85 years earlier, Galileo was buried in secret.

AFTER NEWTON: KIRCHHOFF AND THE CHEMISTRY OF THE STARS

By now, the changes follow on ever more quickly. To the century of Newton also belong, amongst others, Fermat, illustrious mathematician; Romer, who quantifies the speed of light; Grimaldi who studies diffraction; Torricelli, who demonstrates the existence of vacuum; Pascal and Boyle who define the physics of fluids!... The precision of telescopes and clocks increase notably, and an increasing number of astronomers dedicate themselves to establishing exactly the positions of the stars, compiling ever more rich stellar catalogues in the attempt to understand how the Milky Way is made. The nature of the celestial bodies remains outside their interest: if one can determine shape, distance, sizes and movements of the celestial objects, instead understanding of what they are made is surely out of human grasp. For example, in the early nineteenth century, William Herschel (1738-1822), an astronomer capable of studying and deducing the shape of the Galaxy, builder of the worlds largest telescope and discoverer, amongst others, of Uranus, believes without feeling ridiculous that the Sun is inhabited. But within a few years is born astrophysics: distinguished from astronomy (now

Albert Einstein at the Yerkes Observatory (1921)
The necessary answer for astrophysicists came only after Einstein had once and for all, established that mass was "a mode of being" of energy and vice versa. According to the simple relationship identified by him, mass could transform itself into energy, and vice versa: finally the mechanism which allowed the stars to produce enormous quantities of energy without "consuming themselves" was found.

called "classical" or "disposition"), it is based on laboratory tests. Comparing the light emitted from incandescent substances with that gathered from the stars one struggles with the impossible: to discover the chemical composition and, possibly, the structure and the workings of the celestial bodies. Disliked by the "serious" astronomers, it develops thanks to physicists and chemists who invent new analytical instruments basing them on that which Newton had demonstrated regarding the structure of light. In 1814, Joseph Fraunhofer (1787-1826) performs some basic observations on the lines which Wollaston saw in the spectrum of the Sun: there are over 600, and are equal to these present in the spectra of the Moon and the planets; also the spectra of Pollux, Capella and Procyon are very similar, whilst these of Sirius and Castor not. Perfecting the spectroscope[121] with the invention of the diffraction grating (much more powerful and versatile than a glass prism), Fraunhofer observes in the solar spectrum, the two lines for sodium: the spectral analyses of celestial sources began thus. Whilst in the laboratory John Herschel is first to observe the correspondence between spectra and the substances which produce them, Anders J. Angstrom (1814-1874) describes the spectra of incandescent gasses and absorbance spectra ➢196 and Jean Foucault (1819-1868) compares laboratory spectra with celestial sources. Finally Gustav Kirchhoff (1824-1887) formalises the enormous mass of observations into a simple law which changes, once again radically, the manner of studying the sky: "the ratio between the strengths of emission and absorbance for the same wavelength is constant for all bodies found at the same temperature". It is 1859. Even this, which ties the exploration of the sky to atomic physics, is an empirical law, but it is the keystone which allows entry into the chemistry and structure of the celestial objects and the stars: de facto, the spectrum of a Star is sufficient to understand the composition. And precisely using spectroscopy Kirchhoff and Robert Bunsen (1811-1899) demonstrate that in the Sun are present many metals. The observations of the Sun, the closest stars, becomes the interest of a great majority of astrophysicists who analyse phenomena and zones. Sometimes the identification of some lines is difficult, and leads to the discovery of a new chemical element: they begin to think that the Sun has a temperature much higher than that believed. At the same time, the emission lines[194] in the spectra of stars and nebulae demonstrate that around 1/3 of the objects studied are gaseous. Not only: thanks to the work of Johan Doppler (1803-1853) and Aemand H. Fizeau (1819-1896), who demonstrate how the moving away or moving closer, with respect to the observer, of a sound or light signal source provokes the increase or decrease of the wavelength of the signal [219], they begin to specify the movements of the furthest objects. The sky changes again: even the "fixed stars" move!

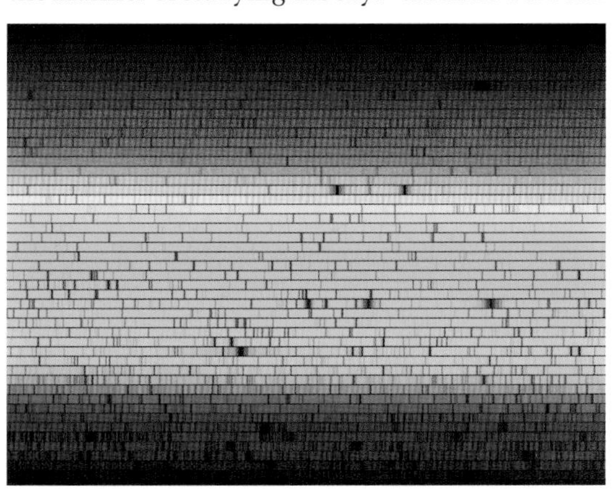

THE HR DIAGRAM: TOWARDS THE FUTURE

By now the collected data are many: father Angelo Secchi (1818-1878), noting that many stellar spectra have similar characteristics, based on the general appearance of the spectra ➢196 classifies the stars into five types. The sequence selected is the correct one: the passage of colours white-blue to red-dark indicates the progressive reduction in temperature, and the temperature is exactly the principal parameter which determines the appearance of a stellar spectrum. New discoveries make astrophysical understanding advance: Johan Balmer (1825-1898) shows that the order at the basis of the wavelengths of the hydrogen lines are summarised

Solar spectrum
The impact that the law of Kirchhoff has had on astronomy of the 19th-20th centuries is similar only to that which the laws of Kepler had on the astronomy of the 17th-18th centuries.

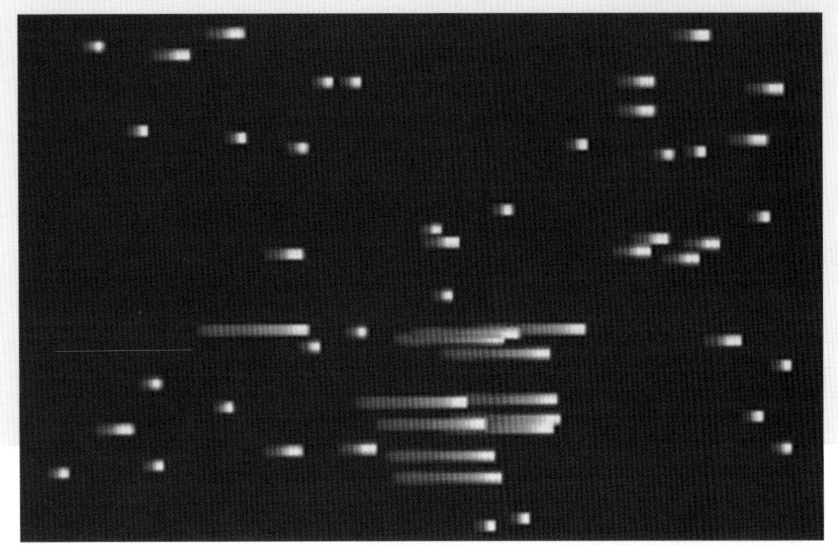

Spectra of stars and galaxies
Starting from the mid 19th century, photography allows the registration, in a faithful and indelible manner, of phenomena until now observed with the naked eye and, in the best of cases, drawn by an astronomer. A photograph, furthermore, can be examined many times, by many people and, with different techniques, even years later, and can highlight invisible objects: thanks to ever increasing exposure times, ever more detailed information is fixed onto the photographic plate, which can then be analysed calmly and with precision. The adjacent image shows some spectra from a group of stars.

with a simple mathematical expression; Pieter Zeeman (1865-1943) notices that a more or less intense magnetic field influences the spectral lines of a source subdividing them into a number of lines proportional to its intensity. That allows the measurement of the magnetic fields present in the stars. They are still empirical discoveries: the theoretical support comes only after having understood the structure of the atom, of the atomic nucleus and the elementary particles. The collected data accumulate while the physicist and the chemist do not come up with instruments sufficient to elaborate hypotheses and exhaustive theories: Faraday and his concept of "field" as "state" of the space surrounding a "source"; Mendeleev and his table of the chemical elements; Maxwell and his theory of electromagnetism; Becquerel and his discovery of radioactivity; the research of Pierre and Maria Curie; the experiments of Rutherford and Soddy on a, b, g rays, studies on black bodies[195] which lead Planck to determine the universal constant; Einstein and his work on the quantization of energy which explains the photoelectric effect; Bohr and his quantistic model of the atom; the theory of special or restricted relativity of Einstein which unites mass and energy in a very simple equation: these are the discoveries that will give the possibility of explaining stellar energy and the enormous lifetime of the stars, to elaborate a timescale much wider than had ever been imagined, to elaborate hypotheses on the evolution of the universe. In 1911, Ejnar Hertzsprung (1873-1967) constructed a graph in which were compared the "colours" and the "absolute magnitudes[➢190] of the stars, and he observes the association between the two magnitudes. In 1913 Henry Russell (1877-1957) creates another using spectral class[➢196] in place of colour, and finds the same distribution. The Hertzsprung-Russell diagram (HR diagram[➢197]) indicates that colour (i.e. temperature) and the spectrum of the stars are linked, and that the spectral type is linked to luminosity[➢190]. But since this depends also on the dimensions of the Star, from their spectra one can draw precise information also on the real dimensions of the Star observed.

We lack an explanation of cause and effect which connects the observations together in a general framework of laws, but the discoveries of chemistry and physics resolve the situation: the reckonings of the atomic model by Bohr reproduce, amongst others, the frequencies of the lines of the Balmer series for hydrogen. Astrophysics has finally the key for the interpretation of the spectra, and the atomic binding energies explain also the origins of stellar radiation and the reason for the enormous amount of energy produced by the Sun. The spectral lines depend on the number of atoms which produce them, but also on the temperature of the gas, the pressure, the chemical composition and the ionisation state: the relative abundance of the elements in the stellar atmosphere are determined, a method which today allows highlighting even very small chemical differences, correlatable with the different ages of the stars. Thus it is discovered that the chemical composition of the stars is almost uniform: 90% hydrogen and 9% helium (in mass: 71% and 27%). That which remains is comprised of all the other elements, already also observed on

Albert Einstein
Even though he was occupied by purely physical problems such as the photoelectric effect, Albert Einstein reached conclusions which served to definitively clarify the most debated problems in the field of astrophysics of the 19th-20th centuries.

Computers: accelerate the mind

The last, great innovation is the invention of electronic computers with the possibility for calculus and ever wider elaborations. For example, they allow the realisation of models of the inside of a Star calculating the values of the various physico-chemical parameters (from chemical composition to temperature, from pressure to density) for each distance from the surface. In a relatively short time, fixing the laws which we want to take into account, the computer shows us the characteristics of the spectrum which we should observe: and the proposed model will be much better the more the "theoretical" spectrum resembles that which is really observed.

Earth. The development of physics also allows the perfection of the theoretical models realised, to explain in a coherent manner what is a Star and how does it work. And the new models, in turn, suggest new observations. New types of Stars are thus discovered: the recurring novae, or the supernovae[204], the pulsars[203] with very short periods of variation... it is also discovered that the stars have an evolution, that they form in groups which are then disaggregated by the galactic tide forces. Further information on our Galaxy comes from radioastronomy, the new "branch" which allows the reconstruction of the structure of the Milky Way exceeding the limits of optical astronomy. New fields of study are opened: the galactic bodies, the globular masses[218], the nebulae[214], the movements of the Galaxy[208], their characteristics, are studied with the help of ever more sophisticated instruments. And the more observed, more numerous are the unknown objects which are discovered, the more the questions grow to which one must give an answer. New and different types of galaxies outside ours are identified; they discover, by examining the Doppler effect[219] that they are all moving away from us. Or better still, the further away they are, the faster they flee...
We are just noticing that the universe does not finish at the confines of the Milky Way, and already it is dilated to excess, filling up with galaxies and ever more stranger objects: up to a thousand billion galaxies, only on the horizon of Hubble. And the discoveries continue: from the galactic centre is observed an antimatter jet which rises, perpendicular to the galactic plane, for over 3,000 light years; objects such as Cygnus A are observed, which emit radio energy equal to 10 million times that emitted by an entire galaxy such as Andromeda; quasars[224] are studied, which in turn seem much closer to that suggested by Doppler effect measurements; this is "perspective effects" which could distort the conclusions... Hypotheses, observations, now hypotheses, new observations, objections, doubts...
There is still not a certain, omnicomprehensive answer: an increasing number of researchers are looking in a thousand different directions. New models of Stars, galaxies, celestial objects are elaborated in this way that perhaps only the mathematical imagination of the researchers will manage to substantiate: giving rise to "black holes", the "foam" universe, "strings"...
By now the number of researchers which occupy themselves with problems tied to stellar evolution, astrophysics, cosmogenetic theory have become so large that it no longer makes sense to speak of someone in particular, nor of a single strand of research: like the other sciences, even astronomy has become a work coordinated on an international basis which proceeds incessantly, without revolutions, in a succession of innovations, inventions, new instruments, interpretations ever more elaborate and, frequently, ever more difficult to understand – sometimes even for those who work on the thousands of parallel routes. It is exactly as Bacon wished things to be during the time of Galileo. Astronomy is "super-specialised" and it comes from those who deal with particular problems concerning physics of the stars, who know nothing about planets or galaxies. Even the language it makes use of becomes more technical and the terms can sometimes recapture entire research, becoming harder to translate into easily understandable language. This creates swathes of jargon, that is often misleading, incomplete and incomprehensible to the layman. Research on the planets, the stars, stellar material, galaxy, the Universe and its evolution are practically separate disciplines though they have the same goal. And whilst information on the heavens and its bodies becomes ever more complete with hypothesis proven and discredited, the universe that we known drifts further and further away from that which was known a century ago. And whilst mathematical models describe one, one hundred or a thousand universes that have curiously more to do with philosophy than science, it should not be forgotten that it all started and few hundred thousand years ago when a quasi-animal planted a pole in order to track the Moon.

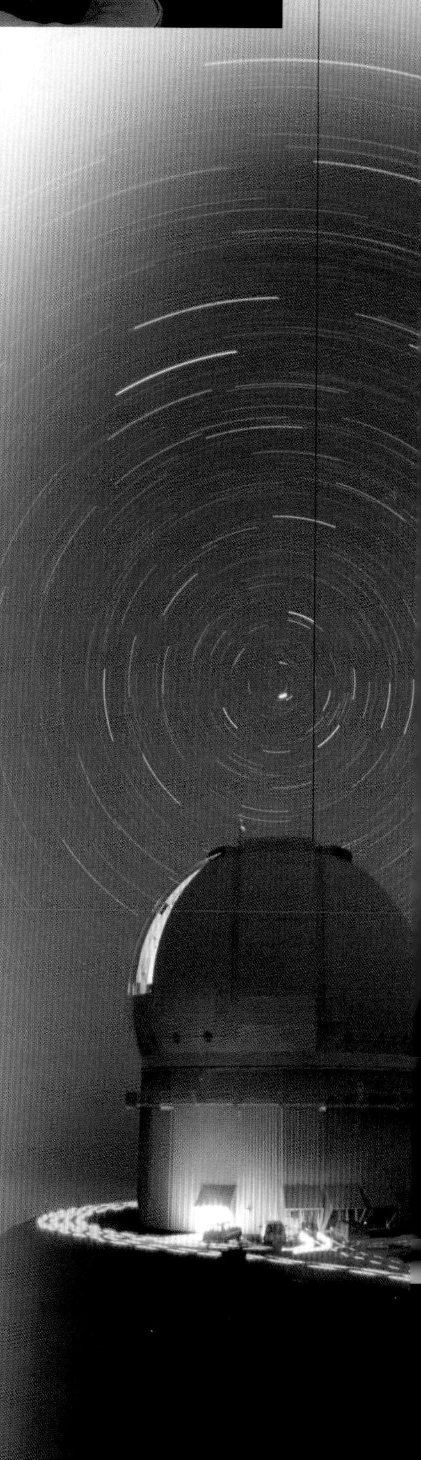

THE EARTH: THE THIRD PLANET OF THE SOLAR SYSTEM OUTWARDS FROM THE SUN. AS THEY SAY: SOMEWHAT CLOSE TO THE RICHEST ENERGY SOURCE IN THE REGION, BUT NOT SO CLOSE AS TO BURN. OUR PLANET, LIKE THE OTHERS, HAS A LONG HISTORY OF AROUND 5 AND A HALF BILLION YEARS, BUT ONLY IN THE LAST CENTURIES – SINCE WHEN MAN HAS BEGUN TO ASK QUESTIONS AND GIVE ANSWERS NOT ALWAYS VERIFIED OR VERIFIABLE – THEY HAVE BEEN AT SIXES AND SEVENS OVER IT. IS IT SMALL, IS IT LARGE, IS IT ROUND, IS IT FLAT LIKE A PIZZA, IS IT CYLINDRICAL, DOES IT REST ON THE BACK OF A GIANT TURTLE, FLOAT ON AN OCEAN OF WATER, IS IT AT THE CENTRE OF THE UNIVERSE ...

AND THEN AGAIN: IS IT HOLLOW, IS IT SOLID, IS IT FORMED FROM A LAYER OF ROCK AND, INSIDE, FROM A SEA OF LAVA, HAS IT ALWAYS BEEN LIKE THIS, HAS IT CHANGED APPEARANCE SEVERAL TIMES, IS IT SWELLING LIKE A BALLOON...

EVEN THE MOON, THE EARTH'S COMPANION IN ITS CELESTIAL VOYAGE, HAS HAD IT'S PART: THE FLIGHTY STAR, IT HAS ALWAYS BEEN ASSOCIATED WITH WOMAN, THE CYCLE OF FERTILITY, WITH SOWING, WITH THE SEASONS. SYMBOL OF CONTRASTS, IT HAS REPRESENTED BIRTH AND DEATH, PURITY AND SENSUALITY, FIDELITY AND FICKLENESS, IMAGINATION AND RATIONALITY. BUT IT HAS ALSO BEEN AN IMPORTANT CELESTIAL INSTRUMENT FOR THE MEASUREMENT OF THE PASSING OF TIME, FOR SCANNING THE SEASONS, THE RHYTHMS OF CULTIVATION AND FISHING. AND TODAY WE KNOW THAT IT'S INFLUENCE IS NOTABLE: IF IT WASN'T FOR THE MOON, THE ORBIT OF THE EARTH WOULD BE DIFFERENT, OUR SEAS AND ACEANS WOULD NOT HAVE TIDES AND THE EARTH'S ROTATIONS WOULD NOT BE PROGRESSIVELY SLOWING, NOR WOULD WE OBSERVE THE ECLIPSES...

WE SHALL SEE, THEREFORE, IN BRIEF, WHICH ARE THE CHARACTERISTICS OF THE CELESTIAL BODIES BEST KNOWN TO US, AND THEIR PRINCIPAL INTERACTIONS.

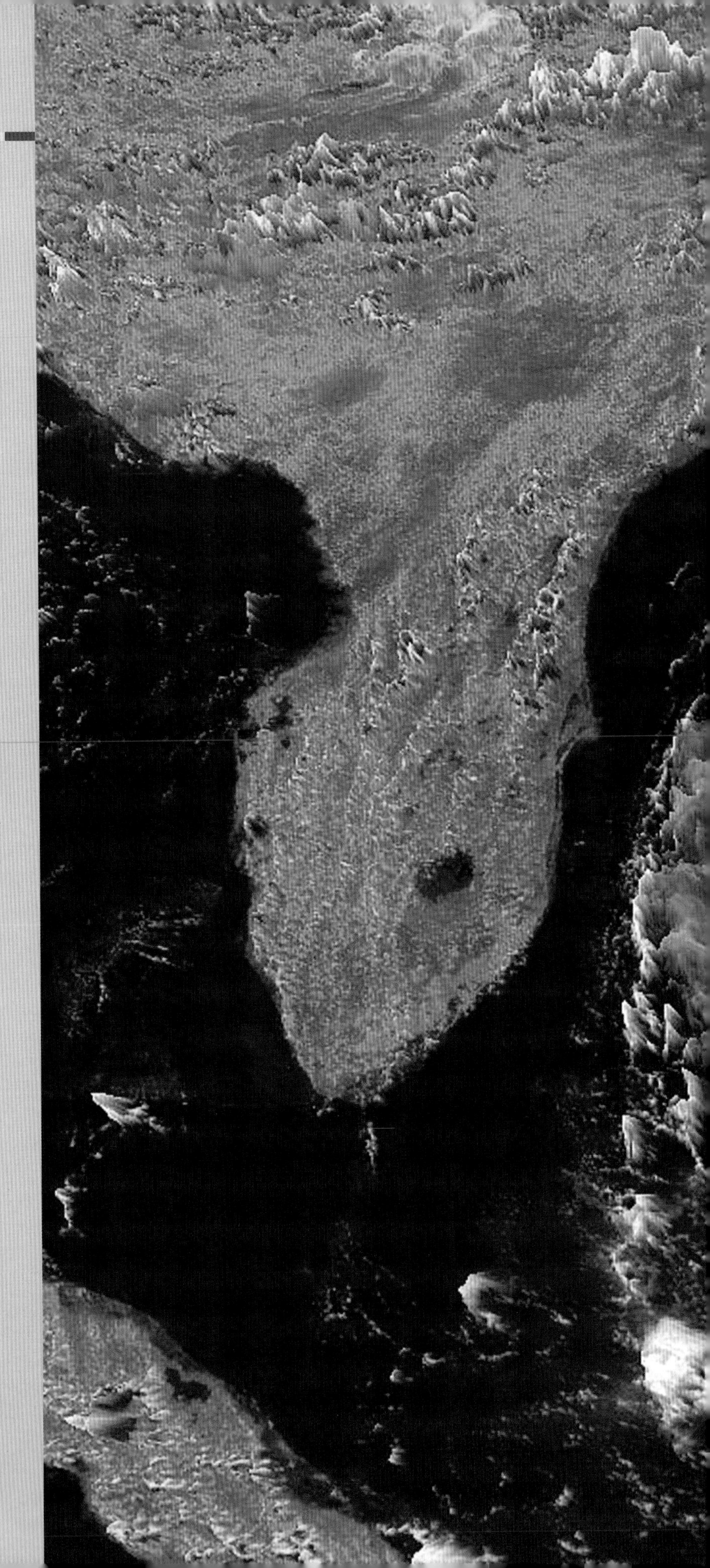

From the Earth To the Moon

THE BLUE PLANET

IN RECENT, EVEN VERY RECENT TIMES, THERE WERE ATTEMPTS TO CLEAR THE FIELD OF BELIEFS ON THE STRUCTURE AND THE DYNAMICS OF THE EARTH. NUMEROUS SCIENTISTS OF VARIOUS DISCIPLINES, PROCEEDING WITH CARE ON THE BASIS OF CERTAIN PROOFS, OF TECHNOLOGICALLY ADVANCED INVESTIGATIONS, OF RELIABLE OBSERVATIONS, OF CAREFULLY EXAMINED AND VERIFIED HYPOTHESES, THEY HAVE PROGRESSED THE UNDERSTANDING OF THE EARTH IN GIANT LEAPS. THAT NOTWITHSTANDING, WE STILL KNOW VERY LITTLE OF SOME ASPECTS OF OUR PLANET: LITTLE IS KNOWN ABOUT IT'S INTERIOR, DIFFICULT TO REACH, BUT FOR STUDYING, FOR EXAMPLE, THE OCEAN DEPTHS, WE ARE DEALING WITH LESS DEEP PROBLEMS THAN THAT DESTINED FOR RESEARCH INTO THE SURFACE OF VENUS, IS IT REALLY OUR FAULT.

The Earth is a solid body which, seen from space, appears perfectly spherical. "Seems": in reality, in fact, measuring it with the most accurate instruments mounted on geodesic satellites, one observes that it is not. Not that it's flat, lets understand! However, if from far away the height of the mountains is insignificant with respect to the entire diameter of the planet, and each difference in level is "flattened" with respect to the immensity of the globe, there exist some areas where the profile varies in a slight, but significant manner, from that of an ideal sphere to which reference is always made. The diameter passing over the poles, for example, is a little shorter than that which passes around the equator, and the radius of the Northern hemisphere is less than that of the Southern hemisphere. Then there are zones in which the surface is more flattened with respect to others: the shape which most approximates to that of the Earth is the shape of the pear, an almost spherical pear, but still a pear. The scientists, greedy for difficult words, have called it geoid, literally "Earth shaped". The Earth, then, is the shape of the Earth... who would have ever said it! Why our planet has just this shape is not known for certain: the most accredited hypotheses consider, among the principal causes,

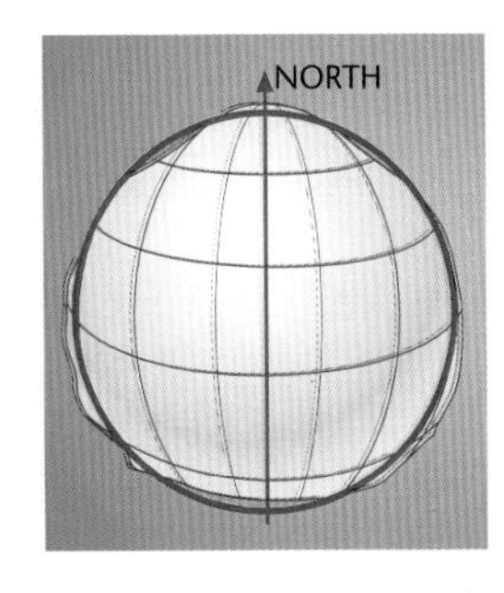

THE EARTH FROM SPACE
Seen from afar, our planet seems indeed spherical: in reality measuring with precision the different heights of the terrestrial surface with respect to the hypothetical surface with equal radius, significant differences are noted. Above, the geoide reconstruction shows these most evident.

hydrosphere
crust 4 - 40 km
inferior mantel
external nucleus 2,290 km
internal nucleus 1,820 km
1,600 km
superior mantel 630 km

THE EARTHS STATISTICS

AGE (IN 10^9 YEARS = MYA)	4.5
EQUATORIAL RADIUS (IN KM)	6,378.388
• MEAN (IN KM)	6,367.65
• POLAR	6,356.912
LENGTH	
• OF THE MERIDIAN (IN KM)	40,009.152
• OF THE MAJOR MERIDIAN AXES (NORTH, IN KM)	6,388
• OF THE MINOR MERIDIAN AXES (SOUTH, IN KM)	6,356.912
• OF THE EQUATOR (IN KM)	40,076.592
• OF THE ORBIT (IN 10^8 KM)	9.42
SURFACES (IN 10^8 KM^2)	5.1
• OF THE EMERGED EARTH (IN 10^8 KM^2)	3.55
• OF THE OCEANS (IN 10^8 KM^2)	1.55
VOLUME (IN 10^{12} KM^3)	1.083
MASS (IN 10^{20} KG)	5.9742
MEAN DENSITY (IN G/CM^3)	5.5
SURFACE GRAVITY (IN cm/s^2)	980.655
INCLINATION OF THE EQUATOR ON THE ECLIPTIC	23° 27¢
DISTANCE FROM THE SUN	
• MINIMUM (PERIHELION, IN 10^8 KM)	1.411
• MEAN (IN 10^8 KM)	1.496
• MAXIMUM (APHELION, IN 10^8 KM)	1.521
ESCAPE VELOCITY (IN KM/S)	11.2
VELOCITY	
• ANGULAR ROTATION (IN °/H)	15
• MEAN ORBITAL (IN KM/S)	29.8
PERIOD	
• OF SIDEREAL ROTATION (IN MEAN SOLAR H)	23.93
• OF REVOLUTION (IN MEAN SOLAR D)	365.26
MEAN SURFACE TEMPERATURE (GROUND, IN K)	295
ATMOSPHERIC PRESSURE AT SEA LEVEL (IN Pa)	101,325
ALBEDO	0.37

RELATIVE MASS OF THE PRINCIPAL SPHERES OF THE EARTH (IN 10^9 KG)

ATMOSPHERE	5.3×10^9
HYDROSPHERE	1.4×10^{12}
BIOSPHERE	1.0×10^6
LITHOSPHERE	6.0×10^{15}

the rotation➢40 of the globe around it's own axis and asymmetric distribution of the continents, but the problem is still open. In any way, for every daily need the Earth continues to be considered a sphere: the difference 21,476 km that exists between the polar radius and the equatorial radius, in fact, is equal to 0.34% of the mean radius; how can one say a square table is not really square because instead of having sides measuring 1 m it measures 100.3 cm with the other side measuring 0.997cm? No one would ever know, not even a skilled carpenter! One continues therefore to talk of maximum circles (equators, meridians), of radius (which, unless specified otherwise, is always "mean"), of diameter (itself also "mean"), of centre angles and so on.

THE PLANET OF WATER AND ITS FOUR "SPHERES"

The Earth, above all on the surface, is very rich in water: covering more than 71% of the surface of the planet, colouring it blue. Due to the particular characteristics of this substance and the typical temperature conditions of the Earth, water can be found, at the same time, in the solid, liquid and gaseous state: a condition which has considerable setbacks both on the thermal equilibrium (natural greenhouse effect), and on the origin and the distribution of organisms within the various environments.

The water shapes the rocks and conditions the life: the seasonal changes, the erosion produced by rivers, oceans and glaciers, the sedimentation in the great marine basins, atmospheric phenomena... are all aspects linked to the presence of water, frequently also visible from space.

It is exactly the world of water which constitutes one of the four "spheres" into which the planet Earth (or geosphere) is subdivided in order to better study it: such as the atmosphere (sphere of vapours, from the Greek atmos); the biosphere (sphere of life, from the Greek bios); the lithosphere (sphere of stone, from the Greek lithos), also the hydrosphere (sphere of water, from the Greek hydros) has a grossly uniform composition and physico-chemical structure and is further subdivided into zones with even more homogeneous characteristics. Let us briefly look at the principal characteristics of each 'sphere".

WATER
The Earth is the only planet to have water on its surface in the liquid, solid and gaseous states. Water is the first essential limiting factor for life, the element which maintains the planetary temperature within a restricted range and the principal factor in the external dynamics of the Earth: expanding when transformed into ice and solubilising almost all the minerals, it is the most effective erosive agent in continuously modifying the appearance of our planet.

IT IS THE MOST EXTENDED AND UNKNOWN PART OF THE EARTH: MASSES OF ROCK IN THE SOLID AND FLUID STATES EXTEND FOR A DEPTH OF OVER 6,000 KM TO THE CENTRE OF THE PLANET MAINTAINING A POWERFUL ENDOGENOUS DYNAMIC ACTIVE

THE LITHOSPHERE

The only certain information we have on the Earth is purely concerned with the surface: at the most, with the deepest drillings, we arrive at a depth of approx. 15 km: a little over 2 thousandths of the mean radius of the planet. That which is hypothesised on what is to be found below is a theoretical elaboration based on the analyses of the movement of seismic waves through the globe. In fact, when an earthquake occurs, the accumulated energy in the rocky layers is suddenly liberated causing the rock to move. Starting from the hypocentre, numerous seismic waves of various types, propagate in every direction then progressively diminish. In other words, the Earth shaker everything, for several minutes, following each strong earthquake shock.
The shocks are registered by the seismographs distributed over the entire surface of the planet and scientists verify the type, the strength and the speed.
It was exactly through studying the seismograms registered by his Zagreb observatory that, in 1909, Andrija Mohorovicic noticed that the signal of an earthquake taking place over 200 km away arrived at his location faster than that produced by a nearby earthquake. An absurdity, apparently.
Mohorovicic found a convincing explanation: the waves had to pass through rocks with different densities: in the less dense, more superficial zones, they had lower velocity with respect to that reached in the denser, deeper zones. And only the waves produced by far away earthquakes could reach deeper depths.
Thus began talk of crust and mantle, and the surface of discontinuity which separated these two zones took the name of the discoverer: today we know that the Mohorovicic discontinuity (or Moho), where the velocity of the seismic waves changes, has a thickness equal to approx. 0.78% of the terrestrial radius and is found at a variable depth of 6-40 km: it is deeper under the continents, more superficial under the oceans.
The idea of Mohorovicic' was liked: a few years later, the German Beno Gutenberg noticed not being able to register the S waves from earthquakes taking place in the other hemisphere: since this type of wave does not propagate in fluids, he thought that the core of the planet must have a "soft" nucleus" where they become dispersed.
Many others dedicating themselves to the study of earthquakes (even these produced by underground nuclear explosions): notice this way that the interior of the Earth is much more stratified than they had ever believed. Taking into account the values which pressure and temperature can assume at the various depths, considering gravimetric data, the magnetic characteristics of the various zones of the crust and many other variables, structural models of the planet have been realised ever more conforming to the observations. And for the moment the Earth is made as shown in the figure to the right. Until proof to the contrary.

epicenter

SEISMIC WAVES
In the movements of the rocky layers shocked by an earthquake, various types of seismic waves are distinguished, recognised through the traces from the seismograph. They are characterised by the preferential movements of the rock (sussultatory, compressive,..) and have different speeds.
The compression or longitudinal or prime waves (P) are the first to arrive: they propagate with as velocity of 5.6 km/s in the continental crusts; 6.5 km/s in the oceanic crusts; 81 km/s within the first 100 km of the mantle and 7.8 km/s further below.
Slower for the transverse or secondary waves (S; 4.4 km/s), and the Love waves, purely superficial, (3.3 km/s).

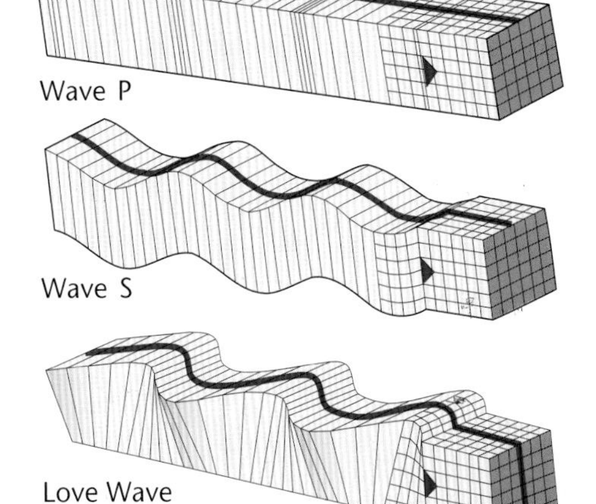

MINERALS, ROCKS AND MOVEMENTS OF THE CRUST

If for the profane saying "minerals" or "rock" has the same meaning, for these who study the Earth the

Mass ratios of the principal parts of the lithosphere (in 10^9kg)

CONTINENTAL CRUST	$1,6 \cdot 10^8$
OCEANIC CRUST	$7,0 \cdot 10^{13}$
MANTLE	$4,1 \cdot 10^{15}$
NUCLEUS	$1,9 \cdot 10^{15}$

Chemical composition of the terrestrial crust

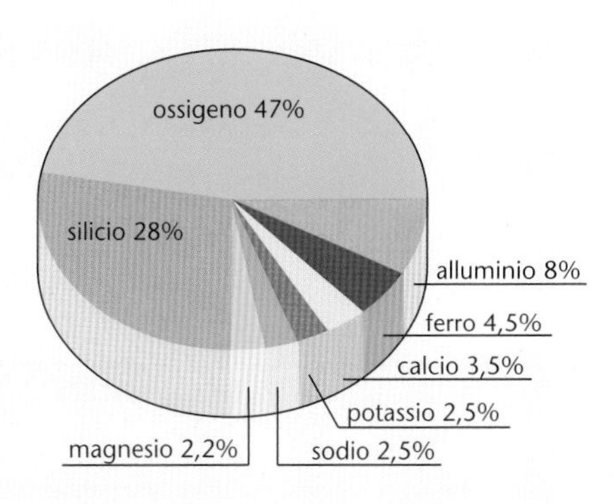

Structure

Below: The Earth is made up of concentric layers of rock with different chemical make up and different temperature and density values, separated by discontinuity surfaces which take the name of the scientists who discovered them. The model reproduced here indicates only some of the principal data on which the scientific community agrees.

distinction is clear: the minerals are stones formed from a single chemical compound which, usually, has a typical crystalline appearance and a precise geometrical structure; rocks are amorphous masses, aggregates of different substances frequently containing crystals or various minerals. For this reason, usually, the minerals are classified according to the main constituent chemical element (minerals of iron, of sulphur or sulphides, of silicon or silicates, of carbon or carbonates...) and the rocks are distinguished according to their origins.

Even the hardest stones are changed if subjected to force: staying below the limits of plasticity the rocky layers can bend but do not break. They form folds: upright, inclined, bedded, ultrabedded if the rock breaks. When the action of the force exceeds the limits of plasticity, the rocky layers break and, if the force is large enough, the fragments move violently away from one another. Fragmented rocks and interrupted layers form the faults and always

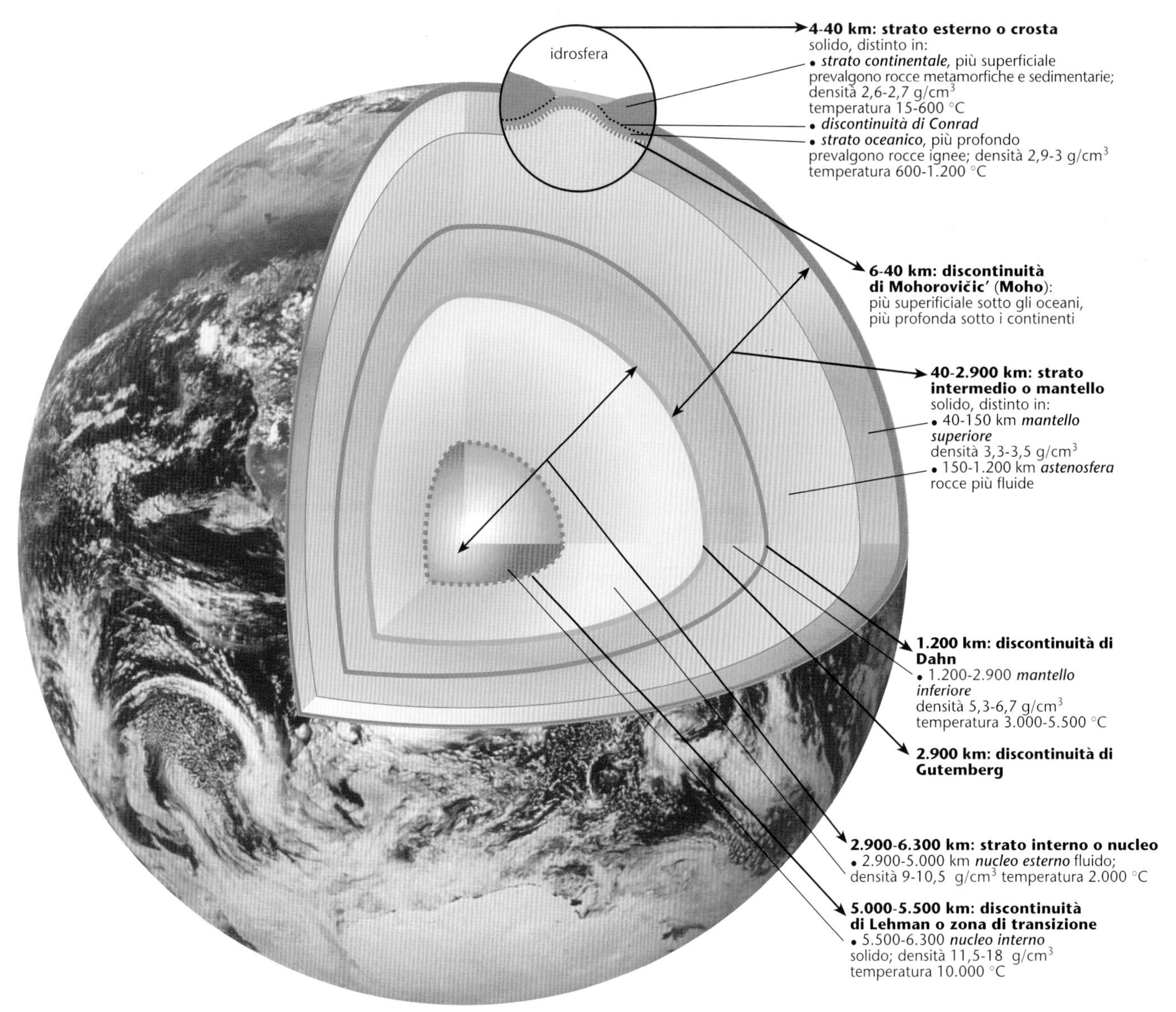

THE SAN ANDREAS FAULT
Perhaps the most famous fault in the world. It marks the edges of the Californian continental plate which runs slowly next to the North American continental plate. The translation al movement is made even more evident by the valleys excavated by the water courses, which suddenly change directions.

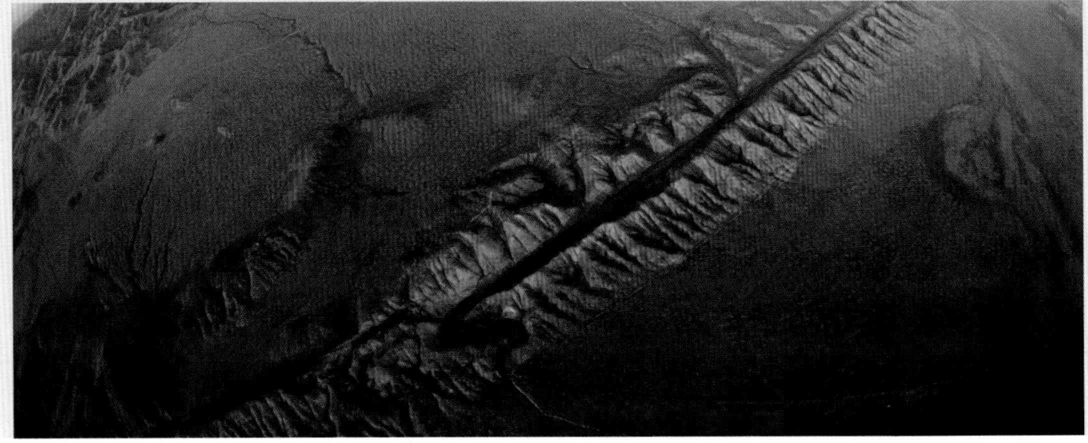

indicate points of the crust in which the forces of compression or stretching are rapid and powerful.

THE DYNAMICS OF THE PLANET

That the terrestrial crust is in continuous motion is clear to everyone. That notwithstanding, up to about a century ago no one was aware: despite observing erosion, earthquakes, landslides, volcanoes... it was thought that the Earth was eternally unchangeable. Beginning in the nineteenth century, using ever more numerous and reliable data on the geography of the entire planet and of the age of the Earth, it was begun to be believed that a mechanism must exist for the formation of the mountains and the recycling of sediments.
The Englishman John Pratt elaborated an orogenetic theory which had great success: according to his isostatic theory, the less dense mountains, "floated" over the denser plains or ocean beds. The crust was made up of prisms, enormous rocky blocks which more or less sank in the astenosphere, the plastic area of the mantle. Isostasis had many supporters, and in some cases it is still valid.
But it was Alfred Wegener, a climatologist passionate about geology, who opened the doors to the modern interpretation of the geological phenomena. He hypothesised that all the continents, once united to form the enormous continent Pangea, were separated from each other like drifting rafts. In Die Entstehung der Kontinente und Ozeune (The formation of the continents and the oceans, 1915) he sets forth his theory on the origin of the continents which is violently refuted by the geological establishment: for what it suggested, it was not based on reliable proofs nor gave any hypotheses on the "mechanism" which defined it. And above all it was not a geologist who proposed it. Wegener dedicated his entire life to the collection of proof, but didn't convince anyone: different to the isostatic theory, in fact, the origin of the continents presupposed the existence of a "motor", a cause: why was Pangea fragmented?

ROCK TYPES

IGNEOUS OR VOLCANIC ROCKS: are the "virgin rocks", produced by the activity of the Earth's endogenous activity. They have origins in the solidification of magma, i.e. the molten rock which is formed above all in some contact areas between the crust and the mantle. They are all constituted by silicates, rich in crystals and frequently full of the bubbles produced by the development of gas during cooling.

They can be differentiated into:
- PLUTONIC the magma solidifies under conditions of high temperature and pressure inside the Earth's crust. That provokes the development of homogeneously sized crystals (for example granites).
- LAVIC the magma solidifies on the surface, at environmental pressure and temperature. That provokes the rapid development of gas bubbles (for example. lapillus, pumice), the formation of heterogeneous crystals and vitrified rocks (for example obsidian).

SEDIMENTARY ROCKS: have their origins in marine environments. In certain points, in fact, under the load of sediment, the foundations give way, they fold: these are the geosynclines where the transformation of incoherent materials into new rocks takes place. Under the weight of thousands of metres of sediments, in fact, the sand and the mud already deposited are pressed, they are compressed, loose water and acquire the consistency of rock. At the same time, grains of calcium carbonate fuse in a slow process known as diagenesis. Over the course of several millions of years, from sand is formed sandstone, from mud marls. It is possible that a geosyncline is pitted up to 10-20 km under the surface of the sea bed: it is at this depth that the pressure becomes sufficient to enable diagenesis. Only sediments accumulated in geosynclines are transformed into sedimentary rocks: these which settle elsewhere are scattered across the sea bed and are continuously moved by the deep currents.

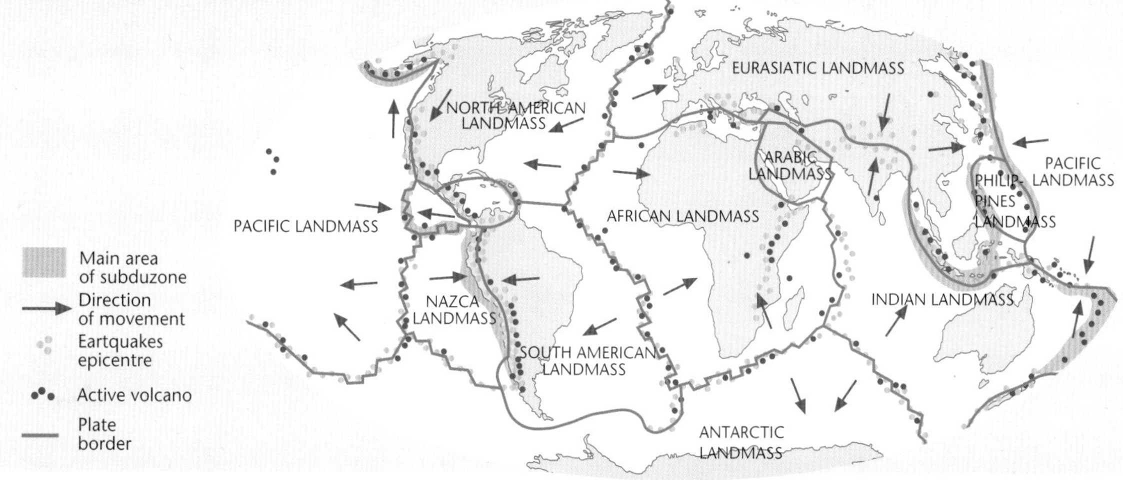

Why did the fragments drift apart? Wegener had no idea.

PLATE TECHTONICS

His revenge came in the sixties from the oceanic abysses. Hess discovered that at the level of the mid-oceanic ridges, out and out volcanic belts, the ocean beds were growing. These, therefore, are the motors for the origin of the continents. On this hypothesis is based the orogenetic theory of the lump or plate techtonics proposed by McKenzie and Parker: the terrestrial crust i san immense jigsaw puzzle of rocky blocks (the plates or lumps) in motion; these rest on the mantle above the Moho discontinuity. Each plates arises from an ocean ridge which, erupting new basalts, is pushed and slides over the mantle exactly as glaciers slide over mountains. Each plate slides along the edges of the neighbouring plates, and that provokes surface earthquakes. Some collide and give rise to an internal mountain chain; others run one over the other creating ocean depths. The rocks of the plate, as they sink, are dispersed into the magma and terrestail coating, creating a chain of volcanic mountains along the coastline. This more acceptable hyposthesis gives a coherent and complete explanation to a lot of geological phenomena, linking the planetary dynamic to natural radioactivity.

A LIVING PLANET
All the Earth is in continuous movement: the air and the clouds, the ocean waters and the glaciers, swarming life everywhere. Even the rock moves: earthquakes make the crust vibrate continuously renewed by the lithosphere dynamics: the rocks, folded, broken, transformed, build mountains which will be flattened by erosion, and fluid lave slips over the rocky surface and renews the appearance of entire regions....
The Earth is the most active of the solar planets.

Sedimentary rocks are recognised by the tendency to flake off into sheets, by the presence both of cavities produced by the disappearance of organic residues, and of fossils (for example travertines, limestones). Although they are all fundamentally made of silica, particular types are distinguished:

- ORGANOGENIC ROCKS are almost exclusively formed from the remains of plankton or animals (for example coral reefs);
- EVAPORITIC ROCKS are the results of the salt deposit produced by the evaporation of a significant mass of water due, for example, to the progressive disappearance of a sea or particularly prolonged climatic conditions. The increase of salt concentration causes various salts to progressively precipitate: first the carbonates (for example chalk), then the sulphates and the chlorides.

METAMORPHIC ROCKS: if the cementation of sediments and diagenesis takes place in geosynclines at depths of 10-20 km, at over 40-50 km, the rocks become fluid, in the intermediate distance (between 20 and 40 km) the temperature and pressure have to high values to form sedimentary rocks and too low values to form magma. Rocks and sediments are deeply transformed, they change appearance, but are not "melted": they thus form a new type of rock, usually crumpled up due to the thermal and barometric forces to which they are subjected and which give rise, it is noted, to the metamorphosis of the sedimentary or igneous rocks. Metamorphic rocks are all silicates with the exception of the marbles; they are frequently recognised by the presence of iridescent sheets (schist's), but have different natures according to the regions which have caused their metamorphosis.

CONTACT METAMORPHOSIS: the rock is transformed due to the proximity of magma which "cooks" the surrounding rocks (e.g. the limestones are transformed into marbles).

PRESSURE METAMORPHOSIS: at depths of 20 – 30 km, away from the magma zone, the pressure also changes the igneous cliffs, being less sensitive to heat, they do not produce new rocks but crystals, such as diamonds.

A FILM OF WATER - ONLY, AT THE MOST, 9 TEN THOUSANDTHS OF THE EARTH'S DIAMETER – ENVELOPES THE LITHOSPHERE. WATER IS EVERYWHERE: IT FLOWS INSIDE THE ROCKS, ERODES THE SIDES OF MOUNTAINS, MAKES PLAINS AND BEACHES... WATER, THE SUBSTANCE MOST CHARACTERISTIC OF THE EARTH, PERMITS LIFE, REGULATES DISTRIBUTION, RULES THE CLIMATE.

THE HYDROSPHERE

A WORLD OF WATER
Even observing the Earth from space, we can be aware of the importance which water has on our planet. But beyond the marine expanses, the frozen vastness of the poles, to the lakes and the rivers, beyond the immense banks of clouds, even the majority of living matter is made of water.

Thus is named the part of the planet occupied by liquid water (oceans, seas, lakes, rivers, underground layers) and solid (polar caps, glaciers, ice packs): a sphere which extends to approx. 8 km in height (the peaks of the highest mountains) to approx. 11 km in depth (the oceanic depths). The water which forms the hydrosphere is in continuous movement: currents, waves and tides agitate seas, lakes and rivers, the glaciers slide over continents, the icebergs float drifting pushed by the winds or transported by the currents, and a thousand rivers filter across rocks, dig caves, dissolve salts whilst they run towards the sea... Thus, the hydrosphere reshapes the lithosphere, eroding it, moving detritus, and accumulating it so as to form new geological structures. Since water has a high specific heat capacity, the hydrosphere represents an enormous reservoir of energy, and influences in a determining manner the climates and winds of the emerged lands.

SEAS AND OCEANS
The seas are large masses of salty water bordered by archipelagos, large islands or peninsulas, or enclosed by land masses relatively close together; there are also particularly extensive great lakes of fresh water, deep or important from the environmental point of view. The oceans, much larger, divide the continents: usually reaching a greater depth to these of the seas, which cover the continental shelf's, i.e. these areas comprised between the continents and the continental escarpments, a submerged precipice which reaches the great abyssal planes. Even if the oceans and seas are linked together, they can differ in salinity, density and temperature: both through strong evaporation, and because salts are dissolved more easily in warm water, tropical waters are more salty with respect to the waters of the cold seas or near to the mouths of the large carrying rivers.
The highest salinity is that of the Red Sea: over 40%; one of the lowest is that of the Baltic Sea: barely 5%.

CORAL REEFS
Characteristic of tropical climates, coral reefs are one of the Earth's most highly diversified biological ecosystems.

CAVES
The subterranean arabesques which one admires in karstic caves are produced by the infiltration of water which first solubilises calcium minerals and then deposits them.

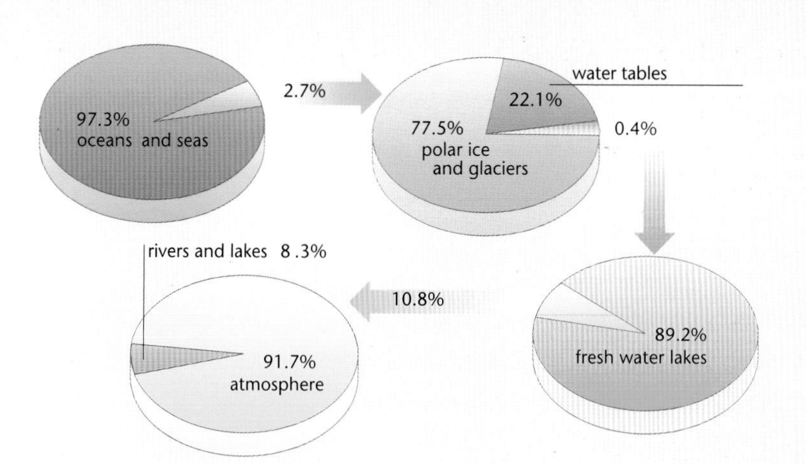

THE HYDROSPHERE IN NUMBERS

OCEAN AND SEA SURFACES WITH RESPECT TO:	
- THE TERRESTRIAL SURFACE	71%
- THE SURFACE OF THE NORTHERN HEMISPHERE	61%
- THE SURFACE OF THE SOUTHERN HEMISPHERE	84%
MASS	$1,4 \cdot 10^{21}$ kg
VOLUME IN KM³	
- TOTAL	$1,4 \cdot 10^{9}$
- SEAS AND OCEANS	$1,4 \cdot 10^{8}$
- POLAR CAPS AND GLACIERS	$2,9 \cdot 10^{6}$
- HYDRAL LAYERS	$8,4 \cdot 10^{5}$
- RIVERS AND LAKES	$2,0 \cdot 10^{5}$

Seas and oceans are subdivided according to the depth in areas with relatively homogeneous characteristics:

• the euphotic zone (from the Greek eu = bene, photos = light) comprises the first layer, not very deep, where sunlight manages to reach; many plants live here and oxygenation is maximal. According to the depth it is divided into the hemipelagic zone (0-50 m depth) and the mesopelagic zone (50-200 m): here only ultraviolet rays arrive, and the plants are reduced to the red-brown algae;

• the aphotic zone (= without light) comprises all the layer of water that remains. Here there is total darkness. According to the depth it is divided into the infrapelagic zone (200-600 m depth) rich in nutrients due to the vicinity to the coasts and to the euphotic zone and populated by numerous types of animals, and the bathypelagic zone (600-2,500 m) which comprises the majority of the waters: here only the large carnivores and animals adapted to the depth live. Finally the abyssopelagic zone (2,500-11.000 m), is almost a desert.

With the increasing depth, the pressure increases by approx. 1.000 hPa every 10m: already at 200 m below the sea surface, onto every square centimetre of surface is loaded 1,023 kg. The temperature decreases irregularly, with notable differences according to the latitude: at the equator, on the surface it measures a mean value of 30 °C, descending to 15 °C at -250 m, to 8 °C at -500 m, to 5 °C at -1,000 m, stabilising at 5-0 °C at -4,000 m. At the equator, therefore, the temperature which at the Poles one finds already at the surface is only registered at 4,000 m depth.

LAKES AND RIVERS

The lakes are important reservoirs of fresh water. According to their origins, lakes are distinguished into circular volcanic, glacial of elongated irregular shapes because they are the remnants from glaciers, techtonic, irregular, produced by the movements of the Earth's crust (e.g. the Dead Sea), coastal also formed from brackish waters, carsic, relics representing that which remains from ancient seas or gulfs (such as the Caspian Sea), alluvional, naturally or artificially obstructed.

Rivers can be fed by atmospheric precipitation (torrents) or glaciers, characterised by a constant flow of water. The amount of water passing every second through a section of a water course is the delivery, and is measured in cubic metres per second (m3/s): it is at a maximum during floods, and is at it's minimum during the period of drought. The tributaries, which make up part of the hydrographical lake, contribute in determining this.

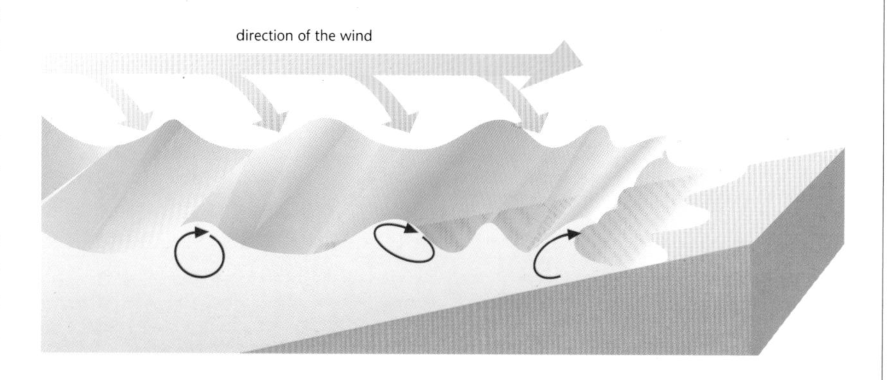

WAVE DYNAMICS
The wind, blowing, transmits rotational movement to the particles of the water surface, (wave). When the sea bed interrupts this movement, it deforms, so as to cause breakers.
BELOW: The main hot (red arrows) and cold (blue arrows) oceanic currents.

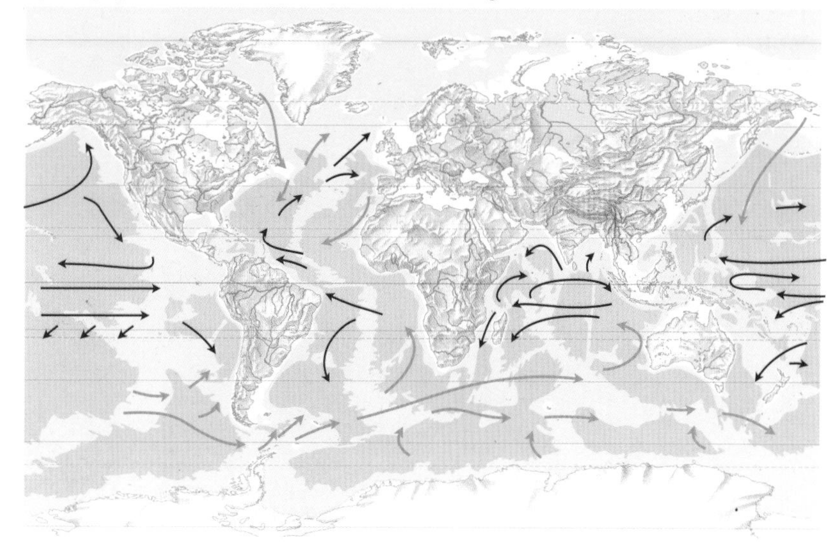

A SPHERE WHICH DOES NOT REACH A THICKNESS OF 20 KM: THIS IS THE SPACE SET ASIDE FOR LIFE. BETWEEN THE SKY AND THE SEA, BETWEEN THE EARTH AND AIR MILLIONS OF DIFFERENT SPECIES SWIM, FLY, RUN, DIG, JUMP, SLITHER...
A WORLD SWARMING WITH ENORMOUS AND MICROSCOPIC CREATURES SURVIVE, CONTINUOUSLY CHANGING, IN DELICATE EQUILIBRIUM.

THE BIOSPHERE

EXTREME ENVIRONMENTS
The areas of the planet in which new rocks are formed, the deserts, the deepest caves, the mountain tops and the marine abysses, represent the extreme environments in which life forms are sparse or even absent.

The biosphere and the hydrosphere are closely linked: water is the essential element for all forms of life, and the distribution of water over the planet (i.e. the limits of the hydrosphere) directly determines the distribution of organisms (i.e. the limits of the biosphere). The term "biosphere", in fact, only recently coined, indicates the collection of the areas of the earth in which life is present. It refers, therefore, to that tight band approx. 20 km thick comprised between the highest mountain peaks and the and the deepest ocean depths: only here, in fact, where the conditions of temperature, pressure, humidity are ideal for the survival of the most disparate organic forms which populate the Earth, is it possible to find forms of life.
Obviously, this "sphere" has uncertain boundaries: it's range coincides with that of the hydrosphere, it overlaps with the lowest layers of the atmosphere and with the most superficial of the lithosphere where it extends, at the most, for around 2 km. However, if for biosphere is meant the area throughout which life is distributed, but also these inorganic parts which are indispensable for life, we must also bring all of the atmosphere into this concept, without whose "shield" against the most energetic radiation no form of life would be able to resist; or the entire terrestrial crust and the upper areas of the mantle, without which there would be no volcanic activity necessary for the enrichment of our surface world with new mineral substances...
The biosphere, therefore, is an ecosystem as large as the planet Earth. An ecosystem in continual change through natural and artificial causes. The natural changes can take place on different temporal scales: can take place over extremely long times determined by astronomical and geological evolution, which necessarily influence the climatic characteristics of the various environments (such as that which happened, for example, during the ice age); or can take place over much shorter times, linked to climatic changes primed by unforeseen geological-atmospheric events (for example the eruption of a volcano, sending large amounts of ash into the atmosphere, can change the climate of vast areas for

Artificial ecosystems
The environment changed by human activity is profoundly different from that of nature, and is in constant precarious equilibrium. Furthermore, the variety of living species is dramatically reduced both due to "extremised" environmental conditions, and for the continual selections made by man.

a considerable period of time).
Instead, the artificial changes, due to human activity, have rapid effects: the deforestation of Africa as a result of the Roman wars of conquest, for example, has certainly contributed to accelerating the desertification process of the Sahara, and there is no doubt that the industrial activities of the last centuries, in rapid expansion, determine rapid and sudden changes in the biological equilibria.

THE LIMITING FACTORS

The biosphere is the point of contact between the different "spheres" into which the Earth is subdivided: it is crossed by a continuous flow of energy originating both from the inside of the planet, and from outside, and is characterised by continual exchange of material, in a never ending cycle which links all environments.
But it is not for this reason that life is found everywhere: it requires particular conditions, without which it would be impossible to find it. There are some physical and chemical elements which "limit" the development of life: the presence and the availability of water is the first and most important. Water is the universal solvent for the chemistry of life, and the primary component of all organisms and without water life is impossible. Not only that: water, passing from the solid state to the liquid and then to vapour, and vice versa, maintains a "natural greenhouse effect" active capable of maintaining the temperature of the planet within the levels compatible with life (i.e. a little below 0 °C and a little above 40 °C).
Also pressure, which must not overly exceed one kilogram per square centimetre (as occurs at a depth of approx. 200 m below the sea), and a wide availability of mineral salts and sunlight, indispensable for plant life, are factors which decisively limit the possibilities for life. There also exists living beings capable of surviving under extreme conditions, in which the temperature, the pressure or the light intensity are far removed from the average values necessary; nevertheless they are, in reality, relatively small in number.

Water and life
Water is the essential element for life, the prime limiting factor. Where it is lacking, only organisms having particular survival systems are found, such as this small plant growing on lave. Where it abounds, life displays itself in millions of different aspects, such as in the case of the rain forests.

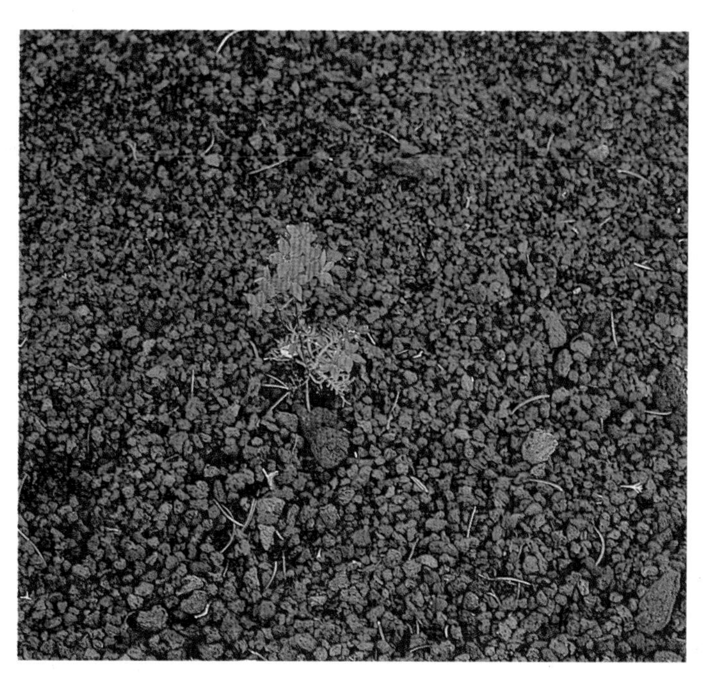

POLAR AURORA AND LIGHTNING, DENSE GREY CLOUDS AND LACES OF ICE, WINDS WHICH BLOW AT 300 KM PER HOUR AND AREAS OF PERENNIAL CALM...
IT IS THE ATMOSPHERE, AN OCEAN OF GAS IN CONTINUOUS MOVEMENT WHICH SHIELDS THE EARTH FROM COSMIC RAYS.

THE ATMOSPHERE

CYCLONES
The troposphere, the lowest and most dense layer of the atmosphere, rich and characterised by more dramatic and visible atmospheric phenomena. It is exactly the vapours, in fact, which condense or crystallise giving rise to the precipitations and the clouds.

Atmosphere is perhaps the most vague term used to identify part of a celestial body: it, in fact, indicates the superficial sheath of a planet or Star constituted mainly of gas and ions (plasma). Put into words it seems simple. The gasses, however, are not like a liquid or a piece of rock for which one can say exactly where are the surfaces which separate it from the surrounding environment: it is impossible to indicate a precise level where the atmosphere finishes and the interplanetary plasma begins. The gasses, in fact, as soon as gravity no longer holds them, "vanish" towards space continuously abandoning the celestial body. In the case of the Earth, then, so close to the Sun, the situation is more complex: to say where the earth's atmosphere finishes and that of the Sun begins (the so called solar wind 148 which, at this distance, is still rather dense), is a problem which can only be resolved theoretically: the external limits of the terrestrial atmosphere is defined as the level up to which it is presumed that the molecules of atmospheric gas cease to be subjected to the earth's attraction and the interactions with the earth's magnetic field. These conditions depend on the altitude and vary with latitude: above the equator the atmosphere ends at a height of around 60,000 km, over the poles at around 30,000 km. But even these values are only indications: the Earth's magnetic field, in fact, is deformed by the solar wind and it's shape can change notably. Changing, as a consequence, even the thickness of the atmosphere. Not only: as happens with the ocean water, even the atmosphere is subjected to the of the rotational movement of the Earth-Moon >54 system and the gravitational interference of the Moon and the Sun:

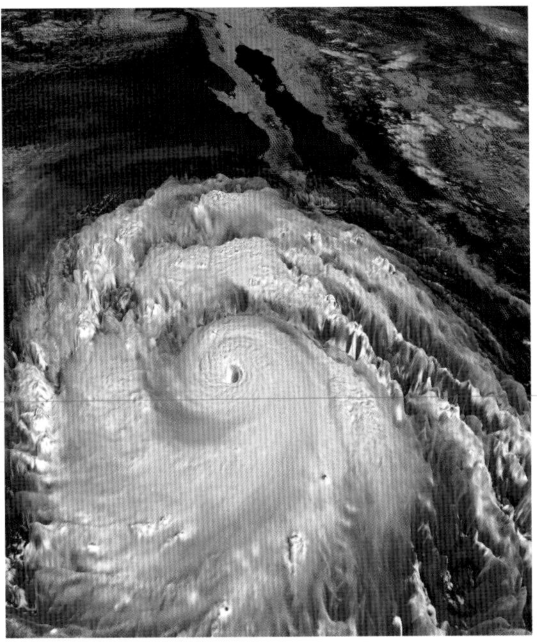

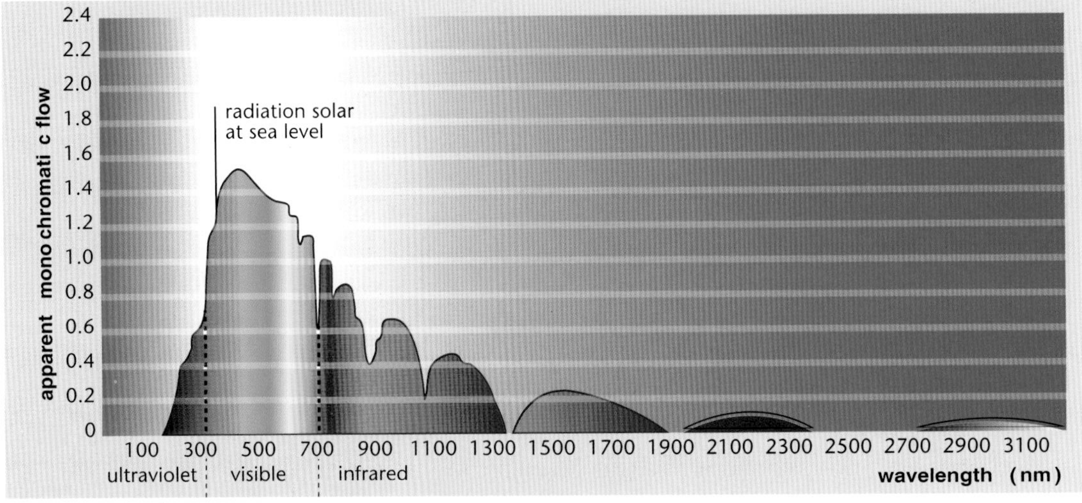

RADIATION
The atmosphere absorbs a large part of the solar radiation which reaches the Earth: the diagram shows how the peak range at sea level is found in the visible and near ultraviolet. More than 70% pf the solar radiation is reflected into space, and around 30% is transformed into heat which feeds the water cycle.

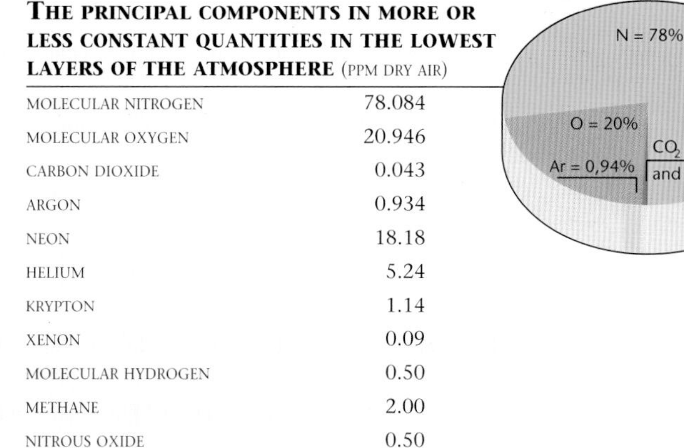

The principal components in more or less constant quantities in the lowest layers of the atmosphere (ppm dry air)

MOLECULAR NITROGEN	78.084
MOLECULAR OXYGEN	20.946
CARBON DIOXIDE	0.043
ARGON	0.934
NEON	18.18
HELIUM	5.24
KRYPTON	1.14
XENON	0.09
MOLECULAR HYDROGEN	0.50
METHANE	2.00
NITROUS OXIDE	0.50

The principal components in varying quantities in the lowest layers of the atmosphere (ppm dry air)

OZONE (FROM UV RADIATION)	0-0.50 ppm
SULPHUR DIOXIDE (FROM INDUSTRY AND VOLCANOES)	0-1.00 ppm
NITROGEN DIOXIDE (FROM INDUSTRY)	0-0.02 ppm
MOLECULAR HYDROGEN (FROM INDUSTRY)	10^{-4} g/m^3
SODIUM CHLORIDE (FROM THE SEA)	10^{-4} g/m^3
AMMONIA (FROM INDUSTRY)	TRACE
CARBON MONOXIDE (FROM INDUSTRY)	TRACE

given that the molecules of gas, lighter and less bound to each other than these of water, have greater possibility for movement, the atmospheric tides are much more conspicuous than these of the oceans. Even though made of gas in continual motion, however, the atmosphere is not homogeneous, not for composition nor for physical characteristics.

STRUCTURE

The atmosphere is not like a solid body in which the molecules are uniformly distributed throughout the volume: 90% of atmospheric gas is concentrated into the part closest to the ground. For this reason, atmospheric density is highest at ground level, and diminishes rapidly moving away from the surface. Thus the pressure, and also the concentration of water vapours, are zeroed out by 45 km. The variations in temperature, instead, are due to the heat sources which are more felt: in the layers closet to the ground, the high temperature is due to terrestrial irradiation, in the outermost layers, and the solar irradiation which predominates. There is cold in the intermediate zones. According to the values assumed for these variables, the atmosphere is subdivided into more homogeneous "spheres":

- **homosphere** (from the Greek homos, equal), 100 km high and with a composition similar to that found at round level;
- heterosphere (from the Greek heteros, different), comprises the air beyond 100 km, with different compositions. Other subdivisions better describe the variations:
- **troposphere**, up to 7 km over the poles and 18 km over the equator, where almost all meteorological phenomena take place,
- **stratosphere**, from 7-18 km to 30-60 km in height,
- **mesosphere**, from 30-60 km to 80-100 km,
- **thermosphere**, from 80-100 km and beyond.

The "spheres" are separated by three surfaces of discontinuity (the "pauses") indicated by a sharp variation in the parameters: tropopause, stratopause, mesopause. Other "spheres" flank these principal spheres:

- **ionosphere**, comprises part of the mesosphere and the thermosphere: here ions and electrons are arranged in different layers during the nights and the days, reflecting electromagnetic waves,
- **magnetosphere**: here charged particles remain confined within the force lines of the Earth's magnetic field (van Allen belt) producing polar auroras, beyond the 1000km range of the force of gravity. They are composed principally of plasma.

A snowflake
All the meteorological phenomena are linked to the presence of water vapour in the air. These however are not characteristic of the lowest atmospheric layers: these are of the troposphere and the lower layers of the stratosphere.

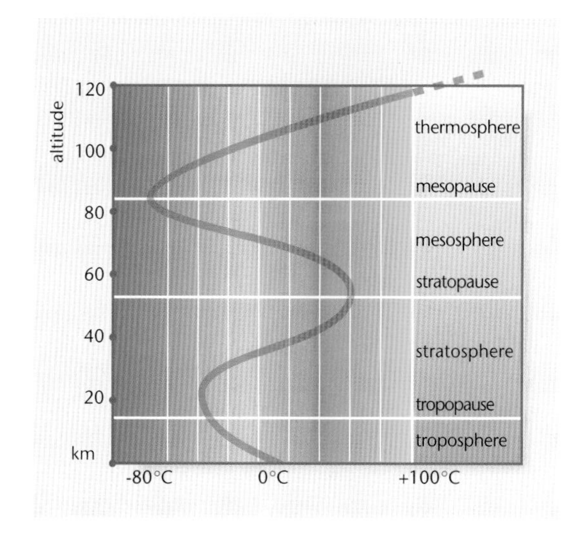

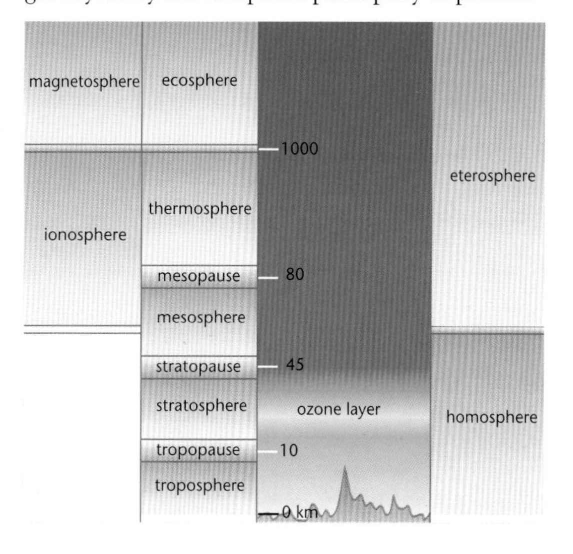

Diagram
The variation of pressure and temperature with increasing altitude allows the easy distinction between the various atmospheric "spheres".

Structure
Starting from the ground numerous "spheres" are distinguished, separated by "pauses" characterised by different mean values of temperature, density and pressure.

THE EARTH, LIKE ALL THE BODIES IN SPACE, MOVES.
BUT IT'S MOVEMENT IS VERY COMPLEX: TO STUDY IT BETTER, IT IS BROKEN DOWN INTO VARIOUS SIMPLER MOTIONS WHICH, RECOMPOSED ACCORDING TO THE LAWS OF PHYSICS, ALLOW THEIR BETTER DESCRIPTION IN A MORE EXACT WAY CLARIFYING ALL THE DYNAMICS.

THE EARTH IN SPACE

THE MOTION OF THE EARTH
The movement of the Earth in space is very complex and is broken down into several simplex movements in order to describe it mathematically: one rotational movement around it's own axis, one of precession, one of rotation around the Sun. Together with the Sun, then, the Earth moves around the galactic centre, and with the galaxy ➢208 it is moved further around space.

ROTATION OF THE CELESTIAL SPHERE
The stars closet to the celestial pole have a declination ➢66 so as to always be visible from the observatories in the same hemisphere, exactly as for the sun at polar latitudes.

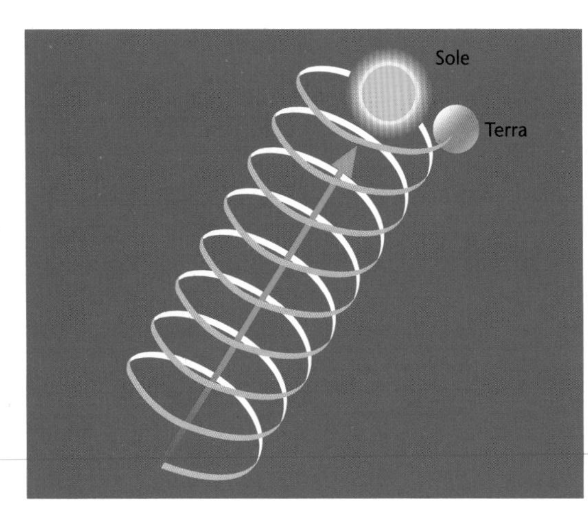

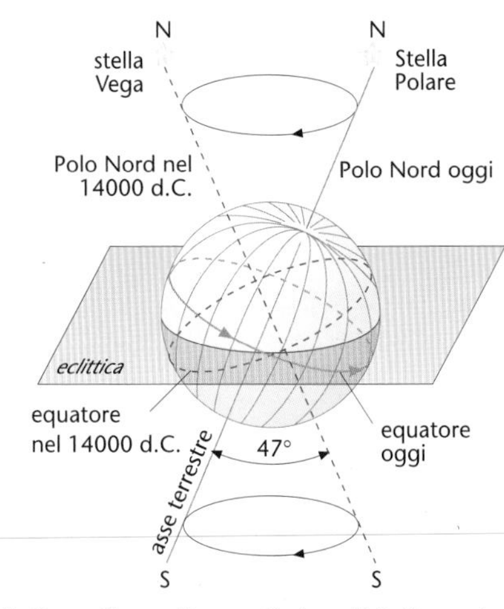

The most important movements into which the movements of the earth are subdivided are rotation around it's own axis, ***precession*** and ***revolution*** around the Sun. Each of these movements, combining with the general characteristics of our planet (fundamentally amongst which, the inclination of the axis of rotation), to produce numerous effects on the climate and on the conditions of life on our planet, in addition to the possibilities for astronomical observations. We will look at each of the essential characteristics, and their principal consequences.

ROTATION AND PRECESSION

The Earth rotates around an axis of rotation which identifies the North Pole and the South Pole on the earth's surface and points in the direction of the Pole Star, a Polaris of the constellation of Orsa Minor[77]. This direction is constant for rather long periods of time but, due to the movement of precession, changes over the course of thousands of years: as with the axis of a spinning top, even that of the Earth describes a complete cone, taking 26,000 years. That also provokes the slippage of the g point➢66 which, in that time, completes a complete revolution of the ecliptic. The axis of rotation is inclined with respect the ecliptic, i.e. the plane at which the revolutions occur, of 66°33' and this inclination is maintained during the entire revolution. Since the earth is solid, the rotation has an equal angular velocity for all the points of its surface (360°/24h = 15°/h), except that where the axis is located (Poles): in a day, each point travels a complete circumference (360°) the length of which depends on the latitude (indicated by ϕ).

THE CONSEQUENCES OF ROTATION AND PRECESSION

Numerous, and of various types.

1. The ***rotation of the celestial sphere***: if the stars, which differing from the Sun have a constant declination➢66 over the course of a year, a have declination of $\delta \geq 90 - \phi$ with (ϕ = latitude of the observer) are circumpolar, i.e. never rise and set. If

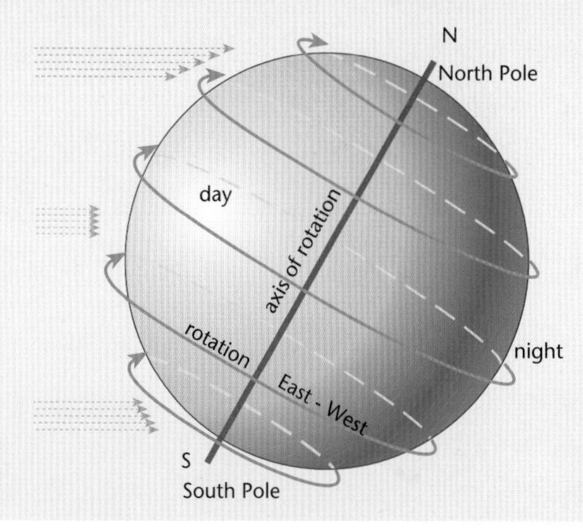

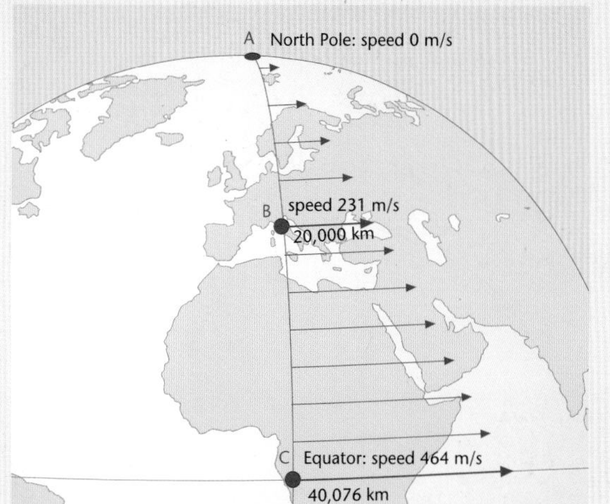
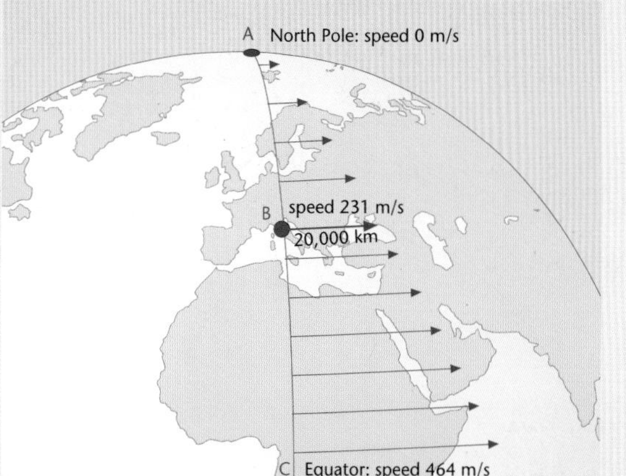

INSOLATION
The height of the Sun on the horizon changes with latitude and the seasons.

TANGENTIAL VELOCITY
A point at the Pole (ϕ = 90°) and on the axis of rotation, has a velocity of 0 km/h. A point with ϕ = 45° travels around 20,000 km in 24 hours at a velocity of approx. 833 km/h. A point with ϕ = 0° is on the equator and travels 40,076 km in 24 hours at 1,670 km/h.

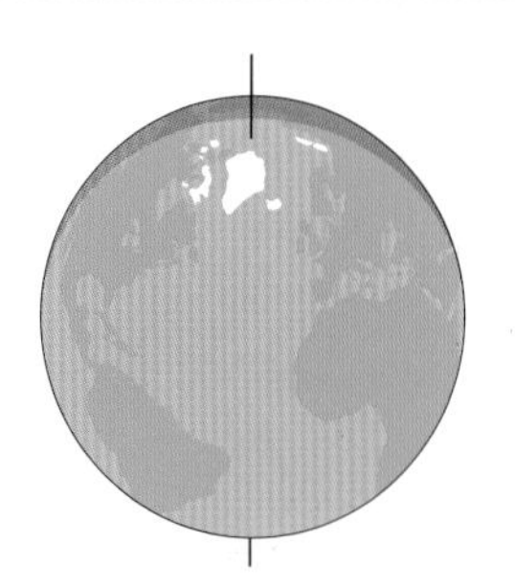

THE SEASONS
The energy which the various zones of the earth receive with insolation is different according to the position of the planet along the orbit and according to the latitude.
Two images of the same Northern hemisphere during the summer (above) and during the winter (below) show this difference.

$\delta < 90 - \phi$, then they rise and set just as the Sun is at our latitudes.
2. The ***alteration of days and nights***: over the course of the year the declination of the Sun which travels the ecliptic varies between 23°27' N and 23°27' S. For places with latitude ϕ < 90° and > 66°33', the >Sun can be circumpolar for a day or more, up to a maximum of six months at the Poles (ϕ = 90°). The alteration of light and dark over the course of 24h takes the name photoperiod: it regulates all the activities of the living beings. In addition, the passing of time is measured on the basis of the apparent rotation of the Sun: the day is subdivided into 24 hours (h), each divided into 60 minutes (min) in their turn divided into 60 seconds (s), each of which is then divided into tenths, hundredths and thousandths.
During the day, therefore, from place to place there is a difference in "hourly time": where the Sun rises, the measure of time is different from where the Sun sets or from where it is already night. For clarity of time zones, the mean solar day and the sideral solar day have been instituted.
3. ***Polar flattening***: the earth is solid but plastic; thus as with the surfaces, the rotation transmits different accelerations at different depths of the planet. Over time, these tangential forces result in progressive enlargement of the equatorial diameter and the reduction of that at the poles.

4. The ***coriolis effect***: since the sites on the surface of the earth complete over the same time a linear path which differs according to latitude, even their velocity of rotation is different. That provokes the *Coriolis effect*: bodies which move along a meridian in the Northern hemisphere are deviated towards the right, in the Southern hemisphere towards the left.

THE REVOLUTIONS

It is the movement which the earth performs around the Sun, considered immobile, at the centre of one of the two foci of the ecliptic. It respects the laws of Kepler>18 and Newton>21: the velocity is greater at the perihelion and lower at the aphelion even if these differences are not relevant since the eccentricity of the orbit is only 0.017. On average, the earth travels around the Sun at approx. 30 km/s.

CONSEQUENCES OF THE REVOLUTION

The ***alternation between the seasons***: is due both to the revolution and to the inclination on the axis. According to the position of the Earth with respect to the Sun, the height of the trajectory followed by the Sun on the horizon changes and, as a consequence, the amount of energy arriving. The climatic conditions depend directly on this phenomenon.

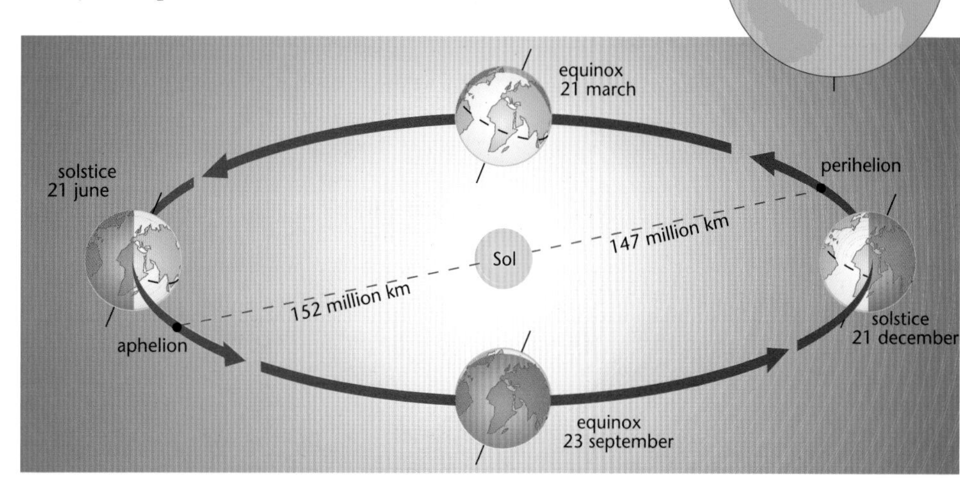

THE MOON
AND THE EARTH-MOON SYSTEM

THE MOON, VERY SMALL IN REALITY, APPEARS AS LARGE AS THE SUN: FOR THIS REASON IT HAS BEEN ADORED AS A GOD OF EQUAL IMPORTANCE. FOR THE REGULAR VARIABILITY OF ITS APPEARANCE, IT HAS BEEN AN INSTRUMENT FOR THE MEASUREMENT OF TIME AND A SYMBOL OF THE CONTRASTING AND COUNTEROPOISING ELEMENTS: ON THE ONE HAND THE FULL MOON, LUMINOUS AND BENEVOLENT, ON THE OTHER THE NEW MOON, DARK AND MYSTERIOUS.

AND STILL TODAY, NOTWITHSTANDING THE ACQUIRED KNOWLEDGE, NOTWITHDSTANDING THAT MAN HAS REACHED IT WITH PROBES AND RESEARCH EXPEDITIONS, NOTWITHSTANDING MANY PROBLEMS HAVING BEEN CLARIFIED, AND IT IS KNOWN THAT THE MOON IS ONLY A BALL OF ROCK SPINNING IN SPACE THERE ARE STILL MANY WHO BELIEVE IN ITS INFLUENCE OVER TERRESTRIAL EVENTS, THE RHYTHM OF THE RAINS AND THE BIRTH OF BABIES.

The Moon, the only natural satellite of the Earth, has an "old, riddled, worn" surface, so said Qfwfq, the imaginary character in the novel "cosmicomiche" by Italo Calvino: beyond the volcanic craters, relatively rare, it is scattered with vast lava plains called seas or basins, which appear darker than the surrounding areas called terrains or plateaux. Bordered by mountain chains with canyons, gorges, depressions and peaks, which are themselves spotted with craters, evidently due to meteoric bombardment from older eras. The overly small mass does not succeed in retaining gas molecules: the absence of an atmosphere and liquids running over the surface, have made it that the rocky lunar formations are conserved intact just as when they formed: the oldest, which date from around 4 billion years ago, still rise adjacent to, or - frequently - below, the most recent.

Mainly, therefore, the Moon is a small world of rock, with sudden changes in temperature of over 200 °C (average temperature in the Sun is around 100 °C, in the shade around -100 °C). The ground, covered by sandy, intangible and very fine dust, vibrates 3,000 times on average each year through very weak endogenous seismic tremors. Alternatively, the falling of a meteorite of a few tens of kilograms can provoke earthquakes which make the Moon shake for several minutes.

OBSERVATIONS

The Moon shines because it reflects light from the Sun: this is a phenomenon already perceived by the most ancient peoples who measured time on the basis of astronomical observations, but it remained almost unknown until Anaximenes, who finally correctly interpreted the dynamics of the eclipses.

According to Parmenides the Moon "forever looks towards the resplendent rays of the Sun", whilst Empedocles thought that the eclipses were due to the interpositioning of the Moon between the earth and Sun, and exploited this to calculate the Moon-Sun distance as being twice that of the Earth-Moon.

Democritus is the first to interpret the lunar markings as being like the shadows of valleys and mountains, whilst Plutarch, observing them, concluded that from the earth, one always observes the same face and that the Moon, like the Earth, has an irregular surface with

SIDEREUS NUNCIUS
Galileo is the first to make a "true" likeness of the Moon, and the emotion of the discovery through the pages of Sidereus nuncius: "Beautiful thing and exceedingly attractive is the ability to gaze at the lunar body. With the certainty that it is given from sensitive experience, one can learn the Moon is not at all covered by smooth and polished surfaces, but rough and unequal, it is in the same manner as the surface of the Earth, showing itself covered in every part by great prominences, deep valleys and gorges."

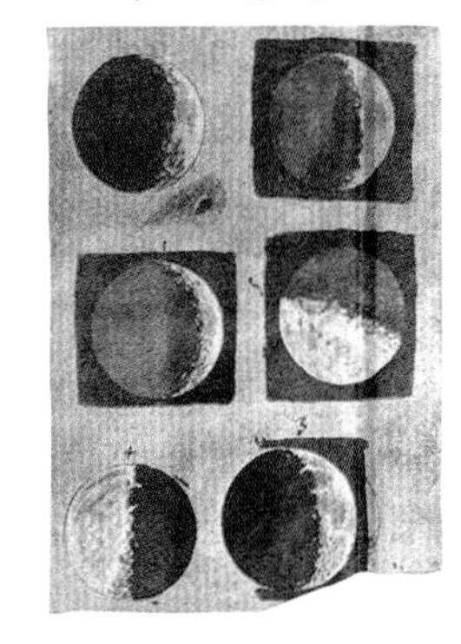

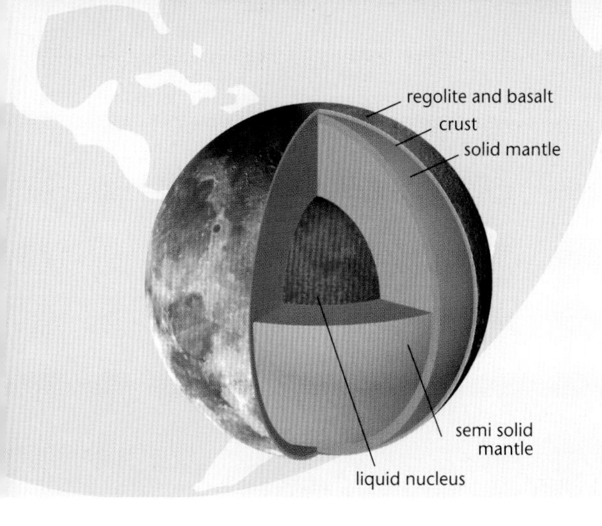

Lunar Statistics		With respect to the Earth
Radius		
• equatorial (in km)	1,738	0.272
• polar (in km)	1,740	0.272
Surface (in 10^7 km^2)	3.8	0.075
Volume (in 10^{10} km^3)	2.2	0.02
Mass (in 10^{24} kg)	0.74	0.012
Density (in g/cm^3)	3.34	0.61
Gravity (in G)	163.38	0.166
Escape velocity (in km/s)	2.4	0.21
Age (in 10^3 mya)	~4.5	
Sidereal month (in mean solar d)	27.32	
Lunation (or lunar month) (in mean solar d)	29.53	
Eccentricity of orbit	0.055	
Inclination of orbit on the ecliptic	5°9'	
Inclination of axis of rotation on the plane of orbit	5°	
Mean distance from Earth (in km)	384,400	

valleys, mountains and oceans which reflect sunlight in different ways. They are the last observations: the philosophy of Aristotle[➢12] blocks all development for hundreds of years. Until Galileo[➢19] who, thanks to the telescope, almost touches by hand that which others have perceived: The Moon is a physical, harsh, rocky object, in all, comparable to the Earth. But Galileo does more: as is his habit, he measures "demonstrating that the unevenness of the Earth is for the large part less than that of the Moon; less, I say, even speaking absolutely, and not only with respect to the sizes of the respective globes". Thus, lunar topography is born.
Newton[➢21] does not make observations: he asks whether the force which makes a projectile fall with a parabolic trajectory is not the same which makes the Moon move in its elliptical orbit, and the answer changes the course of science. But to have further information on the real structure of the Moon and its topography we must wait several centuries: the definitive word will be said with the Russian, American and Chinese space expeditions which will also photograph the hidden face (dark side). It is however probable, that under the soft surface of our satellite new discoveries await.

EARTH-MOON DISTANCE AND OTHER MEASUREMENTS

The first attempts to measure the Earth-Moon distance on the basis of trigonometry and with respect to the terrestrial diameter date from Aristarch and Hipparch[➢12] who obtained values, not very far from the truth. More accurate measurements were performed, following the same logic, starting from the mid eighteenth century, but they have better results when new techniques are used, such as the amount of time a beam of electromagnetic waves takes to complete the Earth-Moon return trip. It is the American Evans, in 1946, the pioneer of this technique, who uses radar. Later laser, pointed at reflectors left on the Moon by the Soviet probes and the astronauts of the Apollo 11, 14 and 15 missions allow the attainment of the lunar distance at perigee, at apogee and at the nodes[51] with a maximum error of 4 x 10^{-10} m allowing the establishment of a mean Earth-Moon distance value of 384,400 km. The precision of this measurement is important: amongst others, it allows the verification of the theory of lunar motion quantifying the energy exchanges between Earth and Moon Luna and the slowing down of the earth's rotation, giving useful information for evaluating the models of the internal structure of the Moon. Furthermore, this type of measurement has allowed the testing of the principle of equivalence proposed by Einstein (which affirms that gravitational mass and inertial mass are identical) and the invariability of the gravitational constant calculated by Cavendish.
The shape of the Moon is very similar to that of a sphere: from the measurements of the terrestrial diameter, mean Earth-Moon distance and mean angular diameter of the Moon (31'5") one easily calculates the diameter, the surface and the volume of our satellite. Due to terrestrial attraction and to its rotation, the Moon also has a hump in the direction of the Earth: the difference between the two axis, however, is insignificant: around 1.5 km, scarcely 0.43%. The mass is determined on the basis of gravitational measurements which assess the land tides or the disruption the orbit of the asteroid Eros[➢162] which is very near to the Earth Moon system.

Earth-Moon System
Whilst the centre of mass is found inside the earth, the "neutral point", where the gravitational force of the Earth and that of the Moon are compensating, is found in space 38,440 km from our satellite and 345,960 km from Earth. Space vehicles launched from Earth which pass beyond this limit enter into the lunar sphere of attraction.

At the time of Galileo it was believed that the Moon was smooth, perfect. The telescope seemed to show oceans and shores, mountains and inhabited plains. Today we know that these shadows we see on the Moon are enormous meteoric craters, extinct volcanoes, immense stretches of lava and powdery plains without life.

LUNAR GEOLOGY

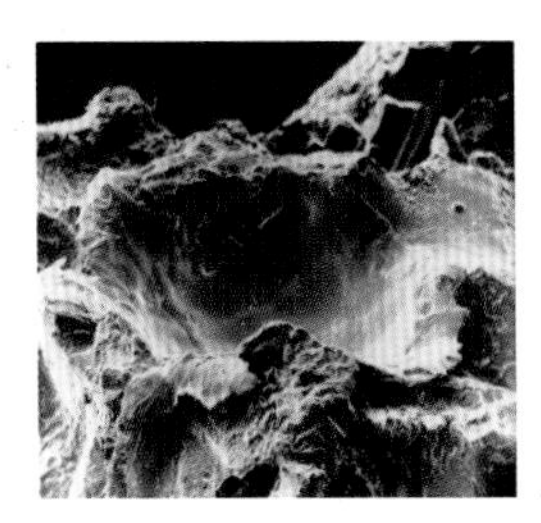

Impact Microcraters
In this microphotograph of a lunar rock a small crater produced by the impact of a microcrater is clearly visible, which has caused the fusion of the rock.

Although the direct exploration of the Moon gave a definite answer to many questions, even observations from Earth supply precious information on lunar geology: from the number of craters, for example, and from their stratification one can extract the relative age of the rocks of a region (the more numerous and overlapping, the older the area is); by measuring the diffusion of sunlight one can say that the lunar surface must be predominantly made up of small and incoherent particles of high porosity; radar techniques can "probe" the lunar crust to a depth of around 1,300 m (using this, pockets of subterranean ice have been found in an area of the lunar South Pole: perhaps the remains of a comet), and so on. Much other information is obtained with teledetection techniques, with telescopic observations and careful study of photographs.
Finally, the gravitational disturbances measured by the probes in lunar orbit allow the evaluation of the distribution, within the moon, of masses with different density: it is with this that have been discovered the mascon, negative gravitational anomalies corresponding with circular basins due, perhaps, to the sinking of enormous asteroids long ago.

THE MAJOR GEOLOGICAL STRUCTURES

SEAS OR BASINS

Called seas because, due to their darker colour, it was thought they were perhaps covered with water, the lunar plains, or basins, are in reality enormous expanses of basaltic rock probably formed in distant eras when the impact of some large meteorites broke the solid crust making, still-fluid material, flow out from inside.
The basins occupy only 15% of the entire lunar surface and are almost totally absent on the face invisible from the Earth (dark side); furthermore, the rocks which form them, characterised by high density (around 3.4 times that of water) are much "younger" than these gathered from other lunar

HYPOTHESIS ON THE ORIGINS OF THE MOON

The age of the Moon is certain. All the rock samples removed by astronauts have been dated and the dates agree leaving few doubts: the oldest regions are the plateaux (around 4 billion years), the youngest the basins (around 3 billion years). Even the Moon, like the entire solar system, originated around 5 billion years ago. But how? Many hypotheses have been made, but the collected proof does not favour any one over the other. Lets look at the most significant. The first has only historical value: to be considered valid, in fact, a hypothesis must explain the given facts for certain, and presuppose that the Moon:

- has the same age as the other bodies of the solar system: the proof, analyses of samples;
- has a density and a superficial mineral composition very similar to that of the crust of the Earth;
- it was located very close to the Earth no more than 2 billion years ago: shown by the dynamics of the Tides;
- it was formed in the same region

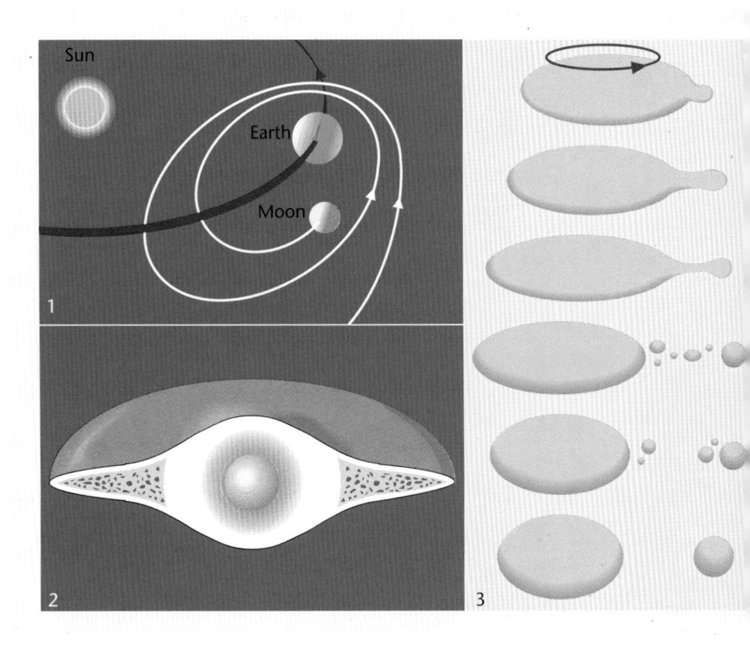

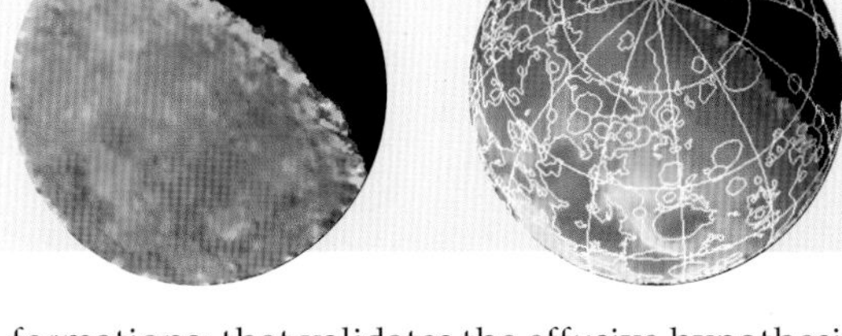

Lunar observations
The observations with polarised light, in the infrared or ultraviolet give data on the composition and nature of the surface rocks.

Gravimetry
The mascons are found both on the visible face (**a**) and on the hidden face (dark side) (**b**).

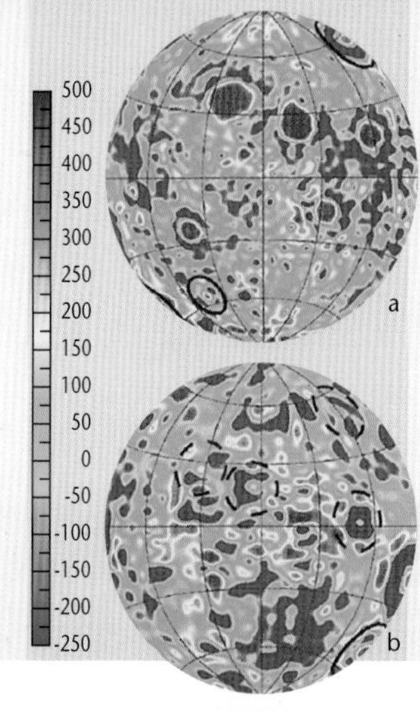

Lunar rocks
Microscopic analyses of a thin section of lunar rock: different colours indicate the presence of crystals with different chemical structures.

formations: that validates the effusive hypothesis of their origin.

CONTINENTS OR PLATEAUX

The continents, called this in the past in contrast to the seas, or plateaux, are the remains of the primitive crust: here the traces of the millenary meteoric bombardment are most numerous. They represent approx. 85% of the lunar surface (70% of the visible face) and their rocks are much less dense than these of the basins (around 3 times that of water) and much older (around 4 billion years).
The crust of the Moon is not, like that of the Earth, divided into moving plates: The lunar mountain chains are not produced by internal dynamics but by meteoric bombardment. The much reduced gravity has then contributed to forming mountains which, in proportion, are much higher than these of the Earth, as Galileo also noted: they frequently reach 6 km and peaks of 8 km are common. On the other hand, that the Moon does not have internal movement is clear: the lunar geological structures do not move by as much as a millimetre and the internal heat flux is too low, because techtonic dynamics can only exist sustained by convection currents.
The measurements made by the Apollo 17 expedition indicate an endogenous heat flux of

Mountains on the Moon
The meteoric origin of the lunar mountains is frequently confirmed also by their arched shape surrounding large basins: collisions of enormous force have smashed the crust making it sink and pushing it sideways. In this image of the Apennine Mountains: they make up part of a series of mountain chains which enclose the enormous basin of the *Sea of Rains*.

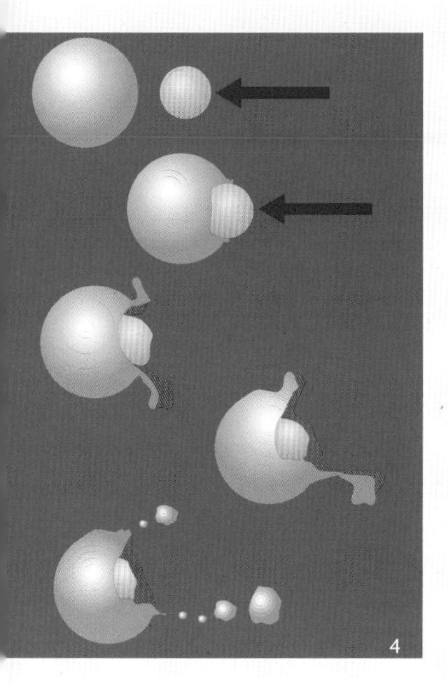

of space in which the Earth was formed: isotopic analysis of samples demonstrates that.
The hypotheses
1. The Moon was a stray body formed at a distance from the Earth (like the comets and asteroids) and captured by chance by the Earths gravitational field.
2. Like the Earth and other planets, the Moon was consolidated from primitive nebulae by the gradual aggregation of " planetesimals". A variation of this theory supposes that the primordial Earth was surrounded by a disk of accretions similar to that of Saturn, extending out to a distance of approx. 3 terrestrial radii: it would have been this material that fed the growth of the protomoon.
3. The Moon is a "drop" of Earth, lost from our planet when it was still semi-fluid, due to the speed of rotation.
4. The Moon had been ripped from the Earths crust by the violent impact of a meteorite.
5. Today, the most accepted hypothesis is a compromise from amongst the previous ones, based on computer simulations which have tried to include all the data acquired since 1986 into a single scenario. When the Earth had just consolidated its rocky crust, a protoplanet of at least 6,000 km in diameter (almost as large as Mars) would have collided with ours disturbing the crust and mantle and provoking, in a few minutes, the formation of two gigantic protuberances of material vaporised in the collision. Within a few hours of impact, the expelled material would have begun to condense in orbit around the Earth. According to this mathematical model, 23 hours would have been sufficient.

CRATERS, RINGS, RILLES AND CREVASSES

Although they are minor geological structures, yet they are these which best characterise the Moon. Furthermore, in some cases, they reach truly remarkable dimensions. We will see what they are made of, how they are classified and to which geological phenomena they are correlated.

Craters and Rings

More numerous on the plateaux, the have very heterogeneous shapes and sizes: the diameter can be comprised of between hundreds of kilometres and a millimetre; they can have a central or off-centred peak or a hilly or completely flat base (in this case they are called rings); they can be surrounded by numerous concentric rocky circles; having circular, elliptical or irregular form; being similar to our gorges, filled with lava almost up to the brim; being well conserved, semi destroyed, fissured by crevasses... The variety is truly enormous!

In the huge majority of cases, they are produced by the long ago impact of meteorites on the lunar surface: The rock analyses demonstrate this, the geological investigations and the fact that almost all the bodies of the solar system have similar superficial structures. It is exactly the examination of impact craters discovered on the Earth that has allowed the outlining of some criteria for the identification of an impact crater with speed and relative certainty:

- it is associated with meteoritic material (metallic and rocky-metallic);
- frequently shows negative gravitational anomalies due, probably, to the subterranean presence of the meteorite;
- it almost always has a circular or stretched shape;
- it has raised edge layers and the stratigraphy is overturned;
- it has a central peak;
- it is surrounded by blocks of material (called ejecta) distributed at a distance from the centre, depending on the degree of violence of the impact. Sometimes the ejecta have conserved energy sufficient to produce secondary craters recognised by the alignment, and the shape being frequently

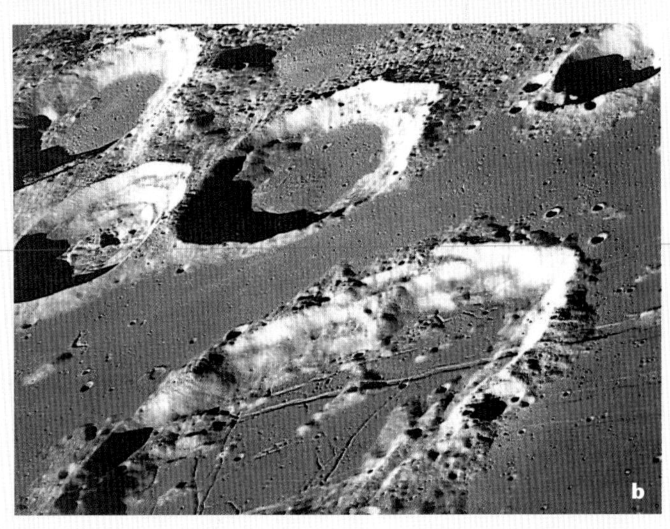

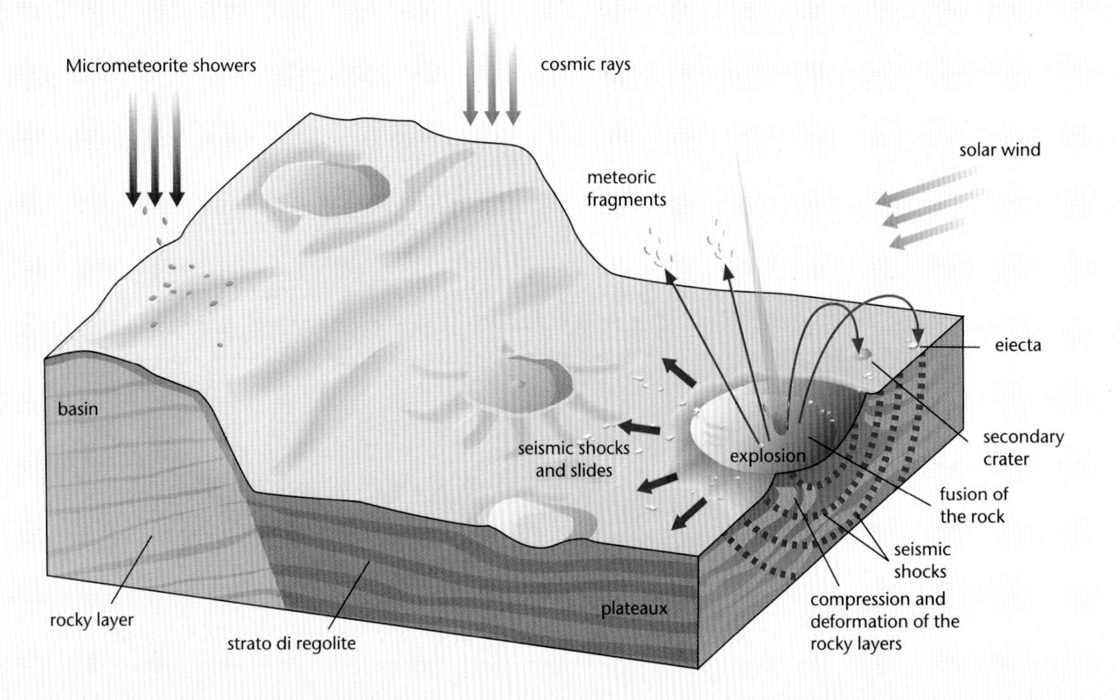

Canals, crevasses and craters
a, **d** and **f**: images of lunar canals. Are probably produced by effusive activity, likely associated to the formation of the basins.
b, **c**, **d**, **e** and **g**: craters. In b are well distinguished the internal crevasses bound to meteoric impacts;
In **c** are identified the internal terracing and the central peak, whilst
d and **e** are small, perfectly circular and relatively deep.
g, the crater Tsiolkovsky, shows its characteristic dark colouration, perhaps due to an effusive process.
NEXT: dynamics of the formation of an impact crater.

elongated, lesser depth and an irregular border. Notwithstanding the common origin, the structure of craters varies with the dimensions.

• LARGE CRATERS: always circular, with a well-defined border and uniform height, having a series of concentric fractures and terraces on the internal walls. On the external walls are seen traces of the flow of molten material and, at a distance, one can be seen a continuous crown of ejecta which follows a discontinuous crown of finer material (light halo).

• SMALL CRATERS: have a diameter of less than 20 km and the smaller they are, the more perfect their circular shape. Poorly terraced pebbly escarpments surround bowl or tunnel shaped depressions.
Some times these craters are volcanic in nature and in this case, they do not usually have central peaks. Proportionally, they are deeper than the large craters.

c

Crevasses and Rilles

The crevasses visible on the bottom of the craters are associated with the splitting of the lunar crust produced by impacts; instead, the rills visible in the basins are produced - perhaps – by lava flow: up to 300 m deep, they stretch windingly for even up to hundreds of kilometres, showing analogous characteristics to terrestrial lava tubes. The low gravity will have greatly enlarged their dimensions, whilst earthquakes and meteorites will have destroyed the roofs.

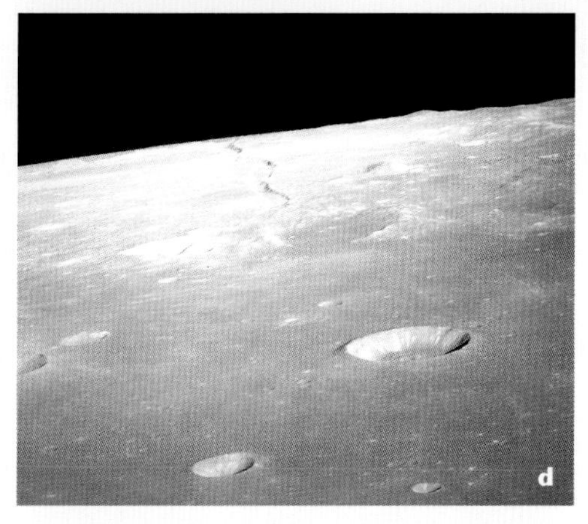
d

e

Folds and Canals

These formations are faults produced by the same phenomena which gave rise to basins, through a series of rapid, isolated and very distanced lava emissions: this hypothesis can be confirmed by the assimetry of the deposits. Steeper on the side of the river basin where they are found.

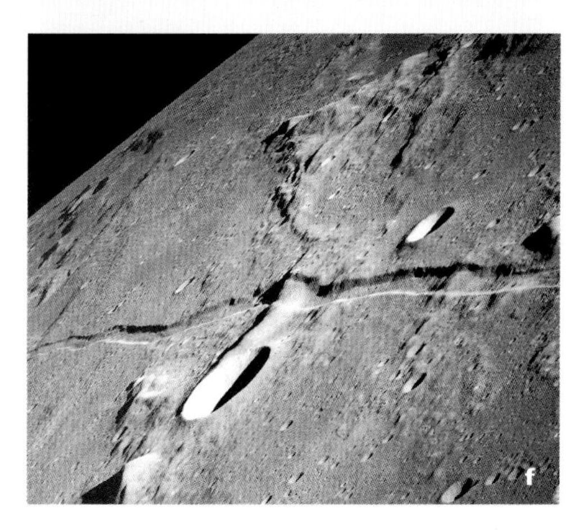
f

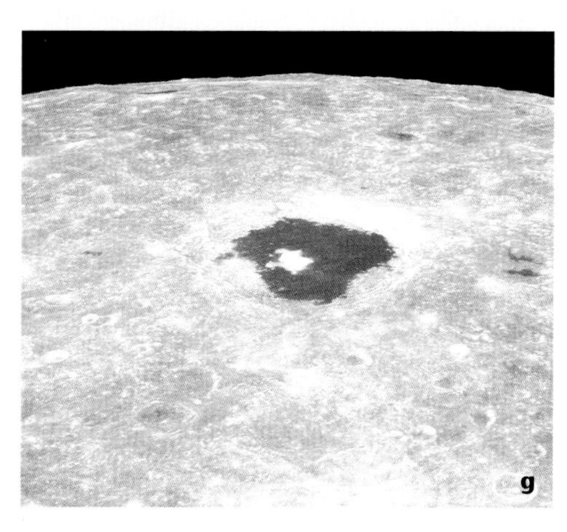
g

Direct examination
Adjacent and on the following page are two moments showing the collection of rocky samples during the Apollo 12 mission.

Regolite
1% of regolite is meteoric material: it is thought however that it is produced by the bombardment and accumulation of meteorites of various sizes.

Centre: a lunar rock brought to Earth by the Apollo missions.

only 2-4 mw/cm2: around a quarter of that of the Earth. Mountain chains talking about precipices which separate two plateaux at different heights: even in this case, one manages to connect the origins to impacts.

MINERALS, ROCKS AND REGOLITE

Even if, on the Moon, in the absence of an atmosphere, water and life, sedimentary rocks similar to these on Earth do not exist, even if all the lunar rocks are almost completely degassed and anhydrous (devoid of water), that notwithstanding the lunar rocks are essentially similar to these on Earth: even their isotopic compositions correspond.
The characteristics of the minerals vary greatly if the rock originates from a basin or a plateau. The basins are formed above all by basalt with a mean density of 3.350 g/cm3, very close to the mean density of the entire satellite (3.340 g/cm3): considering the internal pressure and temperature which they would need to be derived, one can deduce that the Moon cannot be completely basaltic and that the metallic nucleus can only be very small. Although richer in iron, titanium and refractory elements and more scarce in noble elements, the lunar basalts are similar to these on Earth: rich in pyroxenes, they are scarce in alumina which instead abounds in the rocks of the plateaux.
Despite being mineralogically very heterogeneous, the plateaux are characterised by metamorphic rocks: the meteoric impacts, very frequent here, have fused, fragmented and compressed the rocks for millions of years.

Anortosite, a calcium and aluminium silicate rare on Earth, is the rock covering the largest part of the lunar surface; exclusive, are pyroxyferroite, the first extra terrestrial mineral ever discovered, and armacolite, predominantly made up of iron or magnesium and oxides of titanium. But the majority of the lunar surface is not rocky: it is

STRUCTURE OF THE MOON
The hypocenters of endogenous tremors are found between 600 and 950 km in depth: perhaps here is the geological boundary between the crust and the plastic interior. The propagation of the seismic waves also suggests the existence of some discontinuities.

regolite
Homogeneous soil Waves at 100-900 m/s
20 km: discontinuity waves at 6.7 km/s
20-60 km: solid crust, waves at 6.7 km/s
60 km: discontinuity waves at 9 km/s
60-150 km: solid mantle, waves at 7.8 km/s
150-1,000 km, perhaps solid mantle
1,000-1,500 km: fluid central kernel, with a temperature of around 1,100°C

ORANGE SOIL
Detail of samples of "orange soil" brought to earth by the Apollo 17 mission. Vitreous, probably of volcanic origin, like other samples they are rich in titanium (8%) and iron oxides (22%), however it is the exceptional concentration of zinc which characterises them (the dimensions are 20-45 (mm).

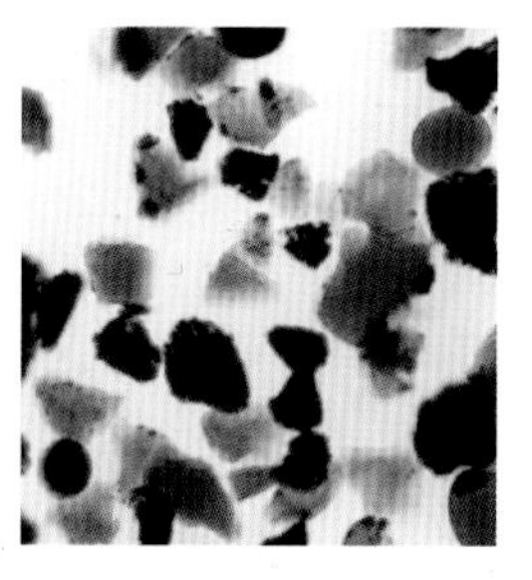

covered by regolite a material made up of fragments with very variable granulometry. The regolite layer has minimal thickness in the basins.

THE MOONS INTERIOR

All the Apollo missions have left instruments capable of measuring seismic events in three dimensions on the lunar surface, identifying, i.e., the point of origin of the waves, the depth of the hypocenter and quantifying the liberated energy. Analogously to that done on Earth, information on the lunar interior has been gathered. Of the approx. 3,000 lunar quakes registered on average in a year, none is superior to the second degree on the Richter scale. They are superficial, due to sudden changes in temperature.

INSTRUMENTS
In the zone of the "orange soil" finding, a sundial with photometric paper serves as a reference to fix the angle of incidence of the solar rays, the scale and colour of the moon.

Like all the celestial objects, the Moon also rotates around its own polar axis. Like all satellites, the Moon also spins around its own planet, the Earth.
But seen from down here, the movement of the Moon is truly complicated, and gives rise to phenomena which have always aroused amazement.

MOVEMENTS AND LUNAR PHENOMENA

The hidden moon
This map, reconstructed right all thanks to photographs taken by the Soviet probes of the Luna series Luna (below Luna 1), show the hidden face of the moon (dark side). The fact that the Earth can see only one face of the Moon is a consequence of the combination between lunar rotations and revolutions.

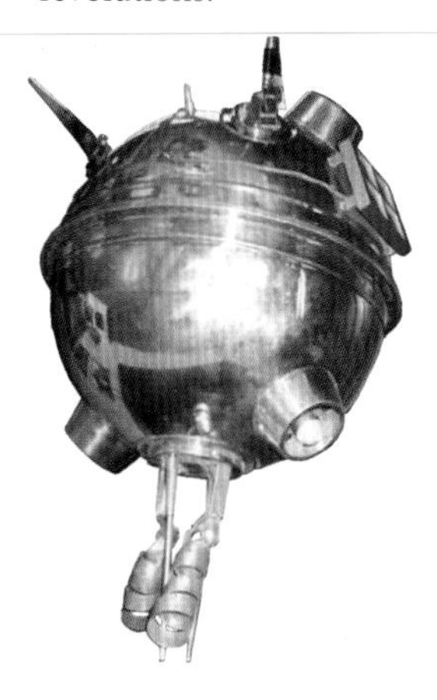

The movement of the Moon in space, like that of all the other celestial bodies, is extremely complicated.
Dominated by the gravitational presence of the Earth, and also very influenced by the Sun→136 and the other planets→154.
Like that of the Earth, the lunar motion has been broken down into simplex motions describable by the laws of Newton[21]: The Moon spins on its own axis with a rotational motion and around the Earth with a revolutionary motion. More precisely, however, one must speak of the motion of the Earth-Moon system around a common centre of gravity which is found at 4,635 km from the centre of the Earth on the conjunction of the centres of mass of the two bodies, i.e. At 1,740 km below the Earths crust. It is this point that travels the elliptical orbit around the Sun called terrestrial orbit.
The Earth, moving around the Sun and rotating around the centre of mass of the system, comes to find itself, with respect to this point, more advanced or more retarded: that determines above all the so called "lunar inequality", a periodic difference in the position of the Sun of 6.4". Regarding lunar motion, however, the necessary calculations to speak in terms of "system" would be extremely complex: seeing that the centre of gravity is found inside the Earth, the almost

consolidated simplification is acceptable.

THE ROTATION AND ITS CONSEQUENCES

The Moon rotates around its own polar axis: with respect to the plane of revolution, the axis is inclined by about 5°. This fact implies that:
• lunar days have only minimal seasonal variations,
• the lunar days have minimal differences in

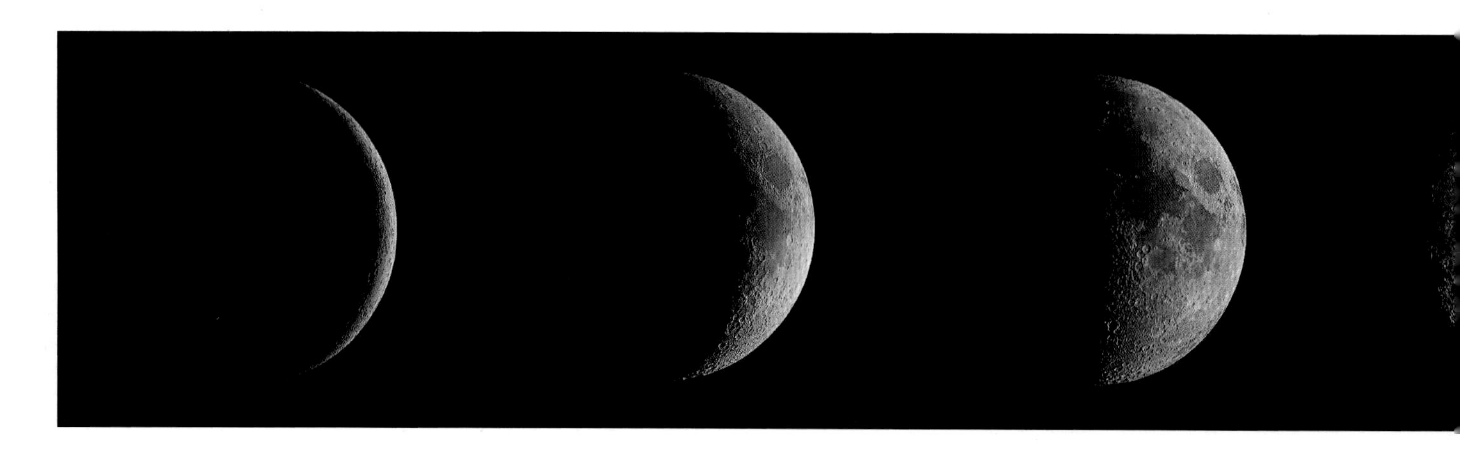

Orbital planes
The angular relationship between the ecliptic and the orbital plane of the Moon. The line of intersection between the two planes and said node line.

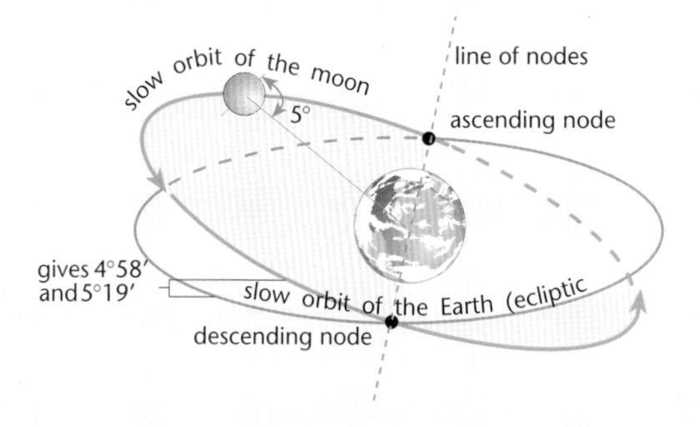

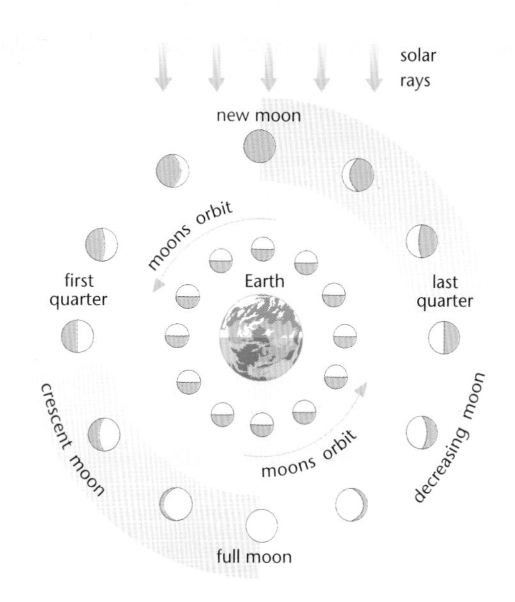

Dynamics
The diagram shows, in the outer band, how the appearance of the lunar face visible from Earth changes with the position of the Moon with respect to the Sun and Earth. In the inner band of the figure, The Moon completes its revolution and is always half illuminated.

duration at different latitudes,

• the Sun always reaches the horizon of the lunar poles.

In addition, seeing that the period and direction of rotation are equal to the period and direction of revolution, the Moon always exposes the same half to the earth: a phenomenon due to the Earths attraction which, as Lagrange demonstrated in 1764, has slowed down the original lunar rotation.

THE REVOLUTION AND ITS CONSEQUENCES

The orbit which the Moon follows around the Earth is approximately an ellipse, of which the Earth occupies one of its foci. The motion takes place from West to East, in the same direction as the rotation and the revolution of the Earth.

The lunar revolution period, i.e. The intervening time between two passages of the Moon through the same point of the orbit, also called sidereal month: it has a duration of 27.32 mean solar days. The lunar plane of revolution is inclined with respect to the ecliptic by an angle which varies (according to the gravitational interference of the Sun) between 4°58' and 5°19'. The node line which the two planes have in common, meets the trajectories of the orbits in two points called nodes. These move along the ecliptic by 19° each year: to travel all the ecliptic takes 18.61 mean solar years. Instead the apsis line, is that which joins the perigeon and the apogee: it also "slides" along the ecliptic, but in the opposite direction to the node line, and to complete an entire revolution takes 8.85 mean solar years.

THE LUNAR PHASES

The illumination of the lunar face turned towards us changes according to the position of our satellite with respect to the Earth and the Sun. The different appearances with which the Moon appears are called phases, and the main ones take various names.

• New Moon phase (or novilunar): is when Sun, Moon and Earth, in that order, are found on the same line; The Moon is invisible because it rises and sets together with the Sun: by day we do not see it because it presents us the non illuminated face, by night it is simply not there.

• the Crescent Moon phase occurs following the new Moon; the illuminated part of the Moon becomes a very thin, ever expanding, segment. Some times, inthis period, the dark part of the Moon can be seen to shine with weak luminosity:

Lunar phases
Our satellite rotates around its own polar axis in 4 weeks: since it takes the same amount of time to revolve around the Earth, it always shows the same half towards us (around 58% of its surface): ours is a synchronous satellite. That means that, if the lunar day lasts 708h, a point on the lunar equator remains in the Sun for 354h and in the dark for 354h. The Moon, however, does not always have the same appearance: according to the position in which it is found with respect to the Sun and the Earth, in fact, changes the illuminated portion which we see. From the left: crescent Moon, first quarter, crescent Moon, full Moon, waning Moon, final quarter. The new Moon is invisible, hence it is missing.

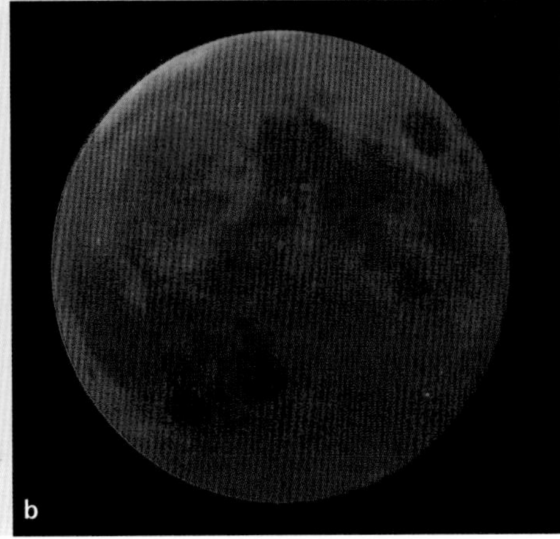

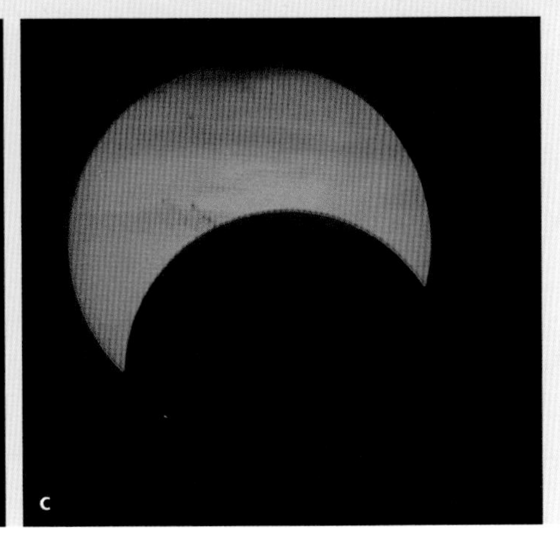

Lunar phenomena
a. Ashen light.
b. Eclipse of the moon.
c. Partial eclipse of the sun.
The eclipses are perhaps the celestial phenomena which have most interested the astronomers of ancient times: Chaldeans, Egyptians and Greeks were already able to predict some.

Lunar months
Whilst the sidereal months close when the Moon again assumes the same position with respect to a fixed Star, so far away that its movements along the Solar orbit are negligible, the sinodic month (or lunation) closes when the Moon again assumes the same position with respect to the Sun. In comparison to sidereal months, therefore, lunation has higher duration.

it is the ashen light. The Moon, in fact, reflects back towards us the light reaching it reflected from the Earth illuminated by the Sun.
• ***First quarter*** phase: the Earth-Moon conjunction is at 90° to the Earth-Sun conjunction; a quarter of the lunar surface is seen (first half of the disk).
• ***Full Moon*** or ***plenilunar*** phase: Sun, Earth and Moon, in that order, are on the same line; all the lunar disk appears illuminated because the Moon rises when the Sun sets.
• ***Waning Moon*** phase: after the full Moon, the illuminated part of the Moon becomes ever less extended, until reducing to a segment.
• ***Last quarter*** phase: the Moon-Earth conjunction is at 180° to the Earth-Sun conjunction; a quarter of the lunar surface is seen (second half of the disk, that which was obscured during the first quarter). The first and last part phases are also called ***quadrature***.
The passing time between two successive moments in which the Moon assumes the same appearance (and in the same phase) is called ***sinodic month*** or ***lunation***, and lasts 29.53 mean solar days. It is one of the first units of time measurement adopted by man: the regularity with which the lunar phases succeed, in fact, is absolute, two identical phases repeat every 29d 12h 44min 3s. The peoples who have used the Moon as a reference for scanning the passing time and organising a calendar have been very numerous, and still today the Arabs establish their year on lunar cycles: in the Islamic calendar, the year lasts 354 days and is composed of 6 lunar months of 29 days alternating with 6 lunar months of 30 days. Naturally the Islamic calendar is as precise as that used by us (based on the apparent movements of the Sun), and perhaps even more precise, but, given that it scans time in a different manner, it does not match ours.

THE ECLIPSES

• If the full Moon phase occurs when the Moon is found in a point of its orbit close to the ecliptic, the shadow cone of the Earth can obscure it: this is a partial eclipse of the Moon
• If the full Moon phase occurs when the Moon is found in a node, the centres of the Sun, Earth and Moon, in that order, are aligned: the Moon completely enters the shadow cone of our planet and we have a total eclipse of the Moon The Moon during an eclipse never completely disappears but assumes a reddish colouration: The shadow produced by the Earth, in fact, is always weakly illuminated by red rays from the Sun which are refracted by the Earths atmosphere as in an immense sunset.
• If the new Moon phase occurs whilst the Moon is in a point of its orbit close to the ecliptic, the shadow cone that it produces can obscure a part of the terrestrial surface: this is a partial eclipse of the Sun. Seen from Earth, the shadow disk of the Moon only partially covers that of the Sun.
• If the new Moon phase occurs when the Moon is found in a node, the centres of the Sun, Moon and Earth, in that order, are aligned: all the shadow cone of the Moon is projected onto the Earth and this is a total eclipse of the Sun. Seen from Earth, the dark disk of the Moon – which has an apparent diameter almost equal to that of the Sun – progressively covers all of the disk of the Sun obscuring

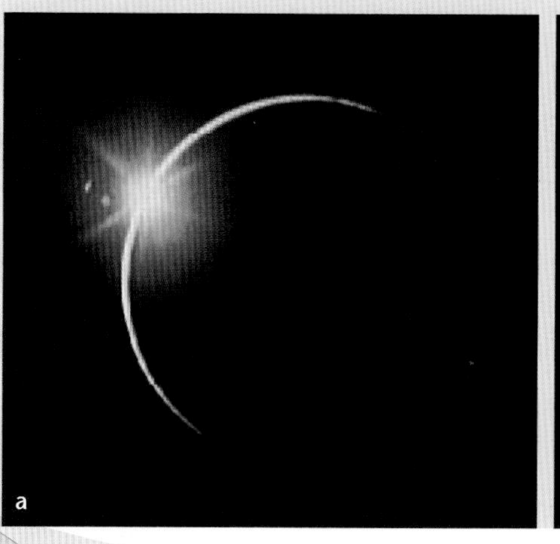

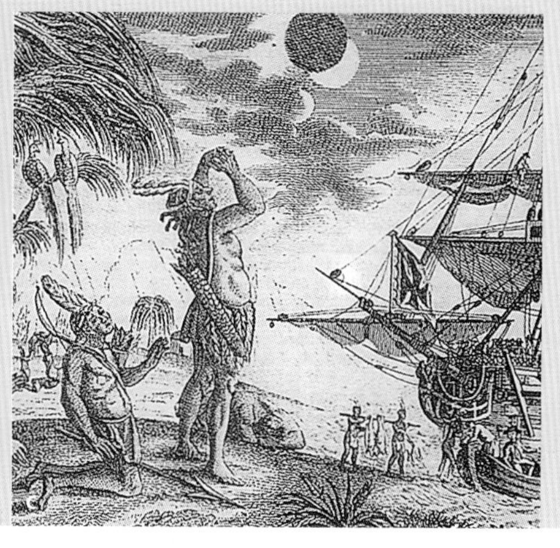

Solar eclipses
a. Total eclipse of the Sun.
b. Annular eclipse.
c. In this period print is depicted the anecdote according to Christopher Columbus, exploiting his astronomical knowledge, he would have predicted a solar eclipse, prevailing so easily over the hostile natives.

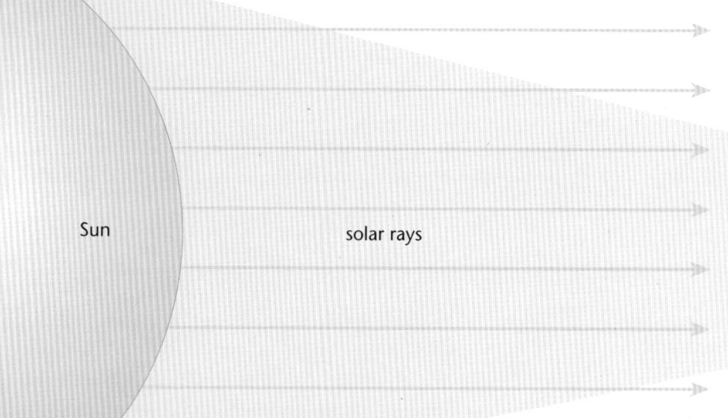

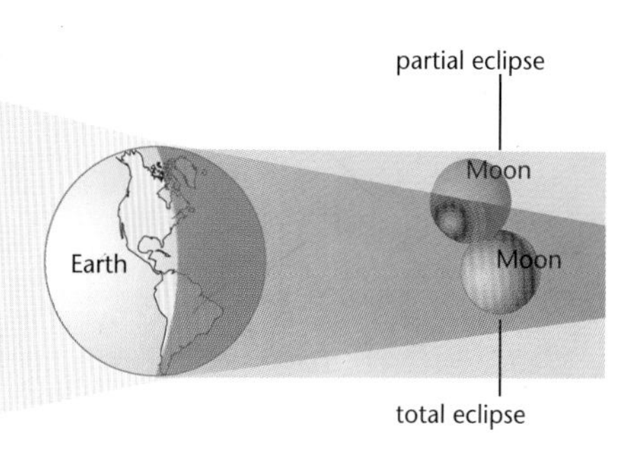

Eclipses of the moon
The diagram shows the arrangement of the celestial bodies involved in a partial eclipse of the Moon and in a total eclipse of the Moon.

Eclipses of the sun
The diagram shows the arrangement of the celestial bodies involved in a partial and total eclipse of the Sun. The photograph taken from space shows the shadow of the Moon on the surface of the Earth during a total eclipse of the Sun.

it for a few seconds. In these moments of total darkness the stars and the parts of the Sun, which are normally invisible, can be seen (for example the corona➢142). The Moon, however, precedes it in motion, and the shadow "sweeps" wide areas of our planet. When the Moon is found at the apogee➢54, it's disk has an apparent diameter a little smaller than that of the Sun: if under these conditions a total eclipse of the Sun occurs, the Moon does not cover all of the luminous disk and we have an anular eclipse. On average, around 70 eclipses occur every 18 years: 42 lunar and 28 solar.

LIBRATION MOVEMENTS

Although the Moon always presents us with the same face, we manage to see a little over 59% of it: it is a phenomenon attributed to apparent oscillations in the lunar globe due to slight changes in our point of view. These oscillations, called librations are of four types:

- latitudinally, it is analogous to the phenomenon of the terrestrial seasons and is due to the fact that the Moon maintains its axis of rotation inclined to the orbit. That allows one to see, in alternate periods, a part of the lunar polar regions beyond the limits of the hemisphere normally facing the Earth;
- longitudinally, is due to the fact that the rotation and revolution are not perfectly synchronous: whilst the rotation is a uniform motion, the

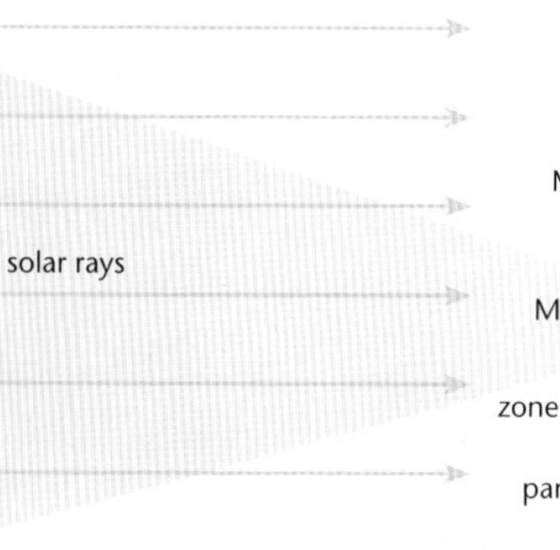

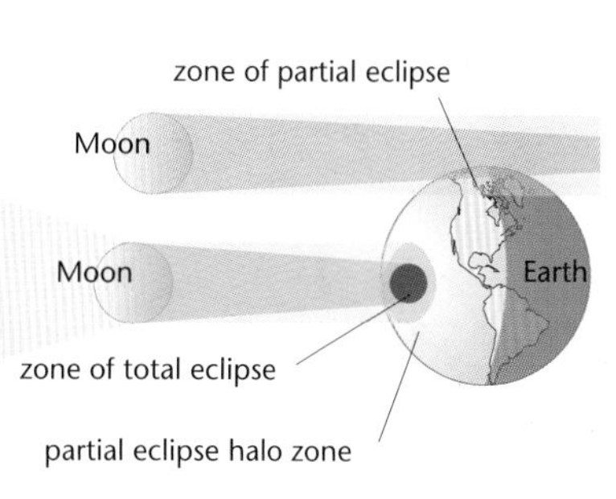

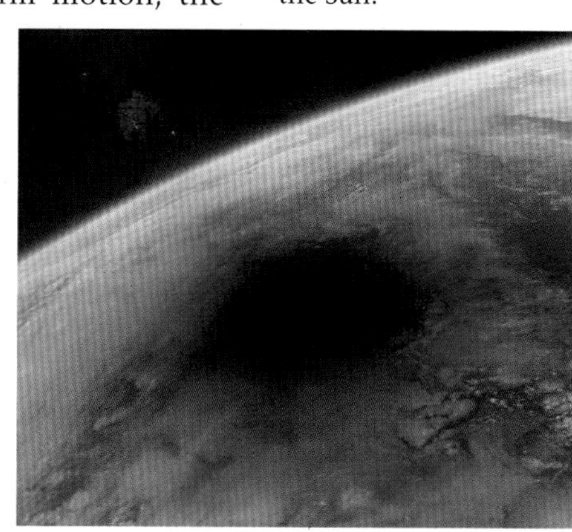

Stonehenge
The study of lunar motion has led to the construction of observation structures from which it should be possible to fix and recognise particular points of the horizon corresponding to the moments of "stasis" in the path of the Moon.

The tides
The gravitational actions of the Moon and Sun and the rotational forces of the Earth-Moon system provoke the phenomenon of the tides. The maximum height of the tides occurs with the new or full Moon; the minimum height to the first or last quarter.
The marine and oceanic tides are very obvious in some coastal areas for their contribution to particular meteorological conditions or the formation of depressions. The tides, however, also effect the continents, the atmosphere and, on the Moon, have established a permanent deformation of the crust and the movement of the centre of mass in the direction towards our planet.

revolution follows the laws of Kepler21 increasing the velocity at perigee and reducing it at apogee. Allowing the view of some oriental and occidental regions beyond the limits of the hemisphere normally facing the Earth;
• parallactic or diurnal, is due to the fact that the observer is not located at the centre of the Earth but on the surface. At Moon rise one can see a little of the surface beyond the Eastern edge, at setting one can see beyond the Western edge;
• physical, is due to real irregularities in rotation. It has negligible effects.
The result of the first three librations is the said apparent libration.

EARTH-MOON DISTANCE

Since the Moon has a relatively small mass, even forces which seem to have no influence over the movement of the earth provoke sensitive variations in lunar movement. The principal disturbances, however, are almost exclusively due to the variations in the Moon-Sun distance: that provokes oscillations in the force with which the Sun attracts the Moon. In the first and last quarter phases, for example, both the Earth and the Moon are subjected to almost equal, but converging, solar attraction, and tend to move closer together. In the new Moon phase, the force which attracts the

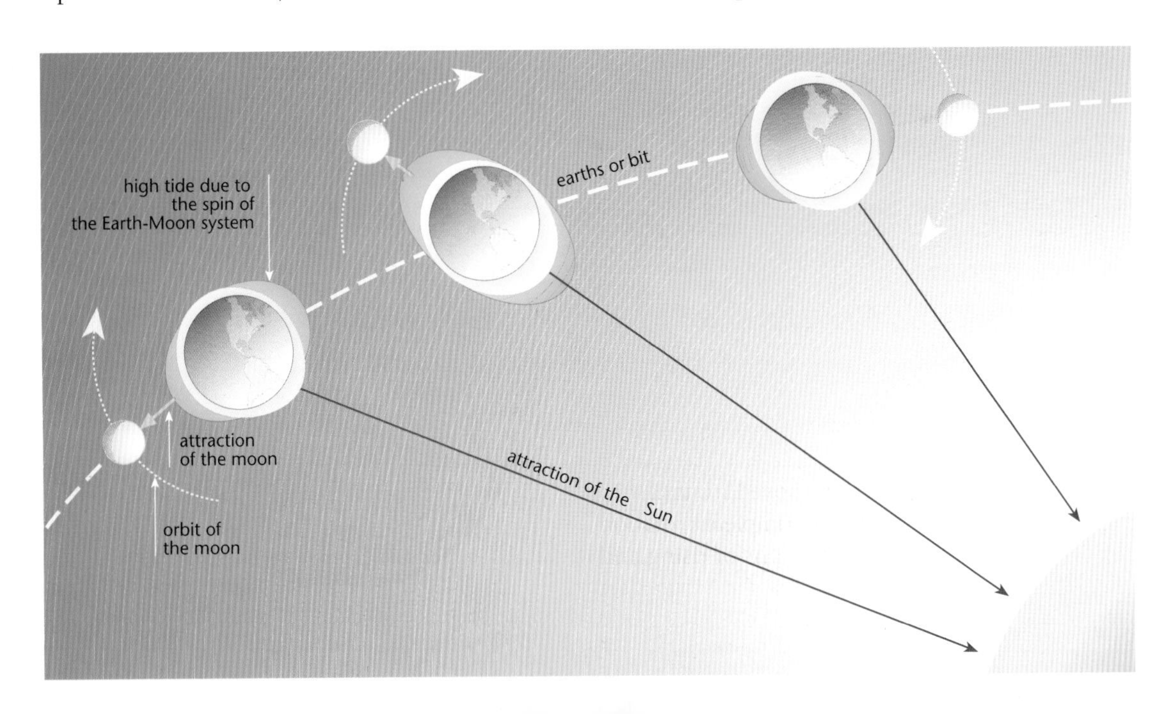

Perigee: less than 356,425 km		**Apogee: more than 406,710 km**	
Date	Distance from Earth	Date	Distance from Earth
15 December 1548	356,407 km	9 January 1921	406,710 km
26 December 1566	356,399 km	2 March 1984	406,712 km
30 January 1771	356,422 km	23 January 2107	406,716 km
23 December 1893	356,396 km	3 February 2125	406,720 km
4 January 1912	356,375 km	14 February 2143	406,713 km
15 January 1930	356,397 km	27 December 2247	406,715 km
6 December 2052	356,421 km	7 January 2266	406,720 km
29 January 2116	356,403 km	18 January 2284	406,714 km
9 February 2134	356,416 km	29 November 2388	406,715 km
22 December 2238	356,406 km	11 December 2406	406,718 km
1 January 2257	356,371 km	21 December 2424	406,712 km
12 January 2275	356,378 km	21 January 2452	406,710 km
26 January 2461	356,408 km	1 February 2470	406,714 km
7 February 2479	356,404 km	12 February 2488	406,711 km

Moon is greater; in full Moon phases it is less. The maximum and minimum distances between the Earth and the Moon, apogee and perigee, however, due to the disturbances in lunar motion can only be determined with an uncertainty of around 20 m: the table shows the distances calculated by Meeus on the basis of a simplification of the theory of Chapront. The maximum apogee and the minimum perigee always occur during the winter in the terrestrial Northern hemisphere because it is exactly in this period that the Earth-Moon system, at perihelium, suffers the greatest gravitational disturbances. For the same reason, the difference between the distances of the perigee (51 km) is greater than that between the distances of the apogee (10 km). Still due to the disturbances in lunar motion, the perigee moves in the same direction as the orbital motion of the Earth: the time which the Moon takes to return to perigee is said to be an anomalous lunar month and lasts 27d 13h 18min 33s.

THE MOONS MOVEMENTS IN THE SKY

If we observe the Moon, repeating the observations at the same time on successive evenings, we would see it change rapidly. It would change shape, because the portion we see illuminated by the Sun (phase) would change; the inclination of the trajectory followed in the sky would change, which – as with the Sun – depends on the season and on the observation point; the point in which it rises (and sets) on the horizon would change, moving every day, and would change the time in which it rises (and sets), rising each day with a delay of around 53 min with respect to the previous day, the time in which the celestial sphere, which due to the rotation→40 of the Earth is moved from East to West by 15° each hour, takes to complete a rotation of 13°. The change of the positions of rising and setting and the delay with which the Moon rises and sets, are a consequence of the rotational motion of the Moon around the Earth exactly as happens, over the course of the seasons, also for the Sun. Over the course of a year, in fact, even the Sun changes the position and time of rising and setting: "revolving" around the Earth along the ecliptic, in fact, in the period comprised of between the spring equinox and that of autumn it always rises more to the North with respect to the point East, and always rises more to the South in the period between the autumn equinox and that of the spring. The Sun, however, performs a complete oscillation with respect to the point East in an entire year. On the other hand the Moon carries out a complete oscillation with respect to the point East in one month.

Epigee and Apogee
The Moon has a different apparent diameter according to the distance from the Earth. Above are compared the scale images of the Moon at perigee and at apogee.

Solar and Lunar motions
Variations in the point of rising and setting of the Sun (in a year) and of the Moon (in a month) at average latitude.

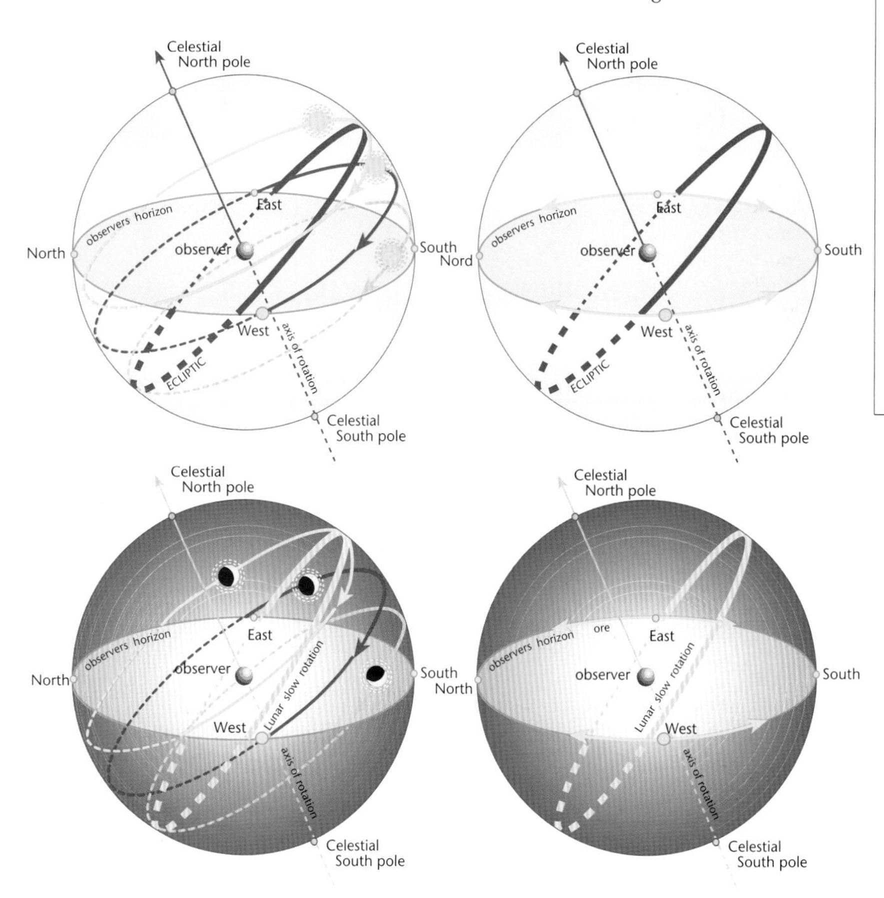

It is thanks to direct space exploration that today we know so much about the Moon.
But the conquest of our satellite has been the fruit of a contested international competition which cost billions of dollars in a few years. And the lives of a few brave men.

EXPLORATION AND CONQUEST

Lunakhod 2
This is the second "lunar rover", the Russian all-terrain vehicle remote controlled from Earth which, disembarked from the probe Luna 21 in the region of the Le Monnier crater, it travelled 37 km in 4 months. Active only during the lunar days, it sent home vast amounts of varied information.

Below, Aside, Centre: The Mercury capsule during preparation and ground testing; the chimpanzee Ham is recovered following his trip on the Mercury redstone rocket

It is the Soviets who started the "race to the Moon". They reached it with probes, ever more perfected: these release remote guided robots onto the lunar surface which succeeded in gathering samples, in performing investigations and in collecting data and images which they sent or physically brought back to Earth. They were the Luna and Zond projects: 32 probes equipped with ever more sophisticated instruments, over the course of 17 years, allowing the creation of the almost complete map of the hidden face (dark side) of the Moon, gathering samples of lunar soil and performing chemical analyses, discovering the lunar magnetic field through specific surveys. That meant perfectly mastering the technology of "soft" landing, of remote control, of the automation which permited performing numerous activities, both in flight, and on the Moon. It was a challenge which the United States could not ignore. And the commitment to make human beings reach the Moon became a point of national pride. An effort which cost hundreds of billions of dollars and the work of over 500,000 scientists and technicians for 15 years: this is the Apollo Program which made the United States the only country to plant its flag on the Moon. But to arrive at just over 300 hours of lunar walking and trips around the craters on special Rovers has cost much more: also used for this were the costly Surveyor and Lunar Orbiter probe programs, which performed accurate preliminary recognitions of the lunar surface, and the Mercury

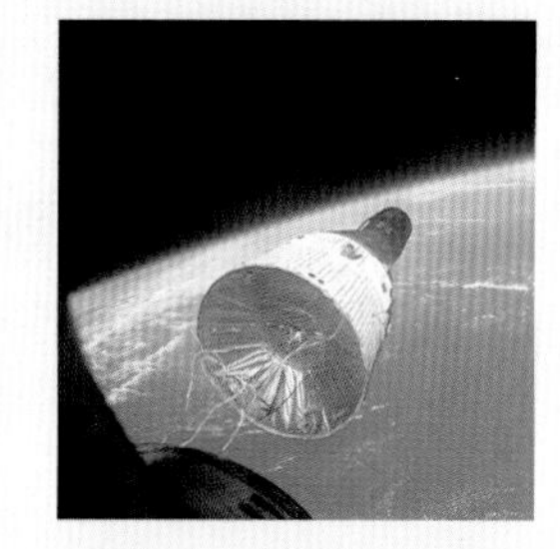

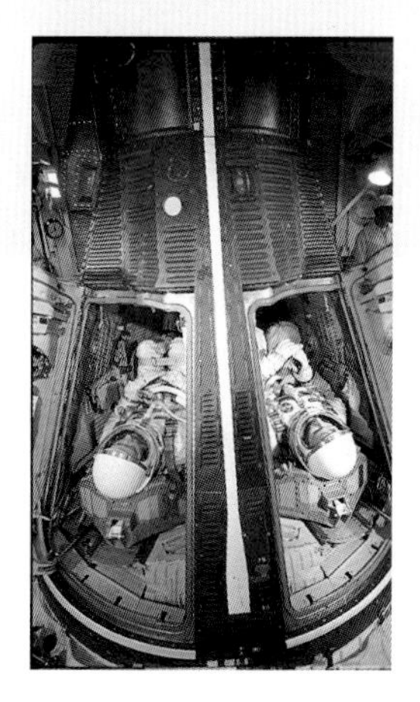

Around the Earth
The first EVA (Extra Vehicular Activity) of a man in space. The Gemini program had to refine all the necessary technology for the landing on the Moon.
Above: the capsule in orbit and inside;
Below: the recovery of the vehicle after splashdown.

and Gemini programs which from 1958 to 1962 developed the necessary technology for the Apollo Program. Lets summarise them:

• the Mercury Program, 393 million dollars and over 2 million people involved, brought the first Americans into terrestrial orbit. It is the beginning of the solutions to great technical problems such as that of the heat shield for re-entry, the stability of the rocket vector, the microclimate of the capsule, the suits...
Only one astronaut at a time is launched a short distance from Earth, and for a short time;

• the Gemini Program, over a billion dollars, doubles the crew, organises space rendezvous and the first exit from the vehicle consolidating the space methodologies acquired.

Is it worth it? From the scientific point of view certainly yes: the protests have been silenced by the copious "spinoffs" which, in more diverse fields, have compensated for the enormous undertaking

ASSAULT ON THE MOON

DATE	MISSION
1959	
2.1	**Luna 1**: passes at 4,000-6,000 km from the lunar surface
3.3	**Pioneer**: passes at approx. 60,000 km from the lunar surface
12.9	**Luna 2**: the first to land on the Moon measures the lunar magnetic field
4.10	**Luna 3**:passes at approx. 7.000 km from the Moon and photographs the visible side
1962	
26.1	**Ranger 3**: attempt at lunar landing, passes at approx. 37,000 km from the Moon
23.4	**Ranger 4**: crashes on the hidden face (dark side) of the Moon
18.10	**Ranger 5**: attempt at lunar landing, passes at approx. 700 km from the Moon
1963	
2.4	**Luna 4**: passes at approx. 8,500 km from the Moon collecting data
1964	
30.1	**Ranger 6**: crashes on the Moon
28.7	**Ranger 7**: crashes on the Moon sending back to Earth more than 4,300 photos taken from a height of 1,500 m over the Sea of Clouds which is then re-baptised Mare Cognitum (the Known Sea). Some photos have a resolution of 40 cm
1965	
17.2	**Ranger 8**: crashes on the Moon sending to earth more than 7,000 photographs with wide panoramas of the Sea of Tranquility: some have a resolution of 2 m
18.11	**Ranger 9**: crashes in the Alphonsus crater sending to Earth over 5,800 photographs
9.5	**Luna 5**: crashes on the Moon
8.6	**Luna 6**: attempt at lunar landing; passes at 160,000 km from the Moon
18.7	**Zond 3**: flies over the Moon sending to Earth 25 photos of the visible side
4.10	**Luna 7**: crashes on the Moon
3.12	**Luna 8**: crashes on the Moon
1966	
28.9	**Luna 9**: first "soft" lunar landing; takes 27 photos and measures lunar radioactivity
31.3	**Luna 10**: is the first artificial satellite of the Moon; repeats the measurements of the lunar radioactivity remaining in orbit
30.5	**Surveyor 1**: lands and sends over 11,100 photos to Earth
1.7	**Explorer 33**: attempt to enter into lunar orbit
10.8	**Lunar Orbiter 1**: enters into lunar orbit; takes 211 photos and performs gravimetric tests discovering the mascon
24.8	**Luna 11**: enters into lunar orbit and collects data
20.9	**Surveyor 2**: attempts lunar landing and crashes on the Moon
22.10	**Luna 12**: enters into a lunar orbit which later changes, takes 422 photos
6.11	**Lunar Orbiter 2**: from lunar orbit takes 184 photos and performs gravimetric tests discovering other mascons
21.12	**Luna 13**: lands and makes some measurements of the ground of the basins
1967	
4.2	**Lunar Orbiter 3**: in lunar orbit takes 182 photos and conducts gravimetric tests discovering new mascons
17.4	**Surveyor 3**: lands, examines the lunar soil and sends over 6,300 photos
8.5	**Lunar Orbiter 4**: from lunar orbit sends to earth 326 photos and, performing gravimetric tests, discovers other mascon
14.7	**Surveyor 4**: attempts landing and crashes on the Moon
19.7	**Explorer 35**: in lunar orbit carries out measurements on lunar magnetism and radiation
6.11	**Lunar Orbiter 5**: from lunar orbit takes 426 photos and, performing gravimetric tests, discovers new mascons
8.9	**Surveyor 5**: lands and carries out a chemical examination of the lunar soil sending over 18.000 photos
7.11	**Surveyor 6**: lands and sends over 30,000 photos; attempts to take off
1968	
7.1	**Surveyor 7**: lands and sends to Earth over 21,200 photos; examines the soil of the plateaux
7.4	**Luna 14**: performs gravitational surveys from orbit
15.9	**Zond 5**: lands gathering soil samples
10.11	**Zond 6**: orbits around the Moon and reports a series of photos
1968	
21.12	**Apollo 8**: return trip for a crew who orbit the Moon
1969	
18.5	**Apollo 10**: return trip for a crew who reach 8 km from the lunar surface
13.7	**Luna 15**: orbits around the Moon and crashes in the Mare Crisum
16.7	**Apollo 11**: first landing on the Moon. The men stay in the Sea of Tranquility 21h 36min then return with 21 kg of stones leaving instruments
7.8	**Zond 7**: orbits around the Moon and reports a series of photos
14.11	**Apollo 12**: second landing on the Moon. The men stay in the Ocean of Storms 31h 31min; bringing to Earth 35 kg of rocks and leave numerous instruments
1970	
11.4	**Apollo 13**: takes a crew around the Moon
12.9	**Luna 16**: lands in the Mare Foeconditatis and collects 100 g of lunar soil
20.10	**Zond 8**: orbits around the Moon and reports a series of photos
10.11	**Luna 17**: lands in the Mare Imbrium, conducts various scientific experiments and sends to Earth over 20,000 photos
1971	
31.1	**Apollo 14**: third landing on the Moon. The men stay in the Fra Maura crater 33h 31min conducting experiments; bringing to Earth 43.5 kg of rocks and leaving numerous instruments
26.7	**Apollo 15**: fourth landing on the Moon. The men drive in an all terrain vehicle to the base of the Apennine Mountains in the Palus Putredinis staying on the Moon 66h 55min. They perform many experiments, put a satellite into orbit and return to Earth with over 100 kg of rocks
2.9	**Luna 18**: sends data from lunar orbit and then crashes
28.9	**Luna 19**: stays in lunar orbit for over 4 years sending data and performing experiments
1972	
14.2	**Luna 20**: lands in the Mare Foeconditatis and collects soil samples
16.4	**Apollo 16**: fifth landing on the Moon. The crew drive in an all terrain vehicle on the Cayley plateaux and build the first lunar astronomical observatory. They stay on the Moon 71h 2min performing numerous experiments
7.12	**Apollo 17**: sixth landing on the Moon. The crew explore the Sea of Serenity and carry out an articulate program of experiments and research. They stay on the Moon 71^{h} 2^{min} bringing back 560 kg of rocks and over 110 kg of stones and powder
1973	
8.1	**Luna 21**: lands in the lemonnier crater and releases the remote controlled module Lunokhod2 which performs experiments and transmits over 80,000 photos
10.6	**Explorer 49**: carries out radio astronomical measurements on the dark side from orbit
1974	
2.6	**Luna 22**: stays in orbit transmitting images for over 1 and a half years
28.10	**Luna 23**: lands in the Mare Crisum and carries out experiments
1976	
14.8	**Luna 24**: in near orbit, collects data for around 3 months before crashing on the Moon
1990	
24.1	**Hiten 24**: in near orbit, collects data for around 3 months before crashing on the Moon
1994	
25.1	**Clementine**: orbits for around 4 months, mapping he lunar surface through various instruments (UV, infrared, visible,)
1997	
24.12	**AsiaSat3/HGS-1**: fly by
1998	
7.1	**Lunar Prospector**: in polar orbit for 2 years maps the surface compositions and searches for ice deposits, measures the magnetic and gravitational fields
2002	
	SMART 1: gathers geological, morphological, topographic, mineralogical, geochemical to answer problems over the origin of the Earth-Moon system, on volcanic activity and lunar techtonics, on the thermal and dynamic processes of lunar evolution
2003	
	Lunar-A: launches penetrators from orbit which sink approx. 3 m into the lunar soil completing research on lunar quakes, on the thermal properties and lunar energy fluxes
	Selene: equipped with 13 instruments (radar, laser altimeters, X, g, and fluorescence spectrometers), collects data from orbit and launches a probe which lands at approx. 3 m/s, destined to collect data on the lunar environment

LUNA 9

THE LANDING OF HUMAN CREWS ON THE MOON HAS CERTAINLY REPRESENTED ONE OF THE MOST SIGNIFICANT MOMENTS IN THE HISTORY OF THE 20TH CENTURY AND HAS PERMITTED A DECISIVE QUALITATIVE LEAP NOT ONLY IN THE UNDERSTANDING OF OUR SATELLITE BUT ALSO IN THE DEVELOPMENT OF SPACE TECHNOLOGIES.

THE APOLLO EXPEDITIONS

Here are some extra details, some curios, some names and some dates: a trifle, with respect to the infinite mass of historical and scientific information which one can trace on this space program which remains in the NASA annals and in the hearts and imaginations of men.

APOLLO 1
Roger Chaffe, Virgil Grissom, Edward White
A dramatic fire which causes the deaths of the astronauts at take off risks blocking the entire program. Thus follow, in secret, the 5 "tests" without crew: Apollo 2, 3,4,5,6.

APOLLO 7
Walter Cunningham, Donn Eisele, Walter Shirra
The first experiment of sending an Apollo capsule into terrestrial orbit with a crew. They experiment the linking in space of two capsules. The duration of the entire mission is around 264^h.

APOLLO 8
William Anders, Frank Borman, James Lovell
The crew orbit around the Moon and the Earth: it is the first time that men see directly, the hidden face (dark side) of the Moon. The duration of the entire mission is 147^h.

APOLLO 9
J.R. Mcdivitt, R.L. Schweickart, D.R. Scott
The program is an experimental flight in orbit around the Earth, during which the lunar module (LEM) and the command module separate and dock. The duration of the entire mission is 240^h.

APOLLO 10
Eugene Cernan, John Young, Thomas Stafford
Young stays in orbit around the Moon aboard Charlie Brown, whilst Stafford and Cernan arrive in the lunar module (LEM) Snoopy up to 8 km from the surface of the satellite. They don't land, but return to the ship, docking the module and returning to Earth. The duration of the entire mission is $192^h\ 3^{min}$.

APOLLO 11
Edwin Aldrin, Neil Armstrong, Michael Collins
Collins stays in orbit around the Moon aboard Columbia, at an altitude of around 110 km, whilst Armstrong and Aldrin land in the LEM Eagle in the Sea of Tranquility.roaming up to 60 m from the LEM, installing a teleseismometer and a laser reflector. They gather 21 kg of stones which they bring back to Earth, television and photographic images. After $21^h\ 36^{min}$ they take off: leaving a commemorative plaque and an American flag.
They undock the LEM, which falls towards the Moon, and they reunite with Collins to return to Earth. The duration of the entire mission is $95^h\ 17^{min}$.

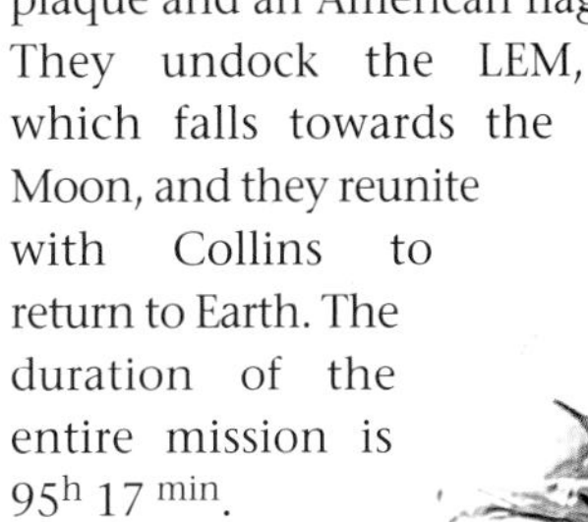

***SPIDER* IN ORBIT**
The lunar module of the Apollo 9 mission orbits around the Earth in lunar landing configuration.

FIRST STAGES
The first stage of the Saturn IB in the assembly area prior to launch.

APOLLO 8
The crew of the Apollo 8 mission. From the left: James A. Lovell Jr. William A. Anders and Frank Borman.

Apollo 11
Historic moments from the Apollo 11 mission, which lands the first crew on the Moon.
ABOVE: few moments from the launch of the Saturn 5 with the Apollo 11 mission, and the detachment of the fuel tanks.

Apollo 13
After almost 100 hours of anguish and frantic work to save the crew, finally the space ship with the three astronauts is recovered following "*splash down*" in the ocean.

APOLLO 12
Alan Bean, Charles Conrad, Richard Gordon
Gordon orbits around the Moon aboard Yankee Clipper, Bean and Conrad land in the Ocean of Storms aboard Intrepid. They roam approx. 400 m from the LEM, retracing Surveyor 3, they remove some parts and gather 35 kg of rocks up to a depth of 70 cm. In their EVA of $31^{h}\ 31^{min}$ they install many instruments amongst which a seismometer which registers the impact of the LEM undocked on leaving. Duration of the mission: $284^{h}\ 30^{min}$.

APOLLO 13
Fred Haise, James Lovell, John Swigert
An explosion damages the oxygen tanks and impedes the completion of the program placing the lives of the crew in danger. The space ship Odyssey, with the LEM Acquarius, rounds the Moon. Then the astronauts head for Earth. The duration of the entire adventure is $134^{h}\ 36^{min}$.

APOLLO 14
Edgar Mitchell, Stuart Roosa, Alan Shepard
Roosa stays in orbit aboard Antares, Shepard and Mitchell land aboard Kitty Hawk in the Fra Mauro crater, where they stay $33^{h}\ 31^{min}$. They roam over 2 km from the LEM and climb approx. 120 m of the Cono crater. Other than a seismometer with which they perform experiments exploding 13 charges, they install a laser reflector, a radio apparatus, a magnetometer which highlights a lunar magnetic field 50 times more intense that envisaged, a solar wind measurer and a plutonium generator for producing electrical energy. They bring back to Earth 43.5 kg of rocks also gathered from a depth of 80 cm. During the return, they perform numerous experiments in the absence of gravity. The duration of the entire mission is $216^{h}\ 42^{min}$.

APOLLO 15
James Irwin, David Scott, Alfred Worden
Worden stays in orbit aboard Endeavour, Irwin and Scott land in Falcon at the base of the Apennine Mountains, in the Palus Putredinis. With the Moon Rover all terrain vehicle they reach over 8 km from the LEM and, in $66^{h}\ 55^{min}$, install many instruments (seismograph, spectrometer, laser reflectors, dust detector, electrical generator), make various experiments and excursions, gather over 100 kg of samples of rocks also from a depth of more than 2 m: amongst these the "Genesis stone", the oldest dated up to now. It is the longest mission of the entire program: over all it has a duration of $775^{h}\ 26^{min}$.

Lunar walks
Buzz Aldrin, pilots the lunar module Eagle, walks on the surface of the Moon during the extravehicular activity envisaged by the Apollo 11 mission.

Plants and plates
The lunar soil is fertile: numerous attempts to cultivate plants, carried out using the samples brought back to Earth by the Apollo missions prove that. It lacks only water.

RIGHT:
the commemorative plaque left on the Moon by the Apollo 11 expedition.

Crucial moments
ABOVE: the LEM of the Apollo 11 mission leaves the spaceship; Aldrin descends from the LEM onto the lunar surface; "splash down" of the Apollo 16 capsule; celebratory parade at the return of the astronauts

Lunar Roving Vehicle
BELOW: The all terrain vehicle from the Apollo 15 mission.

APOLLO 16

Charles Duke, Thomas Mattingly, John Young

Mattingly stays in orbit aboard Caspar, Duke and Young land in Orion on the Cayley plateau. A serious breakdown reduces the envisaged program by 25^h; yet, with the Lunar Rover all the same reaching 6 km from the LEM; mounting an ultraviolet telescope, carrying out experiments that are biological and seismic in nature exploding around 30 charges; launching a small lunar satellite that transmits information for 1 year and gather over 96 kg of rocks: amongst this they find the oldest ever examined to date. They stay on the Moon 71^h 2^{min}, then return to Earth. The entire mission lasts 268^h 16^{min}.

APOLLO 17

Eugene Cernan, Ronald Evans, Harrison Schmitt

Evans stays in orbit aboard America, Cernan and Harrison land with Challenger in the Sea of Serenity, not far from the craters Littrow and Mons Argaeus. This is the last and the most coherent of the expeditions in the Apollo program: in 87^h of EVA, reaching 6.5 km from the LEM, the astronauts gather 670 kg of materials amongst which reddish "glass balls" (orange soil) which probably come from a depth of at least 300 km. They perform gravimetric, seismologic, thermal, electrical measurements, on the lunar micro-atmosphere, on space dust, on neutron emissions from the rocks, take core samples and carry out experiments even whilst in orbit, using a radar probe, a laser altimeter, a radiometer, a spectrometer. The LEM, left to fall to the Moon following the departure of Cernan and Harrison, provokes a quake of a force never before registered on the Moon, equal to the energy produced in an explosion of 700kg of high explosive. The recording is immediately sent to Earth. The whole mission lasts 301 hours.

WE SEE IT EVERY NIGHT, IN INCESSANT MOVEMENT AROUND THE POLE, AND WE HARDLY REALISE THAT IT EXISTS, ITS LIGHT DROWNED BY THE BLINDING LIGHTS OF OUR CITIES THE SKY, BLANKET OF THOUSANDS OF STARS, THE SAME SKY WHICH HAS INSPIRED POEMS AND RELIGIONS, FEARS AND REVOLUTIONS. THE SKY, WHERE THE SUN RISES AND SETS, THE MOON CHANGES PHASE, AND THE STARS SLIDE SLOWLY ACROSS, FROM ONE ZODIAC SIGN TO THE NEXT. AND ALTHOUGH WE ALL KEEP TALKING ABOUT THE SEASONS, WHICH ARE NOT AS THEY ONCE WERE, OUR LIFE IS NO LONGER BOUND TO THESE RHYTHMS, AND VERY FEW AMONG US KNOW WHY THE POSITION OF THE SUN CHANGES DURING THE YEAR, OR WHY THE MOON RISES A LITTLE LATER EACH DAY, OR HOW TO FIND A PARTICULAR CONSTELLATION AMONG THE THOUSAND LIGHTS OF THE NIGHT, WHEN DARKNESS FINALLY COMES.

BUT OUR FASCINATION WITH THE STARRY SKY IS STILL THERE, SOMEWHERE INSIDE, AND WE ALL STILL HAVE THE BASIC NEED TO LOOK UP, AND UNDERSTAND SOMETHING OF WHAT WE'RE LOOKING AT. BECAUSE TODAY IT IS POSSIBLE TO UNDERSTAND A LOT. THE MOON IS STILL MAGICAL, BUT IT'S ALSO THE ONLY CELESTIAL BODY, APART FROM THE EARTH, ON WHICH MEN HAVE WALKED. AND WHILE MARS IS STILL A POINT OF REDDISH LIGHT THAT IS ALMOST IMPOSSIBLE TO FIND, IT IS ALSO THE PLANET WE WILL SOON LAND ON. VENUS, JUPITER AND SATURN CAN BE "REACHED" WITH A SIMPLE PAIR OF BINOCULARS, AND THE "FIXED STARS", NEBULAE AND GALAXIES ARE WITHIN RANGE OF A TELESCOPE. YOU JUST NEED TO KNOW HOW AND WHERE TO LOOK, TO UNDERSTAND A LITTLE ABOUT HOW THE CELESTIAL MACHINE "WORKS" TO HAVE ACCESS TO MUCH MORE THAN GALILEO DID.

Observing the Sky

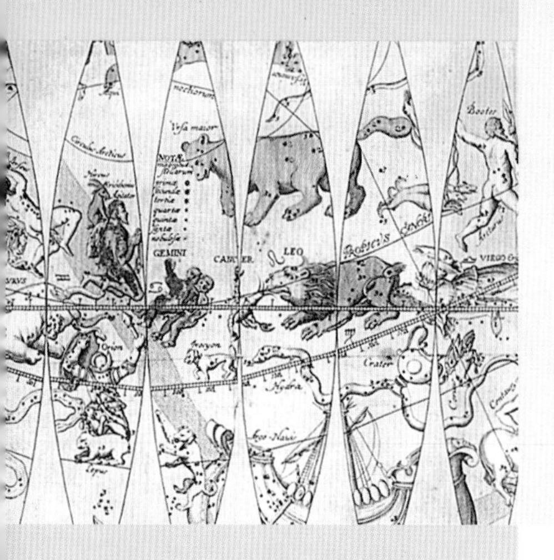

GETTING YOUR BEARINGS IN SPACE

"YOU SEE THAT BRIGHT STAR? AND THAT ONE OVER THERE ON THE RIGHT? IMAGINE A LINE THROUGH THEM, CONTINUING HIGHER IN THE SAME DIRECTION UNTIL YOU COME TO A SMALL AND NOT VERY BRIGHT STAR... CAN YOU SEE IT? IT'S THE POLE STAR, AND WITH THE OTHERS IT FORMS A KIND OF SPOON: THAT'S URSA MINOR. AND JUST TO THE RIGHT.. CAN YOU SEE CASSIOPEIA, THE ENORMOUS W SHAPE? THIS IS REALLY EASY! AND BETWEEN URSA MAJOR AND URSA MINOR, THAT SERIES OF SMALL STARS IS THE CONSTELLATION DRACO...." MANY PEOPLE EXPERIENCE THE THRILL OF DISCOVERING THOSE MAGIC SHAPES WHICH, THE MOMENT WE SEE THEM, STAND OUT AGAINST THE CHAOTIC SPRINKLING OF THE OTHER STARS. ONCE YOU'E SEEN AND RECOGNISED THEM, THE SKY WILL NEVER SEEM THE SAME AGAIN, AND IT WILL BE IMPOSSIBLE NOT TO FIND THEM AGAIN AND AGAIN, IMPOSSIBLE NOT TO KEEP COMING BACK TO FIND THEM.

The distances that separate us from the stars are so high (the "closest" is about 4 x 1013 km) that we cannot appreciate them even with a telescope. To us they all seem to be the same distance away, arranged on the surface of a sphere centred on the Earth: this is the celestial sphere which, because of the motion of the Earth, seems to turn, with a full rotation every 24 hours. But when observing a natural phenomenon, if you want to share or repeat your observations, you need a system of coordinates, so we can indicate the precise position of a phenomenon, whether on the surface of the Earth or in the immensity of space.

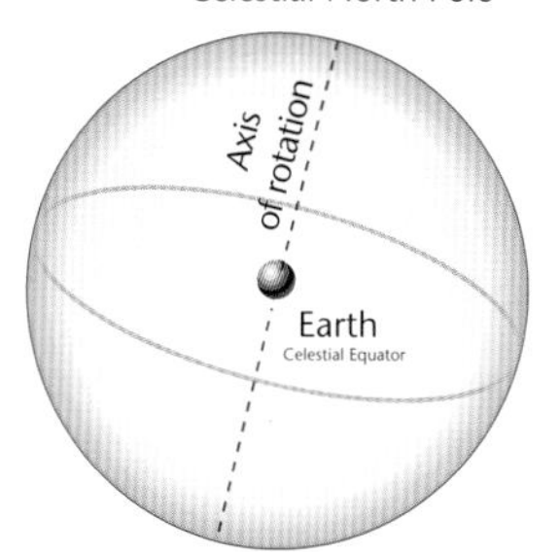

THE CELESTIAL SPHERE
The stars are so far away that to us they all seem at the same distance: this is the celestial sphere, which revolves around us because of terrestial rotation.

GEOGRAPHICAL COORDINATES

On earth the system consists of geographical coordinates: parallels and meridians that intersect at right angles. Since there as many parallels and meridians as angles, there is an infinite number on the earth The largest parallel, used as reference, is the Equator, which corresponds to an angle of 0°, and divides the Earth into two hemispheres: the Northern hemisphere, from 0° to 90° North, and the Southern hemisphere, from 0° to 90° South (by convention the coordinates of the Southern hemisphere are indicated with a minus sign).

THE SKY REVOLVES
As the Earth turns, we can see different sectors of the starry sky every night. The Hubble picture shows the Arches cluster, near the centre of our Galaxy: at median latitudes we would see it rise and set, if we had the sight of the Hubble.

AZIMUTHAL COORDINATES
The azimith is the altitude of a star in the Northern hemisphere.

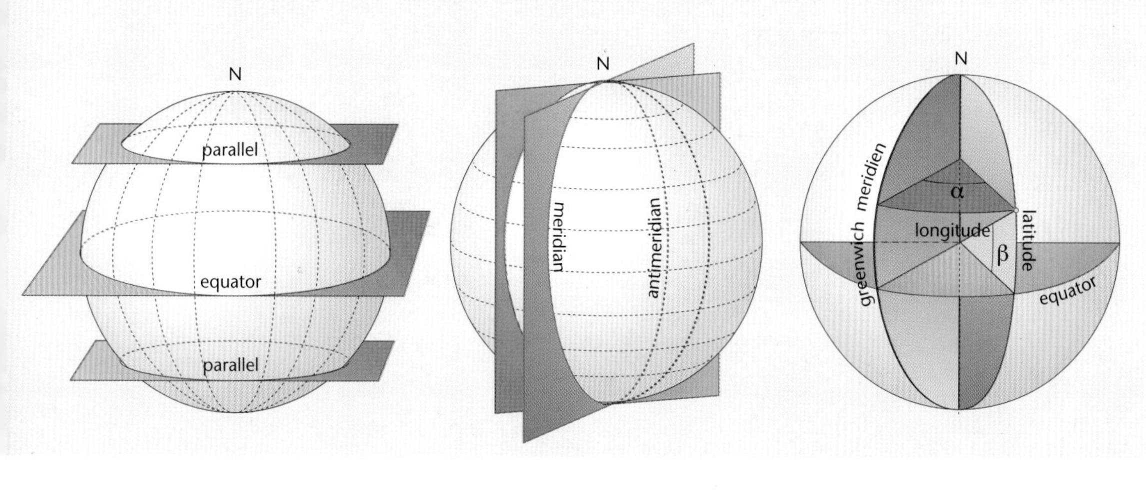

Geographical Coordinates
It was Galileo who called attention to the importance of a reference system for the examination of natural phenomena. On Earth, geographical coordinates indicate the distance of any point on the surface of the Earth from two reference great circles.
- longitude is measured on the parallels,
- latitude is measured on the meridians.

The meridians are all identical. by convention the meridian passing through the Greenwich Observatory, in London, has been chosen as the reference meridian: it corresponds to 0°.
We can always find the position of any point on the surface of our planet by calculating the distance in degrees from the parallel and meridian passing through the Equator and the Greenwich meridian respectively. The first measure is called latitude, and the second is called longitude.

CELESTIAL COORDINATES

As on the surface of the Earth, two great circles also identify any point on the celestial sphere. However, several coordinate systems are used in astronomy, using different pairs of great circles as reference.

ALTAZIMUTH SYSTEM

This takes the celestial horizon and the local meridian as reference circles. The first is the projection of the observer's horizon on the celestial sphere, and the second is the meridian that passes through the North, the Zenith (i.e. the point in which the observer's vertical intersects the celestial sphere, the South and the Nadir (directly opposite the Zenith).
The coordinates are:

• the azimuth (A), which is measured in degrees (with values between 0° and 360°) above the horizontal, from the observer's South point in the direction of motion of the star (East-West);
•the altitude (h), which is measured in degrees, from the horizontal along the altitude circle (i.e. passing through the Zenith, the Nadir and the star). . Altitude values are between 90° and -90°: the value is positive if the star is above the horizon, and negative if it is not. This system can identify a star if the direction of the South cardinal point is known. However, since these coordinates are dependent on the observer's horizon, they change according to the position of the observer on the Earth, and also with time (due to the rotation of the earth).
The only value which always remains fixed is the altitude of the celestial North Pole, in the Northern hemisphere, and of the celestial South Pole, in the Southern hemisphere. this corresponds to the làtitude of the location.

EQUATORIAL SYSTEMS

There are two equatorial coordinate systems: the fixed equatorial coordinate system, dependent on the observer's location, and the mobile equatorial coordinate system, which is independent of the observer's location. .
They both use the celestial equator as reference circle, but the first system takes the local meridian as reference circle, while the

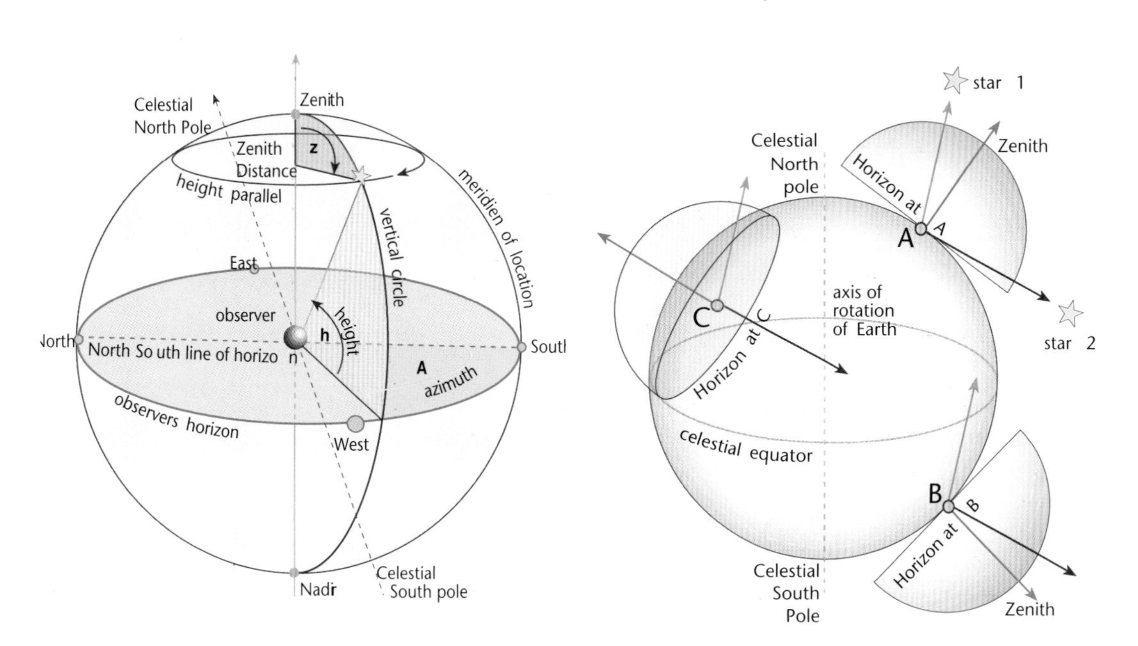

Variations in azimuthal coordinates
Azimuth and altitude change from place to place, and, in the same place, with time. The diagram summarises the situation that occurs observing the same stars from three different points on the surface of the earth. the stars are "at infinity", and so the directions in which the observations are made are parallel.
Each observer has his or her own celestial sphere, with its own coordinates: the altitude of the stars differs according to the horizon.

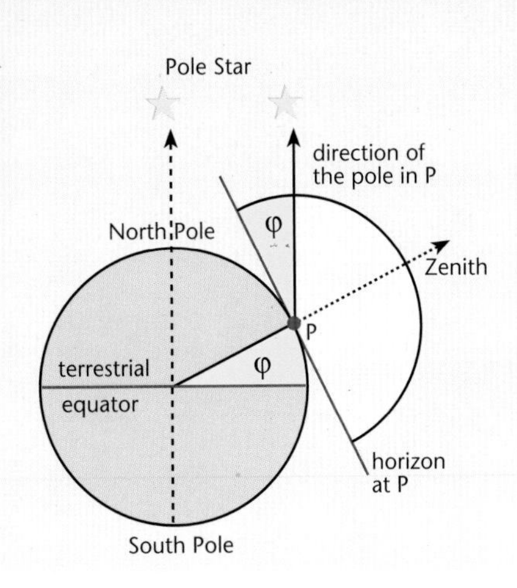

Finding the Pole
It is always important to be able to find the position of the celestial North (or South) Pole, i.e. the point at which the earth's axis of rotation intercepts the celestial sphere.
For an observer in the Northern hemisphere, the altitude of the celestial North Pole on the local horizon coincides with the latitude φ of the location.

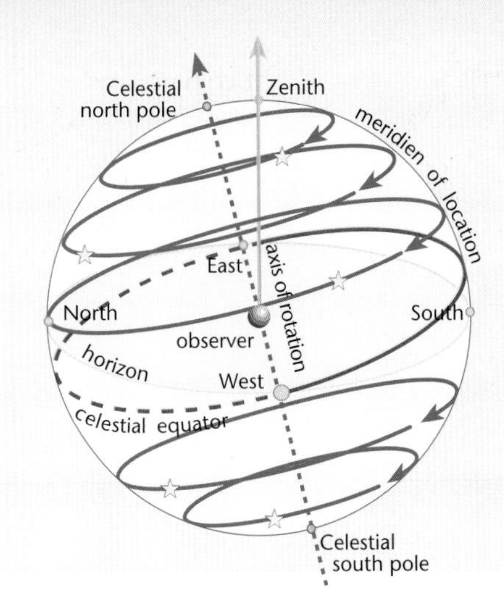

The trajectories of the stars
Every day, each star travels along its own circle parallel to the celestial equator. Apart from tiny movements on the celestial sphere, which take place over very long periods of time, they always move on the same parallel.

second system uses the hour circle (i.e. the circle passing through the celestial poles, which passes through the g point (or Spring Equinox, Vernal Point, or Point of Aries: this is one of the two points at which the ecliptic, i.e. the path of the earth's orbit around the Sun, intersects with the celestial equator).

Fixed equatorial system

In this system the coordinates of a star are:

• the hour angle (t), which is measured on the Equator, from the point of intersection of the celestial equator and the local meridian in the direction of motion of the star (i.e. westward). The value of the hour angle, measured in hours, minutes and seconds, with values ranging from 0 h to 24 h, varies continually during the day. Since the celestial sphere completes a full rotation in 24 h, it follows that: 24 h = 360°; 1 h = 15°; 1 min = 15′; 1 s = 15″, and so on;

• the declination (δ), which is measured along the hour circle passing through the star from the point it intersects the equator. It is measured in degrees, minutes and seconds of an arc, with values ranging from 90° to -90°: it is positive if the star is in the Northern hemisphere, 0° if the star is on the equator, and negative in the Southern hemisphere. The value of δ is roughly constant over time.

Every day each star crosses the local meridian twice along its path: this is called passing the meridian. For a given location, the passages of the meridian of a star correspond to the maximum and minimum altitudes of the star above the horizon: in the former case the star is referred to as in upper culmination or transit, and in the latter it is in lower culmination/transit. The hour angle of a star in upper culmination is always 0 h, and the hour angle of a star in lower culmination is always 12^{h}.

Mobile equatorial system

In this system the coordinates of a star are:

•the right ascension (α), which is measured on the equator from the g point in the direction opposite to the motion of the star (i.e. eastwards). It is measured in hours, minutes and seconds and has values ranging from 0^{h} to 24^{h}.

• the declination (δ), which is measured on the hour circle that passes through the star from the equator in degrees, minutes and seconds, with values between 90° and -90°.

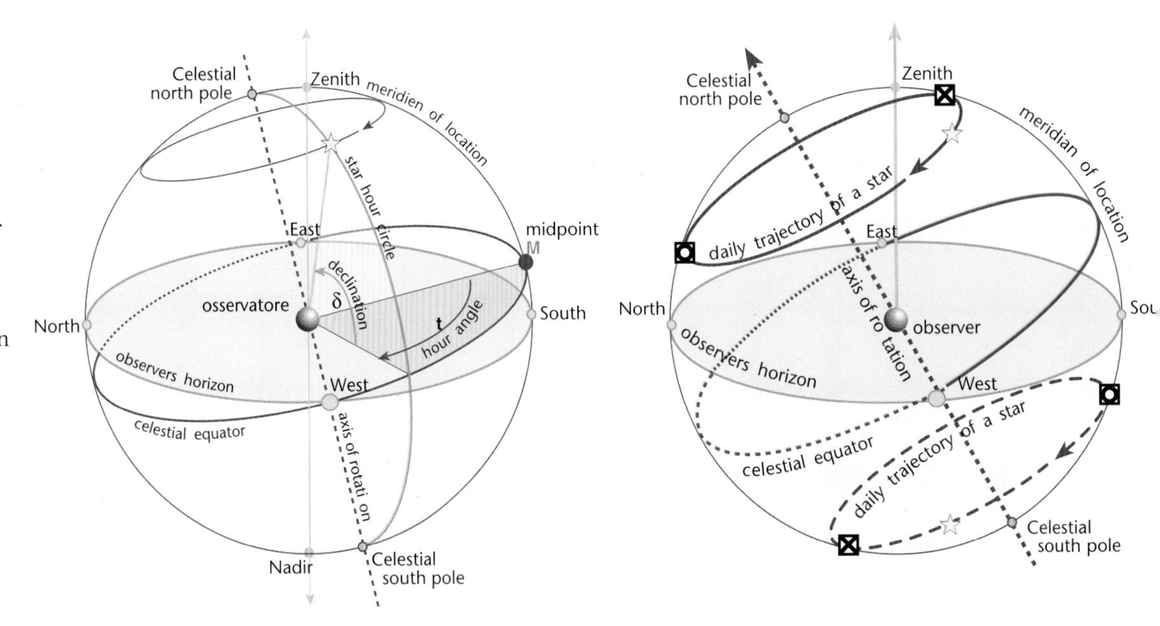

Equatorial systems
On the left: the fixed equatorial system, with hour angle and declination coordinates.
On the right:
◘ Upper culmination and
☒ lower culmination in two stars in opposite hemispheres.

Celestial north pole
Zenith
meridien of loca
latitude φ
East
parallel circle of declination δ = 90° – φ
South
North
observer
observers horizon
West
celestial equator
axis of rotation
Celestial south pole

Latitudes and declination
Top left: a circumpolar star has a declination δ ≥ 90° -M, at local latitude M .
A star with declination δ < M - 90° never appears on the horizon.

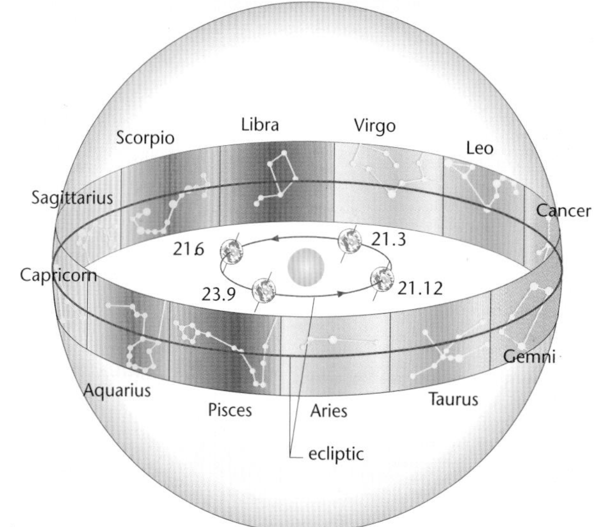

The values of a and d are roughly constant over time and independent of the observer: The γ point turns around the polar axis at the same angular velocity as the stars.
The equator arc between the point of intersection of the celestial equator and the local meridian and the g point is called sidereal time: it indicates how long ago the γ point passed the local meridian.

THE ECLIPTIC AND THE ECLIPTICAL SYSTEM

The orbit of the earth, called the ecliptic, lies on an imaginary plane that intersects with the celestial sphere describing a circumference also called the ecliptic. This term is also used to indicate the circle corresponding to the projection of the path of the Earth through the celestial sphere, or the path of the Sun across the sky.
If we consider the Sun as a celestial body like any other, we can, in fact, imagine that it occupies a specific part of the sky: in this sense the Sun is said to "project itself" in a precise sector of the celestial sphere. However, since the Earth moves every day, then the angle under which we see the Sun changes, and the Sun in turn projects itself in a slightly different area of the sky each day. After a year it "returns" to its initial position:
this "path" of the Sun corresponds to the intersection of the ecliptic with the celestial sphere. However, the axia of rotation of the earth is inclined 66° 33¢ to the plane of the ecliptic. While "viewed from earth" the plane of the ecliptic is inclined 66° 33¢ to the axis of the earth and 23° 27¢ to both the terrestrial and the celestial equator.
The two points at which the ecliptic intersects the celestial equator are called equinoctial nodes.
When the Sun projects itself in one of these it is referred to as being at the equinox: the equinoctial point on which it projects on 21 March corresponds to the g point.
When the Sun is halfway between the equinoctial points it is at the solstice.
So because of the motion of the Earth, the background of stars on which the Sun is projected changes slowly: the winter sky is very different to the summer sky, since the Sun appears to move about 1° per day (360° in 365 days) along the ecliptic.
The constellations ➢86 that are visible change with the season. And because the ecliptic is inclined with respect to the equator, the declination of the Sun, and thus its daily path, also vary during the year.

The ecliptic and the constellations
Because of the revolution of the Earth, the Sun projects itself in different areas of the sky in different periods of the year.
The constellations that are visible are those in the part of the sky opposite that in which the Sun is projected.

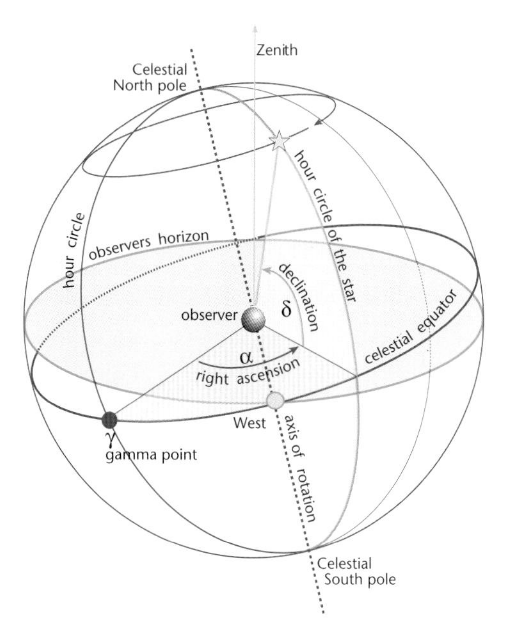

Equatorial systems
On the left:the mobile equatorial system, with right ascension and declination coordinates. When the latitude of a location is neither 0° nor 90°, the celestial sphere is described as oblique: this obviously is the more frequent case.
When φ = 90° (at the Poles) the celestial sphere is parallel: all stars with positive declinations are visible, and all the others are below the horizon.
When φ = 0° (at the Equator), the celestial sphere is normal, all the stars are visible, and they all rise and set.

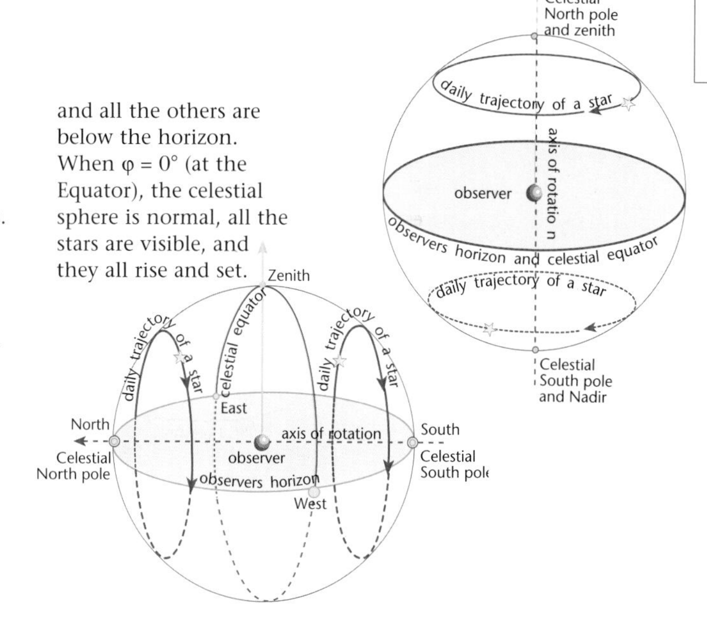

Solar trajectories
Daily trajectories of the Sun with respect to a horizontal with latitude M ≈ 40° during the two solstices (winter and summer) and the two (coincident) equinoxes. The altitude of the Sun varies continually during the year.

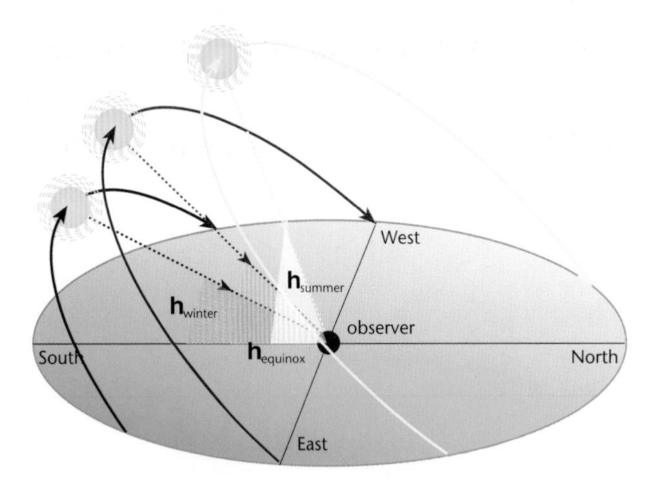

This is the reason why the duration of the day and the night are always different, apart from at the equator and in the day of equinox, when the Sub is in the g point. In the weeks immediately preceding and following an equinox the daily variation in the maximum altitude of the Sun on the horizon is highest: the day progressively lengthens (or shortens) by several minutes each day.
If projected on the celestial sphere, the solstices, i.e. those points on the ecliptic that are half way between the equinoctial points, are at the maximum distance (North and South) of the celestial equator. In these points the Sun has a declination $\delta \pm 23°27'$. In the weeks preceding and following a solstice, the daily variation in the maximum and minimum altitude of the Sun on the horizon is minimum: the day progressively lengthens (or shortens) by just a few seconds each day.

The ecliptical system

This is used primarily to describe the motion of the planets and to calculate eclipses: the reference circles are the ecliptic and the longitude circle passing through the poles of the ecliptic and the γ point. The coordinates are:

•the ecliptic longitude (λ), which is measured on the ecliptic from the γ point in the opposite direction to the motion of the stars to the point of intersection of the ecliptic and the longitude circle passing through the star. This is calculated in degrees, minutes and seconds of arc, with values from 0° to 360°;
• the ecliptic latitude (β), which is measured on the circle of longitude that passes through the star and between the star and the ecliptic; this is also measured in degrees, minutes and seconds of arc, and assumes values between 90° and -90°: it is positive if the star is in the Northern hemisphere of the ecliptic, otherwise it is negative.

THE GALACTIC SYSTEM

This is used in stellar statistics to describe the motions and positions of galactic bodies. The fundamental circles are the intersection of the galactic equatorial plane with the celestial sphere, and the maximum circle passing through the poles of the Milky Way and the apex of the Sun (i.e. the point of the celestial sphere to which the solar motion is directed). The coordinates are galactic longitude and galactic latitude.

The Sun along the ecliptic
Like all stars, the Sun too travels along its own circle parallel to the celestial equator every day.
Unlike the stars, which are too far away for the variations in angulation caused by the revolution of the earth to be appreciable, the Sun follows an apparent trajectory which follows the whole ecliptic, moving about 1° per day.

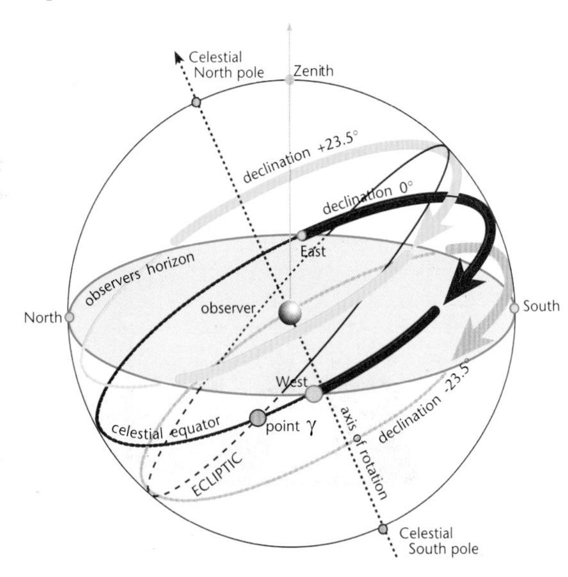

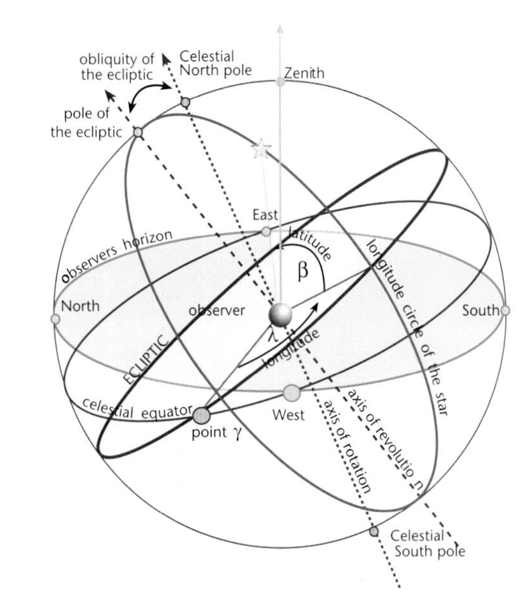

Ecliptical system
On the left: this system independent of the observer has latitude and longitude as coordinates.

UNTIL A FEW YEARS AGO, TO OBSERVE THE SKY WITH A TELESCOPE YOU HAD TO GO TO ONE OF THE MANY ASTRONOMICAL OBSERVATORIES AND BE CAREFULLY SUPERVISED BY AN ASTRONOMER. NOWADAYS PROGRESS IN OPTICS ALLOWS ANYONE WHO'S INTERESTED TO OWN GOOD INSTRUMENTS AND TO MAKE GOOD OBSERVATION. ALL THANKS TO REFLECTOR TELESCOPES.

OBSERVING CELESTIAL OBJECTS

At first you should look at the sky with the naked eye: you'll need to practice some operations, such as orienting yourself, before you can move easily among the constellations, stars and planets. First of all, you need to find which direction is North, using a compass if necessary. Although the direction indicated by the point of the needle is the magnetic North, which is not the geographic North, the angle between these two directions (called the magnetic declination) is just a few degrees, wholly negligible for our purposes.

At medium Northern latitudes, looking straight up from the North point indicated on the compass, the observer can see a region with a not very bright but rather isolated star: this is the Pole star, which here is more or less half-way between the horizon and the vertical above our heads (the Zenith). It's the last star of Ursa Minor, which is easily recognised using the alignment of the two bright stars of Ursa Major: these are stars that are always visible at our latitudes (they are circumpolar).

The main constellations are easy to find. Just follow a series of alignments of the brightest stars, usually indicated on star maps ➢77.

You should choose somewhere not completely dark to start learning your way around the sky, so you can see clearly only the brightest stars.

With a very weak and shielded electric torch you can practice finding and recognising the most important stars and the most well known constellations.

When you then start to observe less bright objects, remember that it's best not to stare directly at the sky in their direction, but to look in their direction out of the corner of your eye, which increases the sensitivity of your eye.

Once you know how to recognise the constellations easily, you can try and see more using binoculars or a small telescope.

WHAT TO LOOK AT

There are lots of things to observe in the night sky. Firstly, the stars: observing and identifying the stars takes just a little practice, although systematic searching for stars, identification of the minor constellations and recognising and observing star clusters and nebulae require more careful and expert study.

You might also observe some stars whose brightness varies in time: these are probably double stars ➢206, variable stars ➢206, novae or even supernovae ➢204 Amateur astronomers, with their dedicated, systematic and careful observations, are often the first to identify such objects, reporting them to the professionals.

But it is unlikely that amateurs could identify new galaxies: although they are extremely bright

OBSERVING THE STARS
Never underestimate how cold it can get when you're observing the stars: even on the warmest summer night you should have a jacket to hand to protect you from humidity and stronger breezes.

Celestial objects
Right: Andromeda (or M31), a galaxy near the Milky Way: this is a giant spiral galaxy very similar to our own, with a diameter of about 200,000 light years. It is also visible to the naked eye, as a hazy dot.

Below, from the top: lunar craters, visible side of the Moon; Mars.

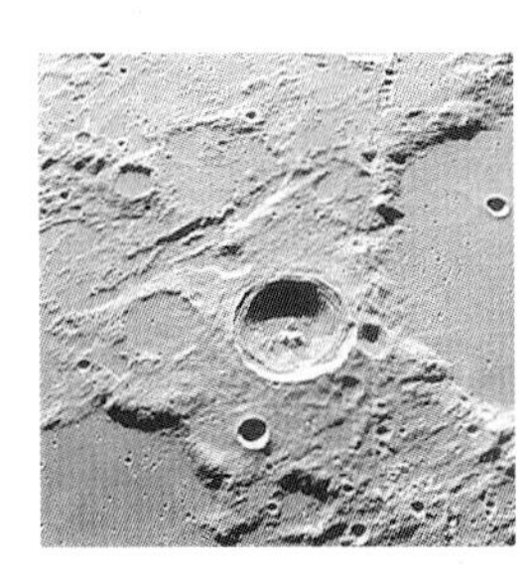

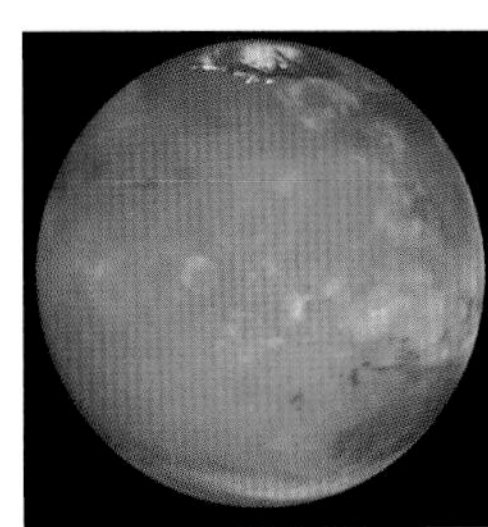

objects, galaxies are all at enormous distances from the earth, so much so that of the 100 billion galaxies in existence, just three are visible (with a degree of difficulty) to the naked eye. Only a trained eye can distinguish the large and small Magellanic Clouds, or galaxy M31. And, in addition, the Magellanic Clouds, first reported in 1519 by the famous Portuguese explorer, are only visible in the northern hemisphere: the Small Cloud is located between the constellations Dorato and Mensa, and the Large Cloud is in Tucana. Galaxy M31 is better known by the name Andromeda: visible in the constellation Andromeda, it is 2.3 x 10^6 light years away, and has two satellite galaxies (NGC 221 and NGC 205), neither of which is visible to the naked eye.
Observing the objects in the solar system is certainly more satisfying: the Moon, first of all, with its changing appearance, and then the rocky planets (particularly Mars and Venus: Mercury, because of its proximity to the Sun, can only be observed with difficulty), and the enormous gas giants, rich in satellites and full of colour: particularly Jupiter and Saturn, with its rings. With a little luck you could also observe an asteroid, or maybe a comet.
The planets are easy to distinguish from the stars, firstly, because they do not shine, but also because of their "steady" light, because they are not points, and therefore the interference of the atmosphere is far less. And the motion of the planets relative to the fixed star is consistent, and observable over just a few days: it is relatively easy to learn to distinguish their motion, and to predict their orbits. Observing a comet is more difficult: you need a lot of patience and a keen eye, to distinguish a comet from the other celestial bodies when it is not yet close enough to the Sun to develop a tail.
"Falling stars", meteorites which fall to earth in showers, on clearly defined dates, are a spectacle everyone can admire: it is impossible to observe them with an instrument, and the only option is to photograph them with a long exposure. Unlike the other celestial bodies, the Sun must never be observed directly: observing it with a naked eye can permanently damage an observer's sight, and using binoculars or a telescope to do so will cause certain blindness. You should not rely on tinted glass or treated lenses: these devices require an observer to look directly at the Sun, and this should be avoided in all circumstances.
The best way of observing the Sun is the same one used in Galileo's day: projecting its image on screen. It is very important to ensure that it is impossible to observe the Sun directly, even accidentally. Care should also be taken to not get burned: the instrument that projects an image of the Sun on the screen, concentrates its radiation, and the temperature can increase to the point of

Images of the sky
Right: image of the comet Hyakutake

Below, from the top: Saturn, the Pleiades. And an image of the Sun in H-α.

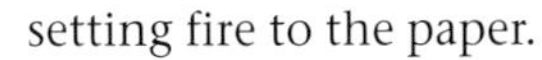

setting fire to the paper.
The best time to observe the Sun is at dusk or dawn: its brightness is not high, and the details of the photosphere can be distinguished more easily.

EQUIPMENT

You don't need much to observe the sky, but a few suggestions may be helpful. Firstly, it's better to be comfortable: pay attention to neck strain and backache, particularly in the winter! A deck-chair or a rug on the ground (or a lilo, at the beach) can be enough, but you should bear in mind that there is often humidity at night: even in the summer you should be adequately covered. A compass and an electric torch are useful. And then, the instruments: you do not need special equipment to make a great many observations: you can discover a lot with the naked eye, or using a pair of field binoculars (magnification 6-8), like those used for bird watching. For example, you can see the moons of Jupiter, some nebulae, and the details of the surface of the moon with a simple pair of binoculars.
Nor is it difficult to photograph celestial objects: all you need is a fixed focus camera and a stable surface to place it on. If the objects you will be photographing are not very bright, you will need to use a special camera attached to a telescope fitted with a device to compensate for the rotation of the earth: enthusiasts will purchase a telescope, which magnifies the planets and bodies of the solar system more than a pair of binoculars.

THE INSTRUMENTS: BINOCULARS AND TELESCOPES

Binoculars for observing celestial bodies must have certain basic characteristics. Firstly, you need to understand the information that summarises the optical characteristics of the instrument. these are engraved numbers such as 8 x 30), which respectively indicate the magnification and the diameter of the front lens in millimetres. Since higher magnifications result in decreasing visual field and brightness, it is useless to use high magnification binoculars to observe a dim star: you will not see it any better than with the naked eye. And for magnifications higher than 10, a tripod should be used. While not only is the image of the object magnified, so are atmospheric interferences: beyond a certain level the distortions become evident, compromising the observation.
Magnification is only useful when looking at non-point objects: the Moon, the planets, the comets, the nebulae. It is of little use when observing the stars: given the distance, whatever magnification is used a star still appears as a hazy point of light. In these cases brightness is more important: if a pair of binoculars does not collect enough light, it is no use for nocturnal observation. And aperture, that is, the diameter of the lens, is the main factor.
Bearing all the variables in mind, the most advisable binoculars would be a 7x50 pair. The field is sufficiently large to not lose the references, and the capacity to see details is such that details of about 15″ of an arc and magnitudes ➤190 of about 10 (i.e. objects 490 times weaker than those at the limit of visibility) can be seen.
The fact that a pair of binoculars weighing more than 1 kg is hard to hold still should also be

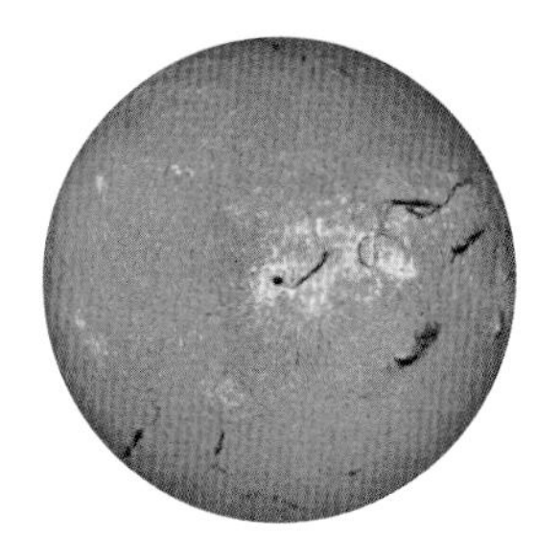

AMATEUR ASTRONOMERS
Many groups of amateur astronomers find valid support for their activities in observatories, and they perform a valuable role, surveying and cataloguing phenomena and celestial objects.

BELOW: often a good pair of binoculars is more than enough to start observing the night sky.

considered: mounting it on a tripod becomes essential, and this is not only expensive, it also impedes movement. In addition, although there are binoculars with magnifications of 4, 7, 10, 20 or 30, there is no great advantage in using the more powerful instruments, since a small telescope can give magnifications of 100 or 200 for the same price.
So the issue of whether or not to buy a TELESCOPE should be carefully considered. But this instrument, too, has its pros and cons: for example, even a small one is bulky and heavy.
of limited size and at reasonable prices.
The best choice could be a Schmidt-Cassegrain with 20 cm diameter mirror: with a weight of about 10 kg, and a suitable magnification, this would allow an observer to see details at around 2″ of an arc and, in particularly favourable conditions, objects at a magnitude of $13^{>190}$, i.e. 600 times weaker than those at the limit of visibility. Since greater precision is limited by atmospheric turbulence, it is not particularly useful to exceed a diameter of 20 cm.
The characteristics of a telescope, whether a refractor or a reflector, are luminosity, magnification, and resolution.
Brightness is a number proportional to the square root of the light energy per second and per unit surface the instrument can concentrate on its focal plane. At a given wavelength, the brightness of a

TELESCOPES

An optical telescope consists of a tube with an objective (i.e. a set of lenses) or a mirror at one end. The objective or mirror provides an image of the object observed, which can be looked at, photographed or analysed by suitable instruments through an eyepiece. If the radiation is collected by a lens (or set of lenses) the telescope is called a refractor, and if collected by a mirror it is called a reflector.
A. Galilean telescope: this telescope is a system consisting of an objective which collects the light from an observed object, and an eyepiece which enlarges an inverted image.
B Keplerian telescope: this is a refractor telescope similar to the Galilean, with a convex lens in place of the concave lens in the eyepiece, which therefore supplies an upright image.
C. Newtonian reflector telescope: concave parabolic mirror which converges the beam of rays from the source onto a flat mirror set at an angle of 45° to the axis of the instrument to form an image on one side of the tube. This made observations much easier. In larger telescopes the flat mirror is replaced by a cabin housing the instrumentation and the observer.
D. Gregorian telescope: this telescope is called after its inventor, James Gregory, and uses the same principle as the Newtonian reflector, using a different mounting which enables the eyepiece to be fitted at the end of the tube.
E. Cassegrain telescope: this is named after Guillaume Cassegrain, and has a structure similar to that of the two preceding telescopes. However, the flat mirror is replaced by a convex hyperbolilc mirror facing the main mirror. This enables the telescope to offer the same performance with a shorter tube, which offers considerable advantages in terms of robustness and stability.
F. Coudé Telescope: from the French for angled, this has the structure of the Cassegrain telescope with an additional flat mirror which deviates the beam so that it always leaves the

a b c d e f g h

Observing the Sun
As in the time of Galileo, the safest way to observe the Sun is to project its image through a small hole onto a light coloured surface in a dark room

For experts
A telescope of this type, one of the most widely used by amateur astronomers, can be fitted with special filters to enable observation of solar protuberances.

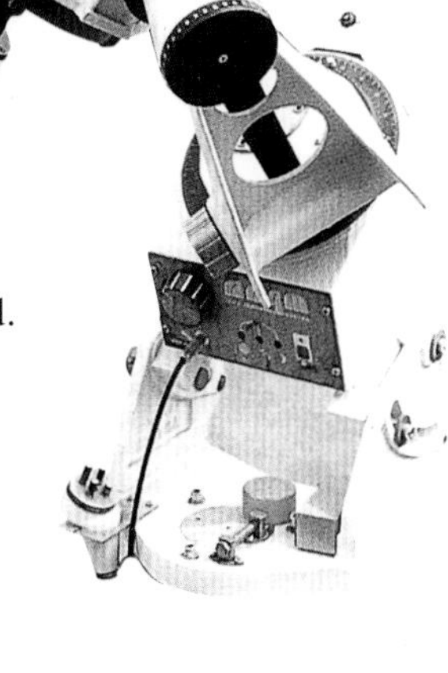

star is given by the square of the diameter of the objective or mirror (called absolute aperture), divided by the focal length, and for extended objects it is given by the objective divided by the focal length (relative aperture). Whether or not "faint" stars can be seen depends on their brightness, not their magnification.
Magnification, which is always angular, is given by the ratio of the focal length of the objective (or mirror) and the focal length of the eyepiece: the shorter the latter, the higher the magnification. It is only relevant when observing extended (non-point) objects.
The resolving power, or resolution, is the smallest angle for which the images of two points are separate and distinct. a high resolving power produces clear and detailed images. Resolving power is directly proportional to the wavelength of the radiation examined, and inversely proportional to the absolute aperture. This quantity too is only relevant for extended objects: However, observation possibilities are limited.

telescope in the same direction, irrespective of the position of the main tube.

CATADIOPTRIC TELESCOPES

These are telescopes which combine lenses and mirrors.
G. Maksutov telescope: this is named after its inventor, and uses a meniscus lens correcting plate to give a spherical aberration to the image which is subsequently corrected by the spherical primary mirror.
H. Schmidt telescope: this too is named after its inventor, and has a corrector plate of complex shape fitted near the centre of curvature of the spherical main mirror. This type of telescope is used primarily for photography.

The mount

The mount of a telescope, i.e. the articulation used to orient and move the tube of the telescope, is also important.
The most commonly used mounting is equatorial, which allows fixed equatorial system coordinates to be used. The telescope can turn around an axis parallel to the axis of the earth and around a perpendicular axis (parallel to the plane of the celestial equator). The declinations can be read on a circle in concordance with the first axis, divided into 360° and fractions of degrees, and the hour angles can be read on a circle in concordance with the second axis, divided into hours and fractions of hours. Once the telescope has been pointed, a clockwork device turns the telescope in concordance with the object observed.

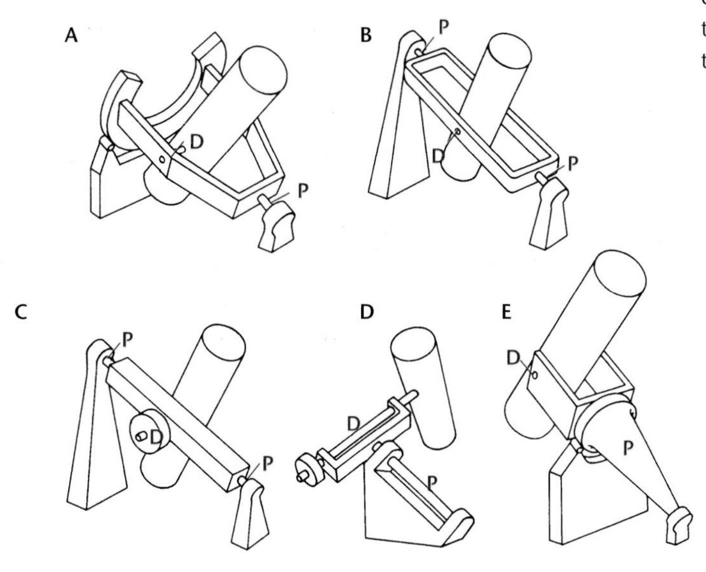

Equatorial mount
Different types of equatorial mount:
A. horse shoe;
B. English;
C. crossed axes;
D. German;
E. fork
P = polar axis
D = axis of declination

OBSERVING THE SKY HAS BECOME THE TRADE OF MANY SCIENTISTS AND THE PASSION OF HORDES OF AMATEUR ASTRONOMERS. WHILE IN THE OBSERVATORIES WORK PROCEEDS PRIMARILY BY THE PROCESSING OF DATA COLLECTED IN SPACE OR PARTICULAR PLACES BY INCREASINGLY POWERFUL INSTRUMENTS, THE AMATEURS MEET IN CLUBS AND SOCIETIES.

PLACES FOR OBSERVATIONS

The oldest astronomical observatories are in large cities. But it has become impossible to observe a night sky full of artificial light: this has led to the building of powerful telescopes in areas increasingly further from inhabited areas, and at high altitudes, to eliminate atmospheric perturbations as much as possible.

In addition to observatories of historic importance, such as the Greenwich observatory, the list below includes some of the observatories with the largest instruments: Zelenchukskaya, in the Caucasus; Mount Palomar, in California; Kitt Peak ➢125, in Arizona, with its famous solar tower; Cerro Tololo ➢123, in the Chilean Andes; Mauna Kea, in Hawai. Here, huge telescopes and long focal length instruments are housed in enormous and almost always circular rooms, with rotating hemispheric domes which can open to the sky, allowing the instrument to be pointed in any direction.

Until not long ago the instruments in these observatories were built by national institutions, and reached a maximum diameter of 6 m.

Nowadays larger and larger instruments are built in the same observatories, and in even more isolated new ones, thanks to new technologies and the considerable investments made by international consortia. This was how the ESO (European Southern Observatory) was built in Cerro La Silla in northern Chile, about 2500 m above sea level, by several nations, including Belgium, Denmark, France, Germany, Italy, the Netherlands, Sweden and Switzerland.

Important centres for radioastronomy studies include Jodrell Bank, in Manchester, which has a large orientable radiotelescope with a metal mirror 76 m in diameter, Arecibo, with a 305 m diameter reflector which converges signals collected on an aerial dish suspended from three towers, Zelenchukskaya, with its RATAN (Radio Astronomical Telescope of the Academy of Sciences), composed of 900 aluminium panels arranged on a 576m diameter circle ; the VLA ➢124 (Very Large Array), in Socorro (New Mexico)with 27 mobile dishes each 25 m in diameter...

DISHES
Traditional radiotelescopes are characterised by a large disc that collects electromagnetic signals and concentrates them on the antenna.
Here the sensors collect them and send them to instruments used to analyse them.

Arecibo
This enormous radiotelescope, built making use of a natural valley in northern Porto Rico (Central America), is also used for the search for signs of intelligent life in space. This radiotelescope can be used to observe objects with declinations between 43° and -6°.

Plate
Attached to the outer surface of Voyager I, sent to explore the outer planets of the solar system and then to proceed into galactic space, the plate below is designed to manifest our civilisation to any intelligent organisms that might live on a planet of another solar system.

The construction of large optical and radio instruments is now accompanied by the creation of instruments to be placed in orbit outside the terrestrial atmosphere ➢124: they record the portion of radiation absorbed by the atmosphere of our planet (infrared, ultraviolet, X rays, gamma rays and cosmic rays), enabling us to study them. Moreover, practically no one observes the sky directly any more: astronomers work on computer-generated calculations or on photographs, while the traditional "surveillance" of the sky is left to the amateurs, who have often identified new bodies of the solar system.

TOGETHER YOU CAN WORK BETTER

There are many places and occasions on which amateur astronomers can increase their knowledge, improve their observation techniques and discuss their results. Societies and clubs often have their own high level instrument, and groups of amateur astronomers increasingly work with the professionals.

IS THERE LIFE UP THERE?

There is no doubt that the enormous commitment of time and money that space investigation requires is in some way connected to the search for life in the universe. For example, there have been more thorough explorations by space probes of those planets in the solar system which rightly or wrongly were thought to be better candidates for life forms: primitive, maybe, but life nonetheless, which might indicate that the Earth is not a singularity in the universe. The search for extraterrestrial life forms, which involves many researchers world wide, has two main directions:

1. Since the existence of higher life forms, and therefore, clearly, of intelligent civilisations, on the other planets of the solar system, researchers are now looking for primitive life forms (such as bacteria) using direct sampling by automatic interplanetary probes.

2. Since it is obviously easier to trace a superior life form (a civilisation like ours certainly makes "more noise" than a bacterial soup, and can be "noticed" at planetary distances by emissions of electromagnetic radiation of clearly artificial origin), and presupposing that life has similar needs in every part of the universe, and that, in particular, that only a narrow range of temperatures similar to those on Earth are compatible with life. Researchers are looking for planetary systems similar to our own, on which the searches for artificial emissions can be directed. This activity includes the use of large instruments such as the Arecibo radiotelescope.

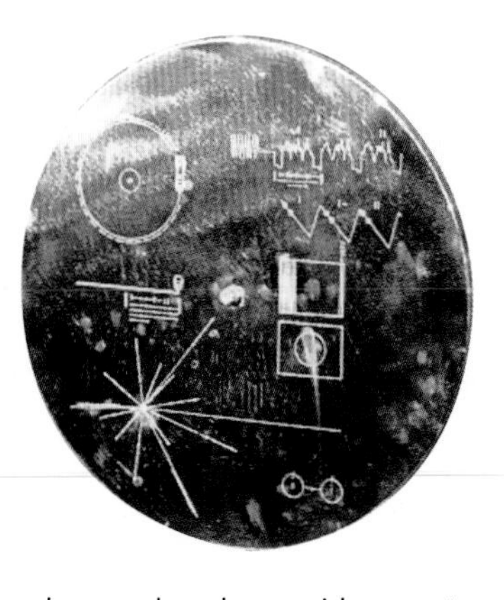

Although neither line of research has produced any evidence yet, some probability calculations indicate that sooner or later some form of life should be found. It is estimated that there would have to be 600 million other planets similar to earth in our galaxy.

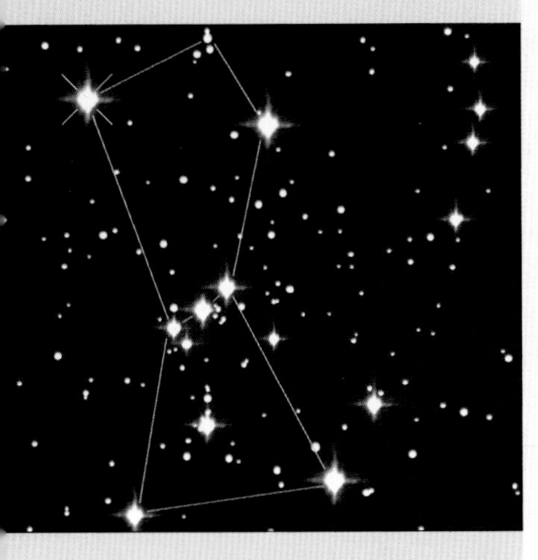

DRAWINGS OF STARS

OK. ALL THOSE ANGLES AND MEASUREMENTS TO FIND OBJECTS ON THE CELESTIAL SPHERE WITH A TELESCOPE ARE USEFUL. BUT SOMEONE WHO'S JUST STARTING TO OBSERVE THE SKY, LIKE SOMEONE WHO'S JUST STARTING TO WALK, MUST EXPERIENCE THE NEW ENVIRONMENT "FIRST HAND". THE CONSTELLATIONS, DELICATE OPTICAL ILLUSIONS WHICH "BRING ORDER" TO THE NIGHT, ALLOW US TO MOVE EASILY AMONG THE STARS ACCOMPANIED BY THE TOUCH OF MAGIC LEFT BEHIND BY THE STORIES OF A BYGONE AGE. THE URSAE, DRACO, CASSIOPEIA, ORION, PEGASUS AND OTHER PEOPLE, MONSTERS, ANIMALS, OBJECTS AND INSTRUMENTS BECOME KNOWN FIGURES. WE WILL SEEK THEM EVEN IN UNKNOWN SKIES, ALMOST AS IF THEY WOULD GUIDE US, AND IT WILL BE IMPOSSIBLE TO LOOK AT THE DARK SKY WITHOUT RECOGNISING THEM: THE NIGHT WILL ALWAYS BE FULL OF FANTASTIC FIGURES, THE FADING DREAMS OF A MYTHICAL PAST.

At first sight it is easy to compare the arrangements of the stars on the celestial sphere to well known images, and men have done this since the beginning of time: who, looking at the stars to the North, has never noticed a large ladle-like figure? It's the constellation Ursa Major. And from this, following alignments and curves, many other figures can gradually be traced. Although we know that the shapes of the constellations are an illusion of prospective, and although we know that they do not correspond to any kind of physical reality, but are merely conventional groupings of stars, the constellations are still useful, because they represent a very simple way of finding our way around the sky.

PICTURES AND STORIES

The ancient astrologers identified the main constellations of the Northern hemisphere, trying to find connections between the stars that might be useful for navigating the sky, and to understand the influences the sky might have on the life of man. Those star figures represented mythological people, animals and common objects, and their origin was explained using myths and legends often linked to religious beliefs. The concept of finding figures in the stars continued until much more recently: the modern constellations, which bear the names of common objects and scientific instruments, are almost all in

A PROBLEM OF PERSPECTIVE
The constellation of Ursa Major is formed of stars that are at different distances from the Earth but which, in the sky, seem aligned, because of perspective. These alignments may be used to find the Pole star, in the constellation Ursa Minor.
RIGHT: the same thing happens with the constellation Cassiopeia. The images show how the constellation appeared 50,000 years ago, and how it will look 50,000 years from now.

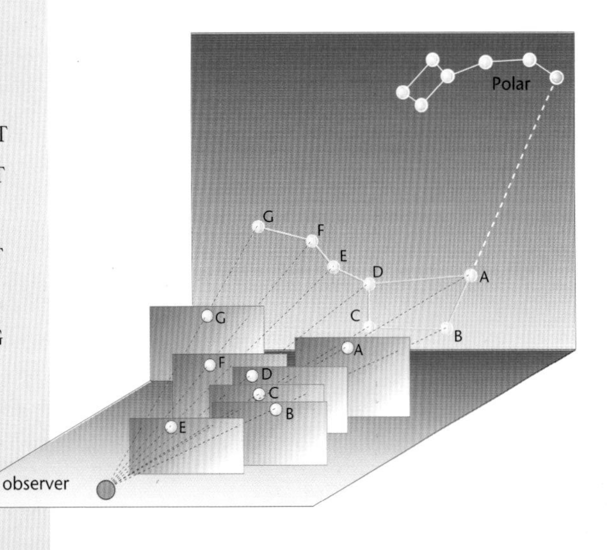

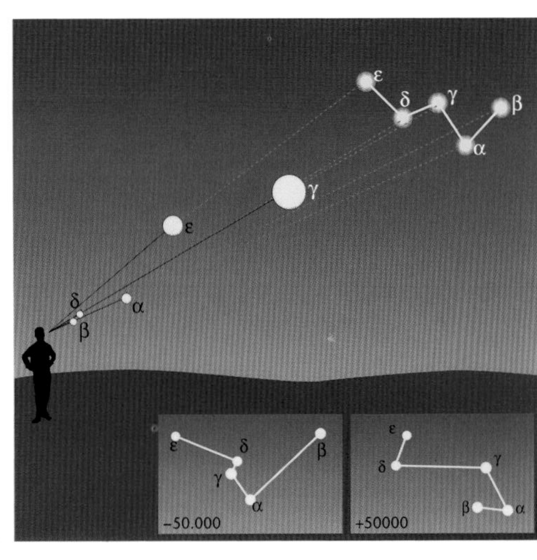

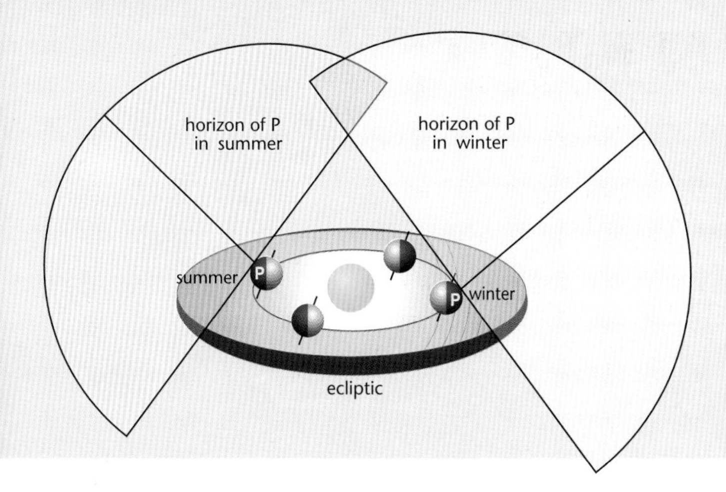

the Southern hemisphere, almost completely unknown until the early 16th century. Not all constellations, in fact, are visible from a particular location, and even those that can be seen in a particular place are not always visible in every season. It all depends on the coordinates ➢62 - 65 of the stars that form them: for example, in a place with latitude j, stars with declinations over φ - 90° never rise. So at median latitudes there will always be constellations that rise and set, constellations that are always invisible, and constellations that are always visible (circumpolar). And because of the rotation and revolution of the earth, the visible constellations change positions according to the time of observation and the season of the year. The visibility of the constellations also depends on the length of the day: for example, when the day lasts more than 12 hours above the Arctic (or Antarctic) Polar Circle, it is more and more difficult to see the stars.

KNOWN AND LESSER KNOWN CONSTELLATIONS

Today, 88 constellations have been officially adopted. 12 form the Zodiac ➢84 ; 27 in the Northern hemisphere and 49 in the Southern hemisphere. The constellations displaying the twelve symbols of the zodiac are certainly the best known: they appear in the night sky in a more or less monthly sequence, because of the revolution of the Earth. The constellations of Ursa Major, highly visible and circumpolar at our latitudes, and Ursa Minor, less evident but also circumpolar, are also well known. Since the Pole star, very close to the North Pole of the celestial sphere, is part of Ursa Minor, these constellations have been used to find the North since the beginning of time.

Apparent motion
Above: the trace left by the stars on the nocturnal photograph is due to the rotation of the Earth.
To the left: The visible night sky changes according to the position of the Earth on its heliocentric orbit.
Left: unlike star **A**, which rises and sets, star **B** is circumpolar for the location, and never sets. Just like the Pole Star.

FOR MILLENNIA THE POSITION OF THE STARS HAS MARKED THE PASSAGE OF TIME, THE RIGHT DIRECTION, WHERE ON THE EARTH THE OBSERVER IS, A SIGN OF FUTURE EVENTS. THESE DAYS, OBSERVING THE SKY IS A SOURCE OF PLEASURE AND STIMULATES FURTHER STUDY.

OBSERVING THE SKY

STAR CHART
Example of a cylindrical star chart:

☆ 1ª
O 2ª
○ 3ª
o 4ª
• 5ª

The stars can be observed anywhere: some can even be seen from city streets. The dimmer light of locations outside the cities, at full Moon, or at dusk, can be used to recognise the main constellations, planets and brighter stars more easily: background light reduces the visibility of the weaker stars, limiting doubts of recognition.
To see all the celestial bodies well, however, you need to find a place where light and smog are at a minimum: in open countryside, in an elevated position (on a hill or roof, for example), far from the bright lights of urban areas.
The stars are easy to distinguish from the planets: they seem to tremble because they are point objects. While the planets, although small, are circular light sources, and dispersed dust, water vapour and atmospheric turbulence interfere less on the light which reaches us from them.

CONSTELLATIONS AND SKY CHARTS

Because of the rotation of the Earth many stars and constellations rise in the East and set in the West: the appearance of the sky changes slowly with time, latitude and season. After several months constellations can be seen in a place and at a time when they previously appeared much later, or could not be seen. Some constellations only appear in certain periods of the year: this is why we speak of a spring, summer, autumn or winter sky. To reproduce the sky of a particular place, then, a single chart will not be valid throughout the night and at every season for every latitude – a thousand charts could be made for every hour of the night and every period of the year. Although they represent different skies, star charts, like geographical maps, are of two types: cylindrical or zenithal.

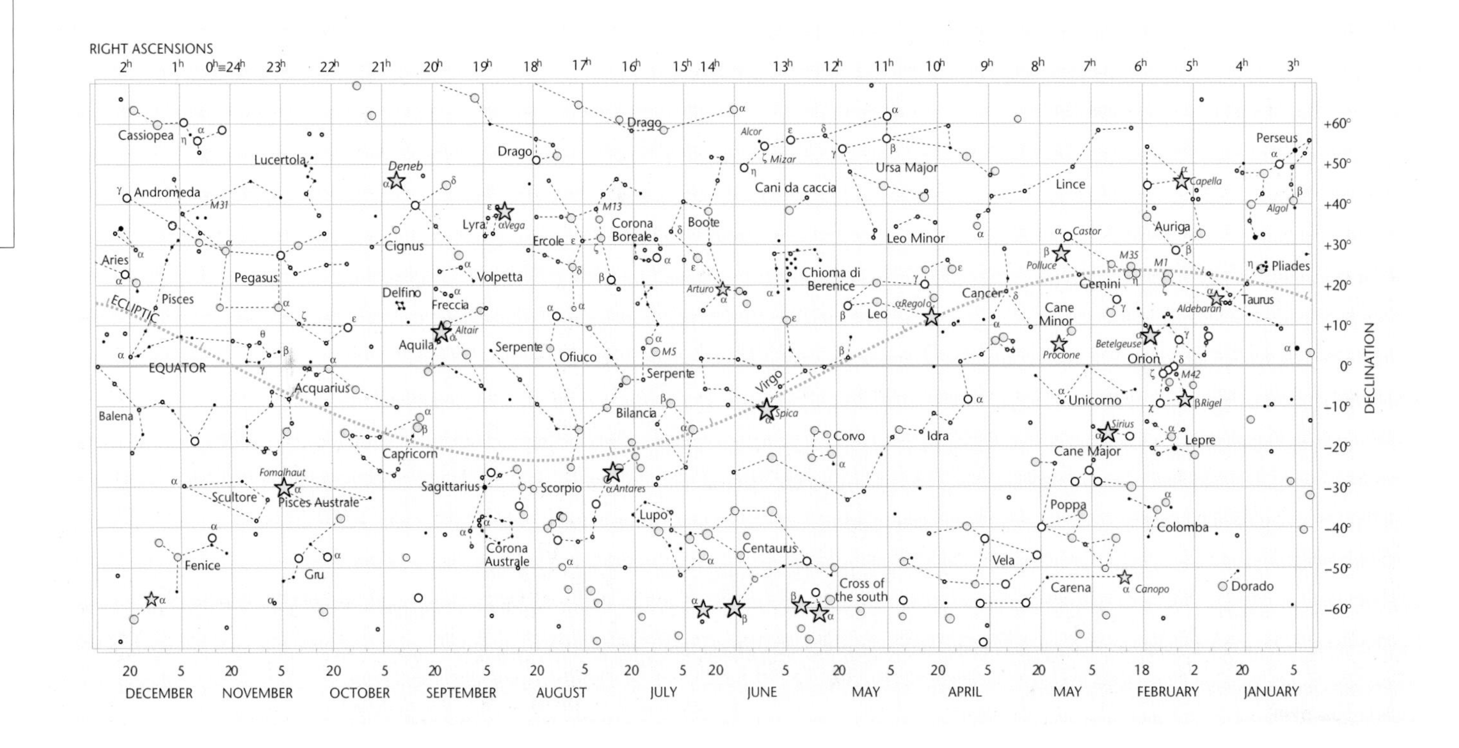

Principal constellations of the northern hemisphere

Name	Declination		Right ascension	
	Extremity of the constellation			
	to N	to S	to W	to E
Aquila	+19°	–12	$20^h 40'$	$18^h 40'$
Auriga	+56°	+28°	$7^h 30'$	$4^h 35'$
Bootes	+55°	+7°	$15^h 45'$	$13^h 35'$
Cassiopea	+78°	+46°	$3^h 40'$	$22^h 55'$
Cepheus	+88°	+53°	$8^h 05'$	20^h
Cygnus	+61°	+28°	$22^h 05'$	$19^h 10'$
Gemini	+35°	+10°	$8^h 10'$	6^h
Lyra	+47,5°	+26°	$19^h 30'$	$18^h 15'$
Orion	+23°	–11°	$6^h 25'$	$4^h 45'$
Ursa major	+73°	+28°	$14^h 30'$	$8^h 05'$
Ursa minor	+90°	+65.5°	Circumpolar	

Principal constellations of the southern hemisphere

Name	Declination		Right ascension	
	Extremity of the constellation			
	to N	to S	to W	to E
Canis Major	–11°	–33°	$7^h 30'$	$6^h 10'$
Carina	–51°	–76°	$11^h 20'$	$6^h 05'$
Centaurus	–30°	–65°	$15^h 05'$	$11^h 05'$
Crux	–56°	–65°	$12^h 55'$	$11^h 55'$
Dorado	–49°	–85°	$7^h 40'$	$3^h 30'$
Eridanus	0°	–58°	$5^h 10'$	$11^h 25'$
Hydrus	–56°	–82°	$4^h 35'$	$22^h 10'$
Octans	–75°	–90°	circumpolar	
Puppis	–11°	–51°	$8^h 25'$	$6^h 05'$
Scorpio	–8°	–45°	18^h	$15^h 45'$
Vega	–37°	–57°	$11^h 05'$	$8^h 05'$

CYLINDRICAL CHARTS

These are made by projecting the entire sky onto a plane: the celestial sphere is "wrapped" in a sheet of paper tangent to the equator, and the positions of the single celestial bodies are projected onto this. A cylindrical chart represents the entire firmament, but the greater the latitude of a constellation the more deformed in shape it will appear. The Poles, for example, become a line. Each night, depending on the time, only a portion of the chart can be observed. This type of chart is more useful to observers with telescopes, since they can be used to obtain the equatorial coordinates of stars immediately.

ZENITHAL CHARTS

These are more widely used, and represent the sky as if it were a dome being observed from below: they represent the sky as an observer sees it on a certain date, in a certain place and at a certain time. However, since the revolution of the earth is much slower than the rotation, the same sky can be observed in the same place at different times on different dates: the same chart can be used on a number of occasions. Those who observe the sky with the naked eye or binoculars will prefer this type of chart because it is amore faithful representation of the appearance of the sky. Zenithal charts also show the four cardinal points, the celestial equator, the ecliptic, and sometimes the zenith, the positions of which are necessary for orientation.

NOMENCLATURE

The stars are grouped in constellations solely on the basis of their apparent position in the sky: often there is no physical link between them, but it is useful to know them to quickly find a star in the sky. The names of the constellations are often taken from Greek mythology, since the Greeks were the first to see the star figures in the sky. Other names are Arabic: many bright stars, such as Betelgeuse, Aldebaran, Rigel or Altair, are evidence of the Arab contribution to astronomy. Other names of constellations and stars are derived from peasant traditions (Virgo, Bootes, "ploughman" in Greek...) while many southern constellations are named after scientific instruments and exotic animals: a trace of the "novelty" of the voyages of exploration which led to their discovery. In each constellation the brightest stars are characterised by a Greek letter which indicates their brightness. The alpha star is usually the brightest star of the constellation, and the alphabetic order is their order of visibility. . Although this order is not always perfect, because of errors or due to variations in brightness.

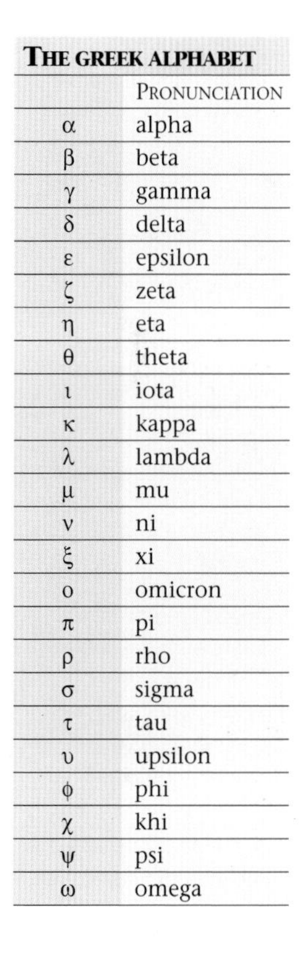

The Greek alphabet

	Pronunciation
α	alpha
β	beta
γ	gamma
δ	delta
ε	epsilon
ζ	zeta
η	eta
θ	theta
ι	iota
κ	kappa
λ	lambda
μ	mu
ν	ni
ξ	xi
ο	omicron
π	pi
ρ	rho
σ	sigma
τ	tau
υ	upsilon
φ	phi
χ	khi
ψ	psi
ω	omega

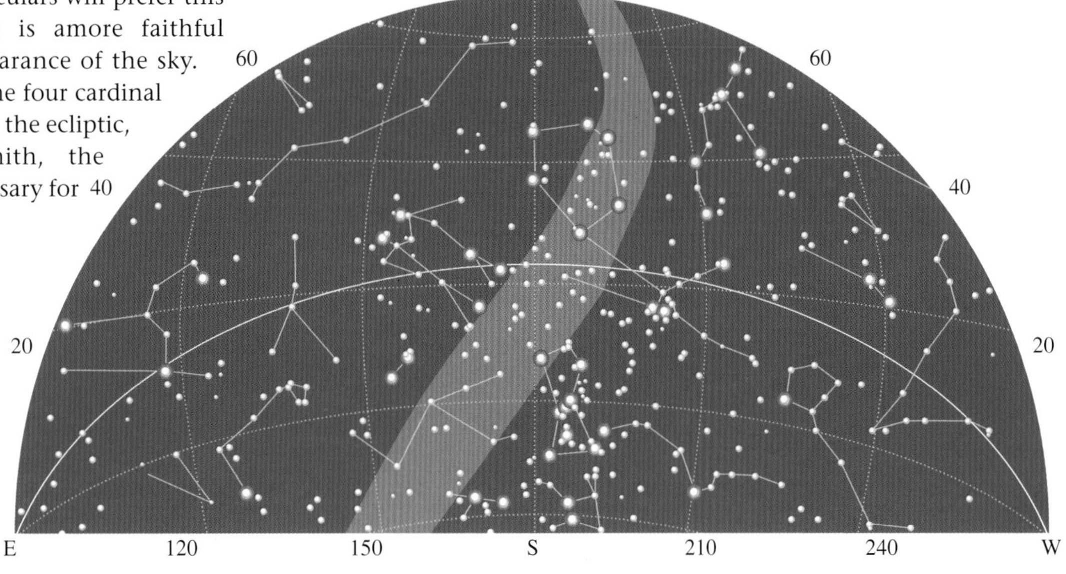

Star chart
Sample of the South side of a zenithal star chart: these charts are divided into two halves (North and South) to make them easier to read.

THE NORTHERN AND SOUTHERN HEMISPHERES OF THE CELESTIAL SPHERE CAN ONLY BE SEEN FULLY, ALL AT ONCE, FROM THE POLES, WHERE THE ZENITH COINCIDES WITH THE CELESTIAL POLE AND ALL THE VISIBLE STARS ARE CIRCUMPOLAR. AND THE CONSTELLATIONS FOLLOW TRAJECTORIES PARALLEL TO THE CELESTIAL EQUATOR.

THE WELL KNOWN CONSTELLATIONS

SAGITTARIUS
This constellation, part of the Zodiac, recalls the mythological Chiron, the wise centaur who tutored illustrious Heroes in antiquity.

DRACO
This constellation was very important for the ancient Chinese.

For regions of the Earth with latitudes greater than 66° 33¢ (i.e. beyond the Polar Circle), the Sun can, for periods of varying length according to the period of the year (i.e. according to its declination) become a circumpolar star: it doesn't rise or set every day, but only when the seasons change. In particular, at the North Pole it rises at the equinox on 21 March, and sets at the equinox on 23 September, while the opposite happens at the South Pole: however, because of the long twilight, the polar night doesn't last 6 months, but much less. In this period all the constellations above (or below) the celestial Equator are clearly visible.

NORTHERN HEMISPHERE

Andromeda: the α star of this constellation is also the *d* of the great square of Pegasus, extending eastwards towards Cassiopeia. The large galaxy M31 is in this constellation, and is visible to the naked eye: through binoculars it may be recognised as a whitish smudge.

Auriga: at median latitudes this constellation appears towards the end of autumn, heralding the arrival of winter. It recalls Phaethon, son of Apollo, the Sun god of the Greeks, who wanted to drive his father's chariot of fire and, after driving too close to the Earth, and almost burning it, was fulminated by Zeus. It may be found from the alignment of delta-alpha in Ursa Major, and the extension of the nose of Ursa Minor, immediately East of Perseus. The main part of the constellation is a pentagon of stars, the brightest of which is Capella, a yellow star of magnitude ➢190 0.06.

Bootes: in this constellation, which represents the driver of the seven bullocks of the Polough of Ursa, is the red giant ➢202 Arcturus, which is about 30 times the diameter of the sun, It is found extending the alignment of alpha-delta in Ursa Major or by extending the curve of the tail (from which it takes its name: arctos aura, tail of the Bear, Ursa). Near Bootes is the small constellation Canes Venatici: a pair of small stars between Arcturus and Ursa Major.

Cepheus: following the alignment of alpha and beta in Cassiopeia leads to the delta star of Cepheus, which has given its name to the regular variable stars called Cepheids ➢207: it varies in magnitude from 3.6 to 4.3 in 5 days. In the year 8000 the alpha star will replace the Pole Star as the star closest to the North Pole, due to the precession of the equinoxes. The myth of Cepheus is related to the story of the Argonauts.

Cygnus or the Northern Cross: immediately East of Lyra is the head of Cygnus, the swan, a third magnitude star almost aligned with Vega (the brightest star in Lyra) and Altair (the brightest star in Aquila). Known also as the Northern Cross, because of its shape, at the tail of Cygnus is Deneb, a 1.26 magnitude star, with the North America nebula visible close by. The first star whose distance has been measured is in Cygnus.

Canis Major: this is the smaller of the two dogs which accompany Orion. Procyon, a yellow star of magnitude 0.5m is the only one visible.

Cassiopea: joining Magrez, the delta star of Ursa Major, with the Pole Star, and continuing in the same direction leads to Caph, the beta star of Cassioneia, a constellation consisting of 5 bright stars arranged in a "W". Near to Schedar (the alpha star) is Achird, a yellow and red double star. The name of this constellation is that of the wife of Cepheus, mother of Andromeda, whose beauty rivalled that of the Neriedes.

Drago or the ***Dragon***: the tail of this constellation is about 10° from the two posterior stars of Ursa Major, and, curving between the two Ursae, it ends with a group of four stars (the head) at about 15° fromVega.

Northern celestial hemisphere
Only the principal constellations are indicated.
The letter a indicates the brightest star in the constellation, and the others follow the order of the Greek alphabet , in order of brightness.

BELOW: The Northern celestial hemisphere in an ancient print.

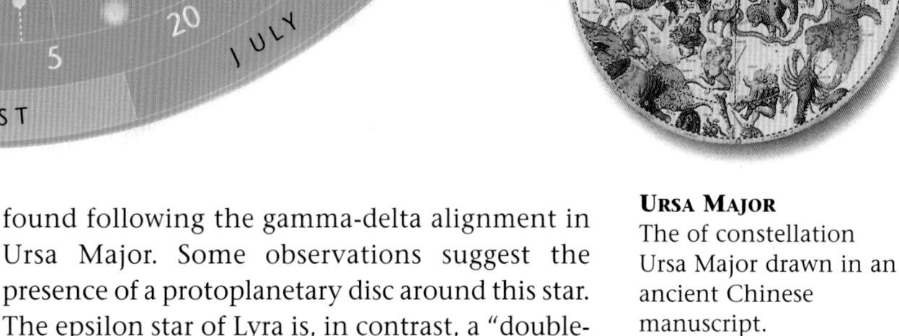

Ursa Major
The of constellation Ursa Major drawn in an ancient Chinese manuscript.

Thuban (the last star of the tail) has been polar: this was the star the Egyptians oriented the pyramids to. Although Draco is a circumpolar constellation, it is most visible in median latitudes in late spring and early summer when it is higher on the horizon. Using a telescope a planetary nebula ➢202 can be seen in Draco. In mythology, the Dragon was the guardian of the garden of the Hesperides, and was killed by Hercules: it is not part of the story linking Cassiopeia, Andromeda and Cetus. This constellation was very important for the Chinese, so much so that it became their national symbol.

Hercules: the principal element of this constellation is a group of 3rd and 4th magnitude stars under the head of Draco. To the West is the M13 cluster: more than 500,000 stars 25,000 light years away, visible with a telescope.

Lyra: this is Orpheus' lute: Vega, the blue-white star to which the Sun (apex) moves, may be easily found following the gamma-delta alignment in Ursa Major. Some observations suggest the presence of a protoplanetary disc around this star. The epsilon star of Lyra is, in contrast, a "double-double" star: the principal pair can be distinguished with binoculars.

Ursa Major, and ***Ursa Minor***: the seven principal stars of the Plough form Ursa Major, the easiest constellation to identify in the Northern hemisphere. Mizar (the zeta star), visually a double, is in reality a system of 7 stars. Following the beta-alpha alignment leads to the Pole Star, in Ursa Minor, which is also a fairly weak star: it is a triple star, the secondary of which, visible with binoculars, is a Cepheid type variable ➢207. The Pole Star is about 350 light years from Earth.

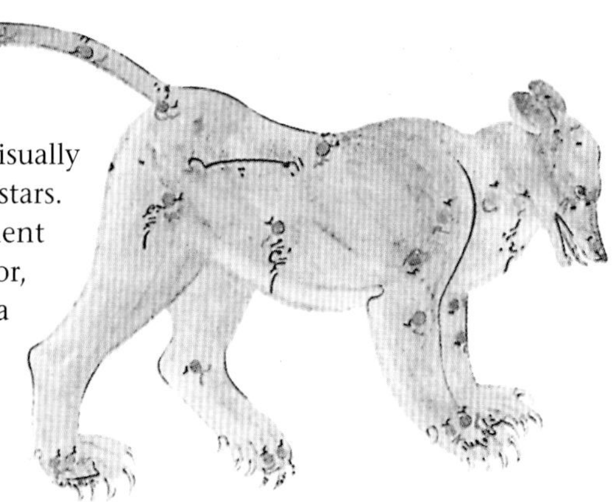

The orion nebula
A detail photographed by NASA's Hubble Space Telescope shows the central part of Trapezium, in the constellation of Orion, about 1,600 light years from Earth. The image shows infrared emissions.

Pegasus: prolonging the line from Ursa Major to the Pole star leads, after Cassiopeia, to the great square of Pegasus: the winged horse is inverted, with the head near the equator, and estends to Cygnus and Delphinus. Pegasus does not only appear in the legend of Perseus and Andromeda, but also in the story of Bellerophon, the hero who rose into the sky and, falling onto the Chimaera, killed it.

Perseus: this constellation is found on the gamma-delta alignment in Cassiopeia, and has a rough "K" shape stretching out towards Auriga. The upper arm ends with Algol (the Demon's Head) or the Head of Medusa: an eclipsing variable star ➢207 which varies in magnitude ➢190 from 2.3 to 3.4. Algenib, or Mirfalk, a yellow giant, is very bright.

Aquila: along the line from the Pole star to alpha Cygni is Altair, a very bright bluish star of magnitude 0.9. The constellation Aquila, the Eagle, stretching towards Sagittarius, recalls Jupiter, who carried off Ganymede to become cup bearer to the gods.

Sagitta or ***Arrow,*** and ***Delfino***: Sagitta is a prominent constellation between Aquila and Cygnus. Farther away to the East, Delphinus, a small rhombus of 4th magnitude stars, forms a triangle with Aquila and Sagitta.

Lyra, cygnus, aquarius and pisces
These four constellations are reproduced in a miniature of the 16th century manuscript, Città di Vita, by Matteo Palmieri.

BETWEEN THE NORTHERN AND THE SOUTHERN HEMISPHERES

Ophiuchus: this is a large, rather dim, constellation under Hercules, crossing Serpens, and is found from the epsilon-delta alignment in Hercules. Ophiuchus is connected to the cult of Aesclepius, the god of medicine. it holds the head of Serpens, the Serpentm in one hand, and the tail in the other.

Orion: half in the North and half in the South. Orion's belt, an alignment of three very bright stars, is exactly on the equator. Formed of a number of bright stars, Orion is often used to find the other constellations. In mythology, Orion is a giant hunter who fought Taurus, the Bull, was killed by Scorpio, the Scorpion, and was brought back to life by Aesclepius. Betelgeuse, the "shoulder of the giant" is a red supergiant star of magnitude 0.9, and Rigel, a blue double supergiant star of magnitude 0.3.
In the sword are the great nebula M42 (16 light years in diameter and 1,600 light years away), and the smaller M43 nebula.

Serpens: this is a very long constellation: its head is near the Corona Borealis (South) and its head is near Aquila (North) In this constellation globular cluster ➢218 M5 is visible to the naked eye.

SOUTHERN HEMISPHERE

Canis Major: following the alignment of the stars of Orion's belt leads to Sirius, the brightest star in the sky a blue-white star of magnitude 1.58. The eye of Canis Major is a double star, companion to a white dwarf ➢202 . Canis Major also includes other double stars, triple stars, and various clusters.

Cetus: consisting of sparse and faint stars (only one 2nd magnitude star), this constellation contains Mira, the first variable star ➢207 discovered: it is an

The southern celestial hemisphere
Only the principal constellations are indicated.

BELOW: The Southern celestial hemisphere and an image of Centaurus, the Centaur, from an ancient print.

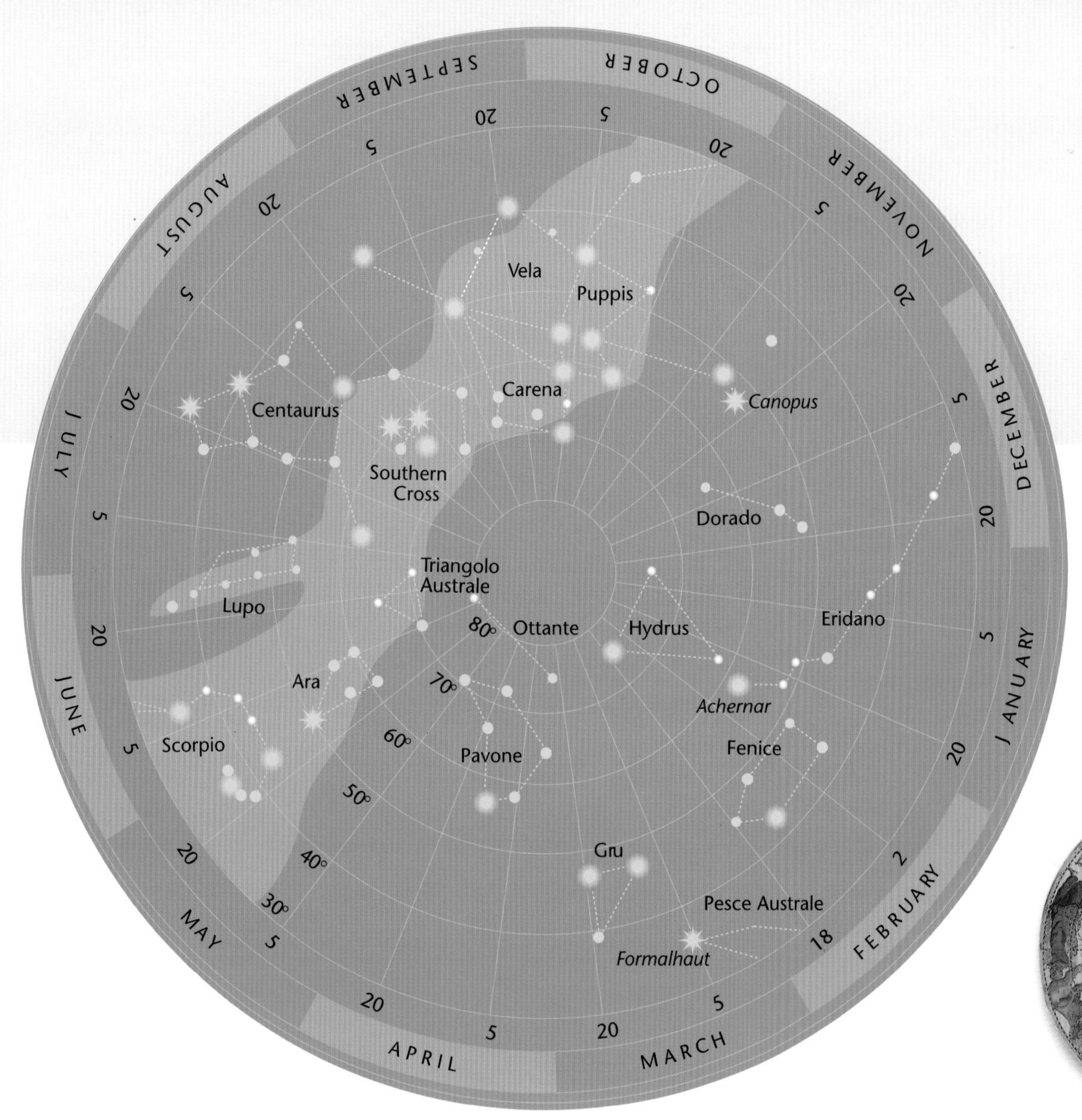

M class ➢196 red star which varies between 8th-10th magnitude (about 2000 K) and 2nd-5th (about 2600 K) in a period of about 331 days. The head, five stars arranged in a circle, is South West of the Pleiades and South of Andromeda. The whole constellation, which extends to the South and West, is named for the monster whale to which Andromeda was satisfied to placate Poseidon.

Lepus and ***Colombia***: these two small constellations are South of Orion. The main part of Lepus is a quadrilateral of 3rd and 4th magnitude stars. Columba, named for the dove which flew from Noah's ark, is even smaller: four stars South of Lepus.

Centaurus and ***Crux***: known since the time of Ptolemy, the constellation Centaurus also included Crux, which has only recently been considered a separate constellation. In Centaurus, the Centaur, a mythological creature half man half horse, the brightest srar is Proxima: this is also the nearest star to our solar system (4.2 light years).

Puppis, ***Vela*** and ***Carina***: these three very bright constellations to the South of Canis Major used to be considered a single constellation, the ship Argo, from the myth of the Argonauts. For this reason they still have some stars in common.
Carina, the Keel, the Southernmost of the three, contains several bright stars: 7 of 1st to 3rd magnitude and 15 of 3rd-4th magnitude. The brightest is Canopus, the brightest star in the Southern sky. Puppis, the Stern, has over 12 stars of magnitude less than 6, and the seta star is of magnitude 2.3, despite being 1,600 light years from earth. There are also many clearly visible nebulas ➢212 in this part of the sky. Vela, the Sail, has an open cluster ➢218 (omicron Velorum) of magnitude 2.5.

THERE'S SOMETHING MAGIC IN THE RHYTHMIC APPEARANCE AND DISAPPEARANCE OF SOME OF THE CONSTELLATIONS IN THE NIGHT SKY. THE ANCIENTS USED THEM FOR ASTROLOGY AND DEVINATION, AND EVEN NOW MANY PEOPLE LOOK TO THE ZODIAC FOR REASSURANCE. ALTHOUGH THE SPRING EQUINOX IS NO LONGER WHERE ASTROLOGERS THINK IT IS.

THE ZODIAC

THE LEONIDS
The shower of falling stars that recurs in early November originates from a point of the sky in the constellation Leo.

The Zodiac is a band of sky 18° wide across the ecliptic: the ancients divided it from the gamma point into 12 signs 30° wide each, and to each sign gave the name of its most representative constellation. As the positions of the Earth and the other celestial bodies change, the Sun, the planets and the Moon are projected onto the Zodiac: during the year the Sun passes through all the signs as it moves along the ecliptic. So do the Moon and the other planets. The constellations of the Zodiac are each visible in a particular period of the year, and they all rise and set following the motion of the sky. It should be emphasised that, about 2000 year ago, at the time of Hipparchus, the gamma point was in the constellation Aries, but the precession of the equinoxes means that since then it has moved about 30° and is now in the constellation Pisces, and in contrast to the affirmations of all horoscopes, at present, in the period between 21 May and 21 June, the Sun is not in the sign of Gemini, but in Taurus.

Aries: this is below and slightly to the East of Andromeda. It is a small constellation, but it has been important in the past, because it used to contain the gamma point.

Taurus: following the alignment of the Pole Star and Capella leads to Nath, which is in both Auriga and Taurus. It is a blue giant much less bright than Aldebaran, the red giant halfway between Nath and the Pleiades. Near the zeta star is the Crab nebula, the remains of the supernova of 1054, at the centre of which is a pulsar ➛203 invisible to the telescopes of amateur astronomers. Near Aldebaran is the open cluster of the Hyades, of which we can see just five or six stars. Taurus represents Zeus, who appeared to Europa to carry her away to the island of Crete.

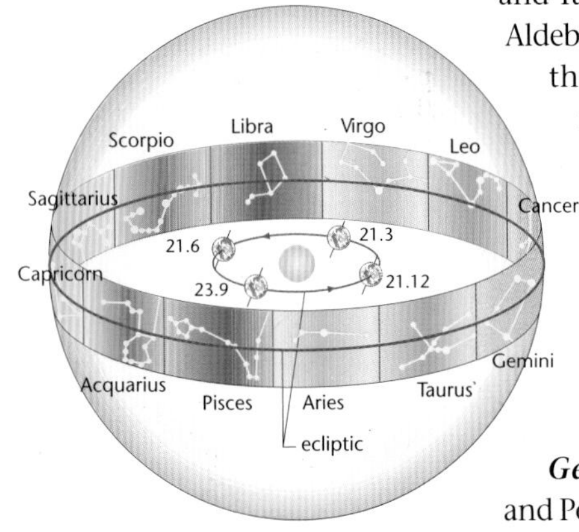

Gemini: : this constellation represents Castor and Pollux, the twin children of Zeus and Leda, who took part in the voyage of the Argonauts. Following the delta-beta alignment in Ursa Major leads to Pollux, a red giant a little dimmer than the bright Castor, a multiple star composed of 6 stars. The open cluster M35 can be seen near Tejat Prior . The summer solstice point (North) is in this sign.

Cancer: this rather faint constellation consists of 4th and 5th magnitude stars, and is located between Leo and Gemelli. At its centre is the open cluster of Praesepe, one of the closest to Earth. The star Asellus Australis is on the ecliptic. Cancer is a giant crab killed by Hercules during his battle with the Hydra of Lerna.

Leo: this is the biggest constellation in the Zodiac, which can be found following the alignment of the Pole star and alpha-beta in Ursa Major. The brightest star is the blue-white Regulus, exactly on the ecliptic. Then come the white Denebola and the yellow double giant Algieba. The falling stars of mid November, called "Leonids", fall from this part of the sky. Leo represents the lion of Nemea, killed by Hercules.

Virgo: Spica, a blue-white binary star, the brightest in Virgo, is found by following the curve passing through the delta, epsilon zeta and eta stars of Ursa Major, and Arcturus, the brightest star in Bootes. Stars of 3rd and 4th magnitude for a "Y" which extends from Spica towards Denebola, the tail of Leo: Virgo is easy to recognise because the sky around it has few bright stars. There is a cluster of hundreds of galaxies in Virgo, about 7 million light years away from us. The name is derived from peasant culture.

Libra: starting from the Pole star and passing through the beta star of Bootes leads to Suben el Genubi, the brightest star in this constellation, about ten degrees below the celestial equator, just to the West of Zuben el Schemali (the beta star),

Scorpio: following the direction from the Pole star passing through the beta star of Hercules leads to Antares, the brightest star in Scorpio, a red supergiant with a diameter 700 times that of the Sun and a

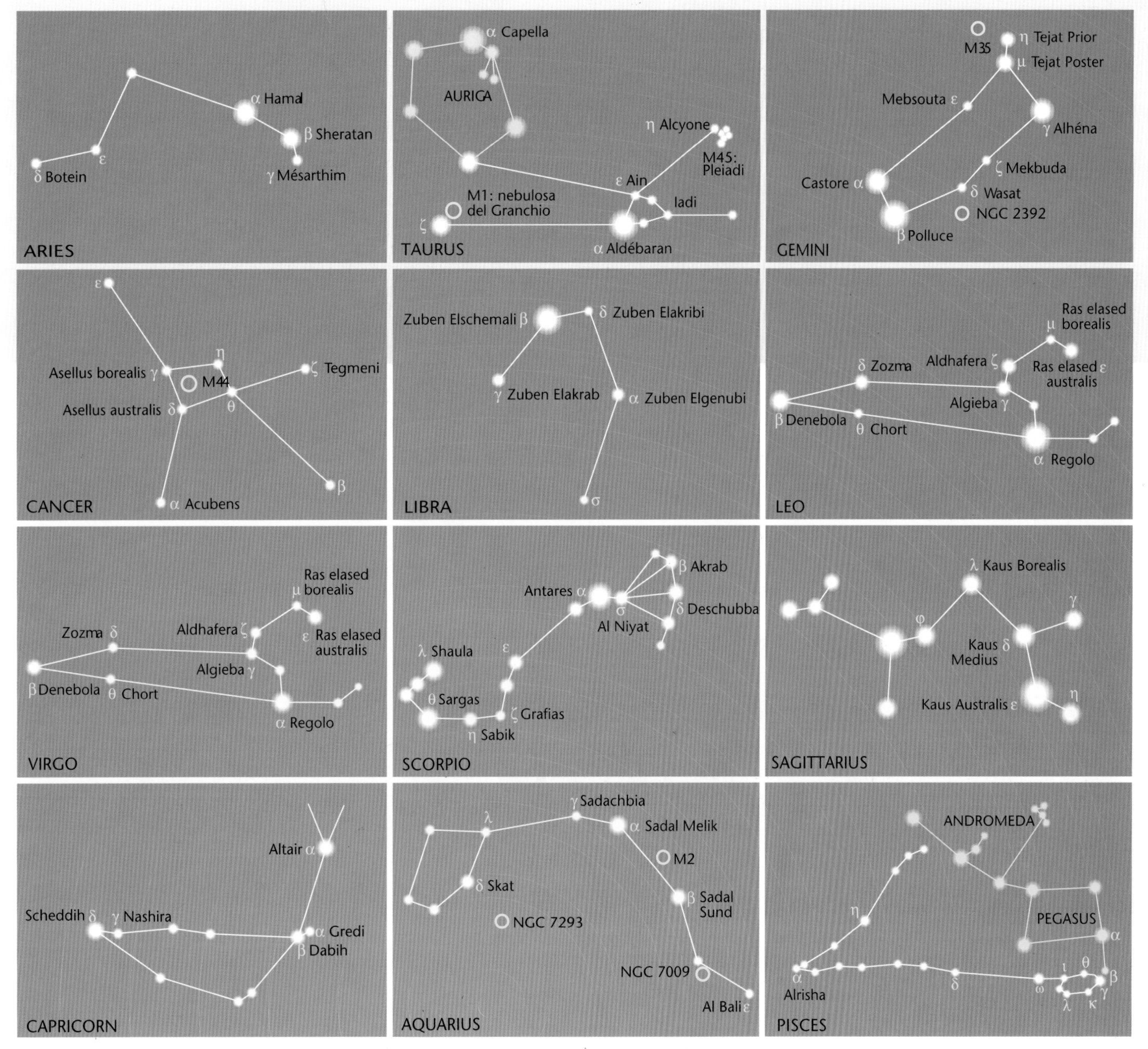

density of one millionth. Antares, which is 250 loight years away, has a weak, greenish companion. There are several not very visible clusters near Antares. This constellation is easy to recognise by its fish-hook shape, and it is named for the Scorpion which killed Orion because he had pestered Artemis.

Sagittarius: proceeding from Altair towards the head of Aquila leads to Sagittarius, about 30° below the celestial equator. Its central part is called the Bear Cub, because it falls a small inverted Bear. Sagittarius represents Cheiron, the wise centaur who taught Achilles, Jason, Aesclepius (god of medicine) and maybe also Apollo. Cheiron killed Scorpio, the Scorpion who killed Orion.

Capricorn: following the direction Vega-Altair leads to Gredi and Dabih. The former is a double yellow star, while all the others in this constellation are quite faint. Capricornus, or sea goat, is a mythical animal with the head of a goat and the tail of a fish.

Acquarius: the alignment of beta-alpha in Pegasus and the beta of Pisces lead to the lambda star of Acquarius. This is a constellation which spreads across the sky below the celestial equator, made up of faint stars. Aquarius represents Deucalion, son of Prometheus, while he pours water from a jar.

Pisces: following the beta-alpha alignment in Pegasus leads to the beta star of Pisces. This is a ring shaped constellation with a long tail which stretches East below Aries, turns North, and ends near to Andromeda. Alrisha (the alpha star) is a blue-white double star. Nowadays the gamma point is in this constellation.

12 CONSTELLATIONS
The zodiac constellations are indicated with lines and white stars, while the yellow stars are stars particularly close to them. The small yellow circles indicate the position of the most important celestial objects found in each constellation.

NORTHERN HEMISPHERE

POLAR REGIONS (63°30'), 12 P.M. WINTER SOLSTICE (21 DECEMBER)

THIS IS THE PERIOD OF LONG NIGHTS: FARTHER NORTH THE SUN WILL ONLY APPEAR AGAIN IN THE SUMMER, WHILE AT THIS LATITUDE THERE IS STILL SOME DAYTIME, ALTHOUGH VERY LITTLE. ON NIGHTS WHEN THERE IS SOME RESPITE IN THE BAD WEATHER TYPICAL OF WINTER, THE CONSTELLATIONS GLITTER BRIGHTLY IN THE FROST.

TO THE NORTH

S A line through the beta and alpha stars of Ursa Major points to Polaris, the North star, while the alignment of zeta and eta Ursae Minoris leads to Etamin, the brightest star in the Draco. Aligning Rastaban and Etamin locates Sadr, the centre of the large cross of Cygnus, and a line from Sadr to Deneb (the "tail" of the Swan) points to Caph the beta star of Cassiopeia, with its clearly recognisable W shape. The alpha-beta alignment points to alpha Cephei, while the gamma-alpha alignment leads to Alpheraz, common to both Pegasus and Andromeda.

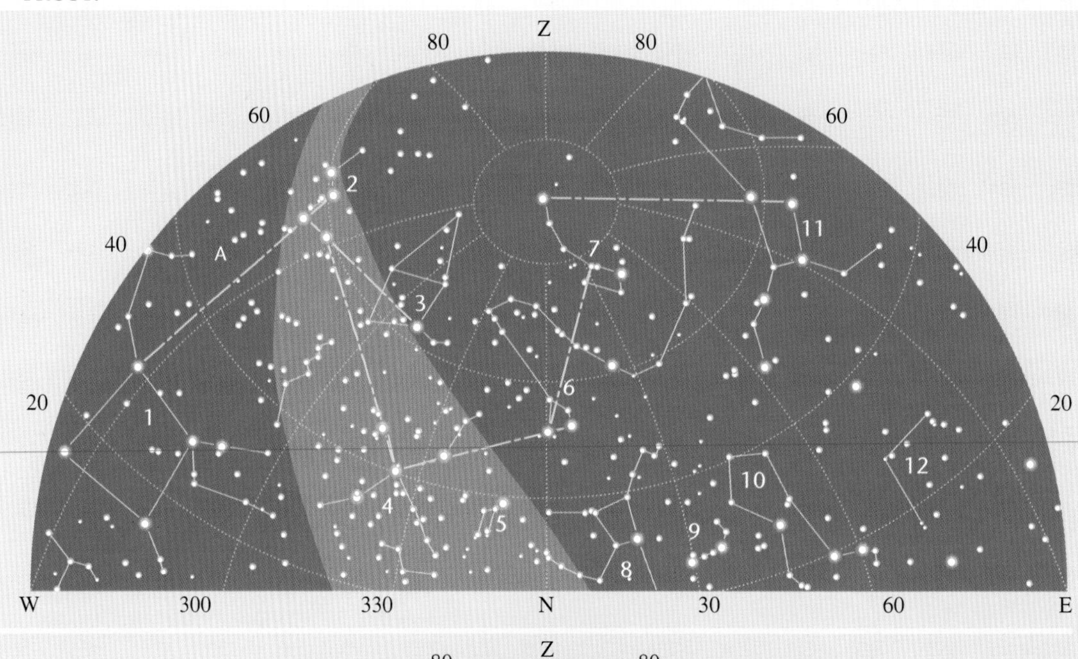

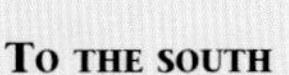

TO THE SOUTH

The constellation Orion is easily distinguished, low on the horizon, by the row of three stars at the equator. Following their alignment to the right (or West) is Aldebaran, the brightest star in Taurus. While low on the horizon on the left (to the East) is Sirius, the brightest star in the sky, in Canis Major. Following the alignment Rigel-Alnilam-Betelgeuse leads to Castor (alpha Geminorum) with its neighbour Pollux, the beta star of Gemini: the beta-alpha alignment points to beta Aurigae, near Capella.

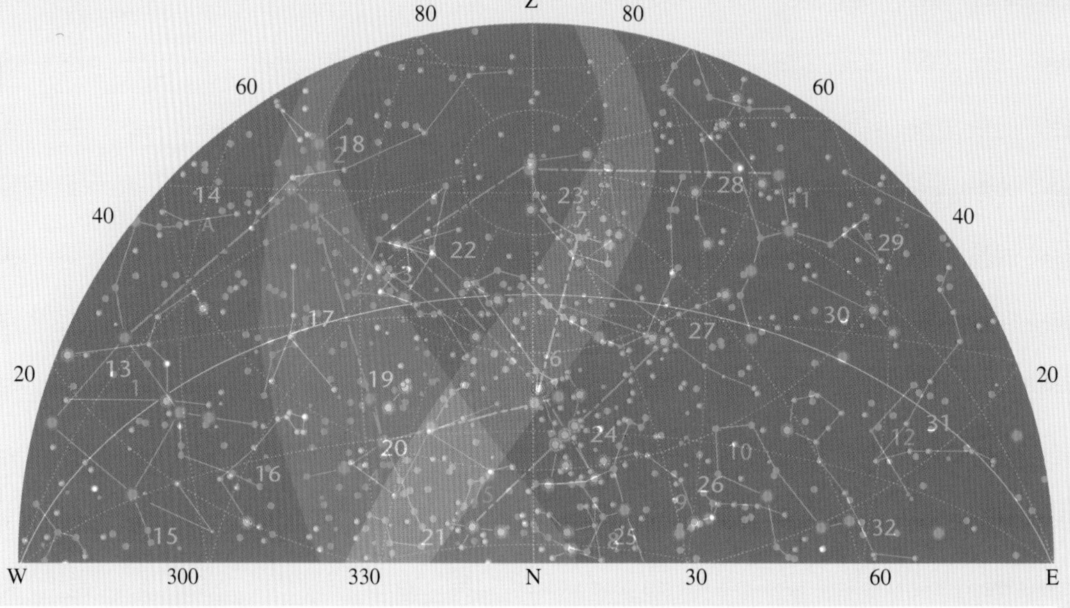

1. PEGASUS
2. CASSIOPEIA
3. CEPHEUS
4. CYGNUS
5. LYRA
6. DRACO
7. URSA MINOR
8. HERCULES
9. CORONA BOREALIS
10. URSA MAJOR
11. BOOTES
12. COMA BERENICES
13. LEO
14. LEO MINOR
15. SEXTANS
16. HYDRA
17. CANCER
18. LYNX
19. CANIS MINOR
20. UNICORN
21. CANIS MAJOR
22. GEMINI
23. AURIGA
24. ORION
25. LEPUS
26. ERIDANUS
27. TAURUS
28. PERSEUS
29. TRIANGULUM
30. ARIES
31. PISCES
32. CETUS

A. GALAXY M31

NORTHERN HEMISPHERE

POLAR REGIONS (63°30'), 12 P.M. SPRING EQUINOX (21 MARCH)

THE NIGHTS START TO SHORTEN MORE AND MORE QUICKLY AROUND THIS DATE THE DURATION OF THE NIGHT VARIES THE MOST. LOOKING SOUTH, MANY GALAXIES ARE VISIBLE BELOW THE CONSTELLATION LEO.

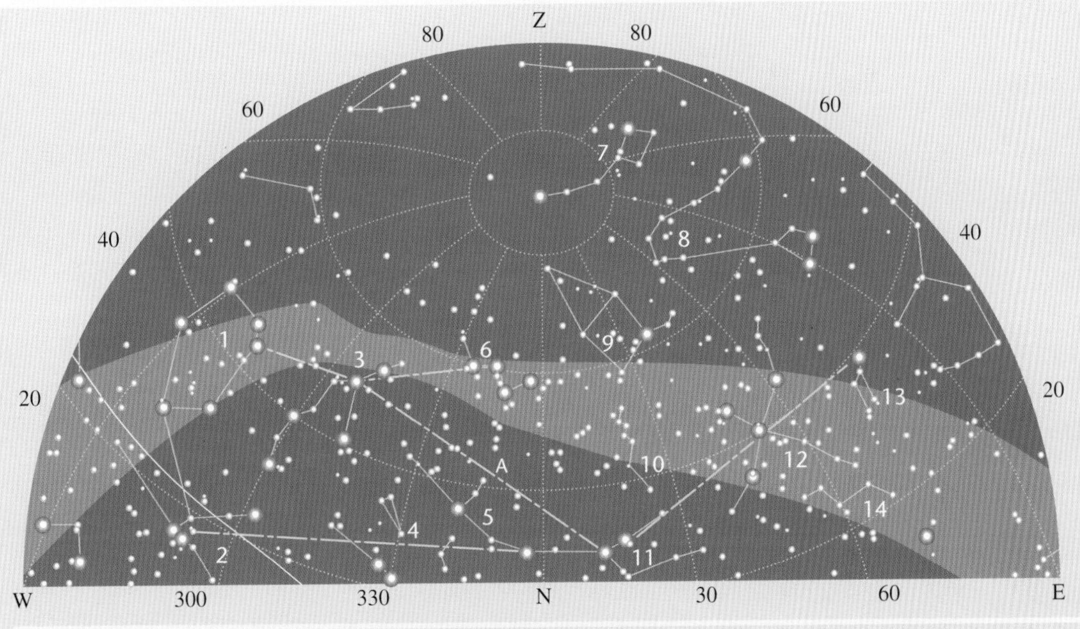

TO THE NORTH
The most recognisable constellations are the large cross of Cygnus, low on the horizon and, higher up, the W of Cassiopeia. Tracing back alignments [78] leads the eye to Draco, Ursa Minor and Pegasus. The alignment of Scheat and Sadr points to Vega, in Lyra, and a line through Scheat and Alpheratz locates Aldebaran (Taurus). The alignment of gamma and delta in Cassiopeia points to alpha Persei, and this, aligned with Scheat, leads the eye to Capella (Auriga).

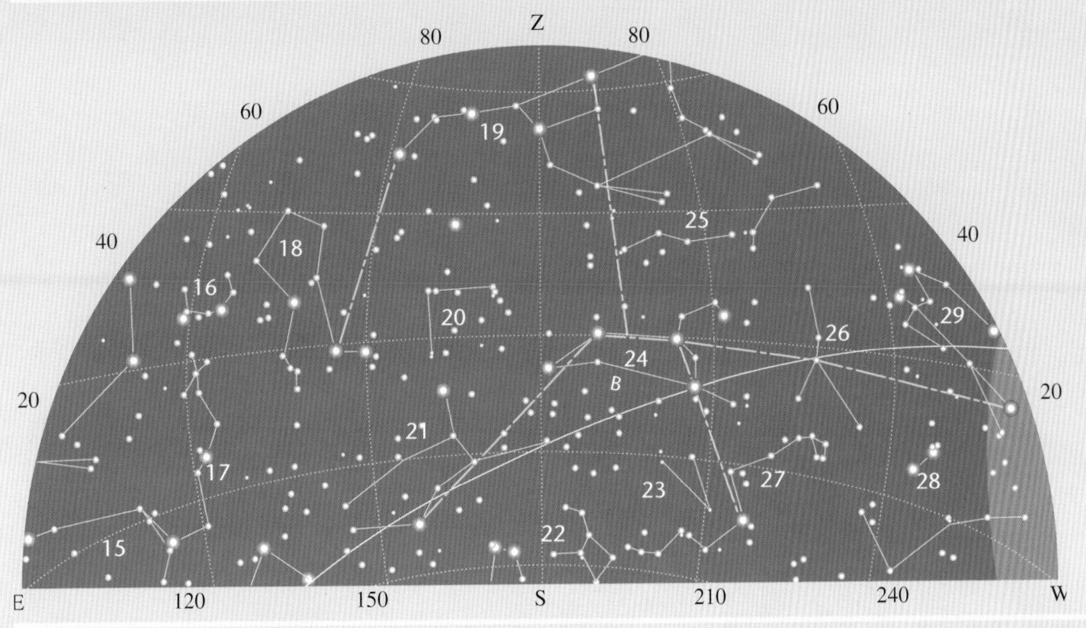

TO THE SOUTH
Ursa Minor, almost at the zenith, leads the eye to Arcturus, the brightest Bootes star in the Northern hemisphere, and to the constellation Leo, halfway between the horizon and the zenith. In Leo, the alignments gamma-alpha, zosma-beta and zosma-gamma point to alpha Hydrae, alpha Virginis (Spica, a white double star) and gamma Geminorum. Sirius can be seen low on the horizon on the right (West).

1. Auriga
2. Taurus
3. Perseus
4. Triangulum
5. Andromeda
6. Cassiopeia
7. Ursa minor
8. Draco
9. Cepheus
10. Lacerta
11. Pegasus
12. Cygnus
13. Lyra
14. Volpecula
15. Corona borealis
16. Serpens
17. Bootes
18. Ursa major
19. Coma berenices
20. Virgo
21. Crater
22. Sextans
23. Leo
24. Leo minor
25. Cancer
26. Hydra
27. Canis major
28. Gemini

A. Galaxy M31
B. Area rich in galaxies

NORTHERN HEMISPHERE
POLAR REGIONS (63°30′), 12 P.M.
SUMMER SOLSTICE (21 JUNE)

THIS IS THE PERIOD OF LONG DAYS: IN THE FAR NORTH THE SUN WILL ONLY DISAPPEAR AGAIN IN THE WINTER, WHILE AT THIS LATITUDE THE DAY IS VERY LONG. THE VERY SHORT NIGHTS ARE ALWAYS ILLUMINATED BY A CONSTANT DUSK, AND ALTHOUGH THE CONSTELLATIONS ARE THERE, IT IS ALMOST IMPOSSIBLE TO SEE THEM.

If the sky were dark, an observer would see these constellations:

1. Leo
2. Leo Minor
3. Ursa Major
4. Ursa Minor
5. Lynx
6. Gemini
7. Auriga
8. Perseus
9. Cassiopeia
10. Cepheus
11. Lacerta
12. Andromeda
13. Triangulum
14. Aries
15. Pegasus
16. Pisces
17. Aquarius
18. Cygnus
19. Vulpecula
20. Aquila
21. Scutum
22. Lyra
23. Draco
24. Hercules
25. Ophiuchus
26. Libra
27. Serpens
28. Corona Borealis
29. Bootes
30. Virgo

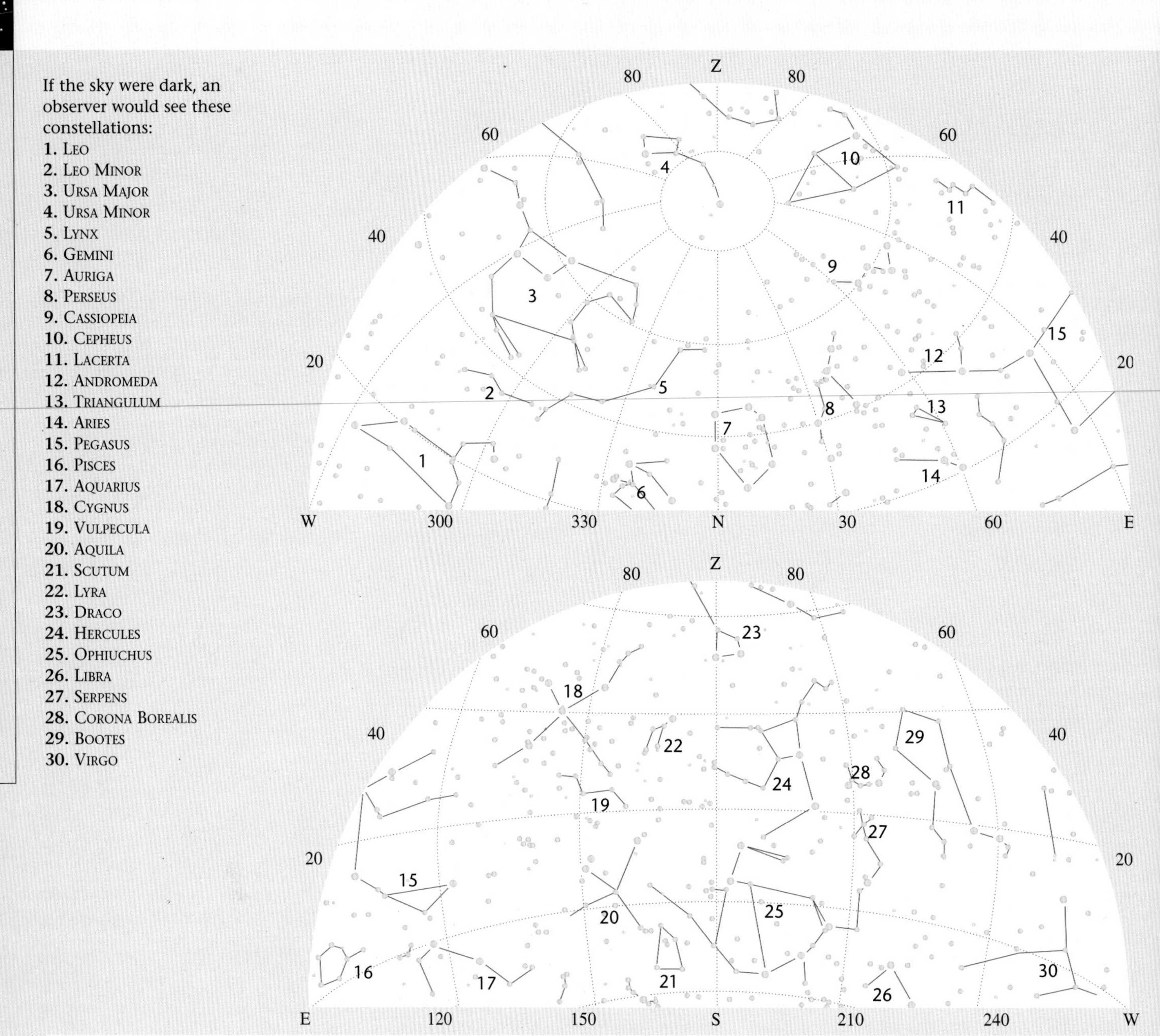

NORTHERN HEMISPHERE

POLAR REGIONS (63°30'), 12 P.M. AUTUMN EQUINOX (23 SEPTEMBER)

WINTER AND ITS LONG STAR-FILLED NIGHTS WILL SOON BE ON US, FOR NOW WE CAN ADMIRE A PART OF THE SKY THAT WILL SOON DISAPPEAR BELOW THE HORIZON.

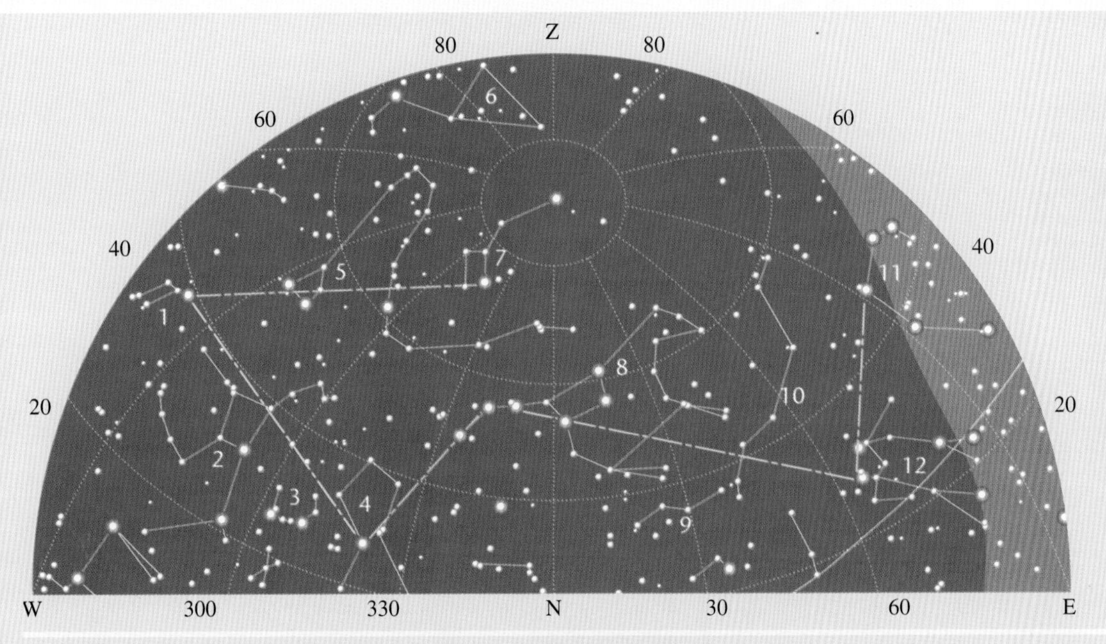

TO THE NORTH
Following the alignment [78] locates Ursa Major, Ursa Minor and Draco, while alignment [79] points to Bootes. On the left, almost to the West, in alignment with beta and gamma Ursae Minoris, shines Vega; diametrically opposite, on the right, almost in the East, shines Capella, which can be located by aligning beta and alpha Geminorum. Pollux, the beta star of Gemini, is found following the alignment of Alioth and Phecda in Gemini.

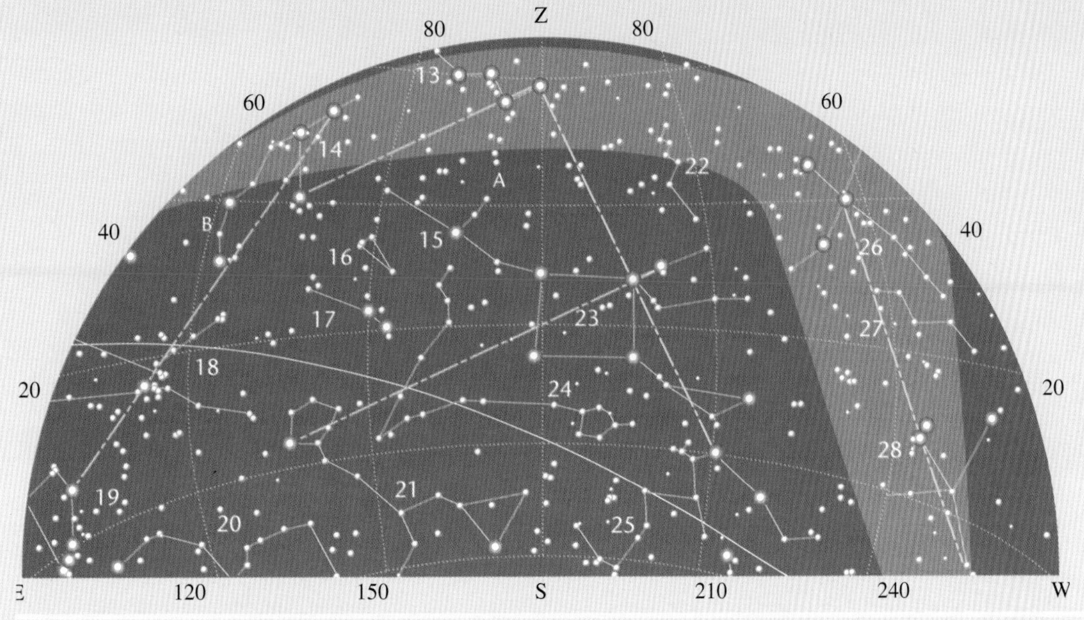

TO THE SOUTH
Pegasus is in the centre, equidistant from the zenith and the horizon, and on the right, we can observe Cygnus shining high on the horizon, with Aquila lower down. On the left (East) is Algol (beta Persei), along the line through beta and alpha Cassiopeiae. Aligning Cassiopeia itself with Aldebaran locates the California nebula and the Pleides. A line down from beta Cassiopeiae to the beta star locates alpha Aquarii, while the alignment of eta and beta Pegasi leads to alpha Ceti. Orion is low on the horizon in the East.

1. Lyra
2. Hercules
3. Crux
4. Bootes
5. Draco
6. Cepheus
7. Ursa minor
8. Ursa major
9. Leo minor
10. Lynx
11. Auriga
12. Gemini
13. Cassiopeia
14. Perseus
15. Andromeda
16. Triangulum
17. Aries
18. Taurus
19. Orion
20. Eridanus
21. Cetus
22. Lacerta
23. Pegasus
24. Pisces
25. Aquarius
26. Cygnus
27. Volpecula
28. Aquila

A. Galaxy m31
B. California nebula

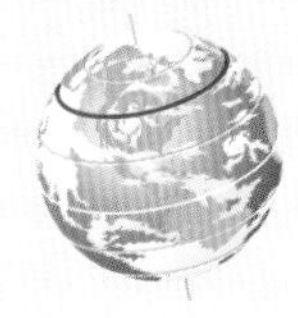

NORTHERN HEMISPHERE

MEDIAN LATITUDES (45°), 12 P.M. WINTER SOLSTICE (21 DECEMBER)

IN THESE LATITUDES WINTER BRINGS NIGHTS WHICH ARE LONG, BUT FAR FROM ENDLESS. ALMOST ALL THE CONSTELLATIONS RISE AND SET, AND THE BEST TIME TO OBSERVE THEM VARIES ACCORDING TO THE MAXIMUM ALTITUDE THEY REACH ABOVE THE HORIZON.

To the north
Following the alignment of the beta and alpha stars of Ursa Major leads to Polaris, the North star. From here, the alignment of zeta and eta Ursae Minoris leads to Etamin, the brightest star in the Draco. Starting from Cassiopeia, easily recognisable by its W shape, lines through alpha-beta, gamma-alpha and gamma-delta lead to the alpha Cephei, Alpheraz, common to Pegasus and Andromeda, and alpha Persei.

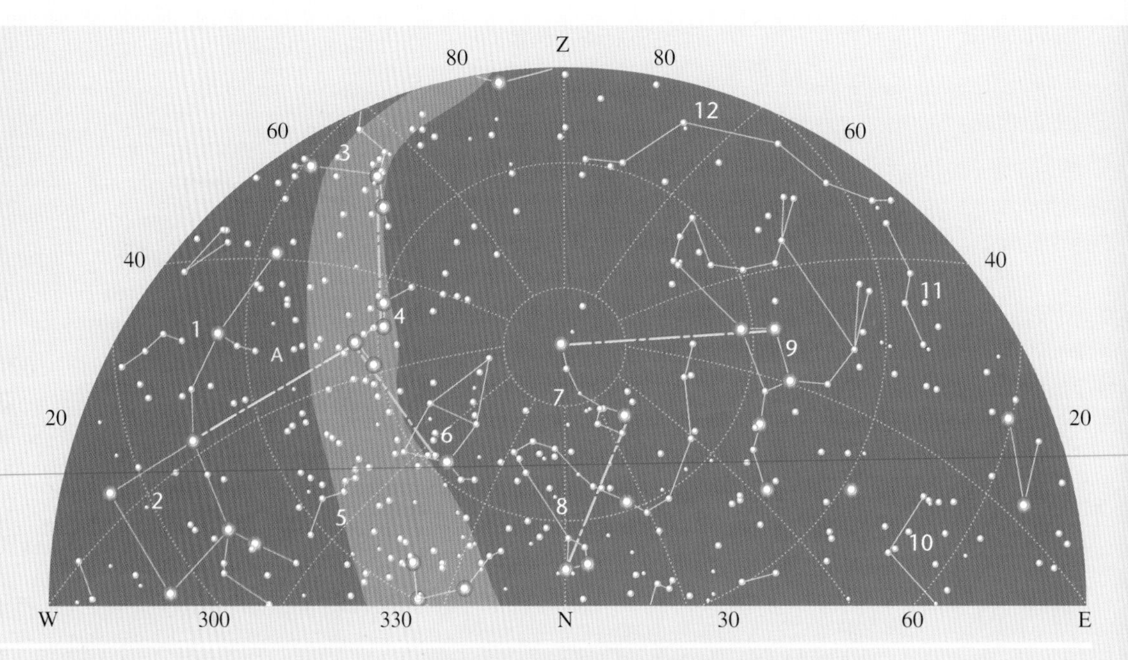

To the south
Orion dominates in the centre. In alignment with the belt is Aldebaran (alpha Tauri) on the right(West) and Sirius, alphja Canis Majoris the left (East). A line through Rigel-Alnitak-Betelgeuse points to Alhema (nu Geminorum). The alignment of beta and alpha Geminorum points to beta Aurigae (in the Northern sector), near Capella, and in the opposite direction, to delta Cancri and Regulus) alpha Leonis) in the South.

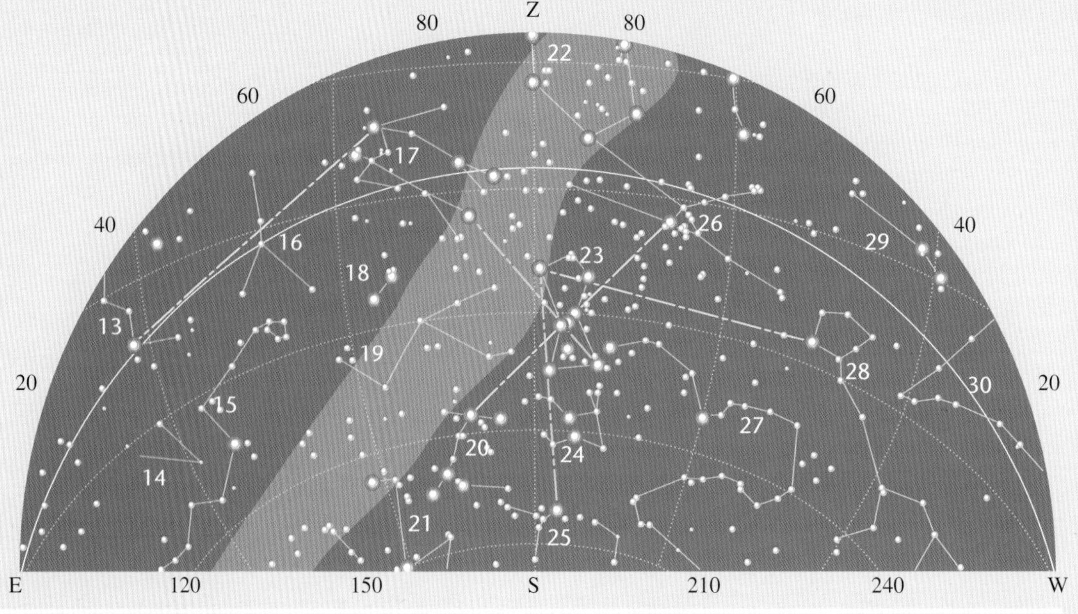

1. Andromeda
2. Pegasus
3. Perseus
4. Cassiopeia
5. Lacerta
6. Cepheus
7. Ursa minor
8. Draco
9. Ursa major
10. Coma berenice
11. Leo minor
12. Lynx
13. Leo
14. Sextans
15. Hydra
16. Cancer
17. Gemini
18. Canis minor
19. Unicorn
20. Canis major
21. Puppis
22. Auriga
23. Orion
24. Lepus
25. Columba
26. Taurus
27. Eridanus
28. Cetus
29. Aries
30. Pisces

A. Galaxy M31

NORTHERN HEMISPHERE

MEDIAN LATITUDES (45°), 12 P.M. SPRING EQUINOX (21 MARCH)

THE SPRING SKY IS CHARACTERISED BY THE USUAL CIRCUMPOLAR CONSTELLATIONS TO THE NORTH, AND A HOST OF ZODIAC CONSTELLATIONS TO THE SOUTH: FOUR ARE CLEARLY VISIBLE, ALL AT THE SAME TIME.

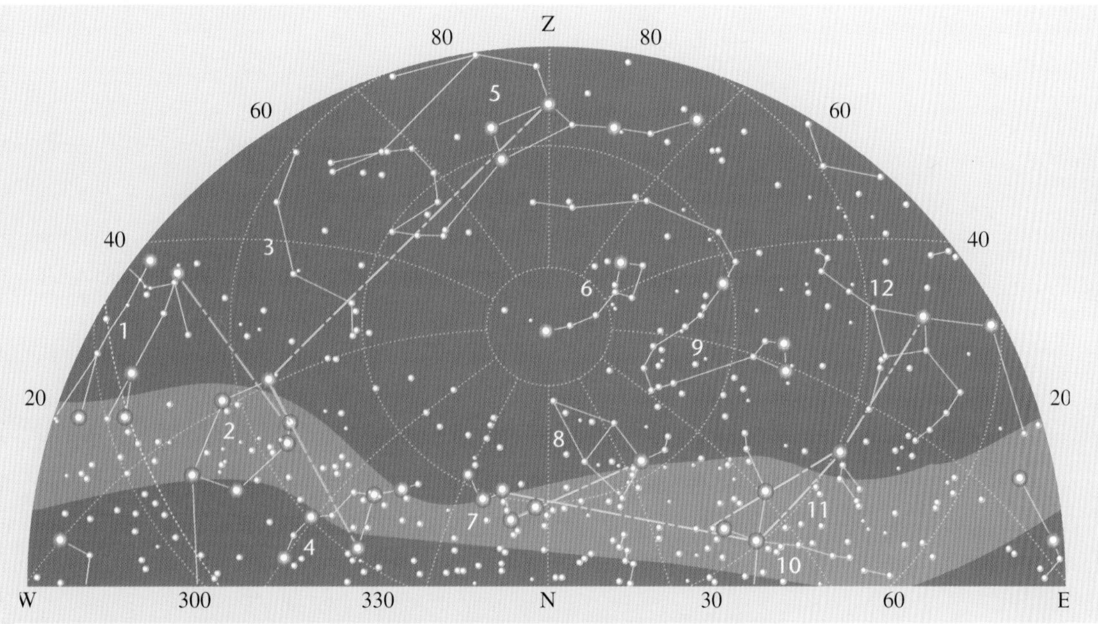

TO THE NORTH
A line projected through Phecda-Dubhe, the gamma and alpha stars of Ursa Major, lpoints to beta Aurigae, close to bright Capella. Following the beta Aurigae-Capella alignment leads to Algol, in Perseus. In the opposite direction (to the west), the line leads to Castor and Pollux, alpha and beta Geminorum. Alderamir (alpha Cephei), and Deneb, alpha Cygni, are found along the alignments of alpha-beta and gamma-beta in Cassiopeia. Vega shines brightly a short distance away, with Hercules further out in the same direction.

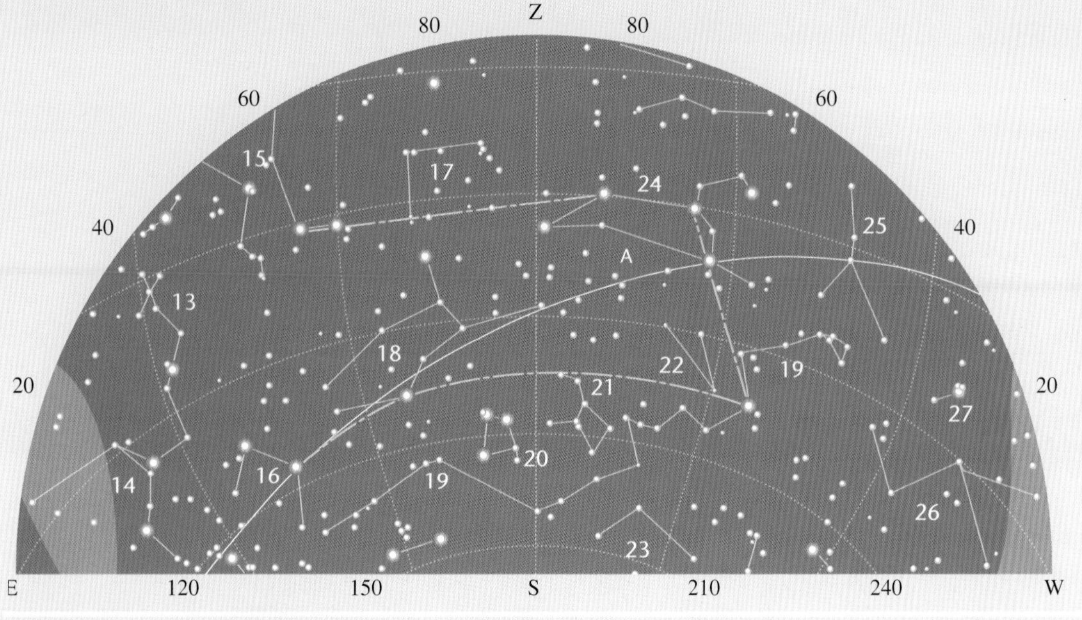

TO THE SOUTH
There are no particularly bright stars on this side, and finding the constellations is quite complicated. The yoke of the Plough in Ursa Major zeta-eta, points northwards to Arcturus, the brightest star in Bootes. A line projected through this to eta star points to the constellation Leo. The line gamma-alpha locates Alphard (alpha Hydrae), and to the East of this along the parallel, lie Spica and Zuben-el-Genubi, the alpha stars of Virgo and Libra respectively, aligned on the ecliptic.

1. Gemini
2. Auriga
3. Lynx
4. Perseus
5. Ursa major
6. Ursa minor
7. Cassiopeia
8. Cepheus
9. Draco
10. Cygnus
11. Lyra
12. Hercules
13. Serpens
14. Ophiuchus
15. Bootes
16. Libra
17. Coma berenices
18. Virgo
19. Hydra
20. Corvus
21. Crater
22. Sextans
23. Antlia
24. Leo
25. Cancer
26. Unicorn
27. Canis minor

A. Area rich in galaxies

NORTHERN HEMISPHERE

MEDIAN LATITUDES (45°), 12 P.M. SUMMER SOLSTICE (21 JUNE)

SUMMER NIGHT: SIX ZODIAC CONSTELLATIONS ARE VISIBLE ALIGNED ALONG THE ECLIPTIC, LOW ON THE HORIZON: LOOKING SOUTH FROM EAST TO WEST, THE STARS COMPOSING AQUARIUS, CAPRICORN, SAGITTARIUS, SCORPIO, LIBRA AND VIRGO CAN JUST BE DISTINGUISHED.

To the north
Nothing of particular note, apart from the usual beautiful and easily recognisable circumpolar constellations: along the alignment beta-alpha of Ursa Major to Polaris, the North star, from zeta-eta in Ursa Minor to Etamin (gamma Draconis); from the alignment of alpha-beta in Cassiopeia to alpha Cephei and on to Veg (alpha Lyrae), or in the opposite direction (gamma-delta) to Mirfaak (alpha Persei), or along the line gamma-alpha (of Cassiopeia) to Sirrah, the alpha star of Pegasus, and part of Andromeda.

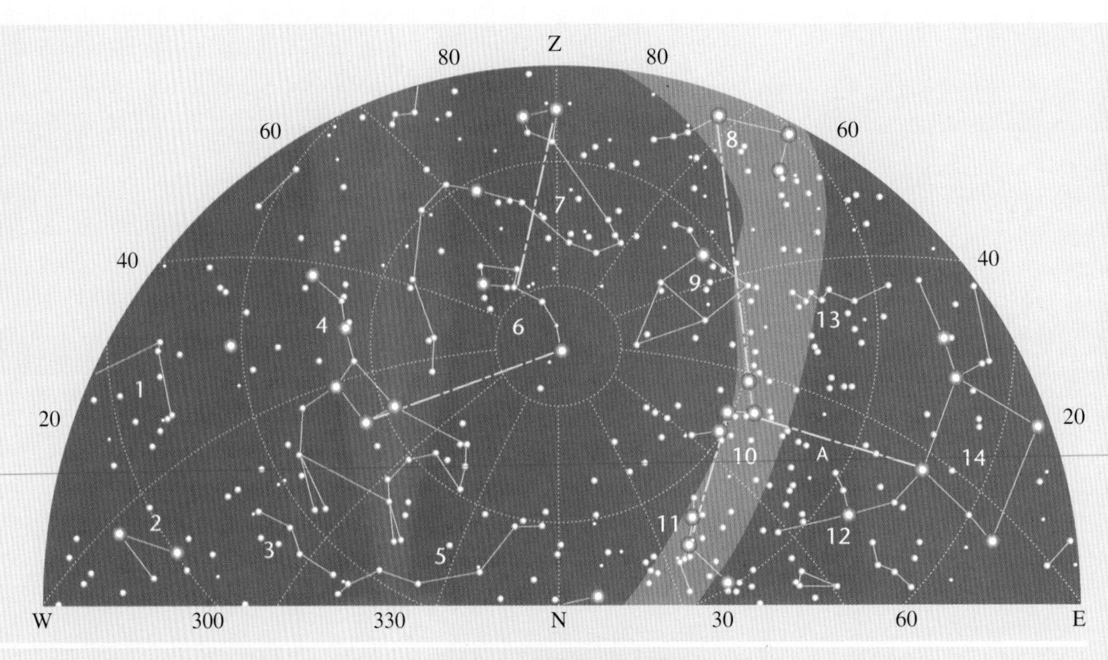

To the south
The most brilliant stars in the Southern stretch are Vega (alpha Lyrae), almost at the zenith, Altair (alpha Aquilae), midway between the zenith and the horizon, but shifted slightly to the left East), and Antares (alpha Scorpii), just above the horizon and shifted slightly to the right (West). A little further away from the South and to the West, are Arcturus, in Bootes, at about the same altitude as Altair, and Spica, just above the horizon.

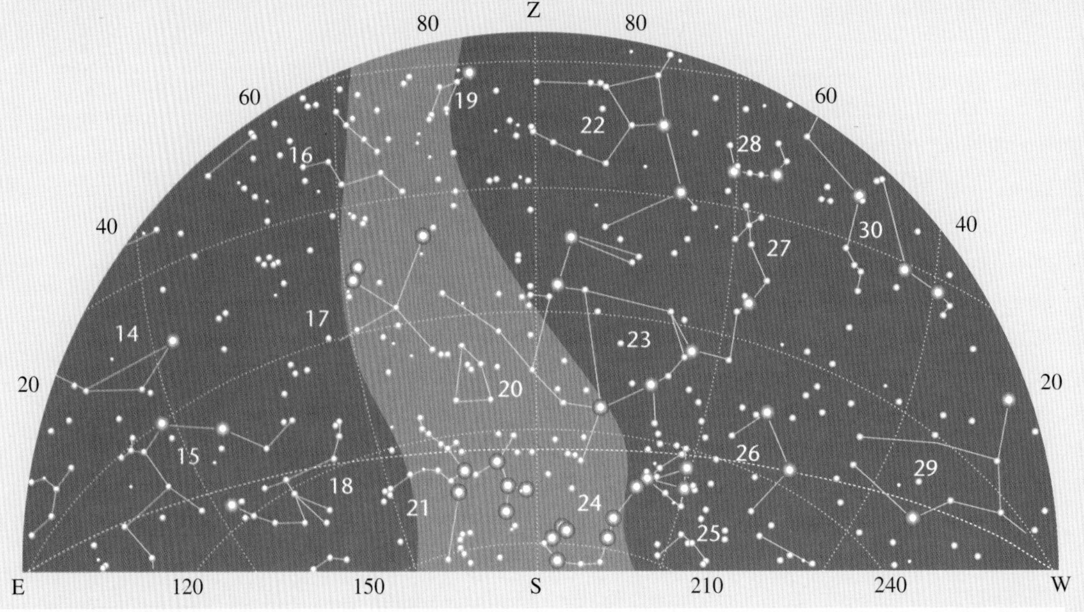

1. Coma berenices
2. Leo
3. Leo minor
4. Ursa major
5. Lynx
6. Ursa minor
7. Draco
8. Cygnus
9. Cepheus
10. Cassiopeia
11. Perseus
12. Andromeda
13. Lacerta
14. Pegasus
15. Aquarius
16. Volpecula
17. Aquila
18. Capricorn
19. Lyra
20. Scutum
21. Sagittarius
22. Hercules
23. Ophiuchus
24. Scorpio
25. Lupus
26. Libra
27. Serpens
28. Corona borealis
29. Virgo
30. Bootes

NORTHERN HEMISPHERE

MEDIAN LATITUDES (45°), 12 P.M. AUTUMN EQUINOX (23 SEPTEMBER)

THE NIGHT OF THE AUTUMN EQUINOX DISPLAYS FIVE ZODIAC CONSTELLATIONS HIGH IN THE SKY TO THE SOUTH, ALONG THE ECLIPTIC. PEGASUS IS ALMOST AT THE ZENITH, AND, TO THE NORTH, URSA MAJOR IS LOW ON THE HORIZON.

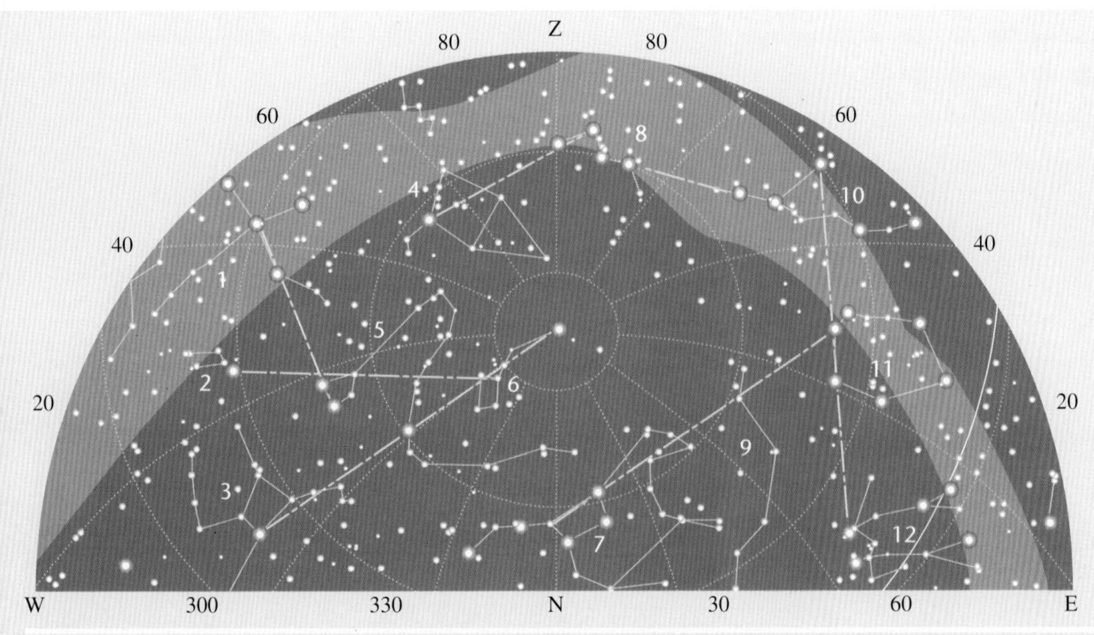

To the north
The circumpolar constellations are the same: Ursa Major and Ursa Minor, Draco, Cepheus, Cassiopeia and Lynx. And, as in the winter, a line through the zeta-eta stars of Ursa Minor points to Etamin, the brightest star in Draco. Now, though, this can be extended out to Vega, in Lyra. The beta-gamma alignment in Draco points to Cygnus, and the the alpha-eta alignment passes through alpha Hydrae and on to Hercules. Perseus is found to the East (right) of Cassiopeia, with Cepheus to the West (left). Capella, in Auriga, can be found following the delta-alpha alignment in Ursa Major, and Castor and Pollux, in Gemini, are found almost on the horizon, below Capella.

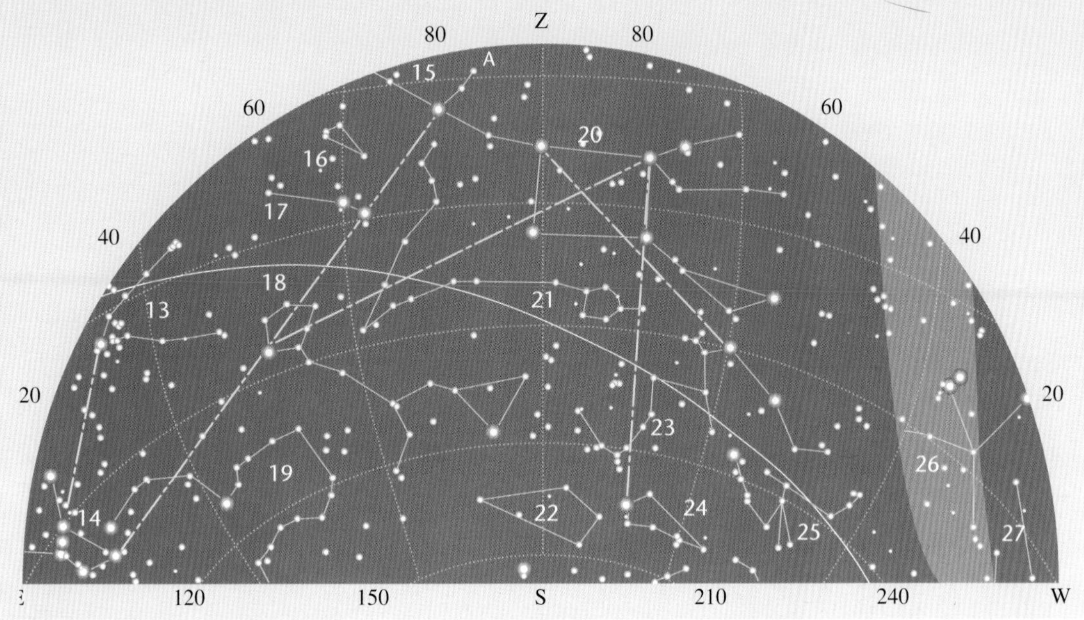

To the south
Pegasus is very high, in the centre, and Andromeda is almost at its zenith, in an ideal position for observing galaxy M31. Altair (alpha Aquilae) is to the West, with Orion low in the East. A line through alpha Orionis and beta Andromedae passes through alpha Ceti, connected to the diagonal of the great square of Pegasus. The other diagonal points to the alpha and beta of Aquarius, and the line through beta and alpha Aquarii leads to the alpha Piscis Austrini.

1. Cygnus
2. Lyra
3. Hercules
4. Cepheus
5. Draco
6. Ursa minor
7. Ursa major
8. Cassiopeia
9. Lynx
10. Perseus
11. Auriga
12. Gemini
13. Taurus
14. Orion
15. Andromeda
16. Triangulum
17. Aries
18. Cetus
19. Eridanus
20. Pegasus
21. Pisces
22. Sculptor
23. Aquarius
24. Piscis austrinis
25. Capricorn
26. Aquila
27. Serpens

A. Galaxy m31

NORTHERN HEMISPHERE

TROPICAL REGIONS (23°30′), 12 P.M. WINTER SOLSTICE (21 DECEMBER)

ALTHOUGH IT'S WINTER, NIGHT AND DAY ARE ALMOST THE SAME LENGTH: THIS IS THE TROPICS, AND THE INCLINATION OF THE AXIS OF THE EARTH ALTHOUGH EVIDENT, HAS LITTLE EFFECT LOOKING NORTH, IN FACT, WE CAN SEE THE CELESTIAL POLE VERY LOW ON THE HORIZON, AND URSA MAJOR RISES AND SETS.

TO THE NORTH
Auriga is almost at the zenith: following the line alpha-beta to the right (East) are Castor and Pollux, the alpha and beta of Gemini, while on the opposite side (on the left, to the West), the line beta-alpha leads to Algol, beta Persei. From Cassiopeia, lines through beta-alpha and gamma-alpha lead to Almach in Andromeda, and to Sirrah, common to Andromeda and Pegasus, while a line through beta-eta of Ursa Minor points to Alderamir, alpha Cephei, and the beta-eta alignment locates Denebola, beta Leonis. The alignment of delta-gamma in Ursa Major points to Algieba (gamma Leonis).

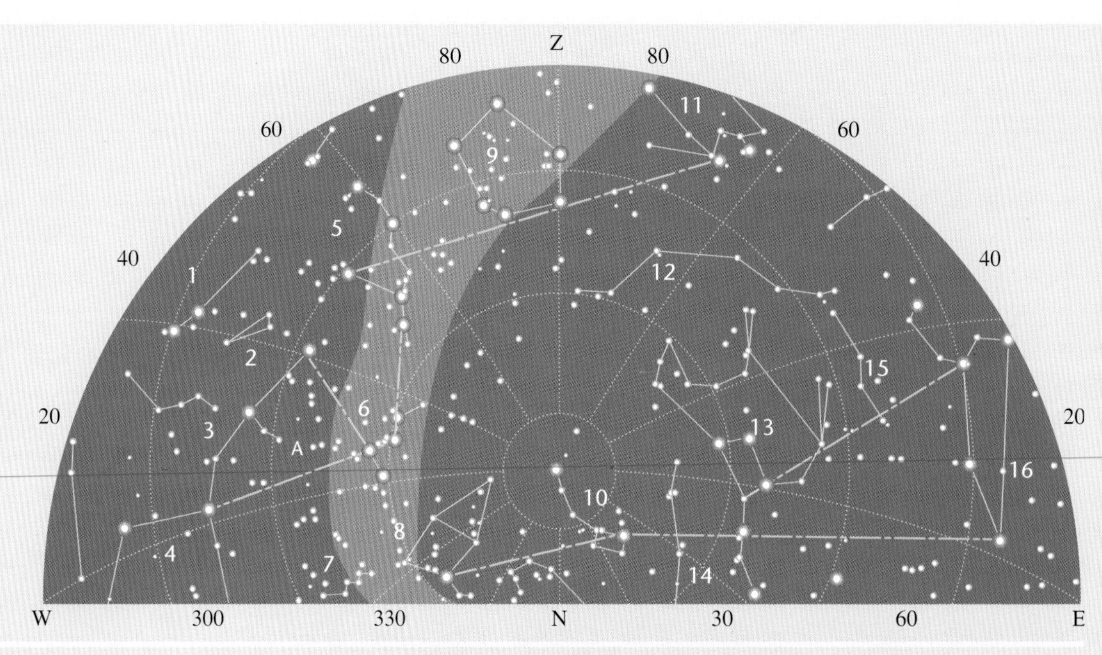

TO THE SOUTH
The sky is crowded with stars. Orion is prominent, high in the centre: following the belt to the right (West), is Aldebaran (alpha Tauri), with the Pleides a little further on, and Sirius (alpha Canis Majoris) on the left (to the East). The alignments beta-upsilon, beta-delta, lambda-zeta-kappa and alpha-kappa point respectively to Mekar (alpha Ceti), beta and gamma Eridani and beyond to beta Ceti, and to Canopus and Alphard (alpha Hydrae.) The alpha direction of Orion-Sirius zeta Hydrae which, aligned with the Southern point, leads to Alsuhail (gamma Velae).

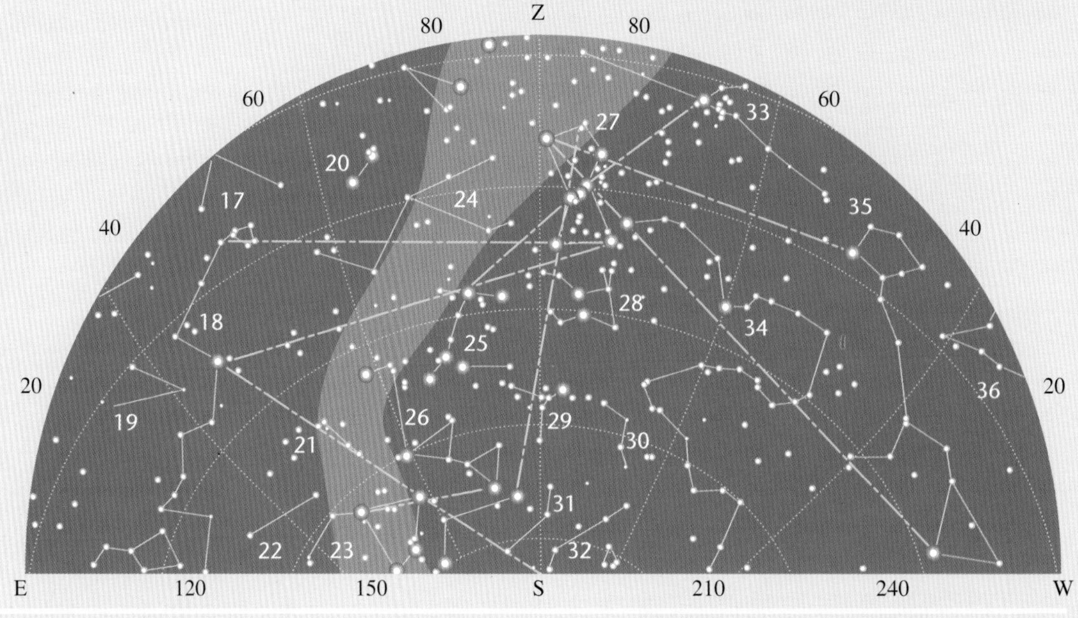

1. ARIES
2. TRIANGULUM
3. ANDROMEDA
4. PEGASUS
5. PERSEUS
6. CASSIOPEIA
7. LACERTA
8. CEPHEUS
9. AURIGA
10. URSA MINOR
11. GEMINI
12. LYNX
13. URSA MAJOR
14. DRACO
15. LEO MINOR
16. LEO
17. CANCER
18. HYDRA
19. SEXTANS
20. CANIS MINOR
21. PYXIS
22. ANTLIA
23. VELA
24. UNICORN
25. CANIS MAJOR
26. PUPPIS
27. ORION
28. LEPUS
29. COLUMBA
30. CAELUM
31. PICTOR
32. DORADO
33. TAURUS
34. ERIDANUS
35. CETUS
36. PISCES

A. GALAXY M31

NORTHERN HEMISPHERE

TROPICAL REGIONS (23°30'), 12 P.M. SPRING EQUINOX (21 MARCH)

The southern cross appears low on the horizon
While polaris, the north star is still visible to the north.
To the south four zodiac constellations can be seen, the highest almost at the zenith.

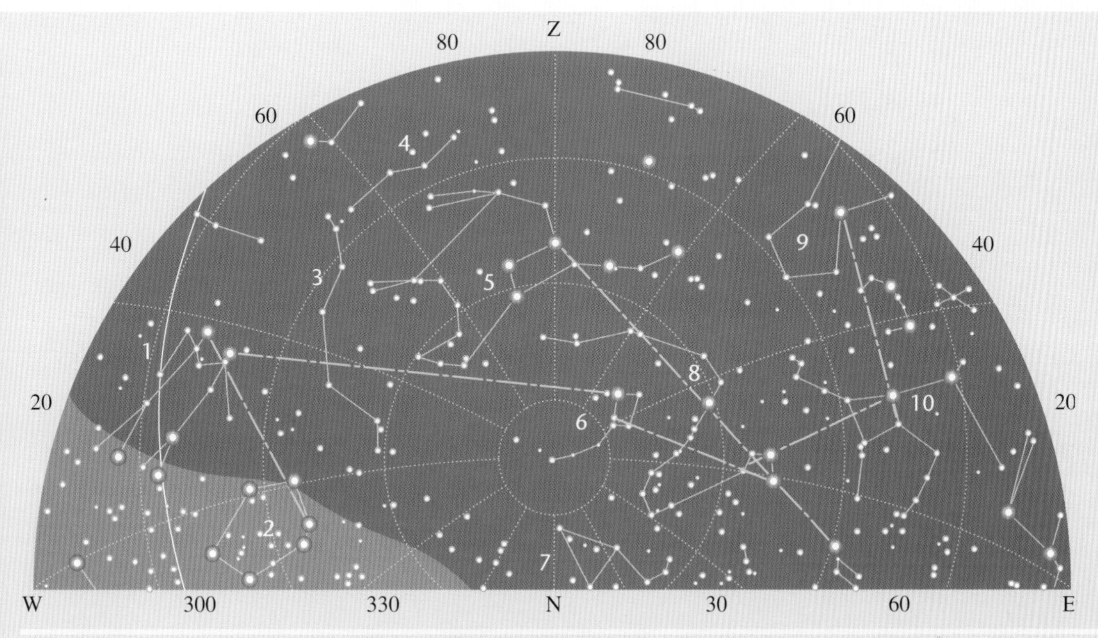

To the north
Ursa Major is directly North, halfway between the zenith and the horizon. A line through gamma-delta leads to Vega, passing through Etamin in Draco, also found along the line through zeta-eta of Ursa Minor. The line through zeta-beta of Draco points to Zeta Herculis which, aligned with epsilon Herculis, points to Izar (epsilon Bootis). On the opposite side of the sky, to the West, a line through gamma-beta of Ursa Minor leads to Castor (Gemini) that, aligned with Pollux, points in the direction of Menkalian and Capella (beta and alpha Aurigae).

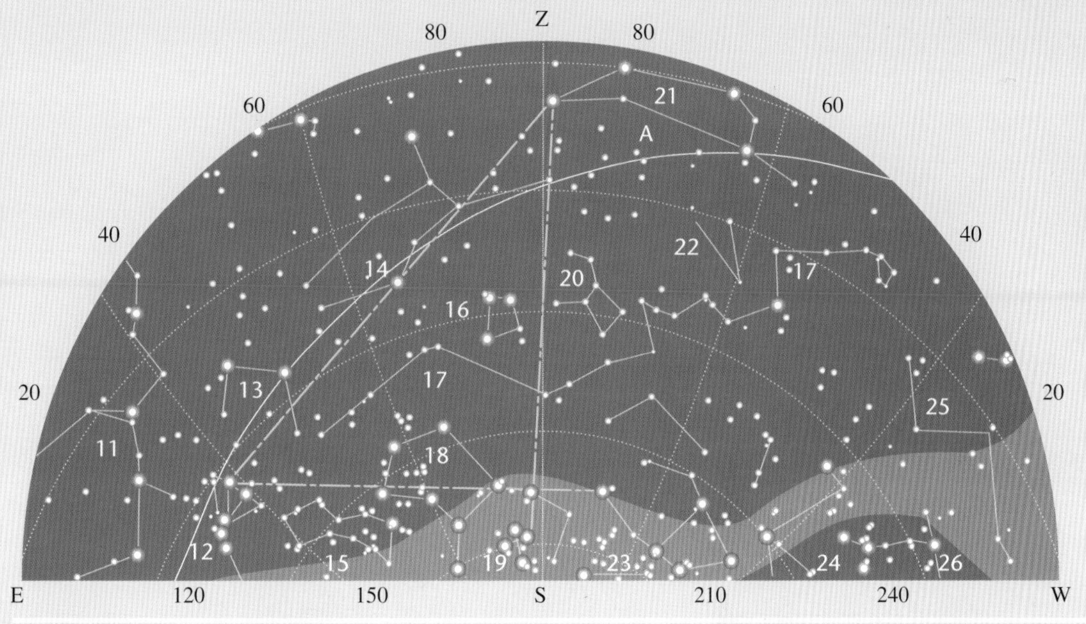

To the south
Leo is almost at the zenith, and Debebola (beta Leonis) is almost directly South: straight down, very low near the horizon are the four stars of Crux. Projecting a line from Denebola to Spica (alpha of Virgo, shining on the left, in the East), and, passing close to alpha Librae, points to delta Scorpii, with Antares below, almost on the horizon. At the same level as Antares there are many bright stars in various constellations: the brightest, almost in the west, is Sirius.

1. Gemini
2. Auriga
3. Lynx
4. Leo Minor
5. Ursa Major
6. Ursa Minor
7. Cepheus
8. Draco
9. Bootes
10. Hercules
11. Ophiuchus
12. Scorpio
13. Libra
14. Virgo
15. Lupus
16. Corvus
17. Hydra
18. Centaurus
19. Crux
20. Crater
21. Leo
22. Sextans
23. Vela
24. Puppis
25. Monoceros
26. Canis Major

A. Area rich in galaxies

NORTHERN HEMISPHERE

TROPICAL REGIONS (23°30′), 12 P.M. SUMMER SOLSTICE (21 JUNE)

THE NORTHERN VIEW IS NOT VERY DIFFERENT FROM THE PREVIOUS ONES, ALTHOUGH THE BEAUTIFUL CONSTELLATION OF CYGNUS CAN NOW ALSO BE SEEN CLEARLY IN THE SOUTH. SIX ZODIAC CONSTELLATIONS ARE CROWDED FAIRLY LOW DOWN ALONG THE ECLIPTIC.

TO THE NORTH
From Ursa Minor to Bootes through Hercules and the Corona Borealis: following beta-gamma UMa, epsilon-zeta Her, through delta-gamma-alpha CrB. Arcturis, in Bootes, can also be found following the "yoke" of the Plough, the most prominent parts of Ursa Major. On the opposite side of the sky, the line through gamma and alpha Cassiopeiae points to Sirrah, common to Andromeda and Pegasus, whose great square is clearly visible. The great cross of Cygnus is easily recognised, shining high on the right (to the East).

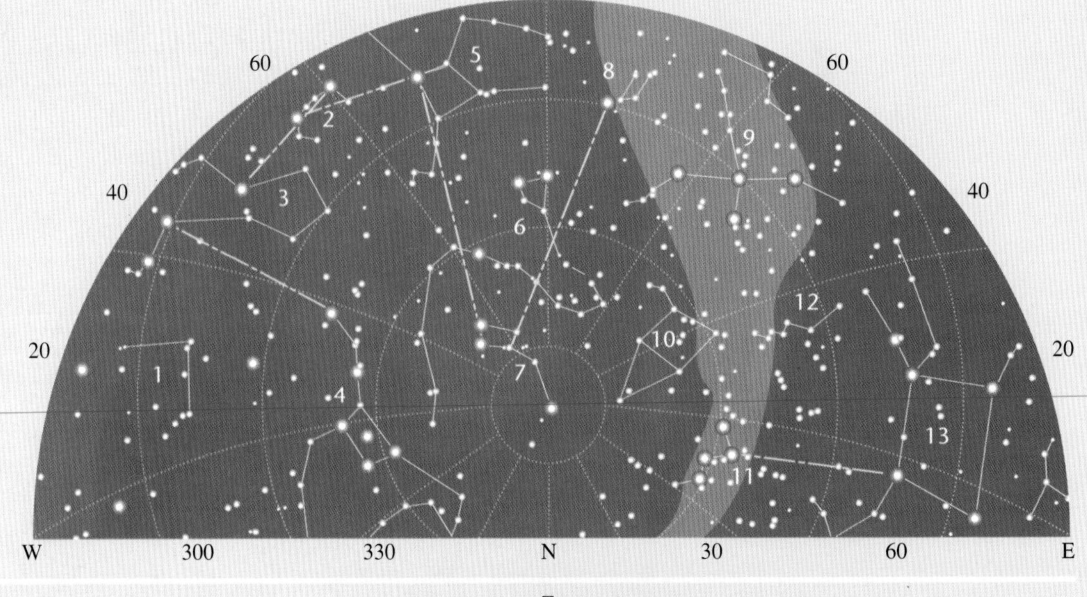

TO THE SOUTH
This part of the sky is full of small constellations made up of stars that are not particularly noticeable. The clearest are, from the right (West, Spica in Virgo and Antares in Scorpio. Continuing in the same direction (roughly the same as the ecliptic), the other zodiac constellations can be located, the brightest of which is Sagittarius, in an area rich in galaxies and nebulae.

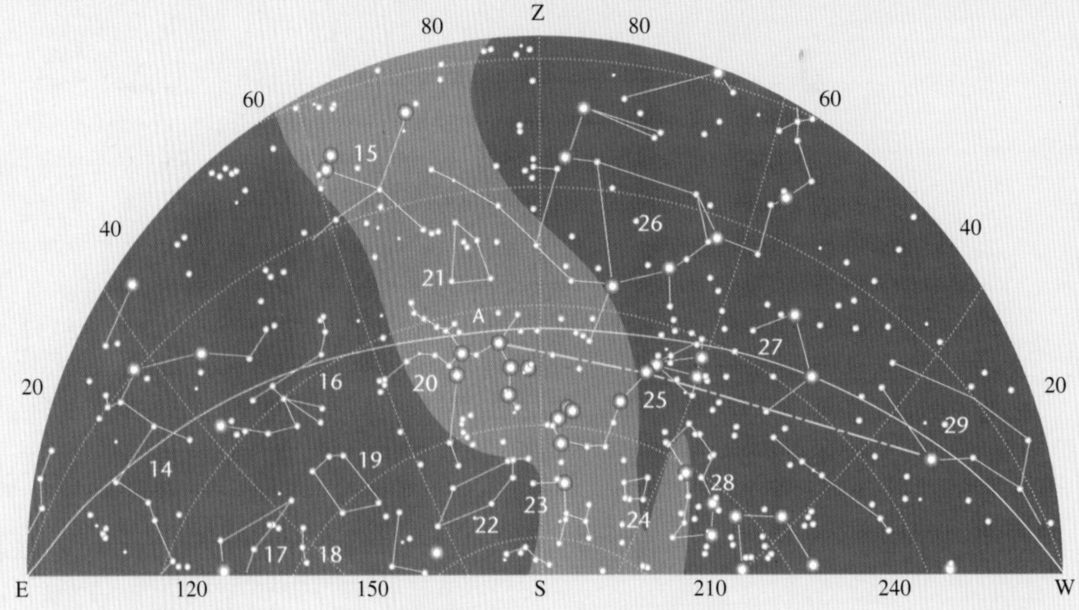

1. Coma berenices
2. Corona borealis
3. Bootes
4. Ursa major
5. Hercules
6. Draco
7. Ursa minor
8. Lyra
9. Cygnus
10. Ursa major
11. Cassiopeia
12. Lacerta
13. Pegasus
14. Aquarius
15. Aquila
16. Capricorn
17. Piscis austrinis
18. Grus
19. Microscopium
20. Sagittarius
21. Scutum
22. Telescopium
23. Ara
24. Norma
25. Scorpio
26. Ophiuchus
27. Libra
28. Lupus
29. Virgo

A. Galaxies M28. M21, M8, M20; trifid nebula, lagoon nebula

NORTHERN HEMISPHERE

TROPICAL REGIONS (23°30'), 12 P.M. AUTUMN EQUINOX (23 SEPTEMBER)

URSA MINOR IS VERY LOW, AND URSA MAJOR IS ACTUALLY BELOW THE HORIZON: TO THE NORTH, WHERE THE ONLY VISIBLE CELESTIAL POLE IS LOCATED, THERE ARE NO CONSTELLATIONS APART FROM THE CIRCUMPOLAR URSA MINOR, WHILE TO THE SOUTH FIVE ZODIAC CONSTELLATIONS CAN BE DISTINGUISHED ALONG AN ALMOST VERTICAL ECLIPTIC.

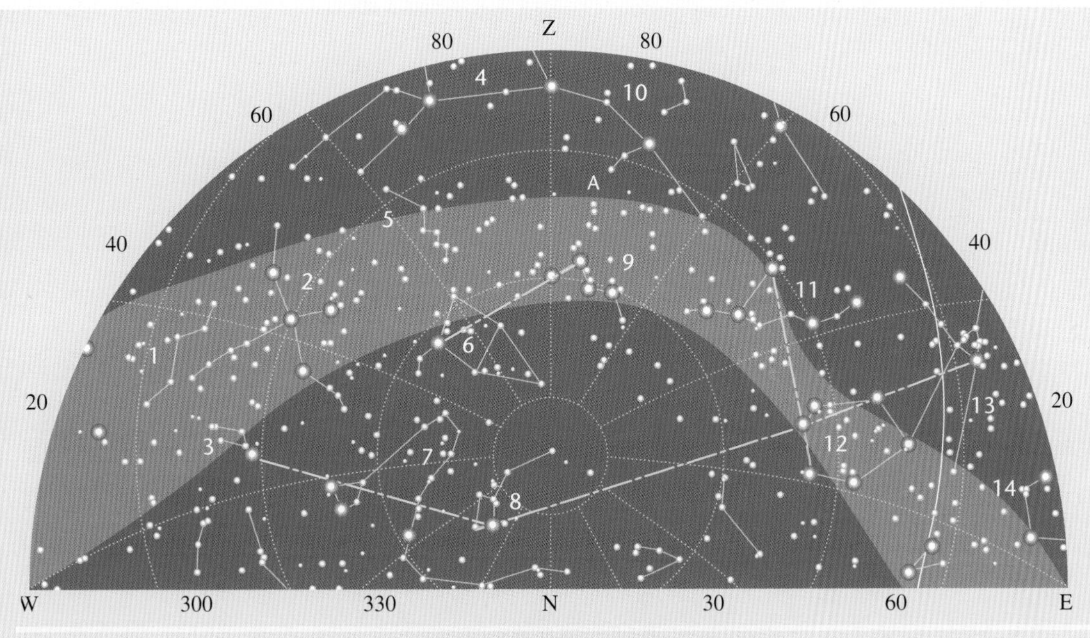

To the north
Pegasus and Andromeda are at the zenith, while Ursa Major is below the horizon. The alignment of gamma-beta in Ursa Minor points to Auriga, and the alignment of beta Ursa Minoris and gamma Draconis leads to Vega, in Lyra. The cross of Cygnus is clearly visible higher up. Along the alignment of alpha-beta Cassiopeiae are Alderamin, or alpha Cephei; low down on the right (to the East) of Auriga, following the alpha-upsilon alignment, is Aldebaran (in Taurus, with Algol (Perseus) along the beta-alpha alignment.

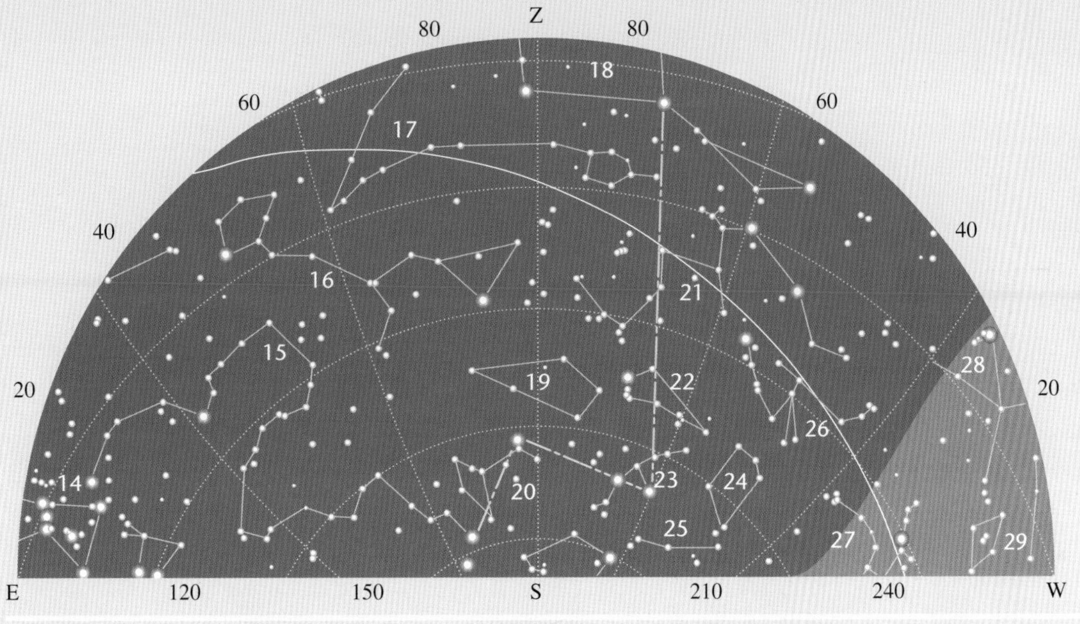

To the south
Pegasus is at the zenith: following the alignment of beta-alpha in the great square downwards, points to Alnair, or alpha Gruis, near the horizon. On its left (to the East) is alpha Phoenicis: aligned with zeta Phoenicis, this points to Achernar, a bright star common to both Eridanus and Hydrus (which is below the horizon. At the extreme East the brightest of the stars in Orion can be distinguished low on the horizon, with Aquila at the diametrically opposite side.

1. Volpecula
2. Cygnus
3. Lyra
4. Pegasus
5. Lacerta
6. Cepheus
7. Draco
8. Ursa minor
9. Cassiopeia
10. Andromeda
11. Perseus
12. Auriga
13. Taurus
14. Orion
15. Ophiuchus
16. Eridanus
17. Cetus
18. Pisces
19. Pegasus
20. Sculptor
21. Phoenix
22. Aquarius
23. Piscis austrinis
24. Grus
25. Microscopium
26. Indus
27. Capricorn
28. Sagittarius
29. Aquila
30. Scutum

A. Galaxy m31

SOUTHERN HEMISPHERE

TROPICAL REGIONS (-23°30′), 12 P.M. SUMMER SOLSTICE (21 DECEMBER)

"BELOW" THE EQUATOR THE ONLY VISIBLE CELESTIAL POLE IS THE SOUTHERN ONE, INDICATED BY SMALL CONSTELLATIONS LIKE CHAMAELEON AND OCTANS. IN THE TROPICAL SUMMER THE ECLIPTIC TO THE NORTH REMAINS QUITE LOW, WHILE SIRIUS, IN CANIS MAJOR, SHINES ALMOST AT THE ZENITH.

TO THE NORTH

The large stars of Canis Major and Orion shine high in the sky. Orion's belt is a pointer to Sirius (on the right, to the East), and Aldebaran in Taurus (to the West). The alignment of sigma-epsilon in the belt points to alpha Ceti. From here, a vertical line down to the horizon intercepts Hamal, or alpha Arietis. A line through alpha and beta Tauri (Nath, near the Crab nebula), leads to Auriga. The line from alpha to beta Aurigae points to the alpha and beta stars of Gemini. The mu-beta alignment in Gemini points to Algieba, or gamma Leonis.

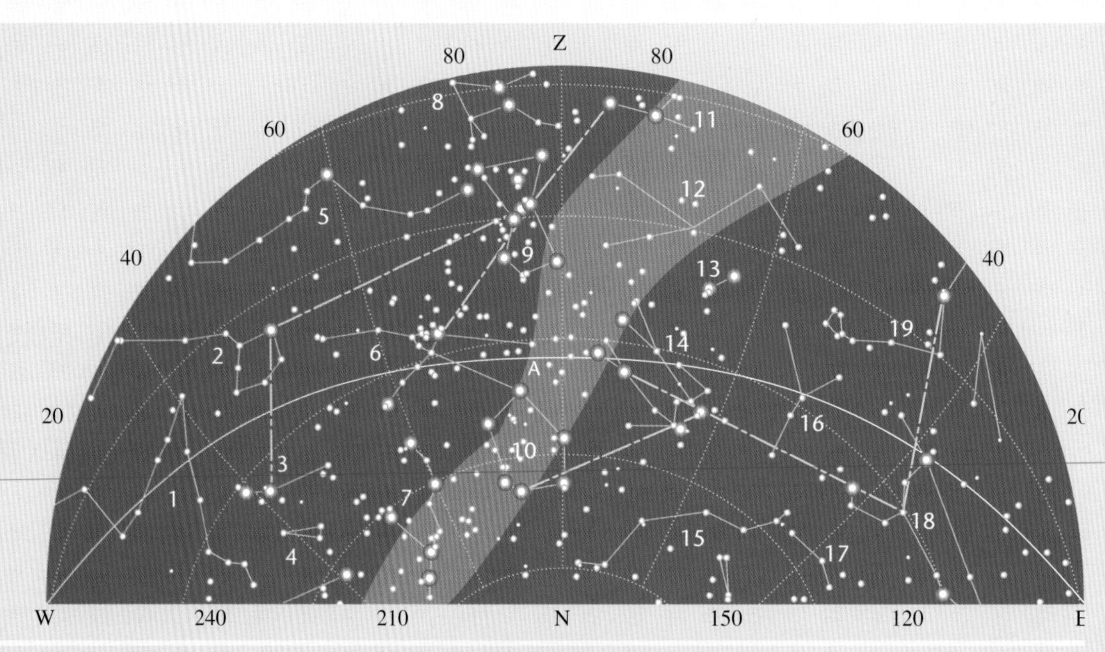

TO THE SOUTH

Canopus, the alpha star of Carina, shines in the centre of the southern sky, directly "under" Sirius and not far from the Large Magellanic Cloud. On the right (to the West) shines Achernar, common to Eridanus and Hydrus, the constellation containing the Small Magellanic Cloud. Low down, almost on the horizon in the south, the alpha-gamma alignment in Triangulum Australe leads to the bright stars of Centaurus and Crux, the longest arm of which points to the celestial pole.

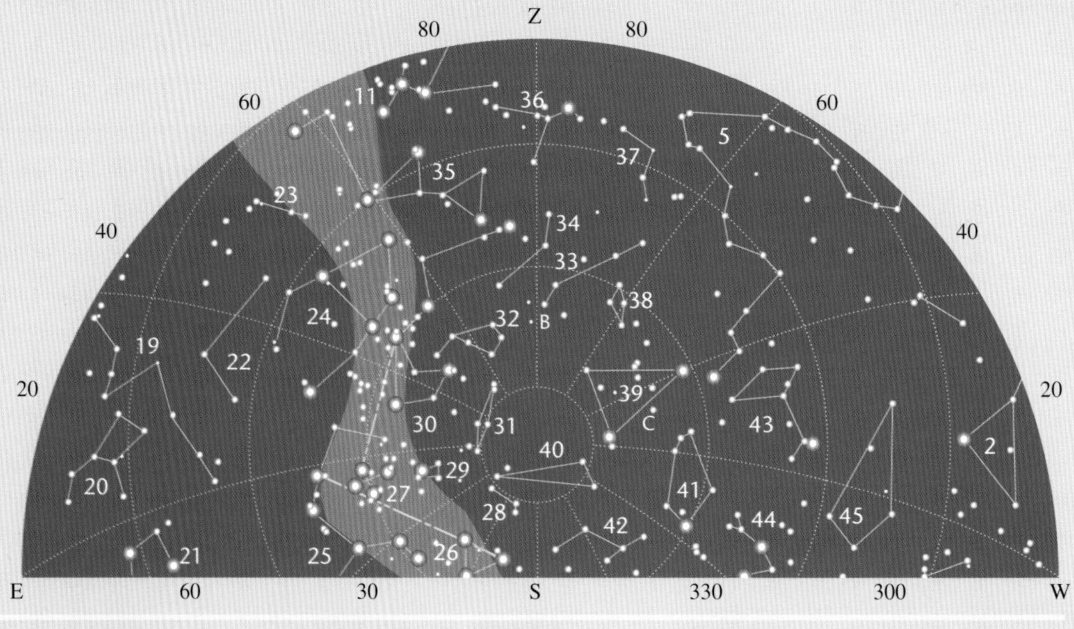

1. PISCES
2. CETUS
3. ARIES
4. TRIANGULUM
5. ERIDANUS
6. TAURUS
7. PERSEUS
8. LEPUS
9. ORION
10. AURIGA
11. CANIS MAJOR
12. UNICORN
13. CANIS MINOR
14. GEMINI
15. LYNX
16. CANCER
17. LEO MINOR
18. LEO
19. HYDRA
20. CRATER
21. CORVUS
22. ANTLIA
23. PYXIS
24. VELA
25. CENTAURUS
26. TRIANGULUM AUSTRALE
27. CRUX
28. APUS
29. MUSCA
30. CARINA
31. CHAMAELEON
32. VOLANS
33. DORADO
34. PICTOR
35. PUPPIS
36. COLOMBA
37. CAELUM
38. RETICULUM
39. HYDRUS
40. OCTANS
41. TUCANA
42. PAVO
43. PHOENIX
44. GRUS
45. SCULPTOR

A. CRAB NEBULA
B.AND C. LARGE MAGELLANIC CLOUD AND SMALL MAGELLANIC CLOUD

SOUTHERN HEMISPHERE

TROPICAL REGIONS (-23°30′), 12 P.M. SPRING EQUINOX (21 MARCH)

WHILE TO THE NORTH THE CONSTELLATIONS ARE STILL EASILY RECOGNISABLE, TO THE SOUTH THE SKY IS CROWDED WITH STARS. CRUX, EQUIDISTANT FROM THE ZENITH AND THE HORIZON, SHINES IN THE SOUTH, AT THE EXTREME EAST AND EXTREME WEST ARE TWO BRILLIANT STARS: ANTARES IN SCORPIO AND SIRIUS IN CANIS MAJOR.

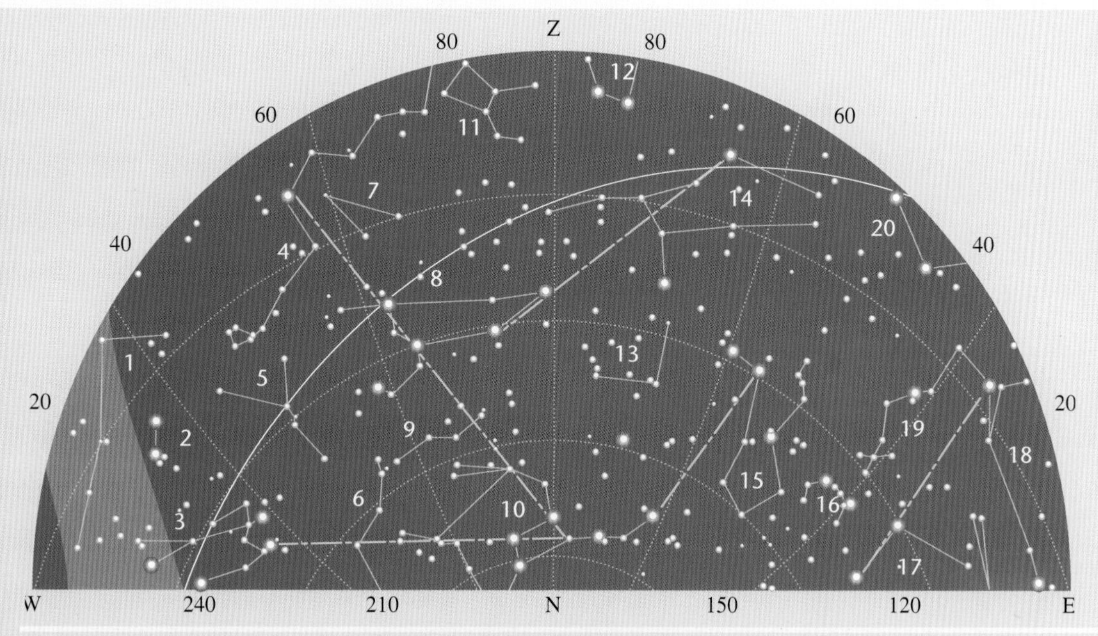

To the north
The celestial North Pole is below the horizon: Ursa Major is just visible above the horizon. The delta-beta alignment points to Castor (Gemini), and the delta-gamma alignment to Algieba and Regulus in Leo, and, then to Alphard in Hydra. While the yole of the Plough points to Arcturus, or alpha Bootis. The alignment of delta-beta in Leo points to Spica, alpha Verginis, and the alignment of eta-zeta in Hercules, above the horizon, leads to alpha Serpentis.

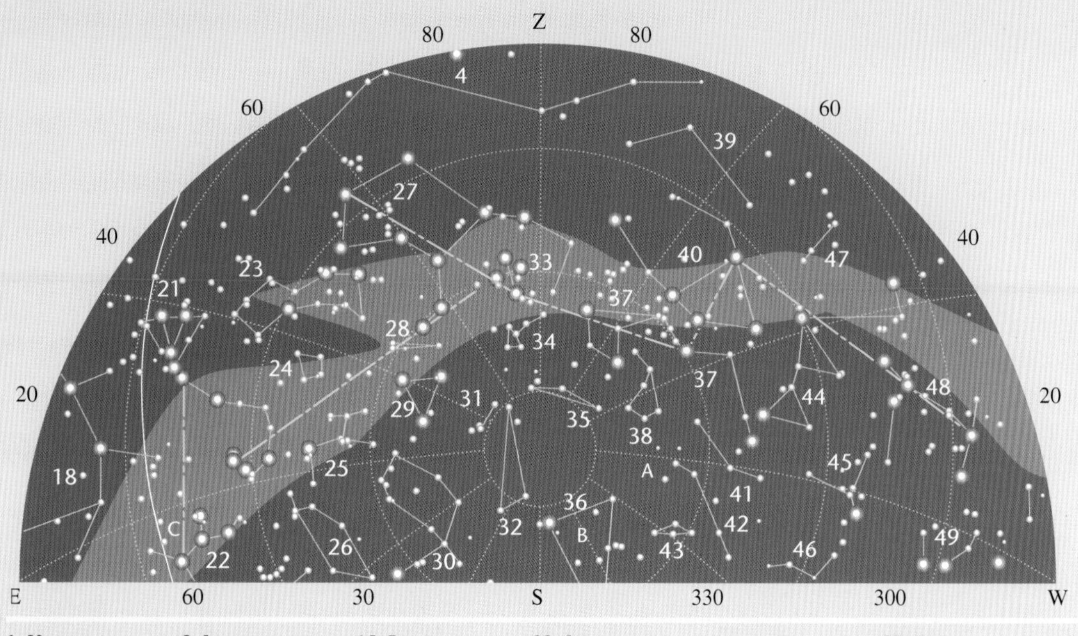

To the south
Hydra, at the zenith, twists from one side of the sky to the other. Directly below Antares (in Scorpio)near the horizon, is gamma Sagittarii, surrounded by nebulae. A line from the tail of Scorpio to beta Crucis passes through the brilliant alpha and beta Centauri, while the alignment of alpha Crucis and beta Carinae points straight to Canopus in Carina, while a line through the backbone of Canis Major (alpha-omega)points to the brighter stars of Vela and Puppis.

1. Unicorn
2. Canis minor
3. Gemini
4. Hydra
5. Cancer
6. Lynx
7. Sextans
8. Leo
9. Leo minor
10. Ursa Major
11. Crater
12. Caelum
13. Coma Di berenices
14. Virgo
15. Bootes
16. Corona Borealis
17. Hercules
18. Ophiuchus
19. Serpens
20. Libra
21. Scorpio
22. Sagittarius
23. Lupus
24. Norma
25. Ara
26. Telescopium
27. Centaurus
28. Pyxis
29. Triangulum australe
30. Pavo
31. Apus
32. Octans
33. Crux
34. Musca
35. Chamaeleon
36. Hydrus
37. Carina
38. Volans
39. Antlia
40. Vela
41. Pictor
42. Dorado
43. Reticulum
44. Puppis
45. Columba
46. Caelum
47. Pyxis
48. Canis major
49. Lepus

A. Large magellanic cloud
B. Small magellanic cloud
C. Nebulae

SOUTHERN HEMISPHERE

TROPICAL REGIONS (-23°30'), 12 P.M. WINTER SOLSTICE (21 JUNE)

IN THE TROPICAL WINTER, THE ECLIPTIC IS ALMOST VERTICAL: TO THE SOUTH THE ZODIAC CONSTELLATIONS OF SCORPIO AND SAGITTARIUS ARE ALMOST AT THE ZENITH, WHILE TO THE NORTH THE CONSTELLATION HIGHEST ABOVE THE HORIZON IS OPHIUCHUS.

To the north
Etamin, gamma Draconis, is now due North. On its right (to the East) shines the cross of Cygnus: looking to the right along the shortest arm, leads the eye to Enif, or epsilon Persei, and Sadal-Malik, or alpha-Aquarii. The cross of Aquila, with brilliant Altair, is clearly visible above Cygnus: a line from Altair to Arcturus in Bootes (on the other side of the sky in the West) passes through the fainter constellations of Hercules and Corona Borealis. While, higher up, a line from eta to delta in Ophiuchus points to Leo.

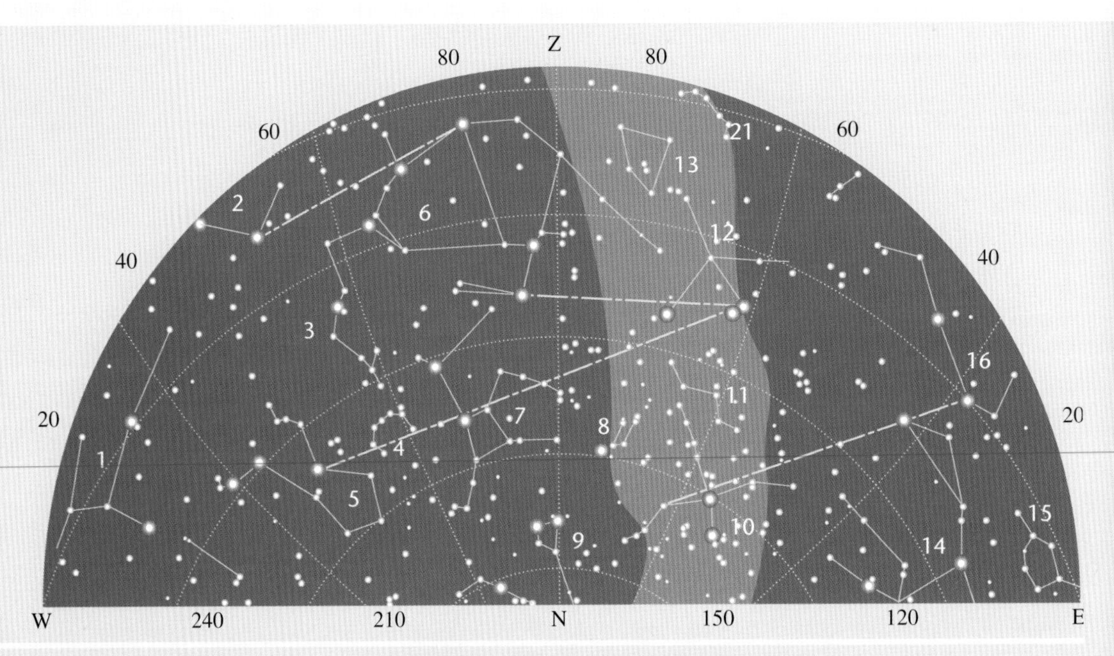

To the south
Scorpio and Sagittarius, almost at the zenith, dominate a sky rich in stars and small constellations. Following the curve of the "tail" of Scorpio leads to Achernar, low on the horizon, a bright star common to Eridanus (below the horizon) and Hydrus. While a line from the tip of the tail of Scorpio to alpha Crucis leads to alpha and beta Centauri. Delta Carinae, almost perpendicularly below Antares, is probably too low to be able to observe clearly.

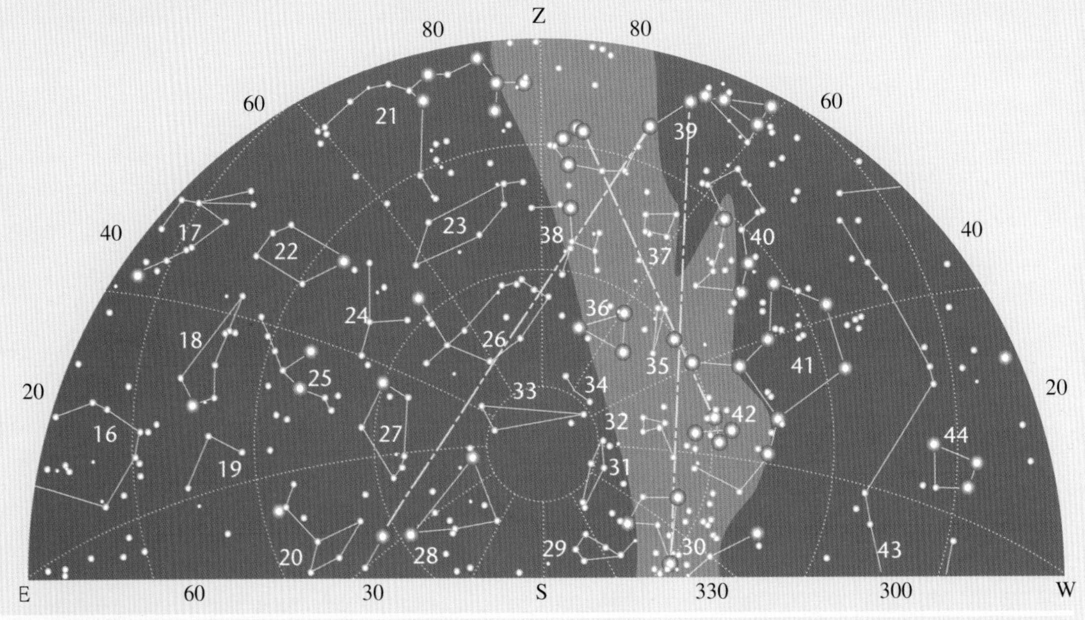

1. Virgo
2. Libra
3. Serpens
4. Corona Borealis
5. Bootes
6. Ophiuchus
7. Hercules
8. Lyra
9. Draco
10. Cygnus
11. Volpecula
12. Aquila
13. Scutum
14. Pegasus
15. Pisces
16. Aquarius
17. Capricorn
18. Piscis Austrinis
19. Sculptor
20. Phoenix
21. Sagittarius
22. Microscopium
23. Telescopium
24. Indus
25. Grus
26. Pavo
27. Tucana
28. Hydrus
29. Volans
30. Carina
31. Chamaeleon
32. Musca
33. Octans
34. Apus
35. Pyxis
36. Triangulum australe
37. Norma
38. Ara
39. Scorpio
40. Lupus
41. Centaurus
42. Crux
43. Hydra
44. Corvus

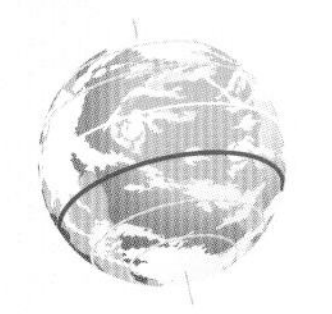

SOUTHERN HEMISPHERE

TROPICAL REGIONS (-23°30′), 12 P.M. SPRING EQUINOX (23 SEPTEMBER)

PEGASUS IS IN THE CENTRE OF THE NORTHERN SKY, WHILE SIX ZODIAC CONSTELLATIONS ARE ARRAYED ACROSS THE HORIZON ON THE ALMOST VERTICAL ECLIPTIC. ACHERNAR, DISPUTED MEMBER OF BOTH ERIDANUS AND HYDRUS, SHINES ALMOST AT THE CENTRE OF THE SOUTHERN SKY.

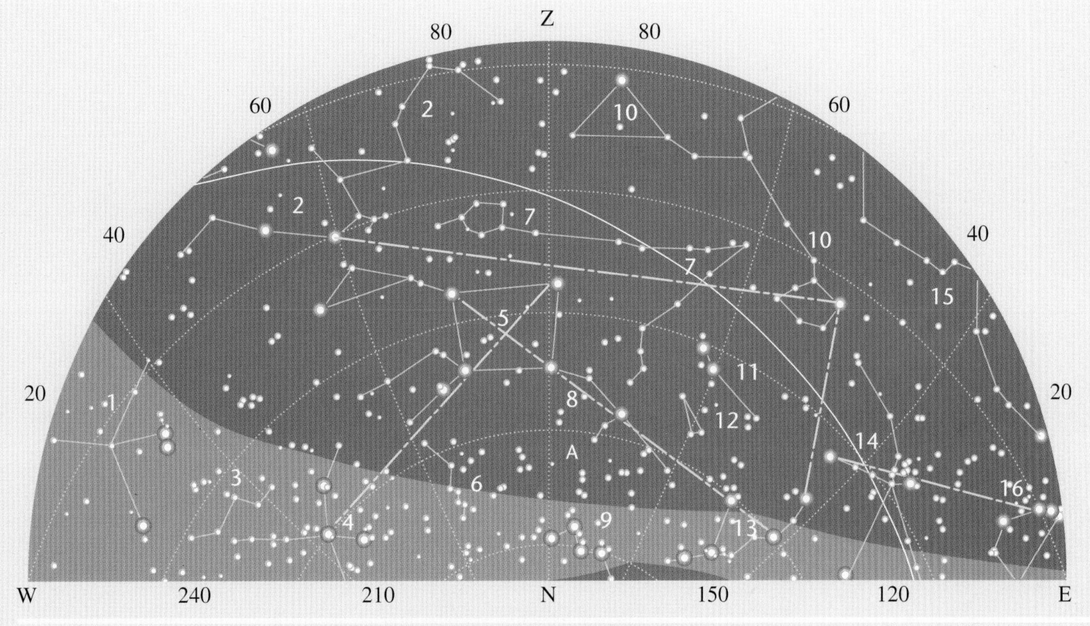

To the north
The great square of Pegasus is in the centre: Algenib and Sirrah are at the top, aligned to the North. Using the diagonals as guides, Mirach in Andromeda, and Algol in Perseus can be found on the right(East), and Sadir in Cygnus to the left (West), in the area of the Pelican and North America nebulae. Further left is Altair, in Aquila. Following the curve of Perseus through the Pleiades in Taurus leads to Menkar, or alpha Ceti, and a line cutting across the pentagon of stars of which this is part leads to alpha Aquarii. Orion is visible low in the east.

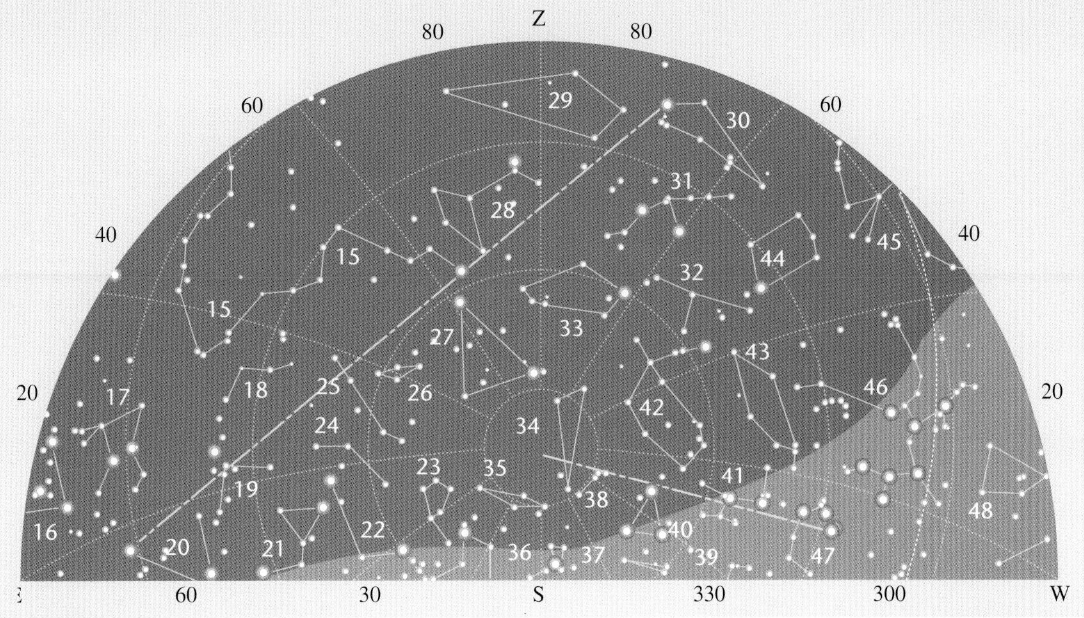

To the south
Sculptor is almost at the zenith, accompanied by Piscis Austrinus, the alpha star of which can be aligned eastward with Achernar, (alpha Eridani) and Sirius (alpha Canis Majoris), very low on the horizon. Aligning the tail of Scorpio (low on the horizon in the West, and Atria, the alpha star of Triangulum Australe, points to the South Pole.

1. Aquila
2. Aquarius
3. Volpecula
4. Cygnus
5. Pegasus
6. Lacerta
7. Pisces
8. Andromeda
9. Cassiopeia
10. Cetus
11. Aries
12. Triangulum
13. Perseus
14. Taurus
15. Eridanus
16. Orion
17. Lepus
18. Caelum
19. Columba
20. Canis major
21. Puppis
22. Carina
23. Volans
24. Pictor
25. Dorado
26. Reticulum
27. Hydrus
28. Phoenix
29. Sculptor
30. Piscis austrinis
31. Grus
32. Indus
33. Tucana
34. Octans
35. Chamaeleon
36. Musca
37. Pyxis
38. Apus
39. Norma
40. Triangulum australe
41. Ara
42. Pavo
43. Telescopium
44. Microscopium
45. Capricorn
46. Sagittarius
47. Scorpio
48. Scutum

SOUTHERN HEMISPHERE

MEDIAN LATITUDES (-45°), 12 P.M. SUMMER SOLSTICE (21 DECEMBER)

SIRIUS SHINES HIGH IN THE SKY, WHILE ORION OCCUPIES THE CENTE OF THE NORTHERN SKY. TO THE SOUTH, THE CONSTELLATIONS VELA AND CARINA, WITH THEIR MANY BRIGHT STARS, ARE PROMINENT IN THE EAST, WHILE CENTAURUS CAN BE ADMIRED LOWER DOWN. ACHERNAR, IN ERIDANUS, SHINES ON THE WESTERN SIDE.

TO THE NORTH
The ecliptic, quite low down now, is marked by the presence of some zodiac constellations, easily found from Orion: a line through Sirius and gamma Orionis points to Aldebaran in Taurus, and Castor (Gemini) is found by aligning Rigel and Betelgeuse (beta and alpha Orionis). Due North is Auriga: a line through theta and tau Aurigae leads to Menkar, or alpha Ceti, passing close to the Pleiades. Higher up, aligning khi and beta Orionis leads the eye to Zaurak, or gamma Eridani.

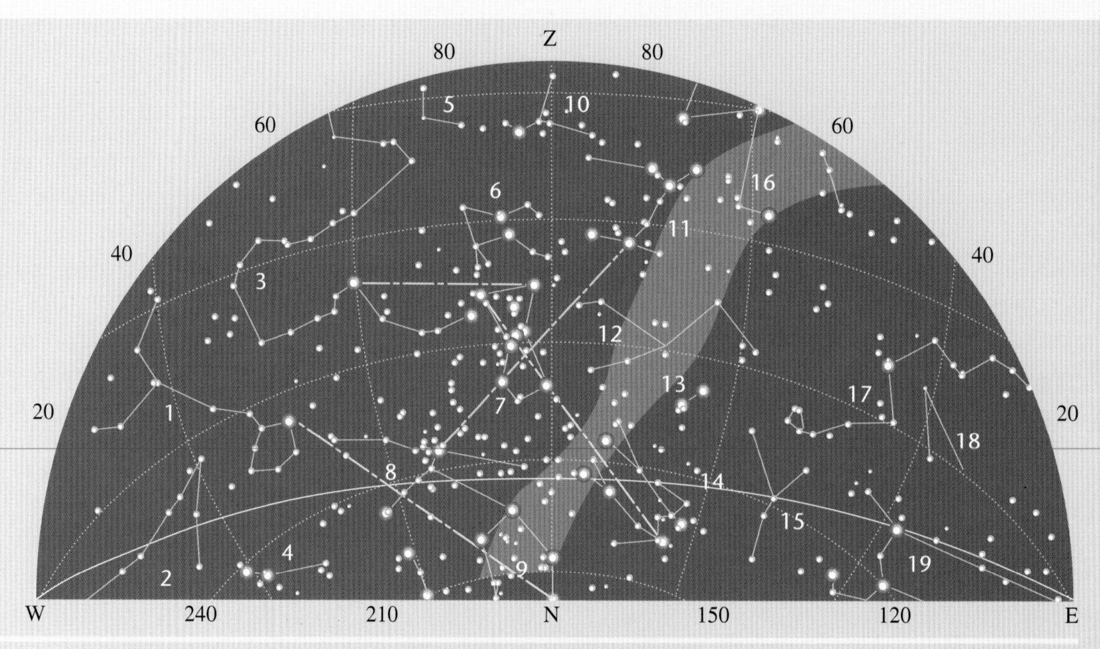

TO THE SOUTH
Some of the brightest constellations in the southern sky can be identified starting from Crux. Its major axis not only indicates the direction of the celestial South Pole, it also leads to the constellation Hydrus. The alignment of gamma and beta Crucis points to Triangulum Australe and alpha Pavonis. Aligning gamma and delta Crucis points to Carina and the minor access identifies Vela above and the two brightest stars in Centaurus (alpha and beta) below.

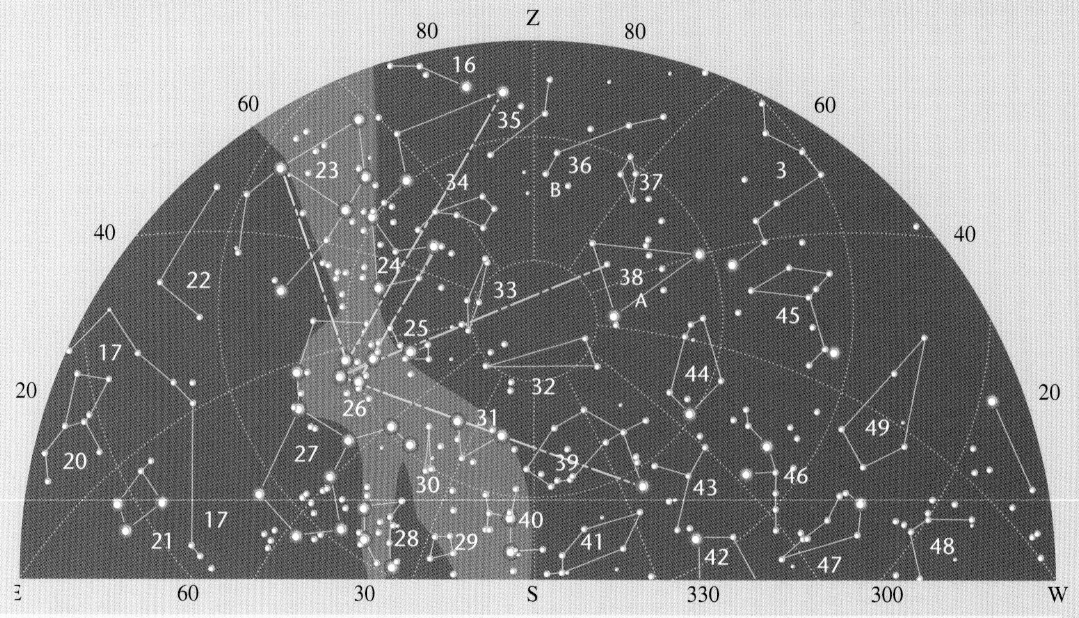

1. CETUS
2. PISCES
3. ERIDANUS
4. ARIES
5. CAELUM
6. LEPUS
7. ORION
8. TAURUS
9. AURIGA
10. COLUMBA
11. CANIS MAJOR
12. UNICORN
13. CANIS MINOR
14. GEMINI
15. CANCER
16. PUPPIS
17. HYDRA
18. SEXTANS
19. LEO
20. CRATER
21. CORVUS
22. ANTLIA
23. VELA
24. CARINA
25. MUSCA
26. CRUX
27. CENTAURUS
28. LUPUS
29. NORMA
30. PYXIS
31. TRIANGULUM AUSTRALE
32. OCTANS
33. CHAMAELEON
34. VOLANS
35. PICTOR
36. DORADO
37. RETICULUM
38. HYDRUS
39. PAVO
40. ARA
41. TELESCOPIUM
42. MICROSCOPIUM
43. INDUS
44. TUCANA
45. PHOENIX
46. GRUS
47. PISCIS AUSTRINUS
48. AQUARIUS
49. SCULPTOR

A. SMALL MAGELLANIC CLOUD
B. LARGE MAGELLANIC CLOUD

SOUTHERN HEMISPHERE

MEDIAN LATITUDES (-45°), 12 P.M. AUTUMN EQUINOX (21 MARCH)

WHILE TO THE NORTH HYDRA STRETCHES ACROSS THE SKY FROM WEST TO EAST ALMOST PARALLEL TO THE ECLIPTIC, MARKED BY SIX CONSTELLATIONS, ON THE OTHER SIDE OF THE SKY CRUX SHINES ALMOST AT THE ZENITH. ABOVE CRUX, THE LIGHTS OF CENTAURUS CAN BE SEEN, AND IN THE WEST THE BRILLIANT STARS OF VELA AND CARINA, WITH SIRIUS A LITTLE FURTHER OUT.

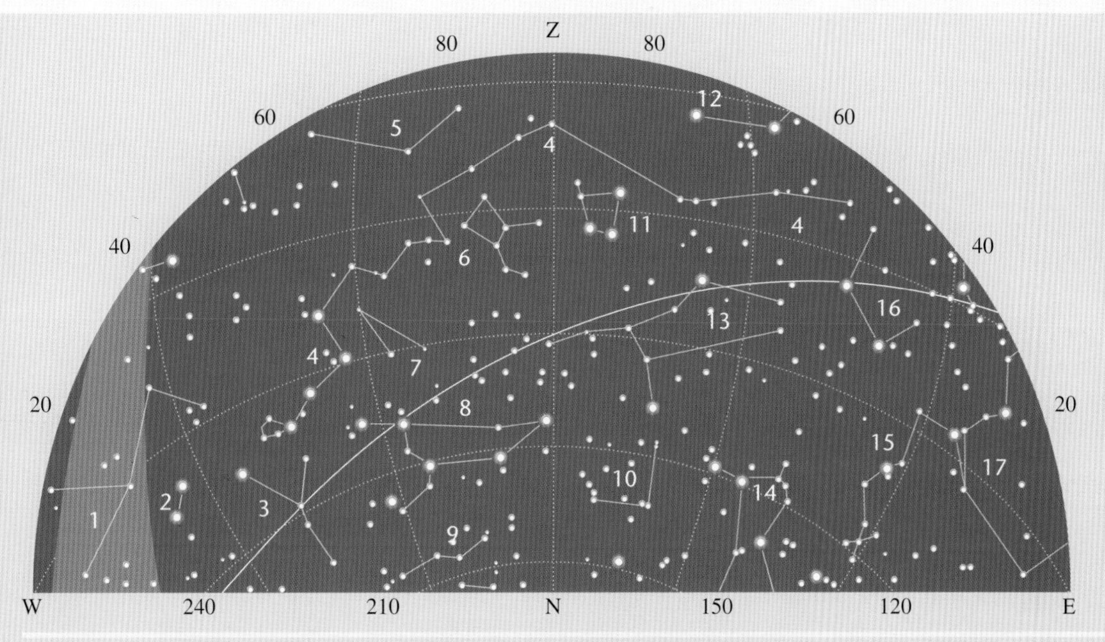

To the north
The brightest stars are distributed along or near the ecliptic: from left to right we can identify Regulus (in Leo), Algieba (gamma Leonis), Denebola (beta Leonis), Spica (Virgo) and Zuben-el-Genubi (alpha Librae). Turning south, the alignment continues, with Antares (alpha Scorpii) and the brightest stars of Sagittarius.

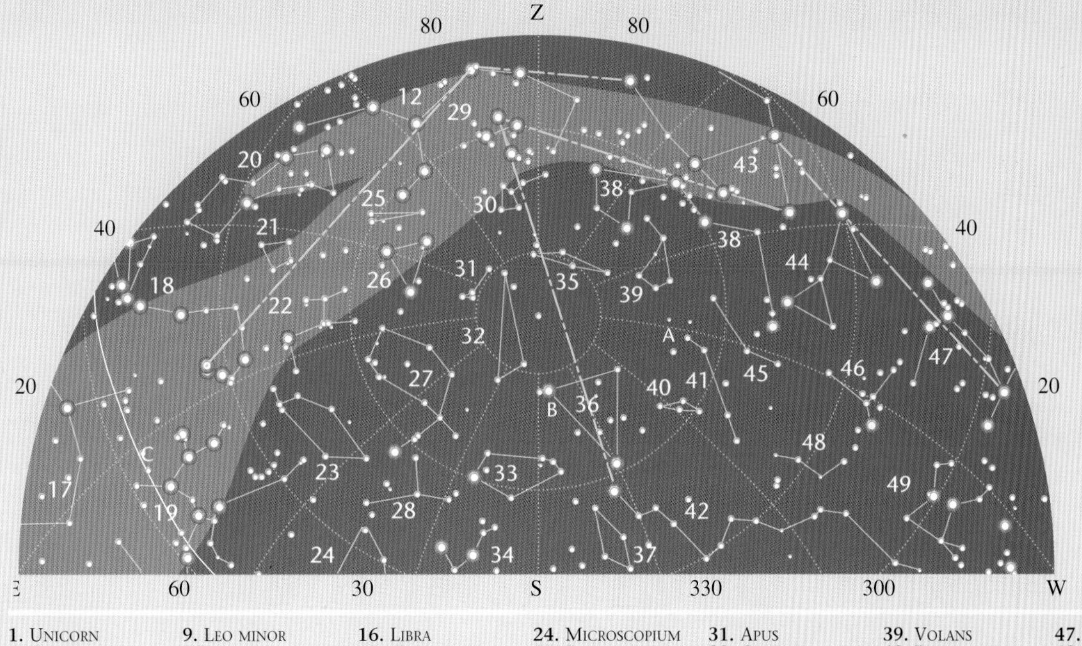

To the south
Centaurus and Crux almost at the zenith can be used to locate Scorpio and Sagittarius, accompanied by the Trifid, Lagoon and Omega nebulae, Hydrus, with the Small Magellanic Cloud, and Dorado, with the Large Magellanic Cloud, all in the east. Sirius shines brightly in the West, with Antares, a little higher, shining in the East. The Canis Major alignment points to Puppis and Vela, the stars of which can also be identified by aligning beta and delta in Crux.

1. Unicorn
2. Canis minor
3. Cancer
4. Hydra
5. Antlia
6. Crater
7. Sextans
8. Leo
9. Leo minor
10. Coma berenices
11. Corvus
12. Centaurus
13. Virgo
14. Bootes
15. Serpens
16. Libra
17. Ophiuchus
18. Scorpio
19. Sagittarius
20. Lupus
21. Norma
22. Ara
23. Telescopium
24. Microscopium
25. Pyxis
26. Triangulum australe
27. Pavo
28. Indus
29. Crux
30. Musca
31. Apus
32. Octans
33. Tucana
34. Grus
35. Chamaeleon
36. Hydrus
37. Phoenix
38. Carina
39. Volans
40. Reticulum
41. Dorado
42. Eridanus
43. Vela
44. Puppis
45. Pictor
46. Columba
47. Canis major
48. Caelum
49. Lepus

A. Large magellanic cloud
B. Small magellanic cloud
C. Area rich in nebulae

SOUTHERN HEMISPHERE

MEDIAN LATITUDES (-45°), 12 P.M.
WINTER SOLSTICE (21 JUNE)

THE ALMOST VERTICAL ECLIPTIC DISPLAYS SEVEN ZODIAC CONSTELLATIONS, AND SCORPIO DOMINATES FROM ITS POSITION ALMOST AT THE ZENITH TO THE NORTH. IN THE SOUTH WEST CRUX CAN BE SEEN TOGETHER WITH A HOST OF BRIGHT STARS, WHILE ACHERNAR IS THE BRIGHTEST IN THE SOUTH EAST.

TO THE NORTH
The constellation Scorpio is almost at the zenith: alpha Librae is lower down and to the left (West) towards the brightest star in that direction (Spica in Virgo). Lower still, near the horizon, are Arcturus of Bootes (on the left) and Vega in Lyra (on the right, looking North). Altair, the alpha star of Aquila is higher up on the right of Lyra. Alpha Herculis is between Vega and Antares.

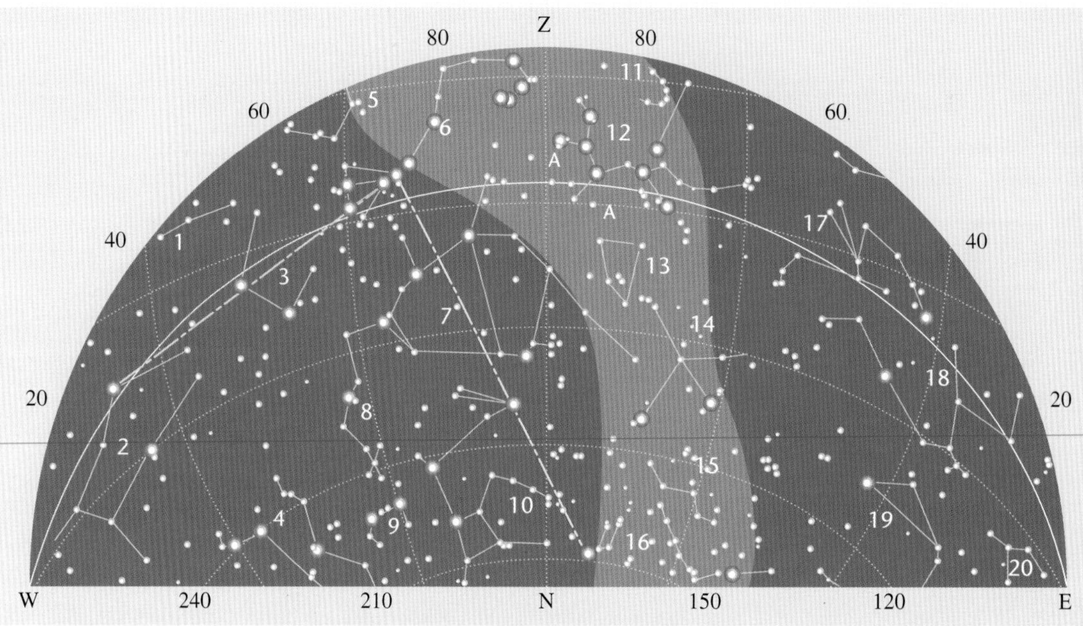

TO THE SOUTH
The brightest stars are massed on the right (West): the stars of the easily recognisable Centaurus and Crux are higher; lower down on the horizon, nearer to the Southernmost point are the starts of Carina and Vela. Canopus is almost due South, just above the horizon. On the left (to the East) is Achernar, between Eridanus and Hydrus, while Fomalhaut (alpha Piscis Austrini) is a little higher. The Small Magellanic Cloud is at the same altitude, but in a more central position.

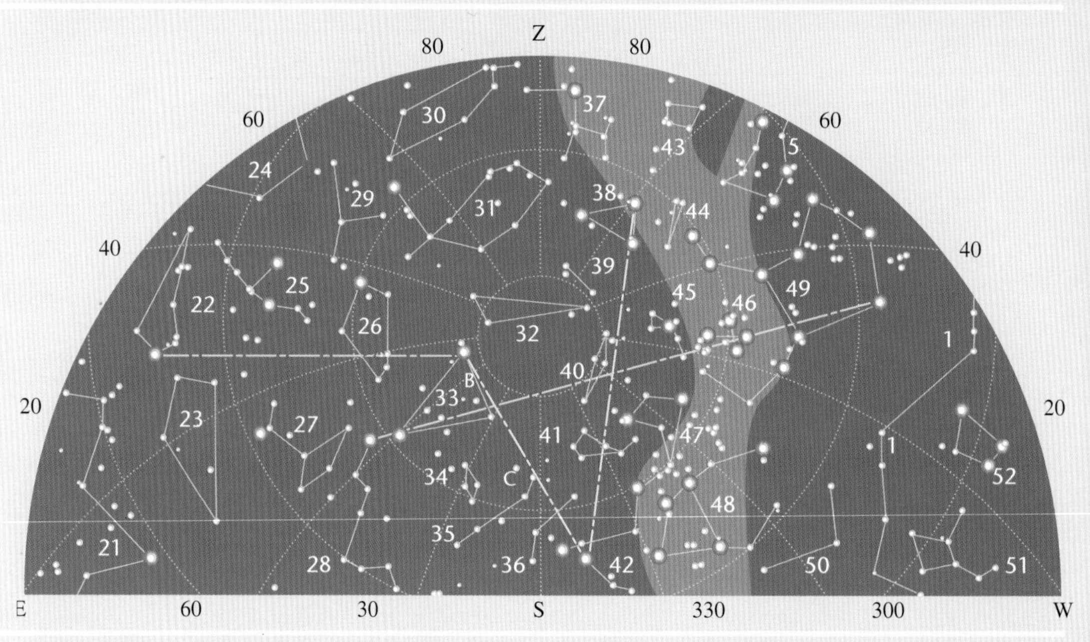

1. Hydra
2. Virgo
3. Libra
4. Bootes
5. Lupus
6. Scorpio
7. Ophiuchus
8. Serpens
9. Corona borealis
10. Hercules
11. Corona australis
12. Sagittarius
13. Scutum
14. Aquila
15. Volpecula
16. Cygnus
17. Capricorn
18. Aquarius
19. Pegasus
20. Pisces
21. Cetus
22. Piscis austrinis
23. Sculptor
24. Microscopium
25. Grus
26. Tucana
27. Phoenix
28. Eridanus
29. Indus
30. Telescopium
31. Pavo
32. Octans
33. Hydrus
34. Reticulum
35. Dorado
36. Pictor
37. Ara
38. Triangulum australe
39. Apus
40. Chamaeleon
41. Volans
42. Puppis
43. Norma
44. Pyxis
45. Musca
46. Crux
47. Carina
48. Vela
49. Centaurus
50. Antlia
51. Crater
52. Corvus

A. Nebulae and galaxies
B. Small magellanic cloud
C. Large magellanic cloud

SOUTHERN HEMISPHERE

MEDIAN LATITUDES (-45°), 12 P.M. SPRING EQUINOX (23 SEPTEMBER)

PEGASUS IS IN THE CENTRE OF THE SKY IN THE NORTH, OPPOSITE CENTAURUS AND CRUX, WHICH ARE AT ALMOST THE SAME HEIGHT IN THE SOUTHERN SKY. SIRIUS AND ANTARES SHINE LOW ON THE SOUTHERN HORIZON.

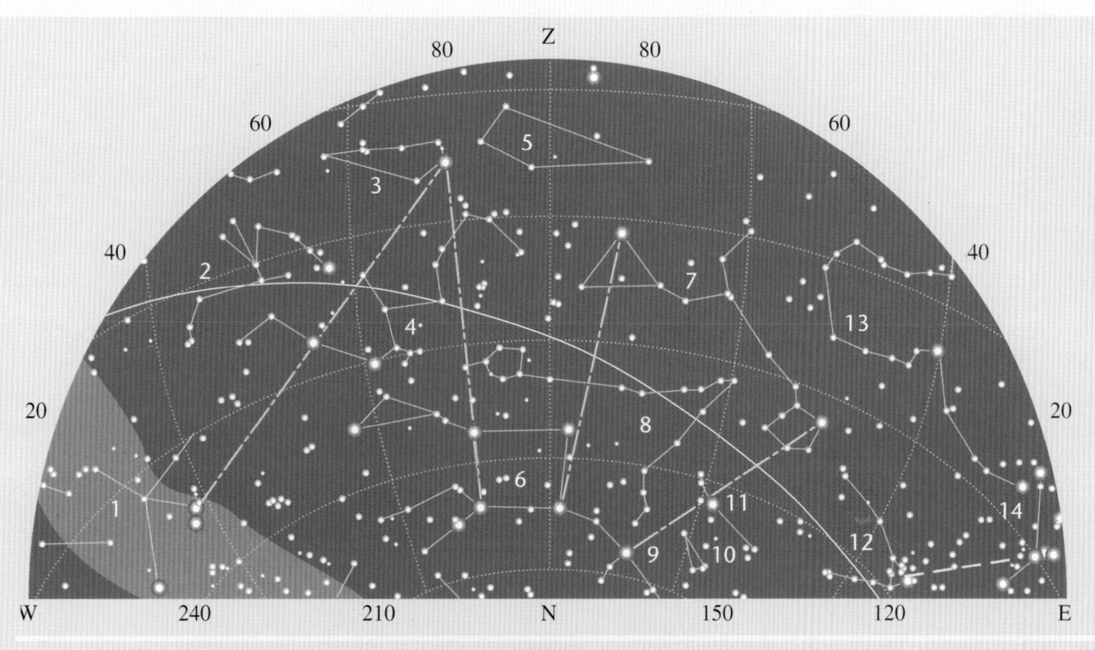

TO THE NORTH
Five rather faint zodiac constellations sketch the ecliptic. The alpha star of Phoenix is almost at the zenith: lower down the stars of the great square of Pegasus can be identified. The sides of the square can be used to identify the following stars, from left to right (i.e., from West to East), Altair in Aquila, Fomalhaut (alpha Piscis Austrini, and beta Piscis Austrini (higher). Lower down, alpha Ceti and Mirach, in Andromeda, identify Hamal (alpha Arietis). Aldebaran, the alpha star of Taurus, and the brightest stars in Orion, are low on the horizon on the right (East).

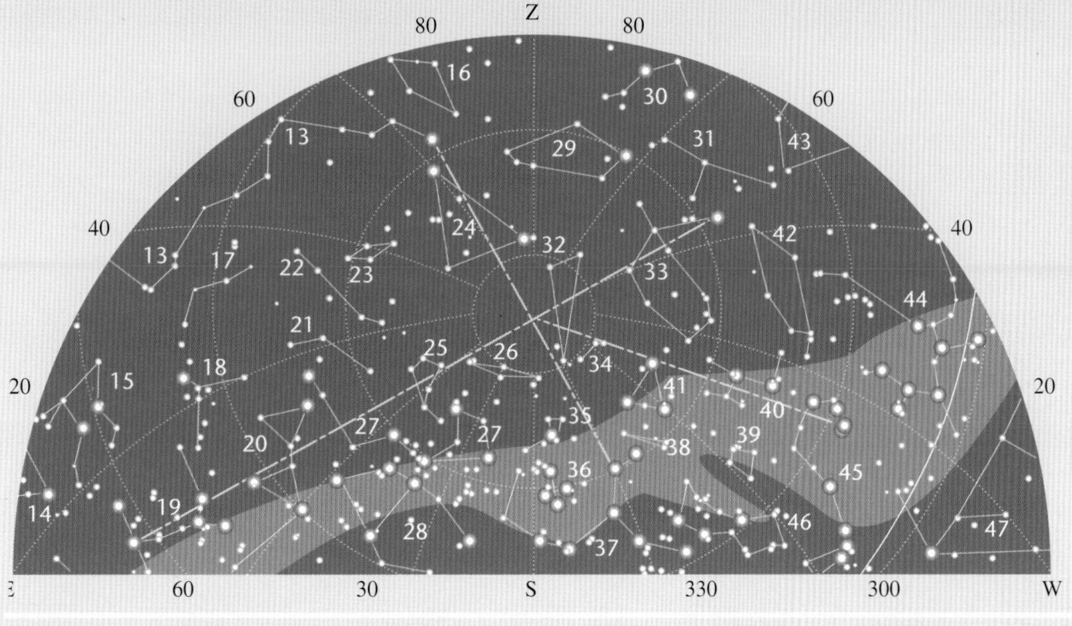

TO THE SOUTH
All the brightest stars are fairly low on the horizon: From East to West we can observe Sirius, Adhara and Wezen, of Canis Major, Alsuhail, Avoir, Aspidiske and Markeb common to Vela and Carina; Crux and Centaurus, with Agena and Proxima, and finally Antares, the alpha star of Scorpio. the tail of Scorpio points to alpha Triangolo Australis and identifies the South Pole: the Peacock alignments of Pavo-Sirius, Achenar in Eridanus and Agena in Centaurus pass through this point.

1. Aquila
2. Capricorn
3. Piscis austrinis
4. Aquarius
5. Sculptor
6. Pegasus
7. Cetus
8. Pisces
9. Andromeda
10. Triangulum
11. Aries
12. Taurus
13. Eridanus
14. Orion
15. Lepus
16. Phoenix
17. Caelum
18. Columba
19. Canis major
20. Puppis
21. Pictor
22. Dorado
23. Norma
24. Hydrus
25. Volans
26. Chamaeleon
27. Carina
28. Vela
29. Tucana
30. Grus
31. Indus
32. Octans
33. Pavo
34. Apus
35. Musca
36. Crux
37. Centaurus
38. Pyxis
49. Norma
40. Ara
41. Triangulum australe
42. Telescopium
43. Microscopium
44. Sagittarius
45. Scorpio
46. Lupus
47. Ophiuchus

SOUTHERN HEMISPHERE

POLAR REGIONS (-63°30′), 12 P.M. SUMMER SOLSTICE (21 DECEMBER)

THE ANTARCTIC SUMMER: THE PERIOD OF THE LONGEST DAY, AND THE FURTHER SOUTH THE LONGER THE DAY THE VERY SHORT NIGHTS ARE ALWAYS LIT BY A CONSTANT DUSK, AND ALTHOUGH THE CONSTELLATIONS ARE THERE, IT IS ALMOST IMPOSSIBLE TO SEE THEM.

If the sky were dark, we could observe the following constellations:

1. Cetus
2. Pisces
3. Eridanus
4. Reticulum
5. Dorado
6. Caelum
7. Lepus
8. Orion
9. Taurus
10. Gemini
11. Cancer
12. Canis Major
13. Monoceros
14. Canis Major
15. Columba
16. Puppis
17. Pictor
18. Carina
19. Vela
20. Pyxis
21. Hydra
22. Crater
23. Sextans
24. Corvus
25. Carina
26. Musca
27. Crux
28. Centaurus
29. Lupus
30. Libra
31. Scorpio
32. Norma
33. Ara
34. Triangulum Australe
35. Circinus
36. Apus
37. Octans
38. Chamaeleon
39. Volans
40. Hydrus
41. Tucana
42. Pavo
43. Indus
44. Telescopium
45. Microscopium
46. Sagittarius
47. Capricorn
48. Piscis Austrinis
49. Grus
50. Sculptor
51. Phoenix
52. Acquarius

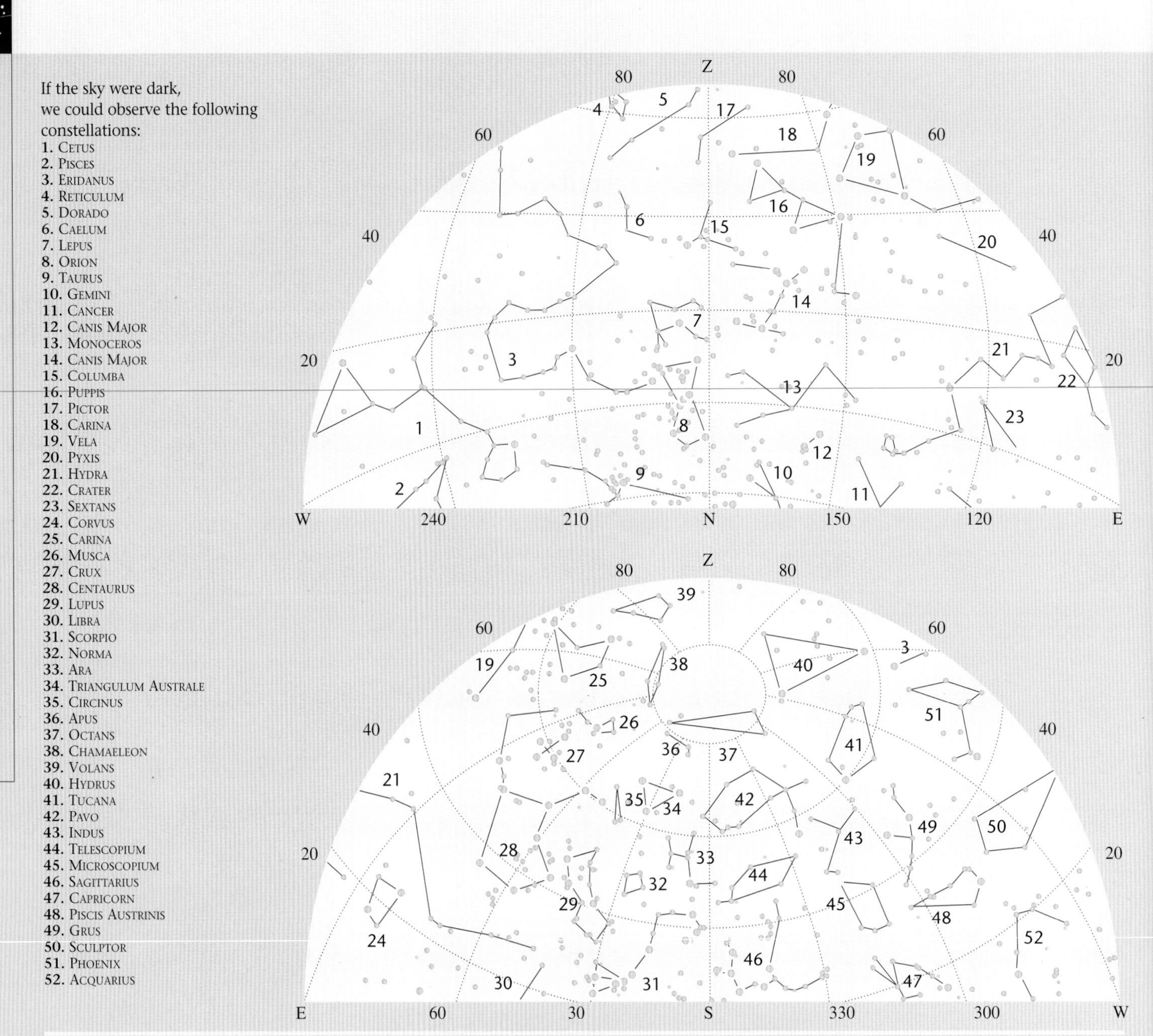

SOUTHERN HEMISPHERE

POLAR REGIONS (-63°30'), 12 P.M. AUTUMN EQUINOX (21 MARCH)

CRUX OCCUPIES THE ZENITH IN THE NORTH AND MUSCA THE ZENITH IN THE SOUTH. THE ECLIPTIC IS QUITE LOW, ACCOMPANIED BY HYDRA THE FULL LENGTH OF ITS NORTHERN SIDE. THE BRIGHTEST STARS ARE ALMOST ALL DISTRIBUTED HIGH IN THE SKY, FAR FROM THE HORIZON.

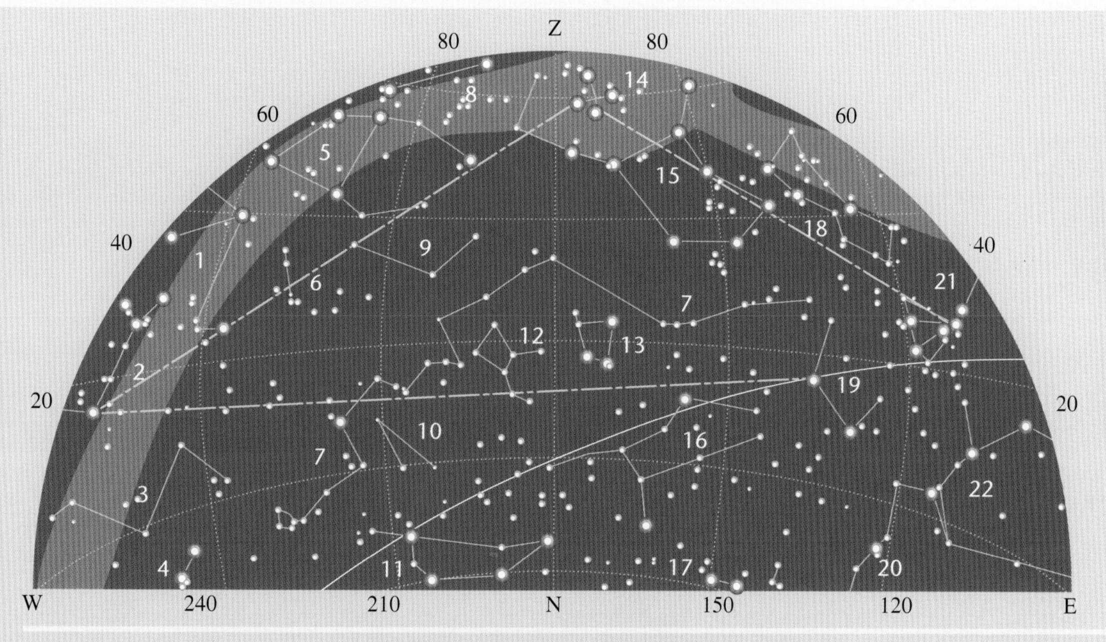

TO THE NORTH

The brightest is Srius on the left (West), then Antares, on the right (East). At about the same altitude as Sirius are Alphard (alpha Hydrae), Spica (alpha Virgo), alpha Libra and alpha Ophiuchus. A line from Sirius to delta1 Crucis passes through Naos, in Puppis, and Alsuhail, Aspidiske and Markeb in Vela, and one joining Antares and gamma Crucis passes through the bright constellation Centaurus.

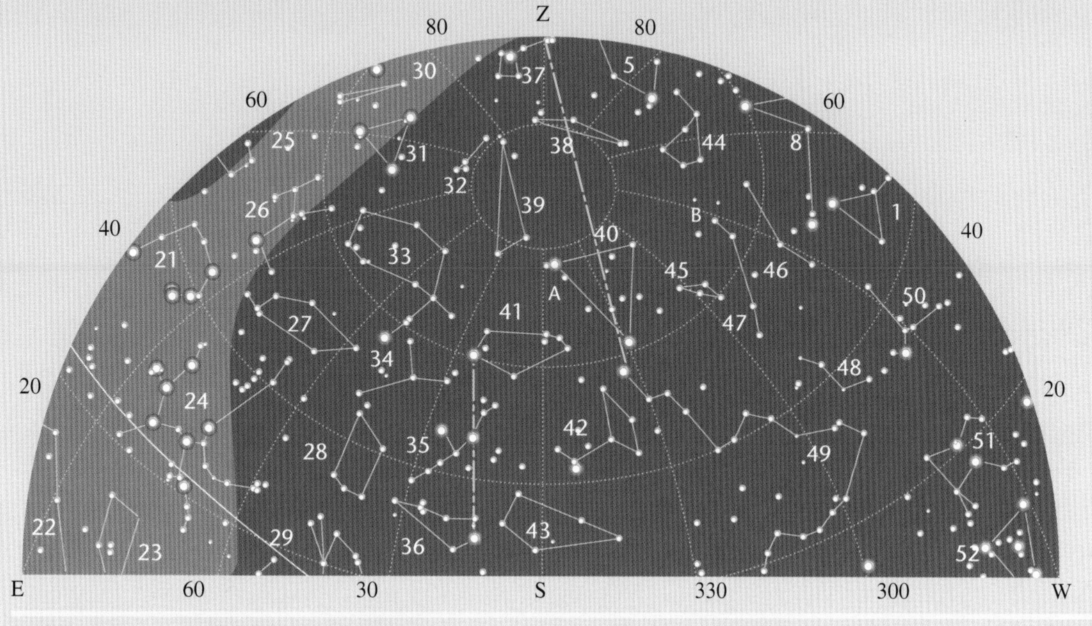

TO THE SOUTH

Above Achernar and just to the left of the constellation Hydrus are the Small and Large Magellanic Clouds to the right (East) of the constellation Dorado, both easy to see. On the right and the left, low on the horizon, are the last zodiac constellations, respectively Sagittarius and Capricorn, partially below the horizon, and Canis Major.

1. Puppis
2. Canis major
3. Unicorn
4. Canis minor
5. Vela
6. Pyxis
7. Hydra
8. Carina
9. Antlia Pneumatica
10. Sextans
11. Leo
12. Crater
13. Corvus
14. Crux
15. Centaurus
16. Virgo
17. Bootes
18. Lupus
19. Libra
20. Serpens
21. Scorpio
22. Ophiuchus
23. Scutum
24. Sagittarius
25. Norma
26. Ara
27. Telescopium
28. Microscopium
29. Capricorn
30. Pyxis
31. Triangulum australe
32. Apus
33. Pavo
34. Indus
35. Grus
36. Piscus austrinus
37. Musca
38. Chamaeleon
39. Octans
40. Hydrus
41. Tucana
42. Phoenix
43. Sculptor
44. Volans
45. Reticulum
46. Pictor
47. Dorado
48. Caelum
49. Eridanus
50. Columba
51. Lepus
52. Orion

A. Small magellanic cloud and

B. Large magellanic cloud

SOUTHERN HEMISPHERE

POLAR REGIONS (-63°30′), 12 P.M. WINTER SOLSTICE (21 JUNE)

At the winter solstice the celestial pole reaches its maximum altitude above the horizon: at the South Pole it is at the zenith, and in this region is at 65°. In the northern sector, Scorpio and Sagittarius blaze in the centre, and in the south Puppis, Vela, Carina, Centaurus and Crux form a trail of light rising from the south into the west.

To the north
At the same altitude and to the right of Antares, alpha Scorpii, are the principal stars of Sagittarius, with alpha Piscis Austrini just above the midline. Aligning Antares with Spica (alpha Virginis) locates alpha Libris: after the trajectory of the ecliptic has been delineated in this way, the constellations of Ophiuchus, Serpens and Aquila, with Altair, can be recognised between this imaginary line and the horizon. Centaurus shines high on the left (West).

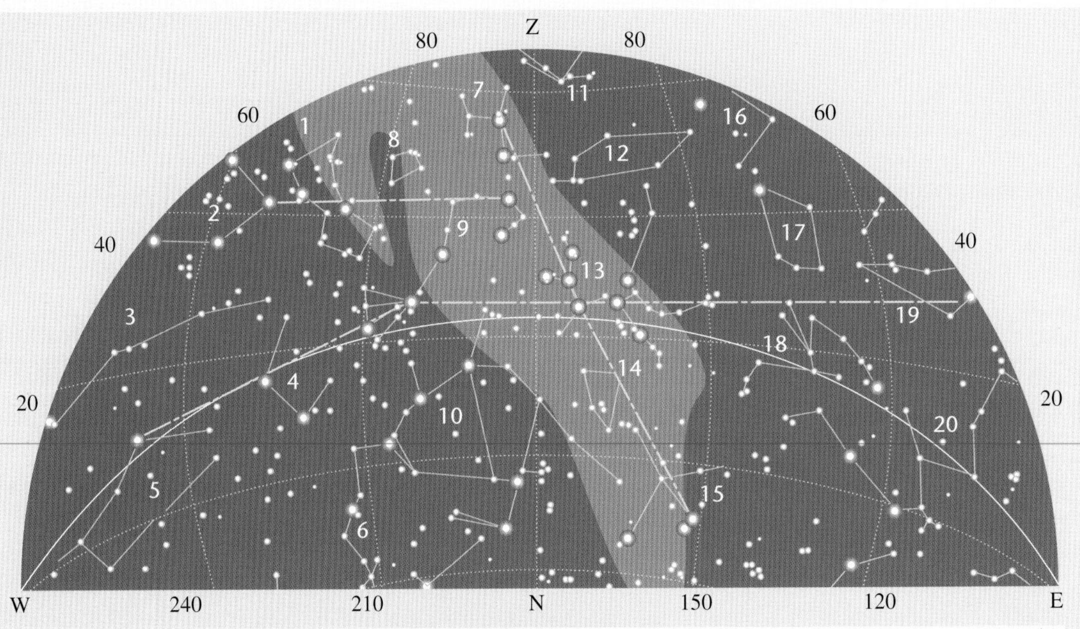

To the south
The Milky Way rises almost vertically, accompanied by the brightest stars in this sector. Canis Major, almost at the Southernmost points, is followed by Puppis with Canopus, Vela with Alsuhail, Aspidike and Markeb, Carina with Avoir and Miaplacidus, Crux with Acrux and Mimosa, and Centaurus with Proxima and Agena. Triangulum Australe and Pavo are almost at the zenith: one to the West (right), and the other to the East (left). Low down on the left is the long constellation of Eridanus, with Hydra low down on the right.

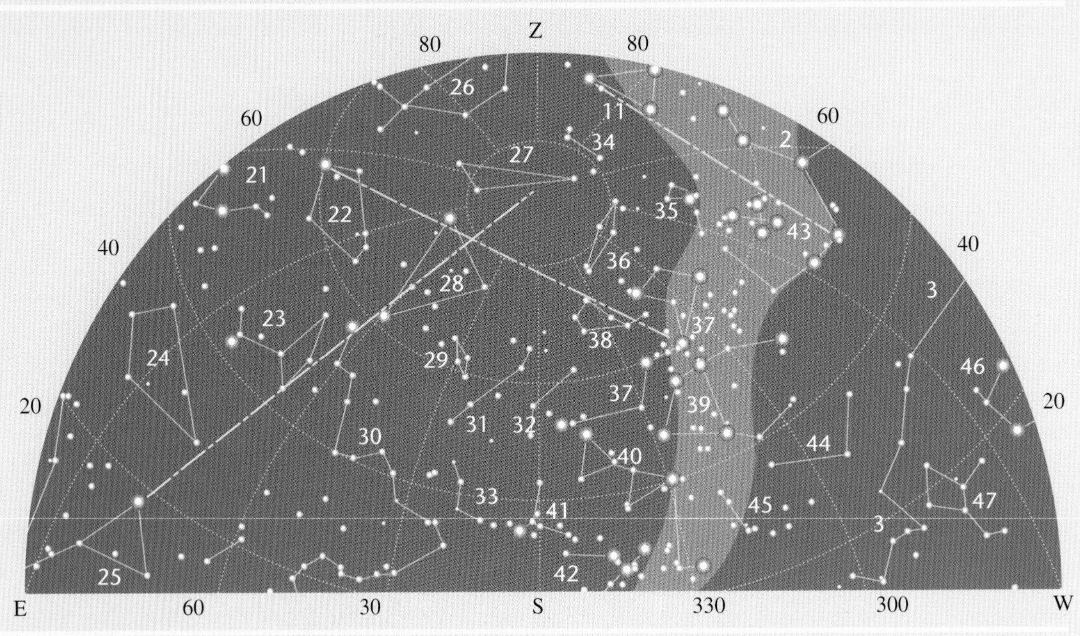

1. Lupus
2. Centaurus
3. Hydra
4. Libra
5. Leo
6. Serpens
7. Ara
8. Norma
9. Scorpio
10. Ophiuchus
11. Triangulum australe
12. Telescopium
13. Sagittarius
14. Scutum
15. Aquila
16. Indus
17. Microscopium
18. Capricorn
19. Piscis austrinis
20. Aquarius
21. Grus
22. Tucana
23. Phoenix
24. Sculptor
25. Cetus
26. Pavo
27. Octans
28. Hydrus
29. Norma
30. Eridanus
31. Dorado
32. Pictor
33. Caelum
34. Apus
35. Musca
36. Chamaeleon
37. Carina
38. Piscis austrinis
39. Vela
40. Puppis
41. Columba
42. Canis major
43. Crux
44. Antlia
45. Pyxis
46. Corvus
47. Crater

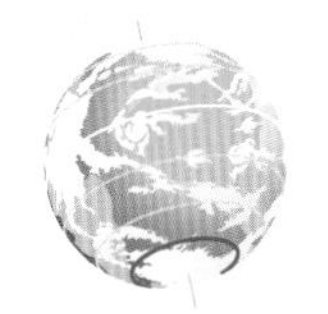

SOUTHERN HEMISPHERE
POLAR REGIONS (-63°30'), 12 P.M. SPRING EQUINOX (23 SEPTEMBER)

TO THE NORTH TUCANA IS AT THE ZENITH, WHILE ERIDANUS STRETCHES EAST FROM THE ALMOST 80° OF ACHERNAR TO THE ALMOST 10° OF CURSA. SOUTH IS A LONG SERIES OF CONSTELLATIONS RICH IN BRIGHT STARS: CANIS MAJOR, PUPPIS, VELA, CARINA, CRUX, CENTAURUS AND SCORPIO FILL THE LOWER PART OF THE SKY FROM EAST TO WEST.

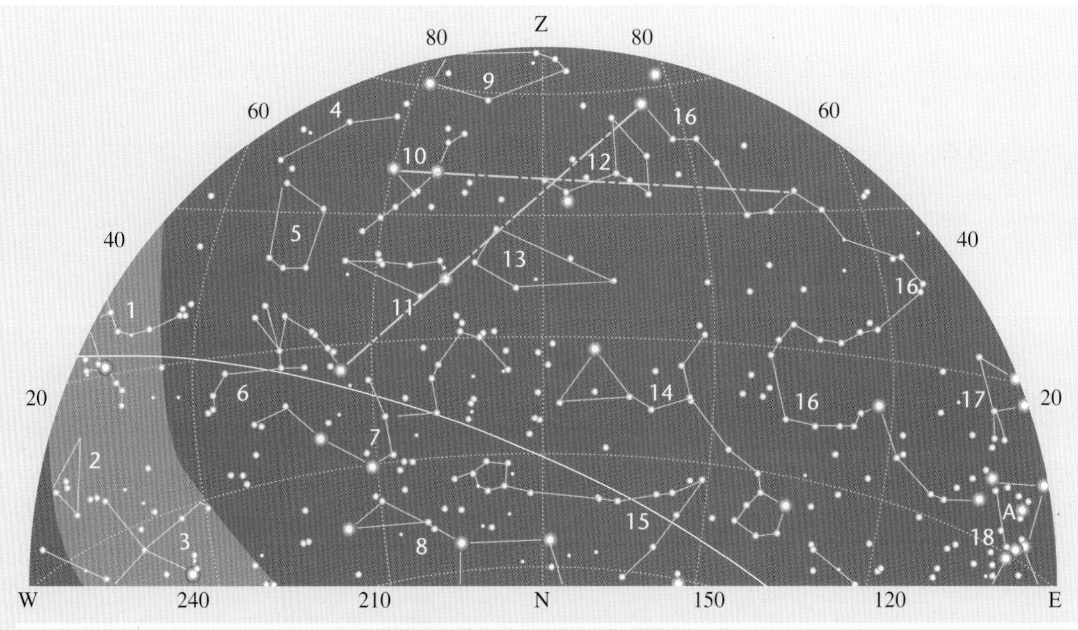

TO THE NORTH
Eridanus is fully distended and visible on the East, where Orion's belt, Rigel, Saigh and the M42 and M43 nebulae ar elocated low on the horizon, visible to the naked eye. A line from Achernar to Formalhaut (alpha Piscis Austrini) leads to Deneb Algiedi (delta Capricorni): tracing this constellation locates Dabih (beta) and alpha Capricorni, a multiple system consisting of two double stars. Aligning the brightest stars in Grus locates epsilon Eridani, which shows evidence of a planetary system.

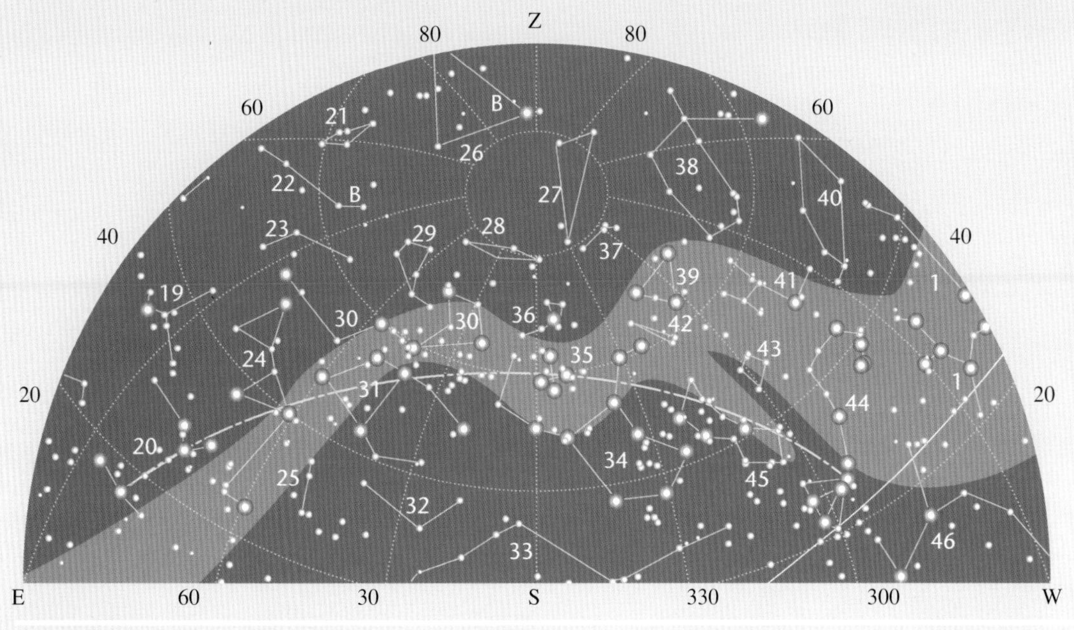

TO THE SOUTH
Here Hydrus is the constellation at the zenith: The Small Megellanic Cloud, near to Hydrus, and the Large Magellanic Cloud, near to Dorado, are clearly visible. From left to right (i.e., from East to West), the brightest stars of the Southern hemisphere may be aligned around an arc centred on Crux, stretching from Sirius, Wezen and Aludra in Canis Major to zeta Puppis and to delta Velorum, surrounded by the other three brighter stars, to the delta and beta stars of Crux, around which the brightest stars of Centaurus make an arch passing through gamma Lupi and ending at Antares in Scorpio.

1. SAGITTARIUS
2. SCUTUM
3. AQUILA
4. INDUS
5. MICROSCOPIUM
6. CAPRICORN
7. AQUARIUS
8. PEGASUS
9. TUCANA
10. GRUS
11. PISCIS AUSTRINIS
12. PHOENIX
13. SCULPTOR
14. CETUS
15. PISCES
16. ERIDANUS
17. CAELUM
18. ORION
19. COLUMBA
20. CANIS MAJOR
21. RETICULUM
22. DORADO
23. PICTOR
24. PUPPIS
25. PYXIS
26. HYDRUS
27. OCTANS
28. CHAMAELEON
29. VOLANS
30. CARINA
31. VELA
32. ANTLIA
33. HYDRA
34. CENTAURUS
35. CRUX
36. MUSCA
37. APUS
38. PAVO
39. TRIANGULUM AUSTRALE
40. TELESCOPIUM
41. ARA
42. PYXIS
43. NORMA
44. SCORPIO
45. LUPUS
46. OPHIUCHUS

A. NEBULAE M42 AND M43
B. MAGELLANIC CLOUDS

Observing the Moon is easy: anyone can do it, you just need a pair of binoculars. It's a large and very bright object, and you don't need to know your way around the sky or to be familiar with instruments:
You don't need telescopes with large apertures or high magnifying power, just look up and fall in love with discovering new and fascinating glimpses of the endless, silent and magical landscape of our satellite.

OBSERVING THE MOON

The visible face of the Moon
Lunar cartography only started in the 17th century.

Whatever instrument you observe the moon with, you'll be experiencing its capabilities to the full: large, luminous and rich in details which change appearance with every change in the light, the Moon is a fascinating object even when viewed through a simple pair of binoculars. However, while a pair of binoculars is sufficient to observe the major formations, a lunar eclipse or the occultation of a star, a small telescope is advisable for observing smaller regions and finer detail. For more advanced observations and studies, and to take high resolution photographs, large aperture and long focal length instruments such as those observatories make available to groups of astronomy buffs have the special equipment (filters, photometers, etc.) needed.

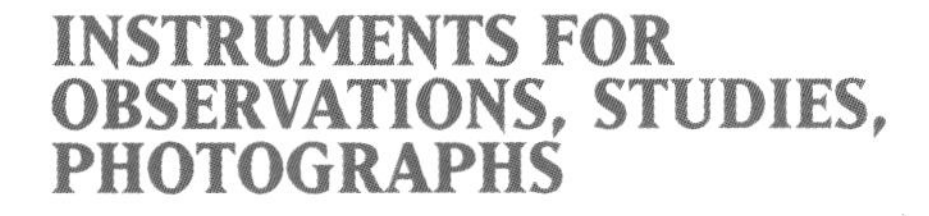

INSTRUMENTS FOR OBSERVATIONS, STUDIES, PHOTOGRAPHS

A few words about the instruments which can be used for lunar observation – binoculars and telescopes – and some hints for photography. But not too much detail: The choice of instrument depends on personal preferences – for a particular type of film, or to use particular techniques to improve the resulting photographs, usually developed by experience, adapting personal knowledge to the instrument available. But more help can be obtained from established astronomy clubs, often affiliated with astronomy institutes and observatories.

BINOCULARS

The most widely used are Porro prism binoculars,

Elger lunar albedo >144 scale

Degree	Examples
0	Black shadows
1	Darkest parts of craters Grimaldi and Riccioli
1,5	Inside of craters Boscovich, Bailly and Zupus
2	Floor of craters Endymion, Le Monnier and Julius Caesar
2,5	Inside of craters Azout, Vitruvius, Pitatus and Hippalus
3	Floor of craters Taruntius, Plinius, Theophilus and Flamsteed
3,5	Insides of craters Archimedes and Mersenius
4	Inside of craters Manilius, Ptolomaeus and Guericke
4,5	Surfaces around l Sinus Medii and the crater Aristillus
5	Surfaces around craters Kepler and Aristarchus
5,5	Walls of craters Picard and Timocharis, rays of crater Copernicus
6	Walls of craters Macrobius, Kant, Bessel and Moesting
6,5	Walls of craters Langrenus, Thaetetus and La Hire
7	Craters Theon< Ariadaeus, Bode B, Wickmann and Kepler
7,5	Craters Ukert, Hortensius and Euclides
8	Walls of craters Godin, Bode and Copernicus
8,5	Walls of craters Proclus, Bode A and Hipparchus C
9	Craters Censorinus, Dionysius, Moesting A, Mersenius B and Mersenius C
9,5	Inside craters Aristarchus and La Perouse
10	Central peak of crater Aristarchus

FIRST OCTANT

THE MAIN LUNAR FORMATIONS OBSERVABLE ABOUT FOUR DAYS AFTER THE NEW MOON ARE MARKED IN THE IMAGE

1. STRABO
2. ENDYMION
3. MARE HUMBOLDTIANUM
4. HERCULES
5. ATLAS
6. GEMINUS
7. CLEOMEDES
8. PROCLUS
9. MARE CRISIUM
10. PICARD
11. MARE MARGINIS
12. FIRMICUS
13. MARE UNDARUM
14. TARUNTIUS
15. APOLLONIUS
16. MARE FOECUNDITATIS
17. MARE SPUMANS
18. MARE SMITHII
19. GUTEMBERG
20. GOCLENIUS
21. LANGRENUS
22. COLOMBUS
23. VENDELINUS
24. COOK
25. PETAVUS
26. SNELLIUS
27. STEVINUS
28. RHEITA
29. FURNERIUS
30. METIUS
31. VALLIS RHEITA
32. FRAUNHOFER
33. FABRICIUS
34. STEINHEIL, WATT
35. VLACQ
36. MUTUS

but these are bulky and heavy: the more expensive but manageable roof prism binoculars are better. Binoculars are ideal for an overall view of basins, mountain chains and the main craters, and to observe lunar eclipses. The essential characteristics of a pair of binoculars are their magnifying power and the objective lens diameter: the greater the diameter and the lower the magnifying power, the greater the luminosity. For the best results the ratio of objective lens diameter (in mm) and magnifying power should not exceed 7-8: in binoculars with a magnifying power of 10 and a 50 mm objective (10 x 50) the ratio is 50:10 = 5, which is optimum for bird watching; in a 7 x 50 pair of binoculars the ratio is 50:7 = 7.1 mm, suitable for most uses, including observing objects in low light at night.

In the case of the Moon though, there's no shortage of light: to see the major craters, the highest mountains and the more extensive valleys and crevices, however, a good magnifying power is useful: an 8 x 30 or 10 x 40 pair of binoculars would be fine. This should not be taken too far: with magnifying power over 10, a tripod must be used. This is the only way 12 x 60, 16 x 70 and 20 x 80 instruments can be used, to provide images with more and more detail.

FIRST QUARTER OR HALF MOON

THE PICTURE, OF THE MOON ON THE EIGHTH DAY AFTER THE NEW MOON, SHOWS THE MAIN LUNAR FORMATIONS AND STRUCTURES WHICH CAN BE OBSERVED ON THIS NIGHT, THE MOST SPECTACULAR IN THE LUNAR CYCLE.

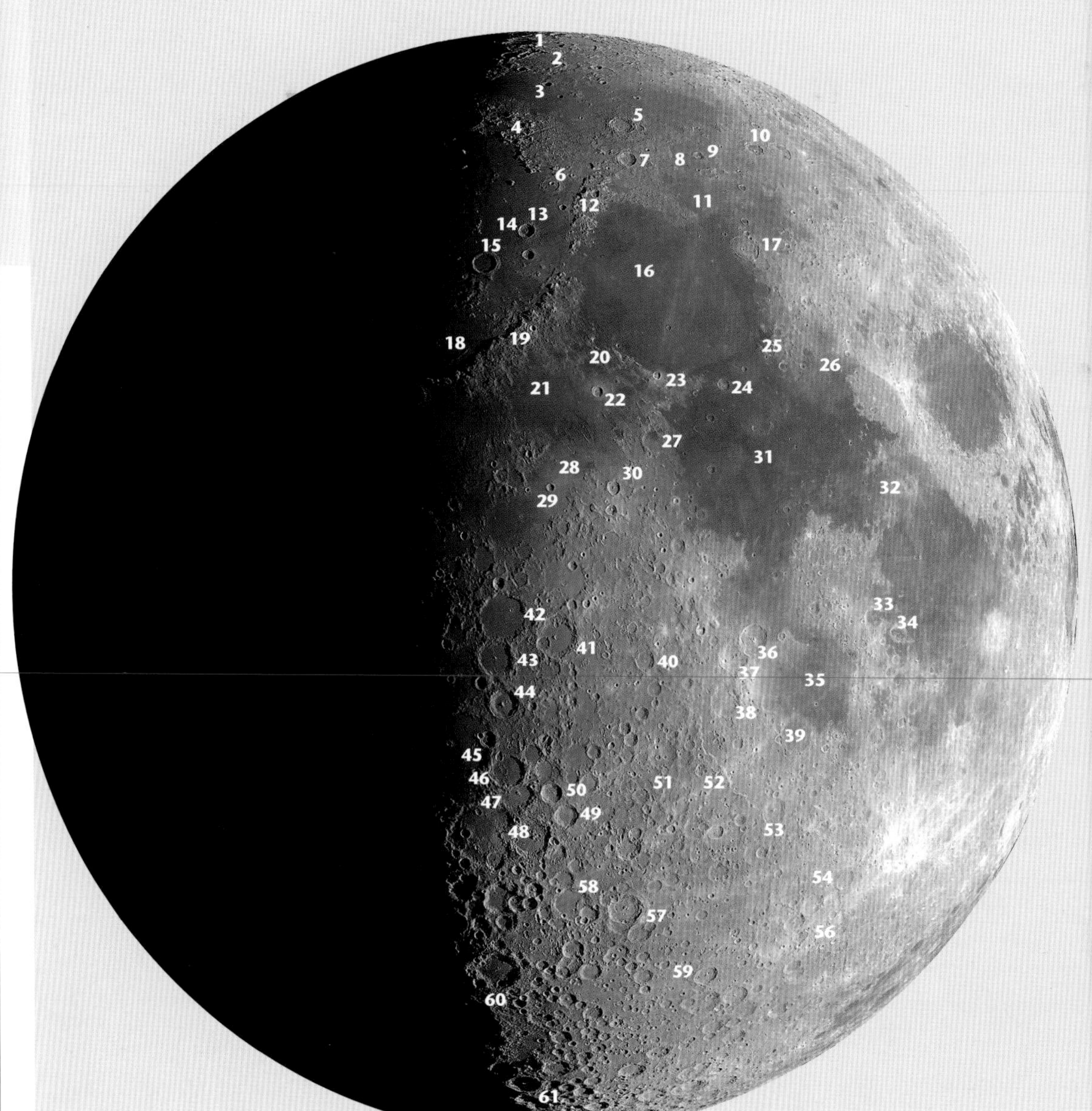

1. Goldschmidt, barrow
2. Bond
3. Mare firgoris
4. Vallis alpes
5. Aritoteles, mitchell
6. Cassini
7. Eudoxus
8. Lacus mortis
9. Bürg
10. Hercules, altas
11. Lacus somniorum
12. Montes caucasus
13. Aristillus
14. Palus nebularum
15. Archimedes
16. Mare serenitatis
17. Posidonius
18. Palus putredinis
19. Montes apennines
20. Montes haemus
21. Mare vaporium
22. Manilius
23. Menelaus
24. Plinius
25. Montes argaeus
26. Palus somni
27. Julius caesar
28. Rima hygunus
29. Triesnecker
30. Rima ariadeus
31. Mare tranquillitatis
32. Taruntius
33. Gutemberg
34. Goclenius
35. Mare nectaris
36. Theophilus
37. Cyrillus
38. Catherina
39. Beaumont
40. Abufelda
41. Albateginus, klein
42. Ptolomaeus
43. Alphonsus
44. Arzachel
45. Thebit
46. Purbach
47. Regiomontanus
48. Deslanders
49. Aliacensis
50. Werner
51. Sacrobosco
52. Rupes altai
53. Piccolomini
54. Rheita
55. Stevinus, snellius
56. Vallis rheita
57. Maurolycus
58. Stöfer, faraday
59. Pitiscus
60. Maginus
61. Curtis

TELESCOPE

If you become really interested in lunar topography, and want to see the smallest details, you need a telescope. Refractor and reflector telescopes: broadly speaking, the former give images less affected by atmospheric turbulence and the latter render colours better. However, amateur enthusiasts mainly use "hybrid" telescopes, a cross between a reflector and a refractor - the catadioptic telescope, an instrument with a mirror, like the reflector, and a corrective plate at the front. This type of instrument is small, light, easy to transport and mains good power: all these aspects are important for anyone living in areas where there is high light pollution, forcing the user to move around to find a clean sky. A 12 cm reflector or a 5-6 cm refractor give good results, and 15-20 cm instruments are already at the limit of normal observation possibilities. Apochromatic refracters give sharper images, but they are very expensive. If the focal length is long enough, the telescope is suitable for observing the Moon: 15-20 times the diameter of the objective lens for a refractor, and at least 8-10 times for a Newtonian reflector. The Cassegrain type, which has very long focal lengths, is best. This is because the magnifying power is calculated by dividing the focal length of the telescope (in

THREE QUARTER CRESCENT MOON

THE PICTURE SHOWS THE MOON AT ABOUT THE TWELFTH DAY AFTER THE NEW MOON. IN THIS CASE TOO THE MAIN LUNAR FORMATIONS WHICH CAN BE OBSERVED ARE MARKED.

1. BOND
2. MARE FIRGORIS
3. CONDAMINE
4. PLATO
5. PICO
6. ARISTOTELES
7. EUDOXUS
8. BAILY
9. MARE IMBRIUM
10. PALUS NEBULARUM
11. ARISTILLUS, AUTOLYCUS
12. ARCHIMEDES
13. THIMOCHARIS
14. MARE SERENITATIS
15. LE MONNIER
16. MONTES CARPATUS
17. ERATOSTHENES
18. MONTES APPENNINUS
19. COPERNICUS
20. SINUS AESTUUM
21. MARE VAPORUM
22. MANILIUS
23. MARE TRANQUILLITATIS
24. SINUS MEDII
25. HERSCHEL, PTOLOMAEUS
26. ALBATEGINUS
27. ALPHONSUS
28. ALPETRAGIUS
29. ARZACHEL
30. BULLIALDUS
31. MARE NUBIUM
32. THEBIT
33. PITATUS
34. DESLANDER
35. CAPUANUS
36. WILHELM
37. TYCHO
38. LONGOMONTANUS
39. MAGINUS
40. CLAVIUS
41. SCHEINER
42. BLANCANUS
43. RUTHERFORD
44. CURTIUS

mm) by the focal length of the eyepiece (in mm): a telescope with a focal length of 1m will have a magnifying power of 100 with a 10 mm eyepiece, and 200 with a 20 mm eyepiece. A long focal length means that eyepieces which are not too large can be used, provides high quality images and is less tiring for the eyes. Eyepieces with focal lengths of 7 mm or more should be used to observe the moon: as for the objective lens, the quality of the eyepiece is important, and an Abbe orthoscopic eyepiece is the most suitable, because there is no distortion, it adapts well to high magnification and it allows the user to make observations without keeping the eye resting against the lens. The higher the magnifying power, the more stable the mounting must be. The best is equatorial: this allows celestial objects to be tracked with a single movement. Movement may be powered by an electric motor, and some "motorised" mountings also take the delay in the Moon with respect to the apparent rotation of the sky into account. if a telescope with a sideral drive performs a complete turn in 23 h 56 min 4 s, a lunar drive telescope takes 24 h 50 min to perform a complete turn, a negligible difference when observing with the naked eye, but a significant one if taking long exposure time photographs.

FULL MOON

THE MOON RISES AS THE SUN SETS, AND DISPLAYS ALMOST ALL THE VISIBLE FORMATIONS IN A SINGLE MOMENT.

1. GIOJA
2. MARE FRIGORIS
3. PLATO
4. SINUS RORIS
5. SINUS IRIDUM
6. MONTES JURA
7. MARE IMBRIUM
8. MONTES ALPES
9. ARCHIMEDES
10. PALUS NEBULARUM
11. PALUS PUTREDINIS
12. MONTES APPENNINUS
13. SINUS AESTUUM
14. COPERNICUS
15. LAMBERT
16. VALLIS SCHRÖTER
17. ARISTARCHUS
18. KEPLER
19. OCEANUS PROCELLARUM
20. GRIMALDI
21. LETRONNE
22. MARE COGNITUM
23. FRA' MAURO
24. SINUS MEDII
25. GASSENDI
26. MARE HUMORUM
27. CAMPANUS
28. MARE NUBIUM
29. PITATUS
30. CAPUANUS
31. SCHICKARD
32. SCHILLER
33. TYCHO

TAKING PHOTOGRAPHS
Observing the Moon is more satisfying than photographing it: the finer details are easier to see because the eye can reach the limit of resolution of the instrument, while photography is hindered by the movement of the mounting, particularly with exposure times longer than a fraction of a second, and by the grain of the film, turbulence and changes in view.
Since the image of the Moon on the focal plane is almost 1 cm for each metre of focal length, a 15 cm focal length telescope creates a 22 mm image of the Moon, and a 35 m telescope gives a 38 mm image. The image of the Moon at the direct focus is small but very bright, and can be photographed with a short exposure time: eclipses, albedo comparisons and occultations can be photographed well like this. For a larger image an eyepiece that projects the image onto the film must be used: depending on the eyepiece and the length of the projection, the dimensions of the image can vary greatly, with corresponding variations in its brightness. Longer exposure times are needed, and the photographs are not as sharp. A telescope with 20 cm aperture and 2 m focal length, with a 6 mm eyepiece 7.5 cm away from the firm is a good compromise: this means that very sensitive film (fine grain) film

WANING THREE QUARTER MOON

THE MOON RISES OVER 2 HOURS AFTER THE SUN HAS SET.
THE CRATERS ARISTARCHUS, KEPLER, COPERNICUS AND TYCHO SHINE BRIGHTLY AGAINST THE DARK BACKGROUND OF THE LAVA BASINS.

1. Philolaus
2. Rupes philolaus
3. Epigenes
4. Pythagoras
5. Sinus roris
6. Harpalus
7. Sharp
8. Montes jura
9. Sinus iridum
10. Mare frigoris
11. Plato
12. Aristoteles
13. Pico
14. Vallis alpes
15. Montes caucasus
16. Palusnebularum, aristillus
17. Mare imbrium
18. Archimedes
19. Palus putredinis
20. Thimocharis
21. Lambert
22. Vallis schröter
23. Schiaparelli
24. Aristarchus
25. Oceanus procellarum
26. Reiner
27. Kepler
28. Copernicus
29. Eratosthenes
30. Montes appenninus
31. Mare vaporum
32. Manilius
33. Sinus medii
34. Fra' mauro
35. Mare cognitum
36. Letronne
37. Grimaldi
38. Gassendi
39. Byrgius
40. Fourier
41. Mare humorum
42. Bullialdus
43. Mare nubium
44. Ptolomaeus
45. Alphonsus
46. Arzachel
47. Thebit
48. Purbach
49. Regiomontanus
50. Theophilus
51. Cyrillus
52. Catharina
53. Lacroix
54. Schickard
55. Pitatus
56. Lungomontanus
57. Tycho
58. Clavius
59. Maginus
60. Maurolycus

can be used, producing high quality images.

FROM NEW MOON TO NEW MOON

As the Moon gradually shifts with respect to Earth, and the portion of surface we see illuminated by the Sun increases, the very sharp shadows become shorter and shorter, and new and spectacular moonscapes become visible to the curious eye of the observer. Every night (since a lunation lasts 29 d, 12 h, 44 min and 2.8 s) we see a segment which is 12° of longitude wider: a segment which, at the Equator, is 350 km wider every day. The details can be seen more clearly after the terminator.

NEW MOON

The Moon looks like this when it is between the Sun and the Earth. The hemisphere facing us is completely dark, and we cannot see anything. Sometimes just the edge can be seen: this is the ideal time to observe the profile of lunar reliefs at the edges. The most notable trough is the *Mare Australis*.

THE MOON RISES AT MIDNIGHT, AND THE OBSERVER HAS TO WAIT UNTIL 3 A.M. AT LEAST TO SEE IT CLEARLY.

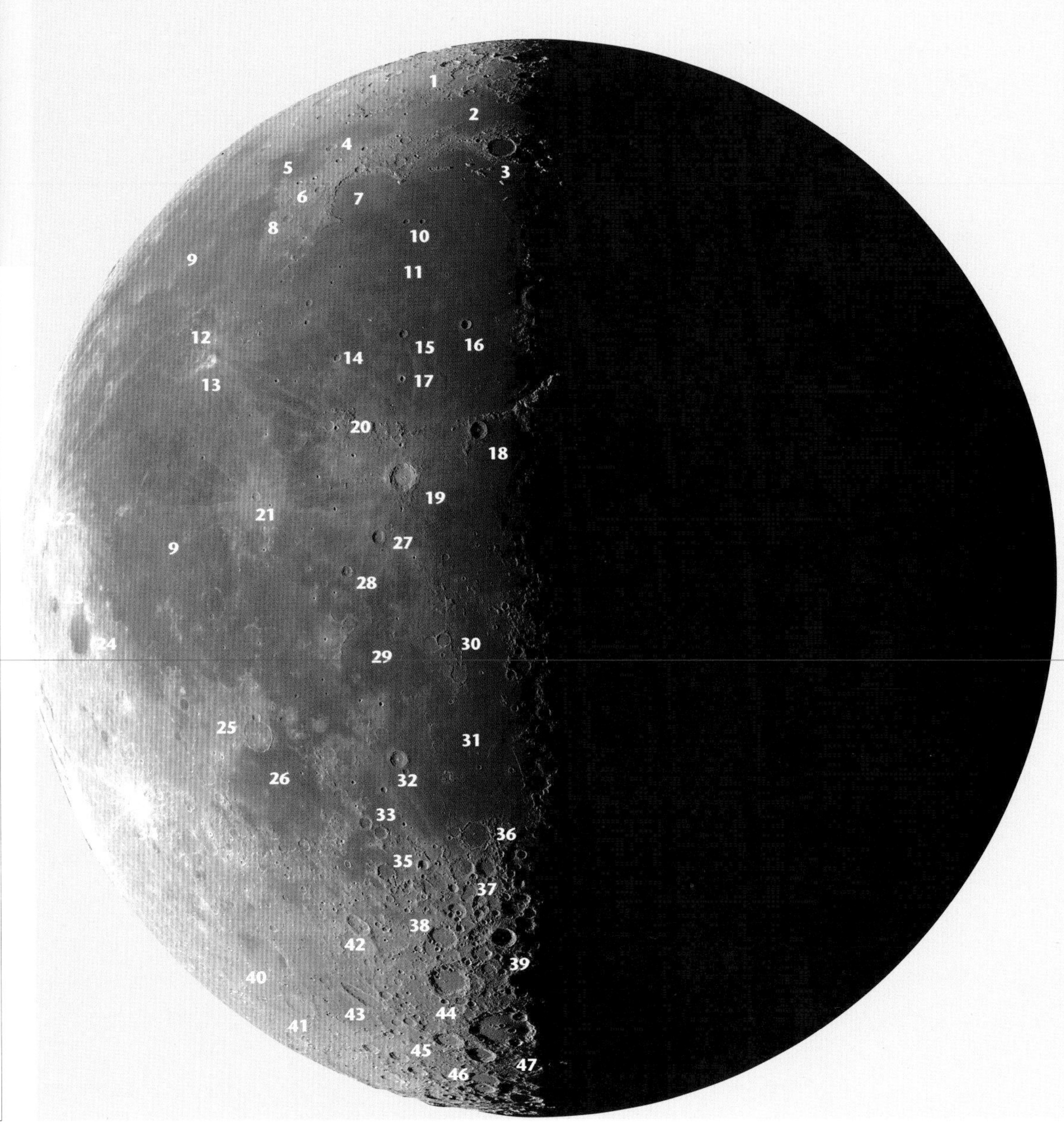

1. Philolaus, rupes philolaus
2. Mare frigoris
3. Plato
4. Harpalus
5. Sinus roris
6. Foucault
7. Sinus iridum
8. Sharp
9. Oceanus procellarum
10. Helicon, le verrier
11. Mare imbrium
12. Vallis schröter
13. Aristarchus, herodotus
14. Euler
15. Lambert
16. Thimocharis
17. Pytheas
18. Eratosthenes
19. Copernicus
20. Montes carpatus
21. Kepler
22. Olbers
23. Riccioli
24. Grimaldi
25. Gassendi
26. Mare humorum
27. Reinhold
28. Lansberg
29. Mare cognitum
30. Fra' mauro
31. Mare nubium
32. Bullialdus
33. Campanus, mercator
34. Fourier
35. Capuanus
36. Pitatus
37. Wurselbauer, Gauricus
38. Wilhelm
39. Tycho
40. Schickard
41. Phocyclides
42. Hainzel
43. Schiller
44. Lungomontanus
45. Scheiner
46. Blancanus
47. Clavius

FIRST CRESCENT
FIRST OCTANT

In particular viewing conditions this may already be visible 20 h after the alignment with the Sun, but usually is not visible for at least 36 h. The crescent is in fact very narrow, and just 12° from the Sun: when it sets there is still too much light to observe it well. Distinguishing any formations on the crescent is even more difficult, but observation conditions quickly become idea for observing earth-light.

FIRST QUARTER OR HALF MOON
By now the Moon sets over 2 h after the Sun has slipped below the horizon, and many illuminated formations may be seen clearly even with the naked eye. Almost a week after the new moon the Moon is at its best for observation.

THREE QUARTER CRESCENT MOON
The Moon now rises later and later, and new formations and more details of the formations previously observed, as they approach the terminator (i.e. the line of demarcation between the dark and the light areas, between "day" and "night"). Observation conditions are ideal for studying the better known and evident lunar formations on the visible face: the craters Copernicus and Tycho, with their enormous rays,

FINAL OCTANT

IT IS INCREASINGLY DIFFICULT TO OBSERVE THE MOON: IT RISES LATE AND ONLY REACHES A HIGH ENOUGH ALTITUDE FOR GOOD OBSERVATIONS NEAR DAWN.

1. PYTHAGORAS
2. BABBAGE
3. OCEANUS PROCELLARUM
4. ARISTARCHUS
5. OLBERS
6. CAVALERIUS
7. HEVELIUS
8. RICCIOLI
9. GRIMALDI
10. FOURIER
11. SCHICKARD
12. WARGENTIN

are visible in more and more detail, with the large spots of the *Mare Imbrium*, *Mare Serenitatis*, *Mare Tranquilitatis* and *Mare Foecunditatis* to the East.

FULL MOON

The Moon rises when the Sun sets. It sometimes appears deformed or squashed by terrestrial refraction, and this decreases as the Moon rises in the sky. Sometimes, depending on how clean the atmosphere is, it can be coloured, anything from pearl white to dark red. When directly compared to reference objects on the horizon it seems larger that it really is: in fact in these conditions its apparent diameter is 2% smaller than its apparent diameter when directly above our heads. Now all the most important formations are visible, although not as beautiful to observe as when they are near the terminator: without the strong shadows they seem flattened, and lose body.

THREE QUARTER WANING MOON – LAST OCTANT

The Moon rises later and later, displaying the large craters Aristarchus, Kepler, Copernicus and Tycho against the dark background of the marias. Observations are increasingly difficult, both because the best conditions occur very late, near dawn, also because the basins now take up much of the surface: since their albedo is much lower, the Moon is less bright.

IF LUNAR ECLIPSES ARE VERY FASCINATING, ECLIPSES OF THE SUN, PARTICULARLY WHEN TOTAL, ARE TRULY MESMERISING, AND ANYONE LUCKY ENOUGH TO SEE A TOTAL ECLIPSE WILL REMEMBER IT FOR THE REST OF THEIR LIFE.

OBSERVING ECLIPSES

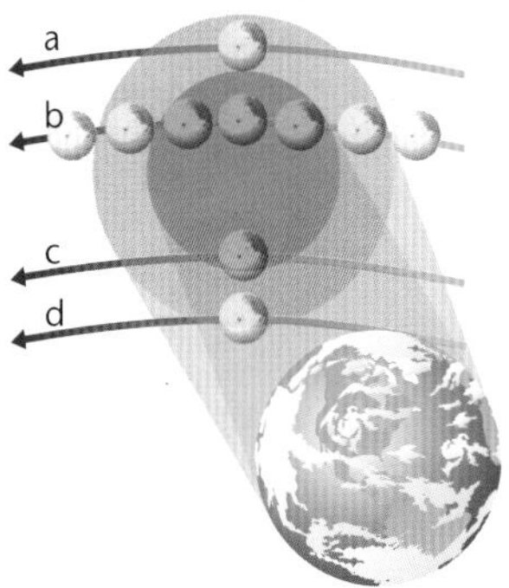

ECLIPSES OF THE MOON
Depending on the trajectory of the Moon (that is, on the distance of the lunar orbit from the ecliptic), they can be:
a. total penumbra;
b. total umbra;
c. partial umbra;
d. partial penumbra.
The closer the Moon is to the nodes, the longer the period of darkness.

ECLIPSE OF THE MOON
The moon is entering the shadow of the earth.

PROGRESS OF A TOTAL ECLIPSE OF THE SUN
A collage of images shows how the lunar disk moves across the Sun during a total eclipse.

The Sun-Earth-Moon and Sun-Moon-Earth alignments ➢52-53 which (respectively) originate the phenomena of lunar eclipses and solar eclipses are not as rare as one might think: in a year, a minimum of four and a maximum of 7 may be seen, combined in different ways (and while there are may be many different numbers of eclipses of the Moon, there are a maximum of two eclipses of the Sun). And while lunar eclipses can be observed from any point on the earth's surface (provided it's night), and for an average duration of about 4 hours in total, eclipses of the Sun may only be seen in the band of earth surface which is "swept" by the shadow of the Moon in the short time the alignment lasts (generally about twenty minutes). Then there is the fact that the surface of the Earth is very large, and it is unlikely that an eclipse of the Sun passes over a band of inhabited countries. In addition, the thermal changes produced by the "loss" of sunshine often causes cloud formation during a solar eclipse: this is a familiar scenario for those who "follow" these phenomena, and find themselves with a dark and overcast sky. So eclipses of the Sun are much rarer, unless the observer goes looking for them. And this is what enthusiasts, who use the opportunity to take a flight to observe moon shade do, as do researchers studying the outer layers of the Sun, although the use of coronographs on satellites has almost completely replaced observation from earth.

ECLIPSES OF THE MOON

While the shadow of the Earth projected on the lunar surface is easily recognisable with the naked eye, through a telescope it can be somewhat confused with the dark spots of the basins. But only for a little while: the progressive advance of the shadow is always visible, even through a telescope. During a total umbral eclipse, the brightness of the Moon is reduced on average to about one ten thousandth of its normal brightness: the phase of total occultation can last up to 2 and a half hours, and is preceded and followed by about an hour of penumbral eclipse. However, even at the darkest moments the Moon is still visible and, in the light of the Sun diffused by the atmosphere of the earth, it assumes different colours according to the pollution in the planet's atmosphere.

ECLIPSES OF THE SUN

Protect your eyes! Looking directly at the sun is really very dangerous for the sight, even using very dark filters.

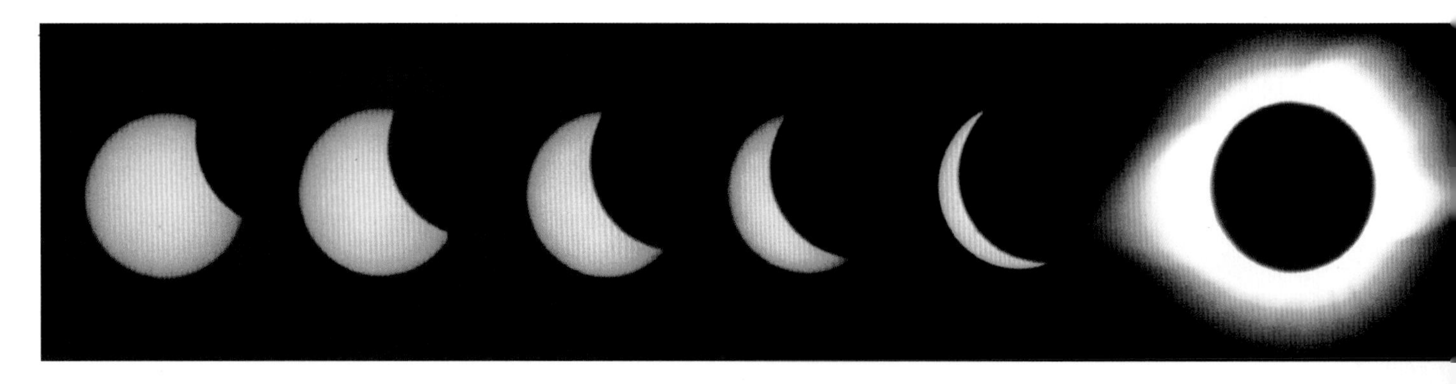

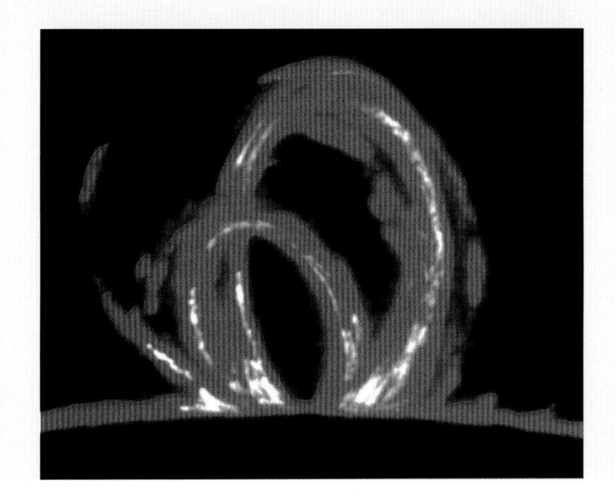

Never point an inadequately shielded telescope or binoculars at the Sun, even if partially eclipses, and never used filters which screw onto the eye piece: the heat of the concentrated rays of the sun can crack them and blind the observer in an instant; never look at the sun without a screen: even if fully eclipses there is a risk that the first rays emerging from behind the Moon could burn the retina. Observers using telescopes should never forget to block the finder: someone could look into it by accident, and be blinded. Neutral filters of density 5 or more and 14-16 degree welding screens can be used. Do not use "traditional" remedies such as blackened photographic or radiographic films and dark glasses. Projecting the image of the sun through a hole onto a cardboard screen or sheet of white paper, just as if the Sun were completely visible, is the best way. During an eclipse of the Sun the luminosity varies enormously over time. But even in the few seconds of totality it varies greatly from the edge of the Moon outwards, increasing from luminosity equal to that of the full moon to one hundred thousandth of that within a distance of just 3 solar diameters. So decide in advance what to photograph if you want to take photographs: A fine grain not very sensitive film is suitable for photographing the edge (the internal corona); a more sensitive coarser grained film is preferable for the outer corona. In this case you need an instrument with a large field (at least 3° and a focal length of at least 1m. The exposure times will depend on the film chosen and the object photographed. Using an instrument with a focal ratio of about 1/10 and a normal film for slides (400 ASA), the exposure can vary from 1s to 1/30s. Variable density filters and eclipse coronographs reduce the internal light and are useful for photographing the external corona. However, to use them properly its worth practicing putting them on and taking them off quickly: making a mistake means wasting the few seconds in which the chromosphere and corona are visible. But remember one fundamental fact: photographing an eclipse is fascinating and useful for all amateur astronomers, but it cannot be combined with observing it as "a spectator", so you will lose the opportunity to admire an unforgettable event like this "live".

Protuberance
The most impressive phenomena of the outermost layers of the Sun may be clearly observed, particularly when the brightness is cancelled by the lunar screen: this shows a beautiful ring shaped protuberance projecting from the edge of the moon during a total eclipse.

Eclipse
Total eclipses of the Sun that can be observed clearly from Earth are rare.

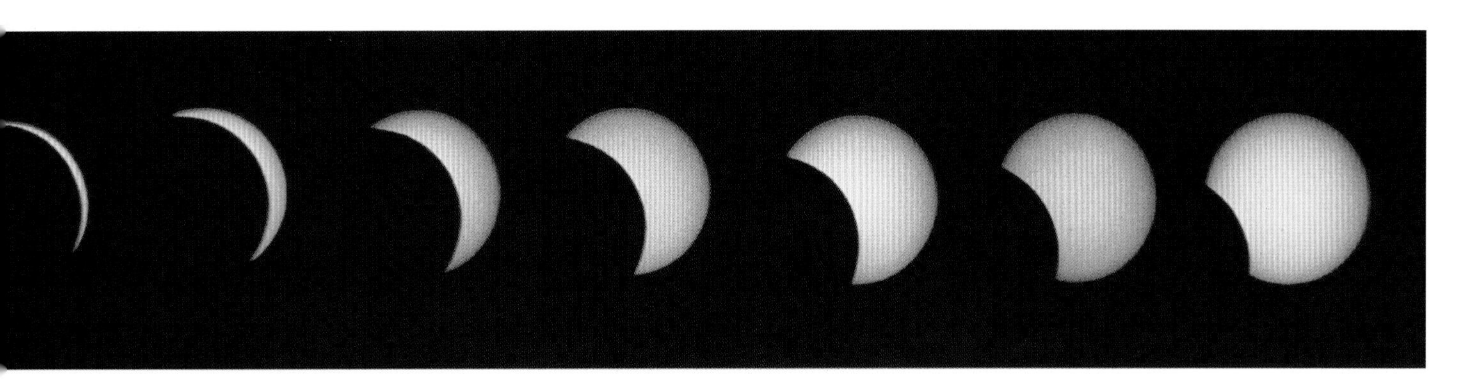

SPACE EXPEDITIONS AND INSTRUMENTS

MAN'S IMAGINATION TRULY HAS NO LIMITS: WITH INGENUITY AND INTELLIGENCE WE HAVE MANAGED TO OBSERVE THE SUN "IN DEPTH" AND THE OTHER PLANETS VERY CLOSE UP, TO WALK ON THE MOON AND TO OBTAIN A HUGE AMOUNT OF INFORMATION ON THE MASS, TEMPERATURE, DISTANCE AND SPEED OF THE MOST DISTANT CELESTIAL OBJECTS FROM JUST A SLIVER OF LIGHT. WHILE THE SPACE EXPEDITIONS HAVE SOLVED MANY PROBLEMS, PARTICULARLY ABOUT THE GEOLOGY OF THE PLANETS AND THEIR CAPACITY TO SUSTAIN LIFE, AND OUR KNOWLEDGE OF THE UNIVERSE HAS MADE ENORMOUS PROGRESS WITH THE ADVENT OF SPACE TECHNOLOGY, WE NO LONGER NEED TO LEAVE THE ATMOSPHERE TO OBTAIN SURPRISINGLY ACCURATE DATA. THE DEVELOPMENT OF COSTLY SPACE EQUIPMENT HAS LEAD TO THE CREATION OF LESS COSTLY BUT EQUALLY ACCURATE INSTRUMENTS ON EARTH.

We can get to know the universe without particularly sophisticated instruments: the great engineers of the past, such as Heratosthenes, Hipparchus and Aristarchus, and Galileo himself, have shown this. However, if elaborate instruments allowing increasingly precise measurements had not been developed, our knowledge of celestial objects would have remained limited to our own solar system at the most.
It is no coincidence that the Copernican revolution needed the sophisticated instruments designed by Brahe, and the telescopes built by Galileo and Newton, to establish itself definitively and irreversibly. And our current knowledge of the stars and celestial phenomena would be unthinkable if our knowledge of optics and chemistry and had not been applied to astronomy, and new analytical instruments invented to examine the "structure" of stellar radiation.
Since Galileo pointed his telescope at Jupiter, instruments have been an extension of man's senses, allowing him to see unimaginable distances - an indispensable instrument, which allows us to develop and test new hypotheses and to design even more sophisticated instruments in a steady progression towards full scientific knowledge of the world that surrounds us.

THE BASIC INSTRUMENTS

The tools needed to collect radiation from extraterrestrial sources are called telescopes. Depending on the type of radiation they can detect, they are referred to as optical, infrared, X-ray and gamma ray or even cosmic ray or neutrino telescopes.

OPTICAL TELESCOPE
With its 1.62 m diameter principal mirror, this instrument is suitable for recording even dim objects. After its installation at the ESO (European Southern Observatory) in La Silla (Chile) it enabled over 12,000 new galaxies (and other objects) to be identified , including some of magnitude 21. And it is by no means the most powerful telescope in existence.

Spectrophotometer
Diagram and picture of the solar spectrum obtained with a spectrophotometer. The light passes through a slit, is focussed by the collimator, dispersed through a prism (or diffraction grating) into a series of monochrome rays. Each ray is then focussed by the objective lens on a photographic plate. Since the light originates from a long narrow slit, each element (dark or light) or the spectrum is in the form of a line.

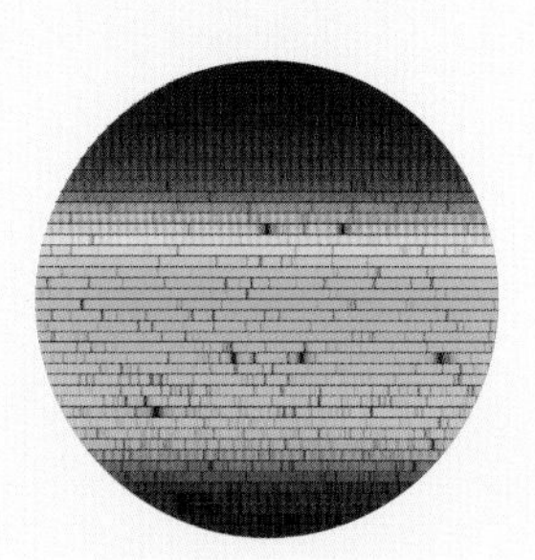

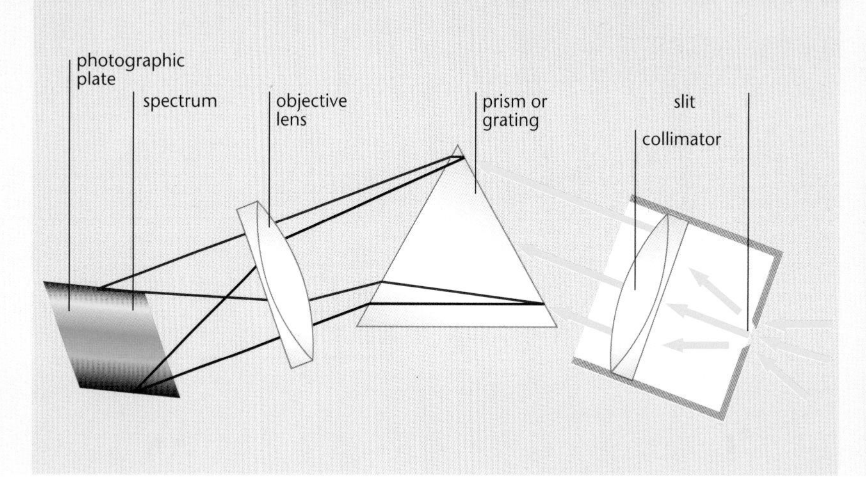

Despite these differences, they all share an essential structure, since they all originate from the refractor and reflector telescopes designed by Galileo and Newton respectively.
An optical telescope collects and concentrates light from a dim source, making it visible or sufficiently bright to be able to photograph it, record it with electronic detectors and analyse it with other types of instrument.
Some telescopes are also able to enlarge the image of the objects observed: this is only useful when observing near objects (in our solar system) or galaxies, since objects which we see as points remain points even when enlarged.
However, when observing non-point distant objects (such as galaxies) the greater the enlargement capacity of the telescope the more important its resolution characteristics, i.e. its capacity to show two objects that are very close together as separate objects, and to identify very distant fine detail.
Other "basic" instruments which are important for astronomical studies are:
instruments to analyse light radiation:
• the spectroscope, which splits light radiation into its spectrum allowing the latter to be observed;
• the spectrograph, which projects the spectrum onto a photographic plate enabling it to be photographed;
• the spectrophotometer, used to compare the light intensity of sources of differently coloured spectra, by directly comparing the respective spectra.
These instruments can analyse any type of radiation: instruments created specifically to analyse infrared or ultraviolet radiation (outside the visible spectrum) use lenses and prisms in special glass that is transparent to the radiation in question, and instruments to collect the spectra generated, such as thermoelectric batteries and bolometers for infrared radiation, and special photographic plates or photomultipliers for ultraviolet. Other instruments have been specifically designed to analyse the Sun, such as the spectroheliograph, used to analyse part of the Sun in a particular wavelength, or the coronograph, used to study the radiation emitted by the outer layers, masking the luminous disk (i.e. artificially duplicating the conditions during a total eclipse of the Sun ➢52).

REFRACTOR OR REFRACTION TELESCOPES

These collect light and enlarge images through an optical system of lenses housed in a tube of adjustable length. The outer lens which points at the object is the objective, and the lens close to the eye is the eye piece. First used by Galileo, the use of this type of telescope was limited by the fact that it concentrated radiation of different colours in separate points, producing a brightly coloured distortion of the images (chromatic aberration). Although this problem has been overcome with the invention of achromatic objectives, and despite the fact that these telescopes were widely used in the astronomical studies of the Nineteenth Century, they are limited by the size of the lens, because, for diameters over a certain limit, their thinness and high weight makes them too fragile. The largest telescope of this type still in use is in Yerkes: it was made in 1897 and has a 1.02m diameter lens.

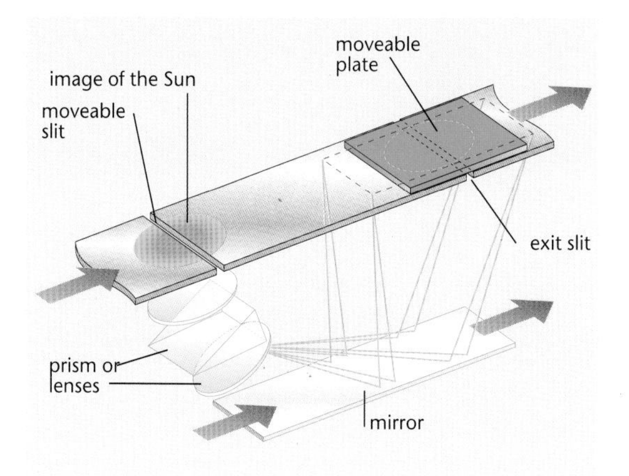

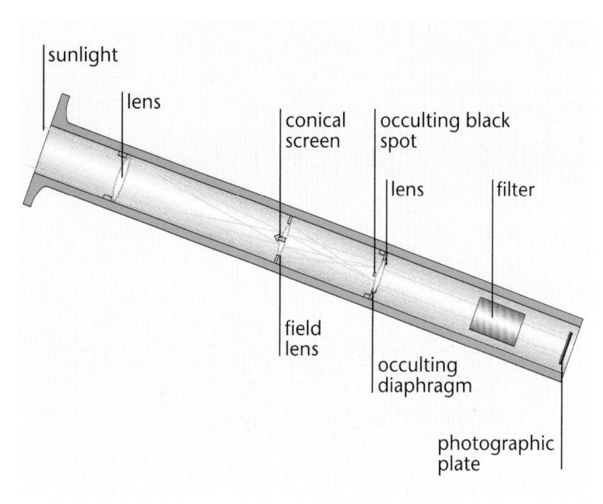

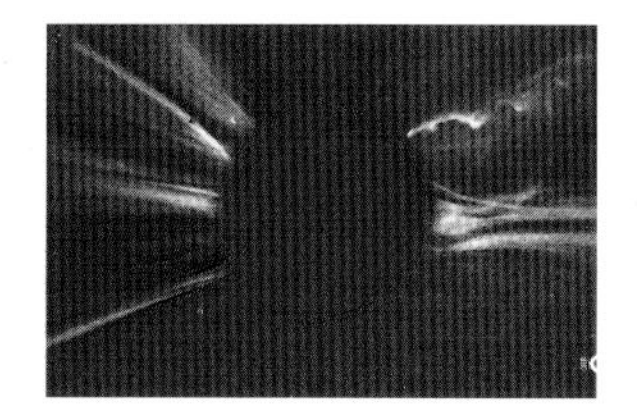

Spectroheliograph
Far Left: Structure.

Coronograph
Left: structure.
Above: photograph of the Sun made using a *SOHO* coronograph

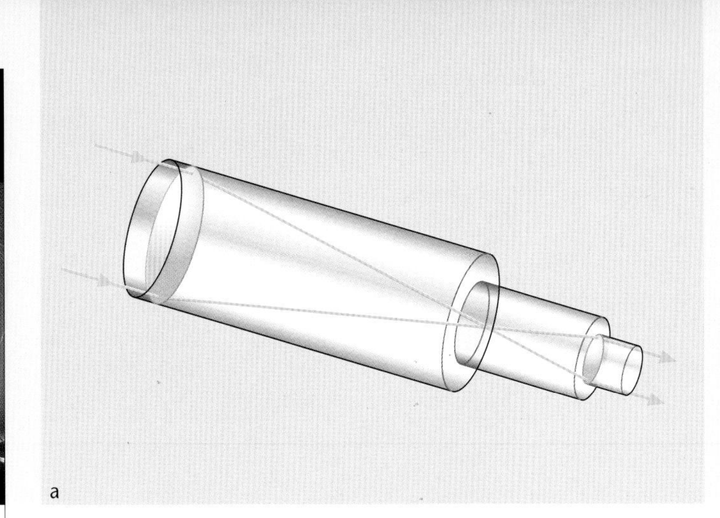

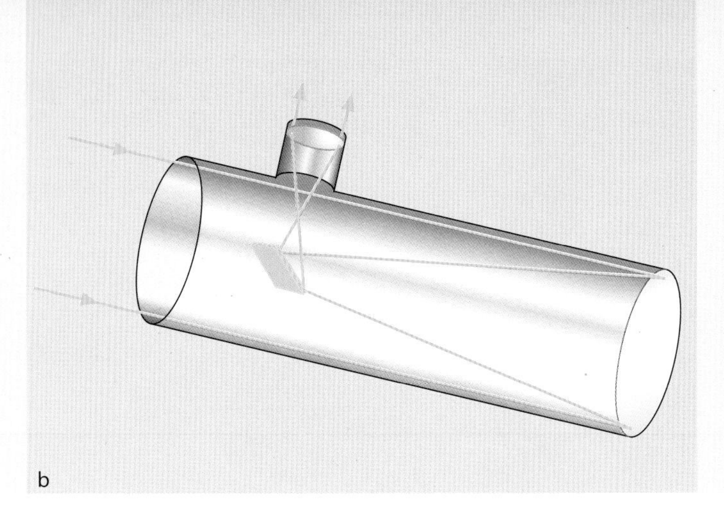

Optical telescopes
Structural diagram of:
a. a refractor telescope,
b. a reflector telescope.

Mirror telescope
One of the four telescopes of the ESO Very Large Telescope during assembly and, on the right, the support of one of these instruments, supporting the secondary mirror.

REFLECTOR, REFLECTION OR MIRROR TELESCOPES

Reflector telescopes are also instruments which collect the light and enlarge the image, but here the central role is played by a mirror instead of a lens. A tube houses the principal or primary mirror, which is convex, reflecting the radiation to converge to a single point. The image is then deviated and possibly enlarged through a system of secondary mirrors and lenses; however, it should be recalled that every reflection or passage through a lens decreases the brightness of the image.

There are a number of different configurations of reflector telescopes, named after their inventors ➢72, but it is the size of the instrument rather than its configuration that is important. The quantity of light that a mirror can collect is in fact proportional to the square of the diameter of the principal mirror: the larger the mirror, the more light is captured, and the more light the better visible objects are distinguished and the more other previously invisible objects are discovered. And in addition: the more light available, the more accurate the spectra obtained from light sources, including the weakest sources.

Moreover, the power of resolution of a telescope is proportional to the diameter of the primary mirror, and the possibility of separating images which are very close together becomes essential in the study of extensive bodies like the solar system, the galactic nebulae or the galaxies outside the Milky Way.

It is no coincidence that the construction of large telescopes in the Nineteenth Century led to the discovery of an unsuspected world beyond our galaxy. However, there are natural limits to the dimensions of a mirror: to not distort under its own

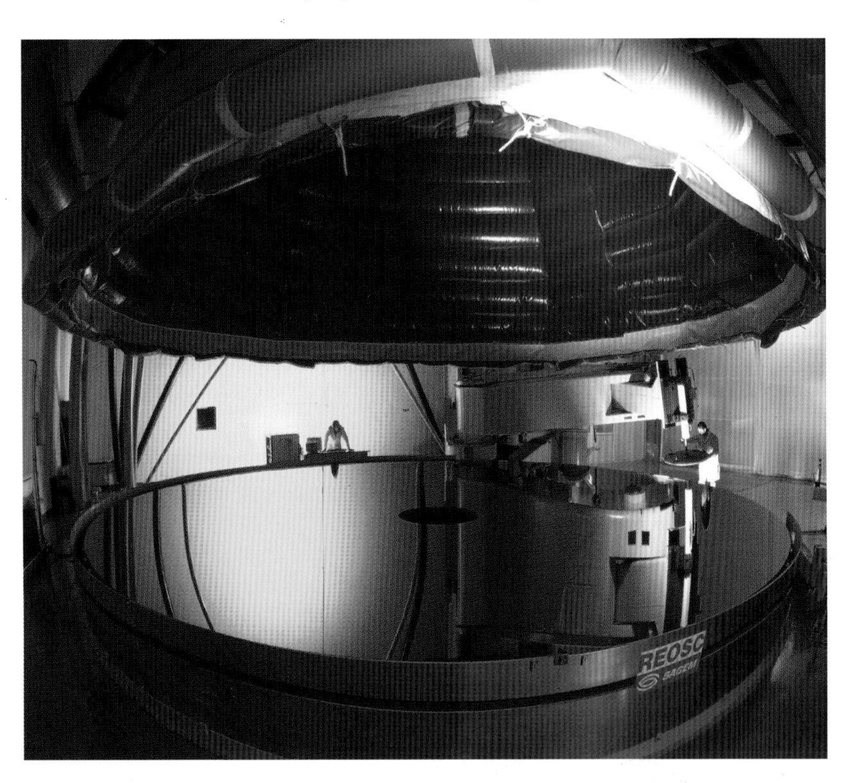

Mirror telescope
Detail of the system of actuator supports under one of the 8m diameter mirrors in each of the 4 identical telescopes making up the ESO (European Southern Observatory) Very Large Telescope.
When all four are in operation it will be equivalent to a telescope with a 16 m diameter primary mirror.

weight its thickness must equal at least 1/5th of its diameter. the maximum dimension, achieved in the Soviet Union, was a primary mirror 6 m in diameter and 42 Mg in weight: it took 2 years just to cool the mass of glass without creating internal flaws.
Fortunately, new technologies, new materials and the development of increasingly complex electronic programmes have allowed these limits to be exceeded. Today telescopes are made in which the principal mirror is ultrathin, or constituted of several mirrors, or of an array of reflective cells. Both the ultrathin and the composite mirrors are maintained in continual movement by a series of sensors and supports which exert different pressures at different points, maintaining the reflective surface in precisely the desired position, and also compensating for the interference of the atmosphere on the trajectories of the light rays collected. All this is made possible by computers, which continuously monitor the atmospheric conditions and collate the information collected by several instruments. Furthermore, in addition to composite mirrors, "composite" instruments are increasingly used. a series of telescopes pointed at the same object, and linked in such a way as to combine the signals collected, can perform the same functions as a single very large telescope. This is the principle used at the ESO.

Observatories
The small picture shows the bleak setting of the La Silla European Southern Observatory (ESO) in the Atacama Desert in Chile, in its early days (1987).

Bellow: the dome of the 2.1 m diameter telescope at Kitt Peak, at dawn.
All modern space observation centres are built at high altitudes, where atmospheric interference is at a minimum.

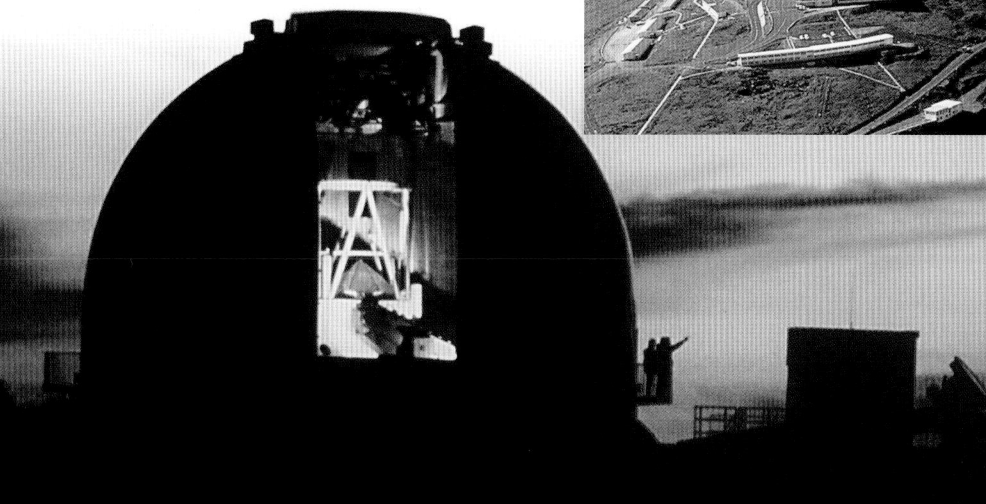

THE BIGGEST OPTICAL TELESCOPES IN THE WORLD

OBSERVATORY	INSTRUMENT	APERTURE (m)	CHARACTERISTICS
MAUNA KEA, HAWAII	KECK	10.0	36 SEGMENT MIRROR
MAUNA KEA, HAWAII	KECK II	10.0	INTERFEROMETRY OPTICAL
MONTE FOWLKES, TEXAS	HOBBY-EBERLY	9.2	SPHERICAL SEGMENTED MIRROR
MAUNA KEA, HAWAII	SUBARU	8.3	
CERRO PARANAL, CILE	VLT	8.2	4 INSTRUMENTS OF THE SAME DIAMETER
MAUNA KEA, HAWAII	GEMINI NORD	8.0	TWIN OF GEMINI SOUTH
CERRO PACHON, CILE	GEMINI SUD	8.0	
LA SERENA, CILE	WALTER BAADE	6.5	TWIN OF THE MMT
MONTE HOPKINS, ARIZONA	MMT	6.5	
NIZHNY ARKHYZ, RUSSIA	BOLSHOI TELESKOP AZIMUTALNYI	6.0	
MONTE PALOMAR, CALIFORNIA	HALE,	5.0	
LA PALMA, ISOLE CANARIE	WILLIAM HERSCHEL	4.2	
CERRO PARANAL, CILE	VICTOR BLANCO	4.0	
COONABARABRAN, AUSTRALIA	ANGLO-AUSTRALIAN	3.9	
KITT PEAK, ARIZONA	MAYALL	3.8	
MAUNA KEA, HAWAII	UKIRT	3.8	INFRARED ONLY
MAUI, HAWAII	AEOS	3.7	MAINLY MILITARY

RADIOTELESCOPES
THE VLA (*Very large Array*), the largest interferometric radiotelescope in the world, is in the Socorro desert (New Mexico): 27 mobile antennas, each 25 m in diameter, mounted on 21 km of track, in which the signals collected by each antenna are combined.

RADIO TELESCOPES

These are instruments that collect and analyse the radio waves emitted by space objects. The most common consist of a metal parabolic dish (called a reflector, receiver, parabola or just dish) that, just like the mirror of a reflector telescope, collect radio waves and converge them onto the feed antenna at the centre. The signal is then sent to a series of instruments that amplify, record and process it to obtained various types of information.

In this case too, the diameter of the dish is fundamental to guarantee the available signal quantity. And here too the limitations on size are technical and engineering in nature: the dish must always be pointed at the objects to be examined, changing its height above the base and the angle of inclination. Until a few decades ago the largest radiotelescope in the world, at Effelsberg in Germany, had a dish about 100m in diameter. But, as for optical telescopes, new techniques also opened the door to further solutions for radiotelescopes: for example, the Arecibo radiotelescope has a 305 m diameter dish the surface of which consists of panels which cover an entire valley in the hills of Puerto Rico ≻75. Since the reflector is fixed, the instrument is pointed by moving the antenna, suspended above the dish between three towers placed at its edges. Given the nature of radio waves, a radiotelescope has a much lower power of resolution than an (optical) telescope: a 30 m diameter radiotelescope, for example, has an angular resolution of for wavelengths around one metre. An instrument of this type can be used to determine the position of a celestial radio source with a precision of 2° if we consider that the apparent diameter of the Sun is about half a degree, it is clear that this "precision" is not precise enough. The problem is resolved by building radiotelescopes up to thousands of kilometres apart (intercontinental interferometers) and connecting their antennae: increasing the number of antennae increases the power of resolution of the instrument is increased far more tha increasing the surface of the dish does. This type of "composite" radiotelescopes are called interferometric radiotelescopes, and their angular

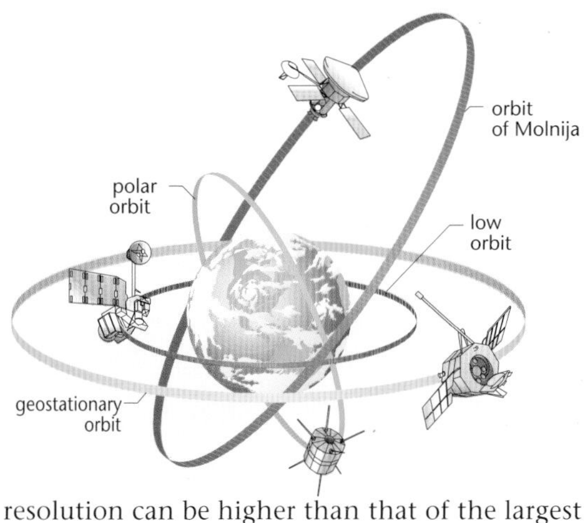

resolution can be higher than that of the largest optical telescopes.

INFRARED TELESCOPES

Infrared radiation (wavelength between 1,150Å and 1 mm) from space is mostly absorbed by the atmosphere: so the largest infrared telescopes are built on the tops of high mountains, installed on special high flying aircraft or balloons, or better yet on satellites orbiting the earth. However, atmospheric absorption is not the only obstacle to analysing this type of radiation on earth: the main problem, which also occurs in space, is to distinguish the signal collected from the "background noise", i.e. from the enormous infrared emissions of the Earth or of the instruments themselves, since any object which is not at absolute zero , (0K = -273.15°C) emits infrared radiation. So everything around the instruments produces "background" radiation, including the telescopes themselves! It is very difficult to produce a "thermograph" of a celestial body without measuring the heat in which the instrument is immersed: in addition to using special photographic film, the instruments must be cooled continuously, by immersion in liquid nitrogen or helium.

ORBITING TELESCOPES

The use of satellites has meant that the trickle of radiations of wavelengths longer (ultraviolet, X rays and gamma rays) or shorter (infrared, microwaves) than those of visible light previously observed has become a flood. Most of these radiations are in fact absorbed by the atmosphere of the earth ➢38, and to observe them the observer needs to be outside the thick layer of gas which covers our planet. Research satellites equipped with telescopes sensitive to different wavelengths can remain in orbit around the Earth or move away from it: Hubble ➢128, for example, is in a geostationary orbit; while SOHO is dedicated to the study of the Sun, and is in a heliostationary orbit. Satellites have enabled observations to be made that would have been impossible from earth: for example, it was discovered that infrared radiation crosses the large clouds of interstellar material more easily than visible light ➢212, and this makes it an important parameter for the study of "cold" space objects.

ORBITS
Shapes and average distances from Earth of possible geocentric orbits.

ARECIBO
A technician checks the enormous reflector of the Arecibo radiotelescope.

KITT PEAK
The solar tower at Kitt Peak, represented in diagram form on the left. Equipped with cooling shields and instruments maintained in a vacuum, it is also suitable for infrared studies of the Sun.
The celostat, present in all fixed telescopes, is a flat mobile mirror which tracks the sun through its trajectory projecting its light into the instrument.

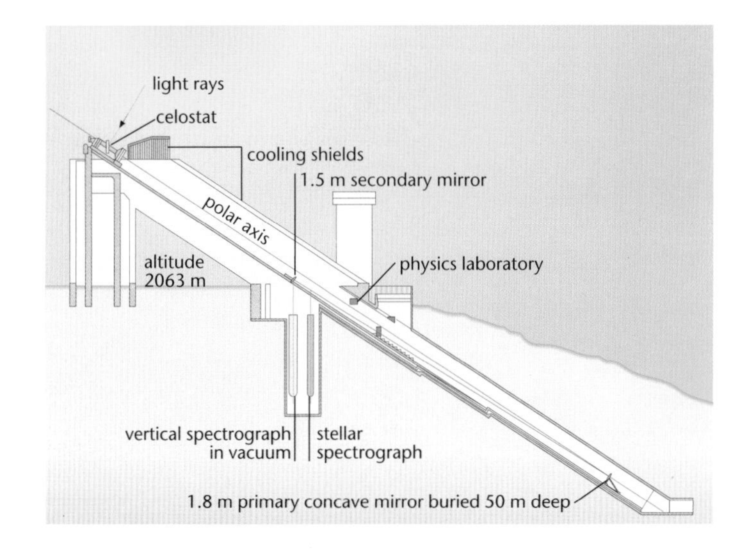

EXPLORATION OF THE SOLARSYSTEM

Lunar orbiter
One of the programme probes. The illustration shows the cloverleaf of solar panels,cameras and transmission antenna.

The space near Earth has never been so crowded with probes, satellites and instruments travelling to a range of destinations. At least 25 probes have left or are about to leave to collect information on the celestial bodies closest to us.
Most of these, predictably, are organised by NASA (National Aeronautics and Space Administration) which has already scheduled many low cost missions (relatively low cost, they still involve hundreds of millions of dollars) which will extend our knowledge of the Moon ➢42 , Mercury ➢158 , Venus ➢160, Mars ➢164, Jupiter ➢170, Saturn ➢174, Uranus ➢177, Neptune ➢179, the nearer comets ➢182 (particularly the soon to return Tempel-1), some asteroids ➢168 and solar wind particles ➢154. But the Russians too have made a considerable contribution to the intellectual conquest of the solar system, particularly in the past, sending probes that collected samples and information to several planets, often using remote controlled landing devices. While in recent years Japanese and European satellites have started to appear in space, although increasingly often large research projects are only accomplished through vast international collaborative efforts.

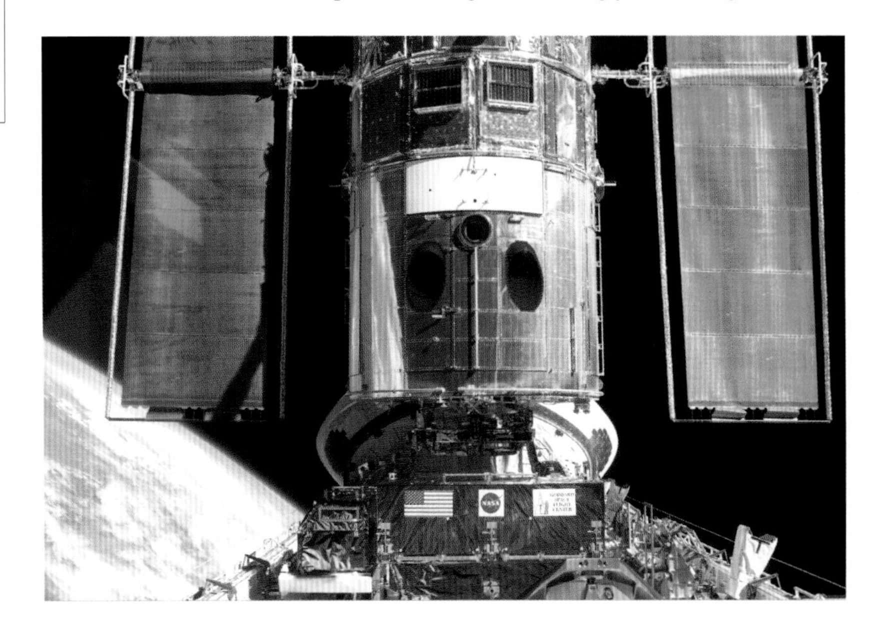

The Hubble
A detail of the Hubble Space Telescope during its installation in orbit around the Earth.

LUNAR PROBES

The Moon has been the "training ground" for all the space expeditions. Although the Apollo expeditions will remain a reference point (for now unsurpassed) in terms of our knowledge of the Moon, a great many probes have been sent to or around our moon. The first were the Russians, with the projects Luna (1956-1976) and Zond (1965-1970). There were a total of 29 missions, 20 of which were successful, and a fair number of firsts: the first probe to reach the lunar surface, the first to fly close to the surface and photograph the dark side, the first soft landing, the first self-propelled vehicle on the moon, the first collection of moon samples, the first space vehicle to orbit the Moon and return to Earth...
In the meantime the Americans did not stand idly by: after the Ranger programme (1964 - 1965) , in which the probes which crashed on the surface of the Moon sent back valuable information, they organized both the Mercury-Gemini-Apollo programme ➢56 and the Lunar Orbiter missions (1966-67), which accurately mapped the surface of the Moon, particularly those areas to be used for the Apollo landings, and Surveyor (1966-1968), which tested soft landings, photographed the Moon from just a few centimetres away, and collected soils samples for magnetic and chemical analysis.
In 1994, long after the glories of the Moon landing, NASA organised the Clementine mission, which continued the lunar mapping project, with greater attention tom the dark side, discovering the ice deposits at the lunar South Pole.
As for the Japanese: the Hiten (1990) satellite, launched by the Japanese space agency (ISAS) followed a very elliptical orbit around the Earth, and in one of the 10 passages it made near the Moon before deliberately crashing on the lunar

Luna 3
Launched on 4 October 1959, this Russian probe flew very close to the Moon and took photographs.

Skylab
The space laboratory in orbit around the Earth, filmed from an approaching module. The solar panels assure the necessary energy for all the instruments, whether for research or for the survival of the crew.

surface, it launched Hagoromo, a small satellite which remains in orbit around the Moon. Unfortunately a problem with data transmission evident before it went into orbit blocks all communications between this satellite and its base. NASA, in preparation for the probable colonisation of the Moon, continued the programme with Lunar Prospector (1998) a probe designed to enter into a low polar orbit and discover if there are subterranean ice deposits on the Moon. Lunar Prospector has also measured the magnetic and gravitational fields, studied the degassing of the lunar soil and undertaken mapping to allow detailed description of the superficial geological composition of the Moon. There are also instruments for astronomical measurements on board, including a gamma ray spectrometer, a neutron spectrometer and alpha particle spectrometer.
New missions have already been planned: SMART 1 (Small Missions for Advanced Research in Technology) built by the ESA (European Space Agency), which has been orbiting the moon since 2002 to test new technologies, Lunar A (ISAS) planned for 2003, consisting of an orbiting element and a drilling probe for geological investigations, and, finally, Selene (SELenological and ENgineering Explorer, ISAS), loaded with instruments and destined primarily to measure the gravitational field of the moon and to touch its surface.
However, the times when a probe was launched to explore a single celestial body, or study a sun or a particular group of phenomena are almost over: we are entering the era of space laboratories such as Skylab or the Hubble, and multifunctional probes like Galileo, which follows in the footsteps of the Pioneer and Viking projects.

SPACE LABORATORIES AND INSTRUMENTS

In reality the most recent space missions to study the bodies of the solar system share one aspect: they are in essence a concentration of technology, a condensation of novelties in the fields of new materials, miniaturisation and automation. Essentially they are small laboratories full to bursting point with instruments which are able to perform a thousand operations at the same time, and to send the results they obtain back to Earth, where someone is always listening (or, more accurately, some instrument is always there to record them). However, the most famous space laboratories are those which, remaining in orbit around the Earth, represent human "outposts" in space: Skylab (1973), and the Hubble Space Telescope (1990), which the professionals call the HST, and the rest of us just call the Hubble, and "Beppo" SAX.

THE SKYLAB

This was a true space laboratory inhabited by the successive crews of scientists involved in the NASA missions until February 1974. Now destroyed, it consisted of 5 functional parts. It was essentially a

Hubble Space Telescope
Two different stages during the assembly of the *Hubble*.
On the left: the final touches during installation in orbit.
On the right: a phase of the assembly on earth, in the enormous NASA hangar.

The *Hubble* in operation
In orbit around the Earth, the Hubble has been carrying out an intense research programme for some time.

space laboratory equipped with everything necessary for a crew to survive 84 days, for manoeuvring in space, and for research activities; every instrumental part was able to function automatically so that, they could be reactivated in orbit and controlled remotely.

THE HUBBLE SPACE TELESCOPE
This represents the first mission in NASA's Great Observers programme: it was in fact designed to develop a research area complementary to that will be made possible by the other elements planned for the programme, which will be built and sent into orbit at a later stage. The Hubble is in essence a 2.4 m aperture Ritchey-Chretien refractor telescope, able to perform observations in the visible, near ultraviolet and near infrared spectra, at a range of wavelengths from 1,150 Å to 1 mm in a fully automatic way.
Placed in orbit by the Shuttle, it was designed in modules so it could quickly be repaired updated or connected to new elements. And this modular concept has already proved its usefulness: after the launch a spherical aberration was noted in the primary mirror, and this was corrected by replacing the wide-angle planetary camera with a second generation device equipped with a corrective eyepiece, and replacing the high speed photometer with a COSTAR (Corrective Optics Space Telescope Axial Replacement), which cancelled the aberration recorded by the other instruments. The solar panels have also been replaced, so that the passage from the shade of the Earth to the light of the Sun no longer causes the pointing instabilities which had previously occurred.
Despite some difficulties, the Hubble works to capacity, and the images it has collected so far are truly spectacular, whether of the solar bodies we are already very familiar with or of the most distant galaxies, at the edge of the universe. And until instruments such as those of ESO had been developed, such precise images were unthinkable; for this reason the Hubble has been hailed as a marvel which, by showing details and phenomena which were previously unknown, has raised new problems and stimulated new hypotheses.

BEPPO SAX
Wholly designed and built in Italy, in collaboration with ESA and NIVR (the Dutch space agency), the SAX (Satellite per Astronomia a raggi X, or Satellite for X-ray Astronomy) can translate electromagnetic radiation from very distant celestial bodies into

Space Shuttle
The spectacular take-off of the Space Shuttle, the spacecraft that allowed space station and space laboratory crews to be easily replaced, and the direct launching in orbit of instruments such as the Hubble and various kinds of satellites.

images. It was the first satellite able to reveal the galactic sources of the whole range of X rays. It has already produced excellent results, the most important of which has been the discovery of a "replay "of the big bang, a phenomenon which liberated enough energy to shine brighter than the sum of all the stars in our universe in a few seconds, inside a galaxy over 12 x 10^9 light years away ➢226 A big bang phenomenon is thought to occur every million years in all galaxies, and is considered an event related to the formation of black holes ➢206. But SAX continues to supply important data to extend these studies.

PLANETS AND SATELLITES, ASTEROIDS AND COMETS

Like the Moon, the planets, some satellites, a few asteroids and several comets have been the destination of a series of specific space missions. In particular, interest in the smaller bodies of the solar system has grown in recent years: they are the NEOs (Near Earth Objects) , asteroids ➢168 which orbit along trajectories between the orbit of Mars and the orbit of Jupiter which are highly eccentric, and intersect the trajectory of the Earth, sometimes approaching it under the influence of the gravitational perturbations produced by the passage of planets and comets. We will briefly examine the most important missions completed to date, and those planned for the future.

MERCURY

The Mariner 10 mission (1973), which supplied almost all of the information currently available on the planet closest to the Sun ➢136, was one of the most proficuous space expeditions ever.
The probe passed within 6,000 km of Venus, so that the planet's gravitational field modified its trajectory and brought it into a heliocentric orbit around the Sun. Mariner 10 approached Mercury three times before it stopped functioning.
The onboard videorecorders collected a collage of images that have allowed the appearance of about 40% of the surface of the planet to be reconstructed.

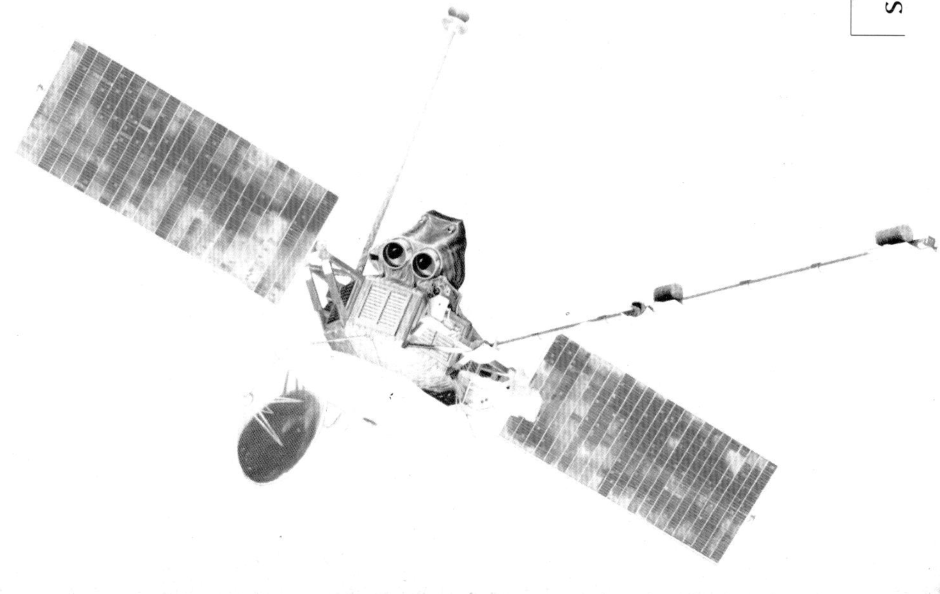

Mariner
The Mariner mission organised by NASA included both a certain number of probes to Mercury, which also collected data on Venus, and probes to Mars.

Magellan
The probe has just been housed in the final stage of the rocket which will transport it out of its terrestrial orbit.
3 days to go to the launch.

Mars Pathfinder
A moment during preparation of the fixed structure of the expedition landing craft: showing the tetrahedral structure destined to open and liberate the Rover, and enable instrumental recordings and photography.

Pioneer Venus
This mission involved two separately launched components, this module, intended to enter orbit, and a Multiprobe able to launch a series of probes onto the planet to collect specific data.

An infrared sensor measured the daily temperature changes, and a magnetometer quantified the presence of a relatively consistent magnetic field, which supported the hypothetical existence of a large metallic planetary "core".
No further information has been obtained since then, there have been no other expeditions: the data collected by Mariner 10 were sufficient for 30 years of studies.
But now that knowledge of the other planets has advanced, there are plans to return to Mercury and examine it with more modern technologies: in the next few years the launches of Messenger (NASA, 2004), Mercury Orbiter (ISAS, 2009 and Bepi-Colombo (ESA, 2009), all probes destined to go into orbit around the planet, are planned. In particular, the Bepi-Colombo mission, named after the Italian scientist who planned the travel times of Mariner 10 and the "slingshot effect" of Venusian gravity to perfection, will not only collect further data on the magetosphere of Mercury, it will also launch a landing module for the physical, optical, chemical and mineralogical analysis of the "soil".

VENUS

The Russians were again the first to become interested in Venus ➢160: during the Venus programme (1967-1983), in addition to collecting data on the planet's atmosphere, remote controlled landing modules were used to transmit precious information and photographs on the environmental conditions under the thick layer of cloud back to Earth. The last probes of the Venus series 14 and 15 (1984), renamed Vega 1 and Vega 2, were destined to fly by the planet and then continue towards Halley's comet. Each probe launched two modules and two balloons to test the atmosphere of Venus about 50 km above the surface. American missions carried out a similar programme of data collection: probes Mariner 2, 5 and 10 (1962-1973) flew close to Venus and collected information, but the Pioneer Venus mission (1978) was the first to conduct more extensive investigations, remaining in orbit and launching small modules to study the planetary surface. This was followed by Magellan (1989), equipped to scan the surface of the planet even through the thick later of cloud, thanks to a radar system which analysts on earth could use to create the most detailed and extensive map of the planet possible. And for the moment no further expeditions are planned.

MARS

This is probably the planet most targeted by probes from Earth: maybe because of the curiosity about it stimulated by the enormous number of science fiction stories about it, maybe because it is relatively close, or maybe

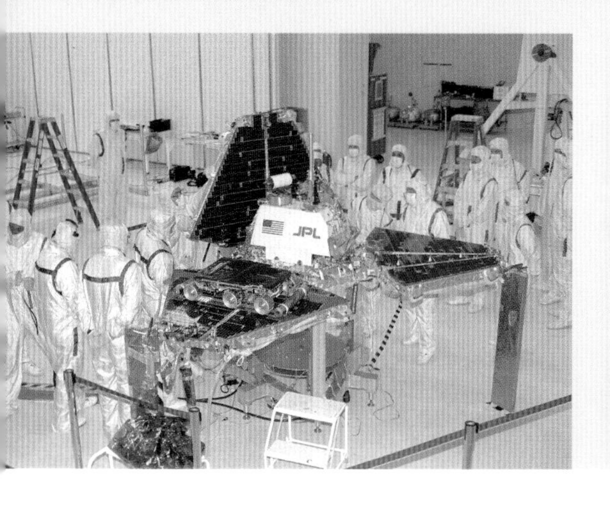

Airbags on mars
The resistance of the "airbag" cover destined to absorb the shock of landing of the Mars Pathfinder module is tested on an artificial Martian surface.

because it is more "credible" as a planet which could sustain life (or might have done so in the past). Whatever the reason, there have been many missions to Mars ➢164, and in future it will attract even more attention, if the colonisation projects go ahead. The United States holds the record on Mars: probes Mariner 4,5,7 and 9 (1969-1971) were the first to get there, passing very close to the planet. The Russians followed – after early failures in 1969, the Mars missions in 1971 and 1973 orbited and landed on the planet, collecting important information.
But it was the United States' Viking programme (1975-1982) that made the real leap forward in terms of the quality of information available on the red planet. Consisting of two spacecraft (Viking 1 and Viking 2), each made up of an orbiting module and a landing module, this mission collected high resolution pictures of the Martian surface, characterised the structure and chemical composition of the surface, and looked for evidence of the presence of life-forms and an atmosphere. The principal contributions of this mission were over 1,400 detailed photographs, a large number of accurate mineralogical investigations and a series of tests which provide clear indications that water was present on the planet in liquid form in the past.
The Russians followed a little later with the Phobos Project, which also involved two spacecraft (Phobos 1 and Phobos 2), fruit of the technical knowledge acquired in the Venus-Vega experience. And although Phobos 1 lost contact with Earth almost immediately, Phobos 2 collected data on the interplanetary medium and, then, adopting an orbit around Mars, on Mars itself and its satellite Phobos ➢167. However, before starting the last phase of the experiment, which was to land a probe, the second Phobos spacecraft also lost contact with Earth.
Ten years after the Viking mission, NASA organised the Mars Observer (1992) , Mars Global Surveyor (1996) and Mars Pathfinder (1996) missions, which collected an amazing amount of data. In particular, the two Pathfinder robots that reached the surface of the planet (a stable automatic base and a semimobile explorer), contributed to collect both scientific data and "logistic"

Panorama of mars
The camera installed on the fixed landing module filmed this 360° panorama of the surrounding Martian landscape. The Sojourner rover, or robotic exploration vehicle, is visible; it has left the module and already travelled several metres to a large rock.

Left: The rover during remote-control tests. The Mars Pathfinder mission cost a total of around 265 million dollars.

Chasing comets
An artist's impression of NASA's Stardust mission to collect material from the tail of Comet P/Wild 2 as it approaches its objective.

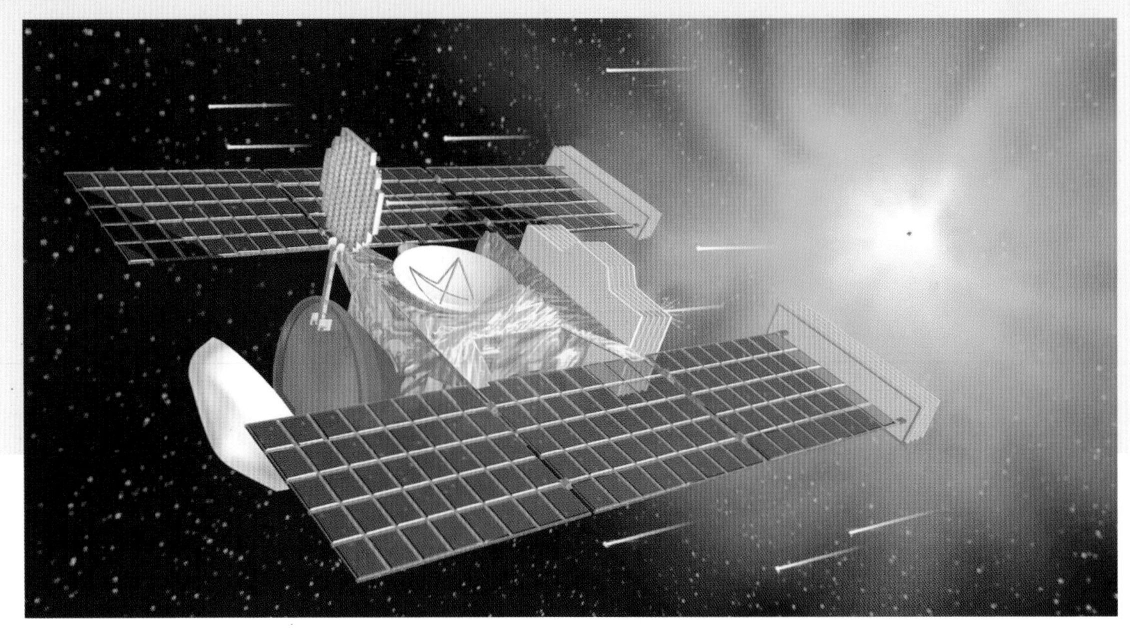

Below: A model of the Giotto spacecraft, which made many observations of the nucleus of Halley's comet.

information, in view of new landings. At the same time, the Russians also undertook a new mission: Mars 96 (1996), consisting of an orbiting module and a landing module, although the latter was not able to enter a trajectory to Mars and crash landed about an hour after its launch.
Further United States missions followed: the Mars Climate Orbiter and Mars Polar Lander, as well as Deep Space 2 (NASA, 1999), were destined to study the Martian climate, the availability of water and carbon dioxide and the characteristics of the soil, using a device designed to penetrate the surface of the planet. But again, the missions failed: they lost contact due to orbit errors as soon as they reached the higher atmosphere of the Earth, or close to Mars. They were followed by the Nozomi (ISAS, 1998), currently in a heliocentric orbit and waiting to meet the red planet at the end of 2003: it will study the higher atmosphere and its interactions with the solar wind. Finally, there is the 2001 Mars Odyssey (NASA, 2001), which will orbit the planet for at least 3 years, producing a complete map of the planet and performing detailed analyses of its mineralogy, radioactivity and environmental conditions. This satellite is also expected to be used for telecommunications for future manned expeditions to the planet.
To prepare for a manned landing, which should take place this century, nine missions have been planned: Mars Express ESA), Mars Surveyor 2003 (NASA, 2003), and Mars Surveyor 2005 (NASA, 2005), all consisting of an orbiting and a landing module, designed to conduct geological and mineralogical investigations, investigate the global atmospheric circulation its interactions with the surface of Mars, the climate and meteorology of the planet, and the logistic conditions of the area in which the explorers' camp will be set up.

ASTEROIDS AND COMETS
Whether or not it is thanks to recent disaster movies, or to studies of the extinction of the dinosaurs, but it's a fact that recently investments in explorations to study the minor bodies of our solar system have increased considerable. Until a few years ago scientists focussed most of their attention on the planets or, at most, on their satellites: in fact, the first mission to explore the NEOs (Near Earth Objects) ➢168, was NEAR (Near Earth Asteroid Rendezvous), organized by NASA in 1996, when we already knew a lot about all the other components of the solar system.
Although not designed for landing, the first probe survived its impact on Eros, one of the more important NEOs and, with its camera and gamma ray spectrometer pointed at the ground, solar panels, antennae and other instruments pointed at the Sun and the Earth, it still transmits data and images: data on the magnetic field of the asteroid and the surrounding space, and on the near infrared spectrum of the radiation collected.
It is envisaged that this probe will continue to orbit with Eros, transmitting to Earth important information on the space travelled through by the asteroid during its long journey around the Sun.The first Deep Space mission left two years after NEAR: this is a probe equipped to measure radiation of various wavelengths and the chemical composition, geomorphology, dimensions, and characteristics of the rotation movements and atmospheres of the asteroid Braille and then of the comet Borrelly.
The comets ➢182 have always attracted more interest than the asteroids, and many probes have been sent to their nuclei or to sample the material in their tails: ICE (NASA, 1978) sent to the comet Giacobini-Zinner; Giotto (ESA, 1985) which reached the comets Halley and Grigg-Skjekkerup; the probes Sakigake and Suisei (ISAS, 1985), which were also

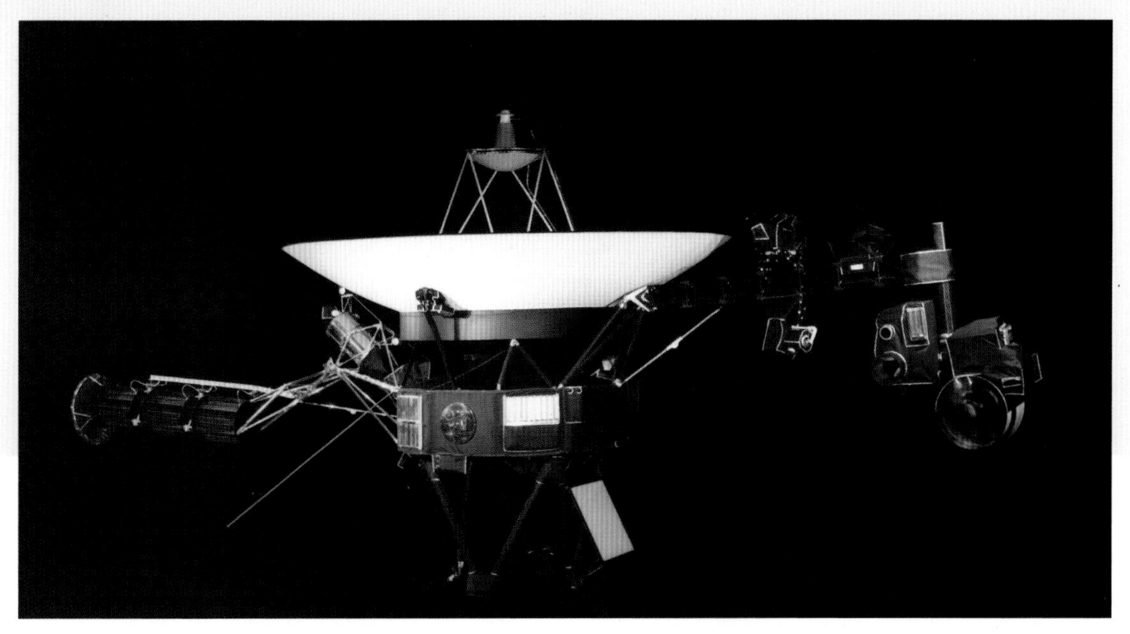

Voyager
A model of the Voyager 2 probe, which has travelled beyond the known frontiers of the solar system.

sent to Halley's comet, and, finally, Stardust (NASA, 1999) destined to collect samples from the tail of Comet P / Wild 2.
Over the next few years a new research expedition to the asteroids and three expeditions to the comets are planned: the NEAP mission (Near Earth Asteroid Prospector, NASA 2002-2005) will go to Nereo, another NEO, and after the departure of the Contour mission (NASA, 2002) which will approach three different cometary nuclei, further missions include the Rosetta (ESA, 2003), which will go to Comet P/Wirtanen, and Deep Impact (NASA, 2004) which will send a probe to comet Tempel 1 and continue to record data until it crashes on the nucleus of the comet, sampling the surface.

OUTER PLANETS

The first missions to the outer planets of the solar system were Pioneer 10 (NASA, 1972) and Pioneer 11 (NASA, 1973): they were the first to cross the dangerous ring of asteroids to reach and photograph Jupiter ➢170, and then continued to the outer reaches of the solar system. Their task is to collect data on the interstellar medium, and on the high energy particles found in the outer heliosphere to determine the distance at which the solar wind is no longer perceptible. Pioneer 11, in its journey to the outer edges of the system, passed close to Saturn ➢174 and collected data and images. The two spacecraft are now travelling in space, towards the constellations Taurus (Pioneer 10, currently more than 14 x 10^9 km from Earth, will reach the first star in about 2 million years), and Aquila (Pioneer 11 will reach this constellation's first star in about 4 million years).
The Pioneer project was followed by the first launch of the Voyager programme (NASA, 1977), recently renamed the Voyager Interstellar Mission: another two probes (Voyager 1 and Voyager 2) are destined to approach the outer planets and leave the solar system.
Designed primarily to study and collect data on Jupiter and Saturn, and their satellites, magnetosphere and the interplanetary medium, their destinations have been modified as work has progressed: while the first probe has approached Jupiter and Saturn and is proceeding to the edge of the solar system, Voyager 2 was redirected so it could also analyse the other giant planets and their satellites, Uranus ➢177 (reached in 1986) and Neptune ➢179 (1999).
Its thanks to the data collected by this probe that we now know a lot about these planets: previously, we knew almost nothing about their temperatures, surface winds or magnetospheres.
Voyager 2, which did not pass sufficiently close to Pluto ➢181 or Caronte to extend our knowledge of these bodies, is now deep in outer space,

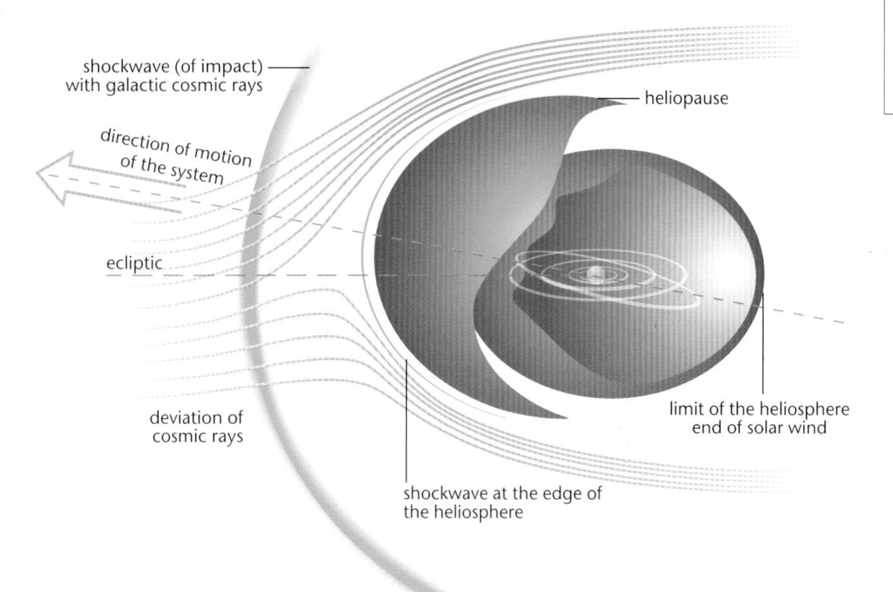

Beyond the solar system
The Pioneer and Voyager probes are travelling towards the exterior of the solar system: their current mission is to check its limits. The image incorporates current knowledge of the galactic environment near the heliosphere, developed by European scientists on the basis of the data collected in these missions and on the decisive data from the Hubble.

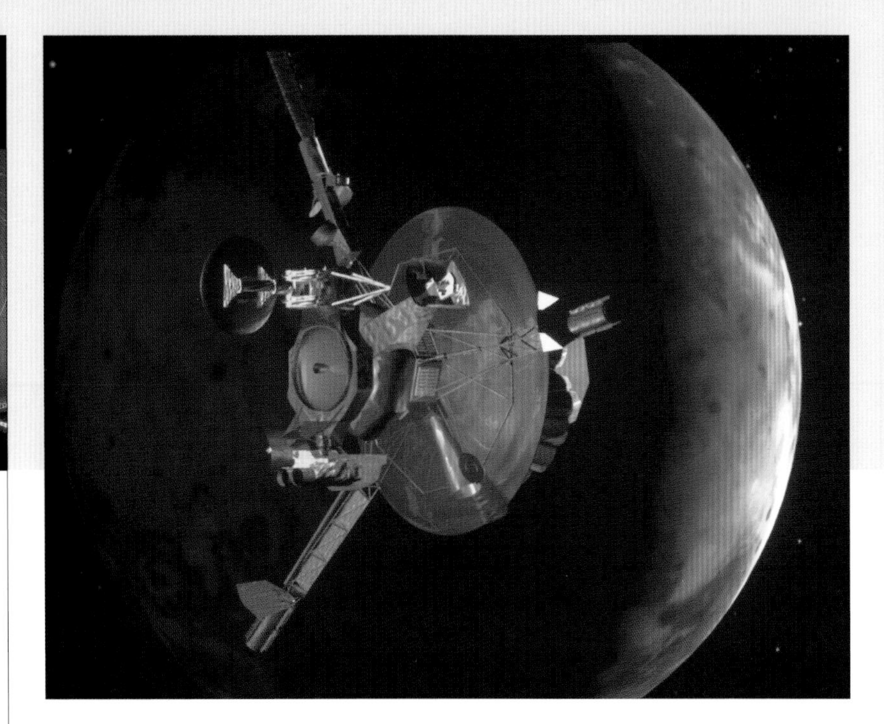

Galileo
An artist's impression of the Galileo mission approaching Jupiter

Cassini-Huygens
This mission, organised by the ESA, will reach Saturn in June 2004.

continuing its task of sampling and studying the solar wind in search of the heliopause.

We will have to wait 12 years before the Shuttle launches another probe to Jupiter: NASA's Galileo mission consists of an orbiting module and a module for atmospheric sampling. However, before it reached Jupiter, Galileo passed close to Venus, using the "slingshot effect" of the two planets to increase its speed. It collected previously unseen images and data on Venus and the Moon, and then on to the asteroids Gaspra and Ida, as it passed through the asteroid belt. And to crown it all, it also collected data on the spectacular crashes of fragments of the comet Shoemaker-Levy 9 on Jupiter. Entering into orbit around the large planet, Galileo released the atmosphere module, which is determining the composition, structure, depth and characteristics of the atmosphere of Jupiter, including its formations, radiation balance, charged particle flows and lightning development. The results have been of fundamental importance, and it is still active.

Another expedition to Saturn and Titan ➢176 was organised by ESA in 1997: Cassini-Huygens, as the orbiting and landing modules are called. While the first module will orbit around Saturn, the other will land on Titan. Their task is to determine the structure and dynamics of the rings, atmosphere and magnetosphere of Saturn, the composition and geology of Titan, and the variability and surface characteristics of its clouds.

***ISEE* probes**
This mission, organised by NASA, collects information, particularly about the solar wind and the inter-relations between the Earth and the Sun.

SOLAR PROBES

In recent years there have many large-scale projects for the exclusive study of the Sun ➢136. We summarise their principal characteristics below.

ISEE

The International Sun-Earth Explorer, launched in 1978, was part of a NASA mission for three different elements to execute a series opf studies from a heliocentric orbit. The first tasks were to examine the inter-relations between the Sun and the Earth, the borders of the earth's magnetosphere and the structure of the solar wind near our planet, and to study plasma motion and mechanisms, conducting investigations of cosmic rays and sun spot emissions. Renamed ICE (International Cometary Explorer) it was launched through the tail of the comet Giacobini - Zinner.

SMM

Solar Maximum Mission, the first satellite repaired in orbit by the Skylab crew, was launched in 1980. After 10 years of activity in which it collected an enormous quantity of information on the solar cycle, recording hundreds of phenomena in various wavelengths (including X rays and gamma rays), it disintegrated over the Indian Ocean because a violent sun storm lowered its orbit. Among other things, the data collected allowed scientists to detect the slight variation in the solar constant during the eleven year cycle of the Sun.

ULYSSES

Produced by the ESA (1990), this probe was destined to measure the properties of the solar wind, the interplanetary magnetic field, the galactic cosmic rays and the neutral interstellar gas, as well as determining the composition and acceleration of the energy particles at different solar latitudes.

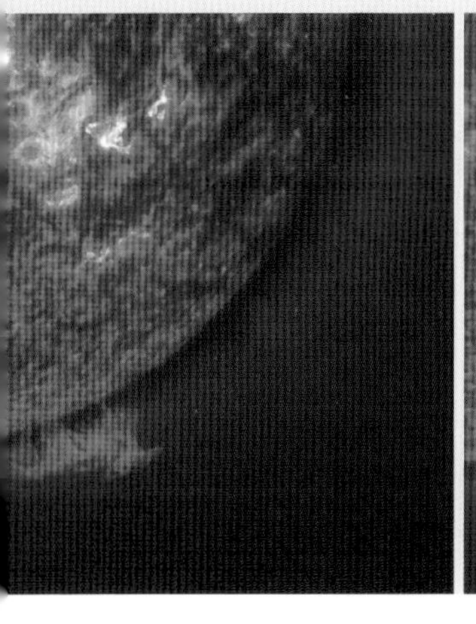
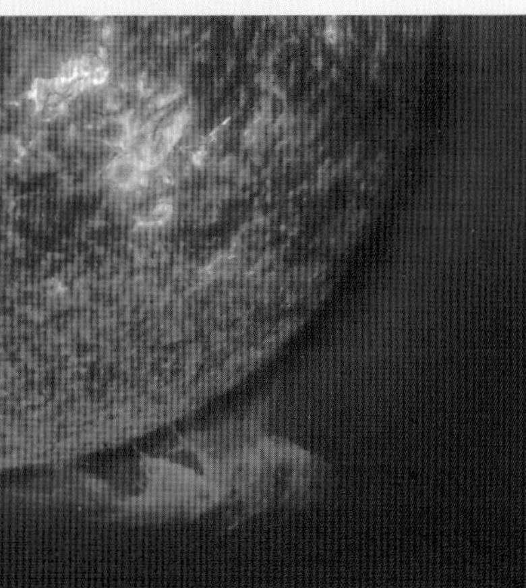
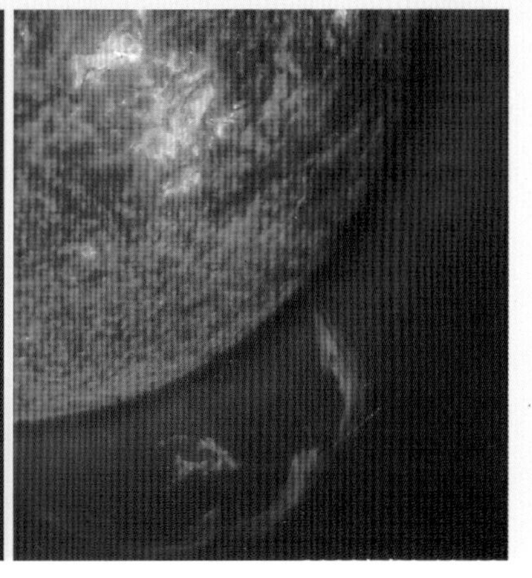

SOHO
An exceptional sequence of images collected by SOHO at intervals of 6 hours, showing the development of a solar protuberance.

Equipped with two magnetometers, two instruments to collect and analyse solar wind plasma, three instruments to analyse charged particles, a sensor for interstellar neutral gas and a sensor for cosmic dust, it would have extended our knowledge of the solar corona and also conducted studies of gravitational waves.
Exploiting the "slingshot effect" of Jupiter, it entered a very steep heliocentric orbit, flying over the South Pole (1994, 2000) and North Pole (1995, 2001) of the Sun at a distance of about 2 UA.

YOHKOH
The result of a collaboration between ISAS and NASA, and launched in 1991, this probe, equipped with two spectrometers and two telescopes, collects data on high energy radiation produced by sun spots and on the calm conditions on the sun's surface which precede a sunspot.

SOHO
A joint ESA-NASA project, SOHO (Solar and Heliospheric Observatory) was launched in 1995. Destined to study the physical processes which form and heat the solar corona, expand the solar wind, and characterize the interior of the Sun, it is equipped with 12 different instruments which produce a continuous flow of data. It has already produced incredible photographs and films which show the solar "earthquakes" produced by sun spots, collected an enormous quantity of data, discovered "rivers of plasma" under the surface of the Sun, and a magnetic "carpet" on its surface which is thought to provide energy sufficient to heat the corona and many comets invisible from the Earth. After some flight problems, and losing contact for a few months, SOHO is again fully active.

GENESIS
The primary objective of this probe, the fifth to be launched as part of NASA's Discovery programme (2001), is to collect samples of solar wind particles and bring them back to earth for further isotopic and chemical analysis. The study of these samples should provide further information to confirm current theories about the origin and evolution of the solar system and the composition of the primitive nebula. About 200 million dollars have been spent to collect 10 to 20 mg of material.

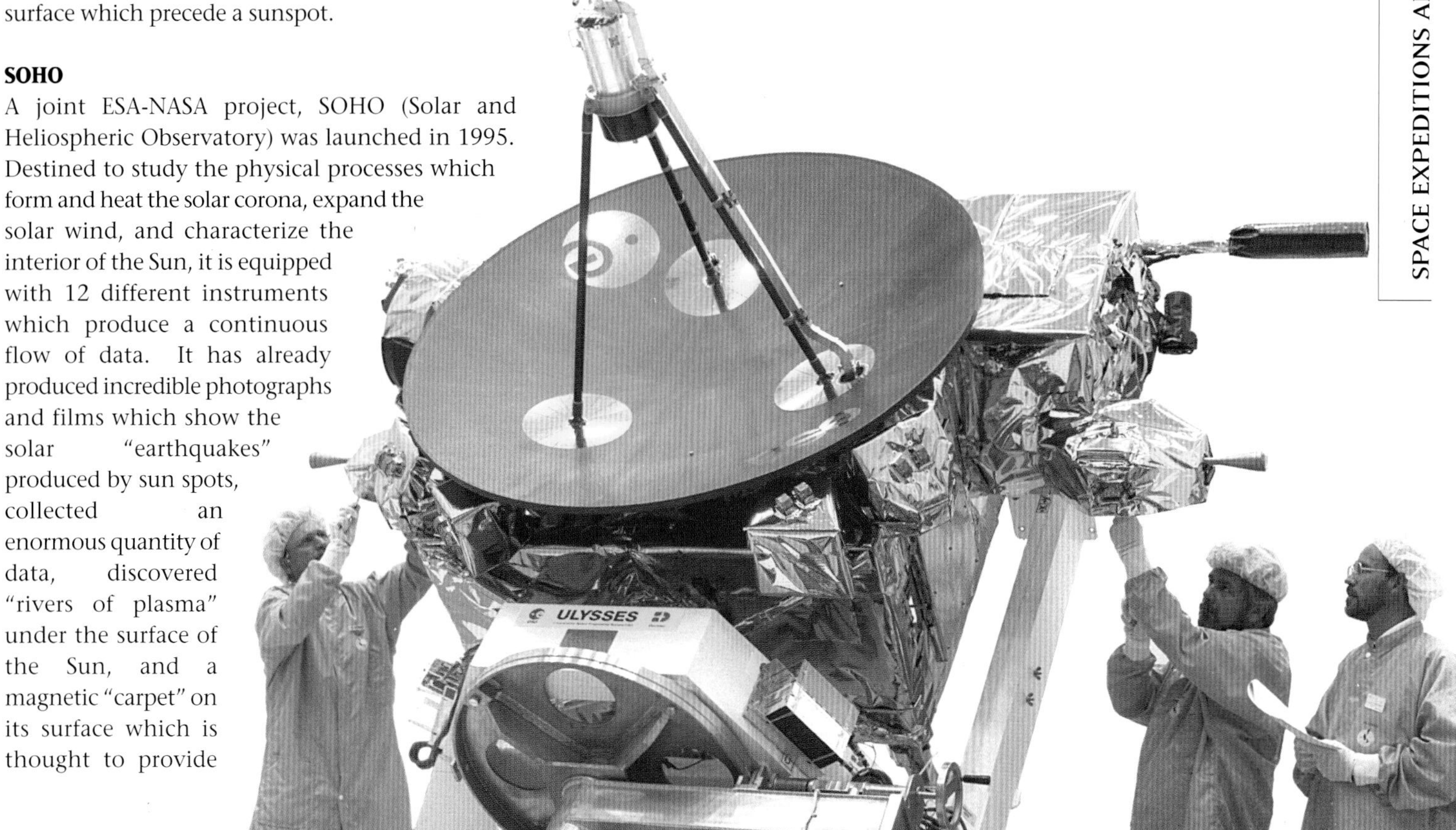

Ulysses
This probe, designed to study the Sun, has already flown twice over the North and South Poles of our star, collecting essential data on its magnetic field.

Since earliest times man has been aware of the fact that the Sun was the origin of life on Earth, and is still. Ancient civilisations, dependent on agriculture, fishing and hunting and therefore on the climate and seasonal cycles, transformed the Sun into one of the most important divinities ever known. The planets were also gods, each having a hidden tie with different human functions and activities, and the course of man's destiny could be gleaned from their motion. For a long time, therefore, astronomical observation was limited to the study of the apparent motions of the celestial bodies and it was only in the seventeenth century that the motions of the Sun and its planets began to be observed in a properly scientific manner. Galileo was the originator of those studious observations. Though he was not the first to observe the heavens, he was the first to do so in a methodical manner. His descriptions of the spectacular sunspots and solar phenomena caused a sensation, as did his vivid descriptions of lunar mountains and plains and the discovery of Jupiter's satellites. Now, barely a few centuries later, with the perfection of observational technologies and the invention of new instruments, we already know a lot more about our solar system. Space probes have truly revolutionised the way in which we can now observe the bodies closest to Earth. From these we know that Venus, contrary to previously held beliefs, is the hottest planet and that Jupiter and Saturn give out more energy than they receive from the Sun. The outer planets have numerous rings and an equally numerous court of satellites, Pluto and Charon are double planets or in fact a cometary body. The confines of the solar system broaden to Kuiper's belt and beyond, carried on the breath of the solar wind.

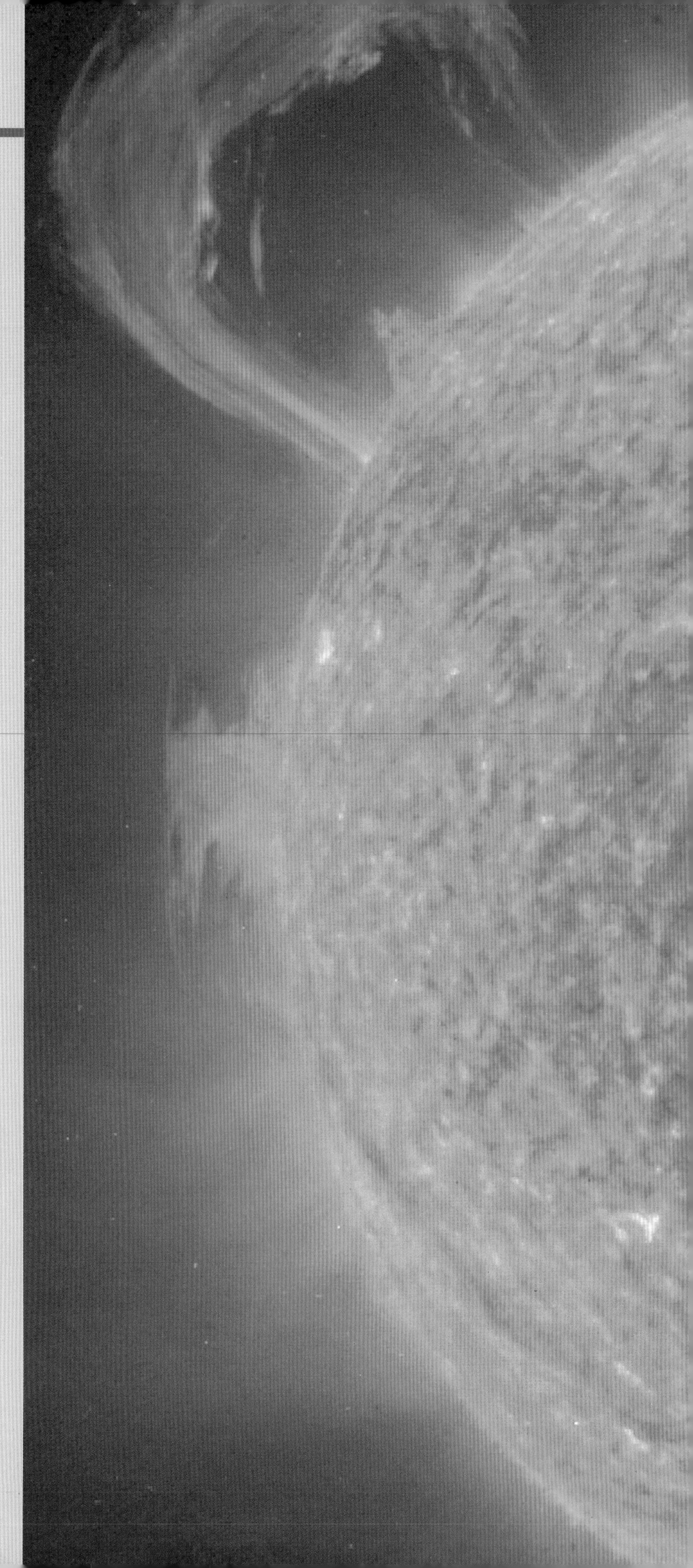

The Sun and its planets

OUR STAR

After Galileo injured his eyes while observing the Sun, our star ceased to remain perfect. It has cycles of dark sunspots; it turns on itself and, on careful observation, it can be seen that it emits very high flares, lightning and light spots... It was not until the 19th century when discoveries regarding the atom were made and Einstein's theory of relativity publicised that the link between matter and energy was established; that solar activity could be studied in a quantative manner and the observations explained by theories. Today much is known about the Sun's dynamics, and even though there is still much to be learned, what we know already helps greatly in understanding all the other stars. Though the Sun holds a special place of fascination for us, it is not in fact that special. It is not huge, like a giant star, neither is it small as a white dwarf, or unstable like a nova...It will continue its placid existence for a long time yet.

The Sun is very large. The visible surface is a sphere some million and half kilometres in diameter with a mass of 2,000 billion billion billion kilograms. It appears small to us (it has almost the same apparent diameter as the Moon, which explains the occurrences of eclipses) only because it is very far away, about 150 million kilometres. For the same reason the smallest details observable from Earth are some 150 km in extent. The details that are to be seen are many. The so-called photospheric or surface phenomena: spots, flares, granulation, prominences, faculae... all of which had already been observed in the past, with some very credible hypotheses to explain their origin. But our firsthand knowledge regarding the Sun is limited to its radiation, registered by the most powerful and modern equipment on Earth, or else registered by yet more sophisticated equipment mounted on astronomical satellites in close orbit around our star. Thanks to the Solar Maximum Mission, or to SOHO (SOlar and Heliospheric Observatory), solar light can be analysed with methods that at one time were simply unthinkable, allowing us to either confirm theories or formulate new ideas about the internal structure, activity and surface dynamics of the Sun.

AVERAGE MASS AND DENSITY OF THE SUN

Mass has been calculated on the basis of Newton's universal law of gravitation. Considering the Earth's negligible mass compared to that of the Sun, knowing the distance between the two bodies ($1{,}496 \cdot 10^8$ km) and the time taken by the Earth to complete its orbit (a sidereal year is about 3.16 · 107 s), the Sun's approximate mass can be calculated. The value obtained in this way is 1.99×10^{30} kg which, taking advantage of the information available from the satellites in orbit, does not require significant adjustment. Once the mass is calculated, knowing the Sun's radius

Sun
An image of the Sun, taken from the Solar and Heliospheric Observatory (S O H O), shows a huge prominence.

SOHO
With the advent of satellites orbiting closely around the Sun, essential data has been gathered enabling the study of the conditions of the solar poles, the Sun's internal structure and emissions which are otherwise screened out by the Earth's atmosphere. The S O H O satellite became operational in 1996.

Atmospheric screen
The chart shows how the Earth's atmosphere progressively absorbs solar radiation at different wavelengths.

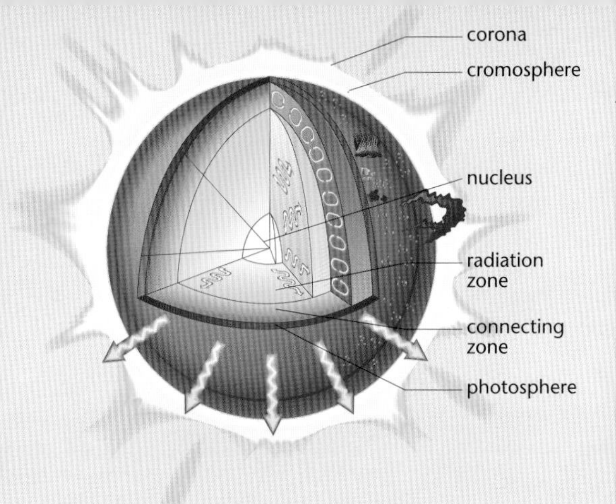

FACTS ABOUT THE SUN

AGE (IN MYA)	5
EQUITORIAL RADIUS (IN 10^5 km)	7
APARENT DIAMETER FROM EARTH	31'
MASS (IN 10^{30} kg)	1.99
• COMPARED TO EARTH	3.32×10^5
AVERAGE OR ESTIMATED DENSITY (g/cm^3)	1.409
• OF THE PHOTOSPHERE	8×10^{-8}
• OF THE CONVECTION ZONE	6×10^{-3}
• OF THE RADIATIVE ZONE	2
• OF THE CORE	160
• COMPARED TO EARTH	0.26
AVERAGE OR ESTIMATED TEMPERATURE (IN K)	
• OF THE PHOTOSPHERE	6×10^3
• OF THE CONVECTION ZONE	6×10^5
• OF THE RADIATIVE ZONE	4×10^6
• OF THE CORE	15×10^6
STELLAR CATEGORY	G2, 2nd GENERATION
AVERAGE OR ESTIMATED PRESSURE (IN kPa)	
• OF THE PHOTOSPHERE	10^{-3}
• OF THE CONVECTION ZONE	10^1
• OF THE RADIATIVE ZONE	3×10^{13}
• OF THE CORE	3×10^{14}
SURFACE GRAVITY (IN 10^4 cm/s^2)	2.7398
• COMPARED TO EARTH	28
DISTANCE FROM EARTH (IN 10^8 km, MEAN VALUE AT INFERIOR CONJUCTION)	1.496
ESCAPE VELOCITY FROM SURFACE (IN km/s)	617.7
ANGULAR VELOCITY (LAT 17° IN 10^{-6} rad/s)	2.87
EQUATORIAL ROTATION PERIOD (IN d)	25 APPROX
EQUATORIAL INCLINATION TO ECLIPTIC	7°15'
LUMINOSITY (IN 10^{33} erg/s)	3.826

(about 700,000 km) the average density can be calculated to be some 1.4 times that of water..

SOLAR RADIATION

The energy produced by internal nuclear reactions within the Sun is propagated throughout the surrounding space in different forms:
• in the form of heat, which is transmitted in solar gas both through the increase of the motion of the particles that make up plasma, as well as through the insurgence of immense convection currents;
• in the form of sound, like the shattering noise of a supersonic plane, as a result of the rapid movement of masses of gas;
• above all, in the form of electromagnetic radiation, a chain of oscillations originated by the intensity of the electric and magnetic fields which disperse in space after being generated within the Sun, like ripples in a pool of water which radiate from the point where a stone has been thrown.

The electromagnetic energy produced by the Sun propagates itself in empty space: travelling in a straight line, it reaches the Earth and travels beyond, dispersing into space. Various groups of wavelengths with various characteristics can be discerned in the radiation given out by the Sun:
• gamma rays (γ) and X rays: the radiations most rich in energy, with wavelength less than 10^{-9} m;
• ultra-violet rays: with wavelength between 4×10^{-6} m and 10^{-9} m;
• visible light: radiations with wavelength between 4 and 8×10^{-6} m: blue and violet light (10^{-6}m), green and yellow light ($5\text{-}7 \times 10^{-6}$m), red light ($8 \times 10^{-6}$ m);
• infrared: with wavelength between 8×10^{-6}m and 8×10^{-2} m;
• microwaves: with wavelength between 8×10^{-2} m and 100 m;
• radio waves: radiations with wavelength greater than 100 m.
Not all the radiations that fall to Earth can be picked up by instruments: the more energetic (γ, X, ultraviolet) and the less energetic (infrared, microwaves, radio) are to a large extent absorbed or reflected by the atmosphere.

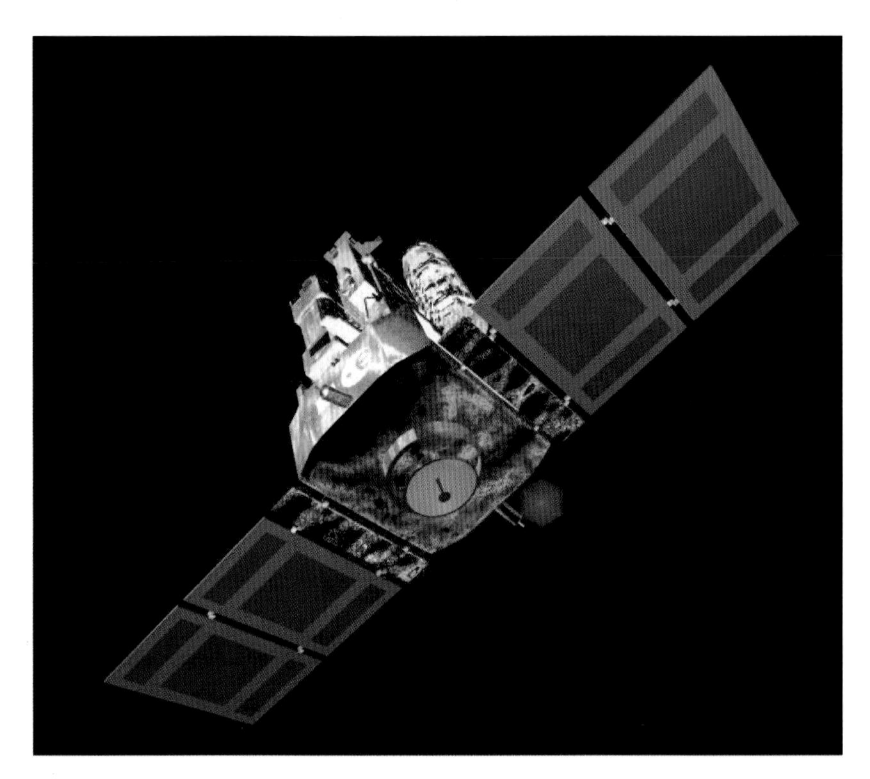

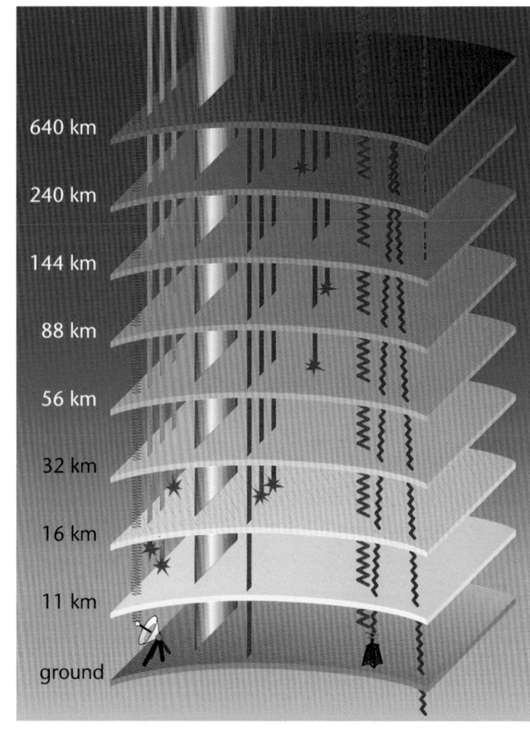

For years the only part of the Sun that was well known was its surface area, stirred by eruptions and flames, perturbed by enormous groups of sunspots and dominated by immense prominences.

Today, thanks to remarkable technical progress, even the deepest recesses of the Sun can be scrutinised anew with new instruments that put theories regarding its internal structure to the test.

STRUCTURES AND PHENOMENA

Based on past observations, spectral studies and the most reliable atomic physics and astrophysical theories, solar astronomers and physicists have developed theoretical models that are able to coherently explain all solar phenomena and characteristics. According to the model most currently in use, various concentric 'shells' can be discerned inside the luminous ball that we call the Sun, with sufficiently homogenous physical characteristics to enable them to be easily identified. Starting from the centre of our star, we can recognise:

- a nucleus, or core, with a radius of some 150,000 km. About 40% of the entire solar mass is concentrated in this area and density is at its maximum (on average about 160 g/cm^3): pressure reaches 3 x 10^{11} kPa and temperatures 1.5 x 10^7 K. It is here that the spontaneous thermonuclear fusion reactions of hydrogen to helium can be triggered: this nuclear furnace has already 'consumed' 40% of the original hydrogen (which constituted about 75% of the mass of the nucleus);
- a radiative zone extending up to 450,000 km from the centre of the Sun, with a thickness of about 300,000 km. Still in the realm of theory, this area is characterised by density and pressure values that are much lower than those hypothesised

THE SUN'S RADIATION

Solar radiation - the only means we have of obtaining information about our star - is made up of a vast range of wavelengths (continuous spectrum [195]) with limited areas in which the radiation is much less intense (absorption lines [195]). The diagram below shows the division of solar radiation in large regions according to the wavelengths, with particular attention to the visible part. The diagram above shows the atmosphere's selective absorption of this radiation. As can be seen, the capture of a large part of the radiation most rich in energy is due to the ozone (O^3), while in the infrareds much of the radiation is absorbed by vapour.

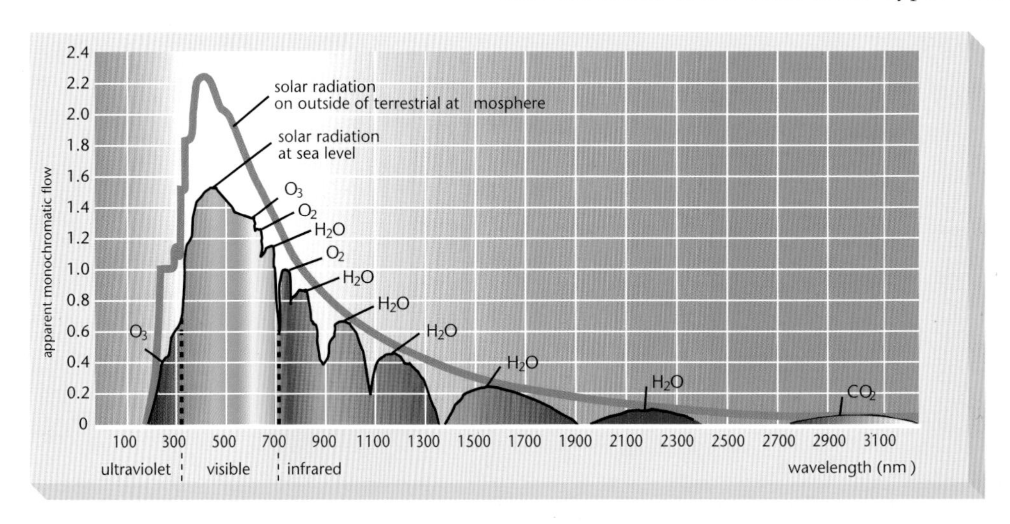

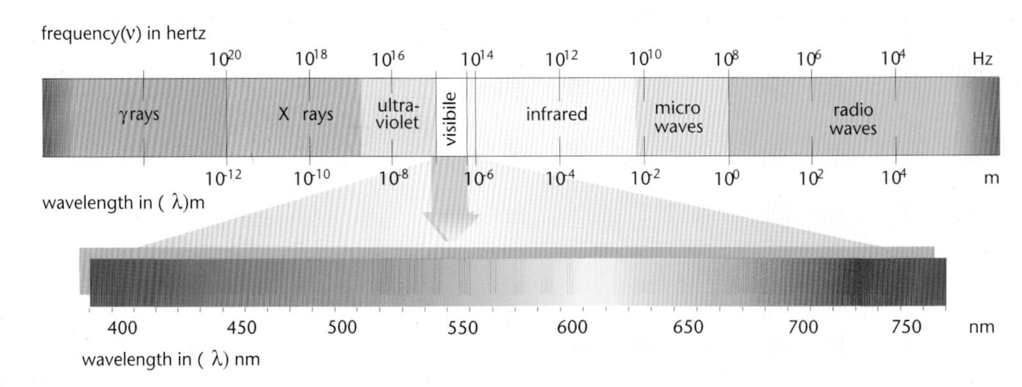

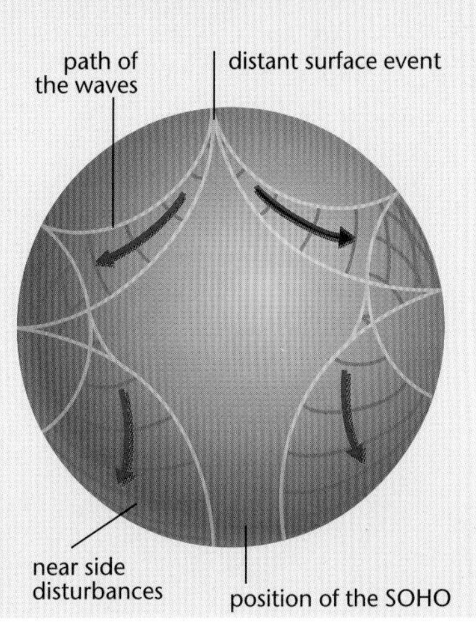

THE SOUND OF EXPLOSIONS
The SOHO probe has been able together and measure the sound waves that travel through the plasma ball which is our Sun. The Sun acts like a resonant cavity inside of which the sound waves and gravitational waves are transmitted at very high speed. Starting from the right, we can see: - a computer image showing how the Sun vibrates internally as a result of the sound waves: the different colours represent opposing movements of the plasma; - a diagram showing how the sound waves spread.
BELOW: the inside of the Sun based on echo-Doppler data.

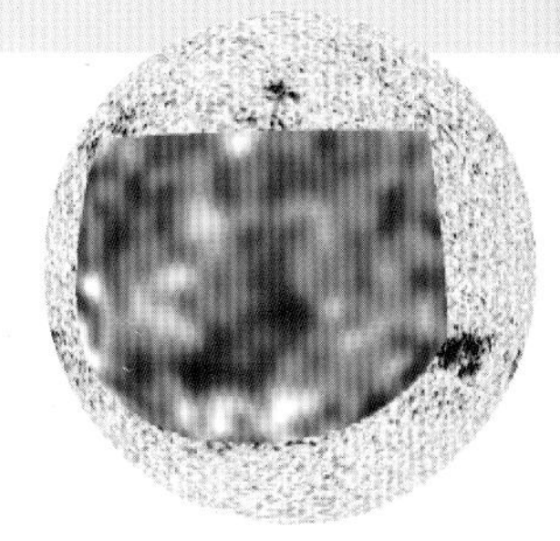

regarding the nucleus: some ten times less. The temperature therefore drops – relatively speaking of course! – to 4 x10^6 K. Here the energy is transmitted through the plasma only by radiation in a sea-saw of absorptions and re-emissions. Freed in the form of photons from the nuclear reactions, the radiation is absorbed and re-emitted millions of times before emerging in the upper layers transformed in Ω, X, ultraviolet, visible and infrared rays (heat);

• a convection zone, that extends for about 250,000 km. Here the density, pressure and temperature values continue to diminish: density reaches 6 x10^{-3} g/cm3, pressure 10 Pa (about 10^{-4} times the atmospheric pressure), the temperature reaches 6 x 10^5 K. In this area, energy is also transmitted to the plasma through high speed convection currents that continually 'remix' the solar material. To explain some surface phenomena, it has been suggested that in this zone profoundly deep giant convection cells slowly produce other smaller cells as they approach the next layer;

• photosphere, literally 'sphere of light' is the part which finally we are able to see. It has a thickness of barely 400 km, a mean density of barely 8 x 10^{-8} g/cm3, an average pressure of only 10^{-12} Pa and a temperature of about 6,000 K. This is the 'solar

THE INSIDE OF THE SUN
A diagrammatic summary of the Sun's hypothetical internal structure, with presumed sizes.

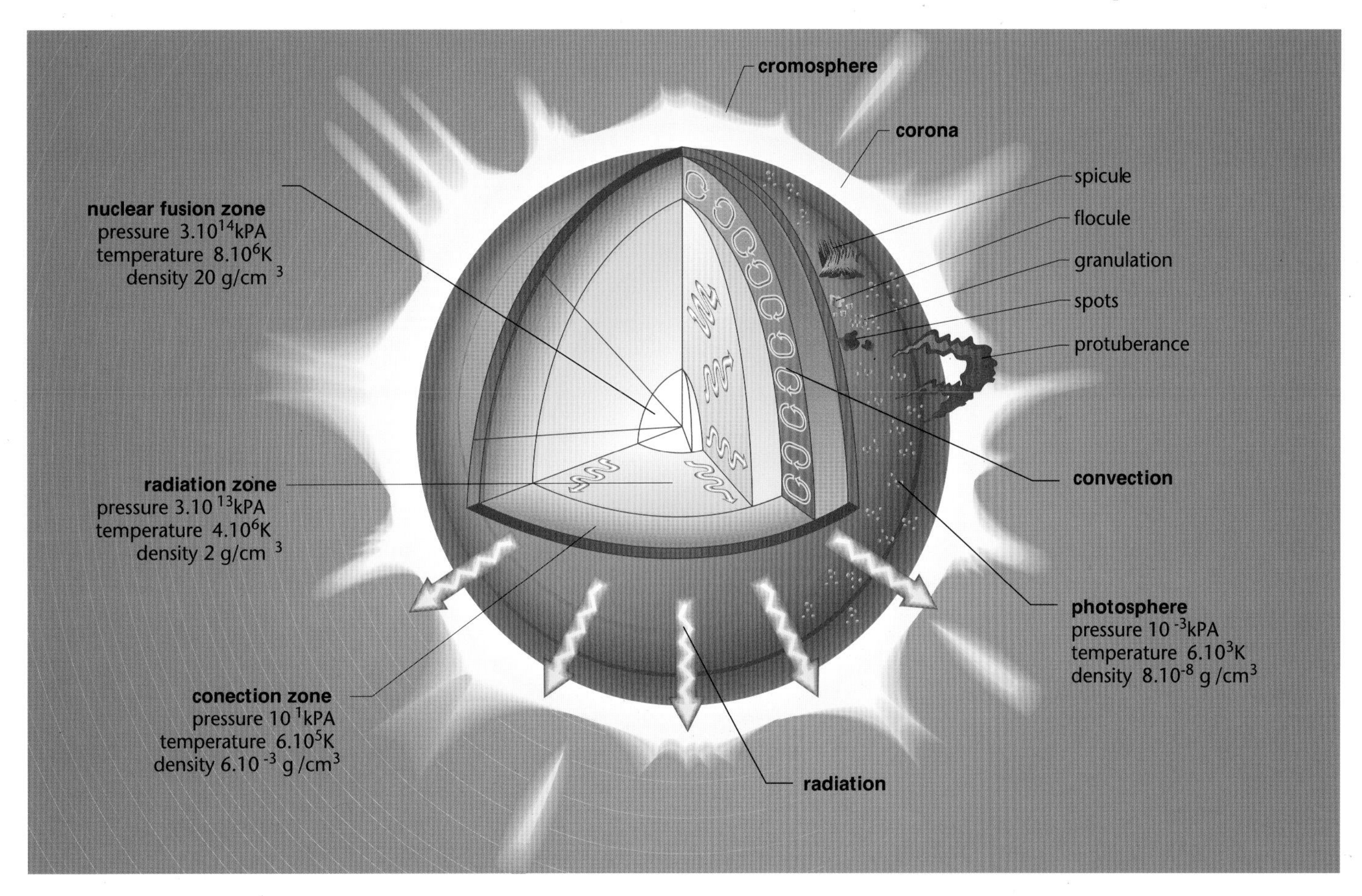

Velocity
The velocity of the gas which gyrates beneath the solar surface has also been measured on the basis of the data gathered by SOHO. From this point of view, the Sun has many layers that compared to the average can be either faster or slower. Yellow-red colours indicate a greater velocity and the darker ones a slower velocity.

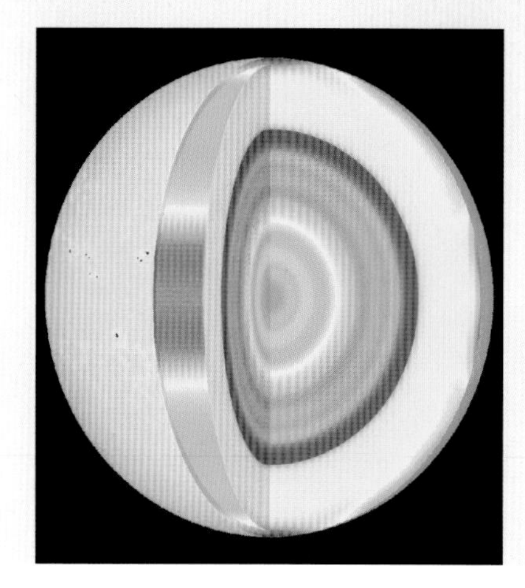

Sunspots at the speed of sound
Three-dimensional Doppler analysis of a sunspot. The upper horizontal layer is at the level of the photosphere, the lower layer is at a depth of 22,000 km. Yellow-red colours indicate a greater speed and the darker ones a slower speed.

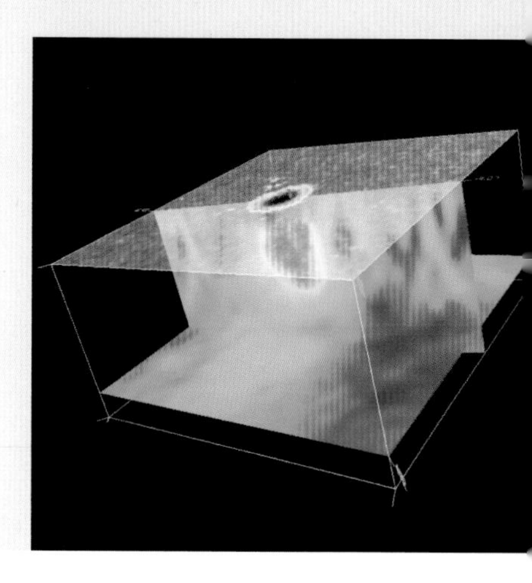

surface' to which reference is being made when speaking of the solar diameter. After a long time, which can be up to 10 million years from the moment in which the radiation is produced by the nucleus, the radiation is finally emitted in a radically altered form, of course, as a result of the journey undertaken. It is also in the photosphere that the sunspots and granulation which are the best-known and studied solar phenomena occur;

• the chromosphere, 'colour sphere' (it appears pink during eclipses), is a plasma layer some 10,000 km high above the photosphere and is considered to be the Sun's lower atmosphere. With a mean density of around 10^{-12} g/cm^3 and a temperature that increases according to altitude, reaching 0.5×10^6 K, it is here that many other solar phenomena occur: spicules, faculae, flocculi and flares;

• the corona: extends beyond the chromosphere dispersing into space as solar wind. It is considered the upper atmosphere and is characterised by temperatures that rise rapidly: in just a few thousand kilometres it reaches 5×10^6 K. It is only visible from earth, even to the naked eye, during a total eclipse and it remains clearly distinct up to an altitude of about 2×10^6 km. The corona is also where the most spectacular phenomena occur: prominences - or filaments - can sometimes reach dimensions that are comparable to those of the Sun itself.

Gravity and light
The Sun is a ball of gas in a state of precarious balance. The nuclear reactions that take place internally tend to make it expand while the gravitational force tends to contract it towards the centre of the mass. With every change in endogenous activity, therefore, there is a corresponding change in contraction or expansion of the Sun.

***SOHO* in the vicinity of the Sun**
The instruments mounted on S O H O have allowed the Sun to be probed to previously unthinkable depths, and information and images that are truly revolutionary to be collected.

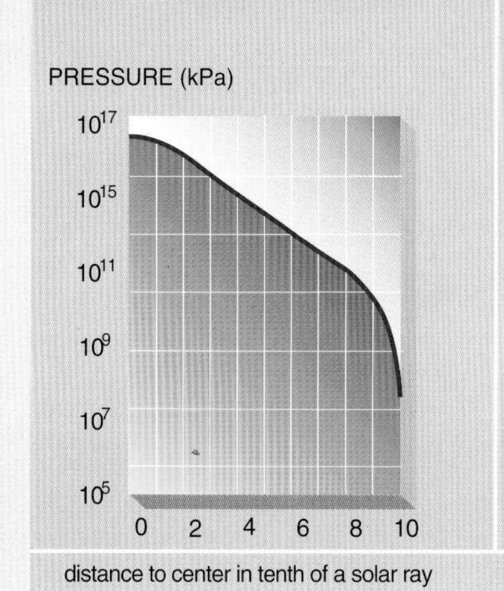

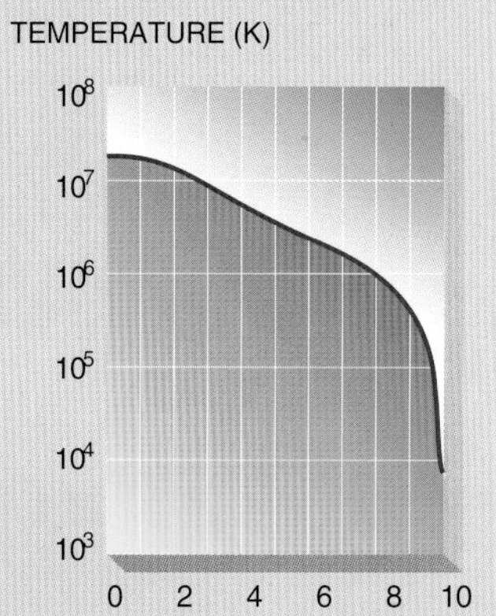

distance to center in tenth of a solar ray

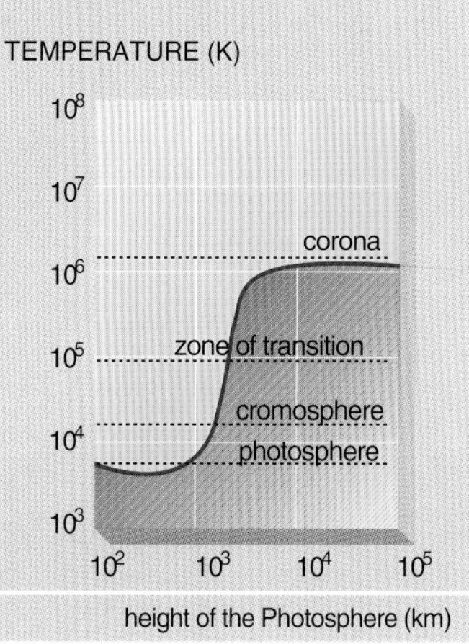

height of the Photosphere (km)

Diagrams
Variations in the pressure and temperature with the varying of the distance from the Sun's centre. In particular, the last drawing illustrates the abrupt change in temperature which comes about when going from the photosphere to the solar corona.

SOLAR PHENOMENA

If we consider that a star is a sphere of gas and that gas is constantly being released into space, it becomes obvious that it makes little sense to talk about the Sun's 'surface'. We have seen how some layers characterised by relatively homogenous conditions of temperature, density, pressure and opacity to radiation, can be identified. The photosphere is different. It is the first 'visible' layer ending where drastic and sudden changes in the physical characteristics of the plasma take place. The change is such as to make the layers which dominate it (chromosphere and corona) completely transparent to visible radiation. It is no coincidence, therefore, that it is here, in the photosphere and in the two outermost layers of the Sun, that the phenomena which are the object of study take place.

GRANULATION AND SUPERGRANULATION

These are characteristic phenomena of the photosphere and represent the visible part of the convection currents. The granulation comprises lighter spots (warmer, ascending currents) and darker spots (cooler, descending currents) that are constantly changing size and form. Each irregular polygonal granule has an average size of between 300 and 1,000 km and remains visible for no more than 5 minutes. The granules at the centre of the solar disk are more regular in appearance. For reasons related to perspective, little by little, the granules become deformed as the solar rotation moves them towards areas that are closer to the borders. Conversely, the granules that are closer to the spots distorted by the magnetic fields assume a more elongated shape.
The difference in temperature between the granules (light zones) and the intergranular lanes (dark zones) is of the order of 100-300 K, equal to a difference in luminosity of 15-20%.
With the help of Doppler analysis the velocity of the gas rising in the light zones has also been calculated. It reaches 300 m/s, three times stronger than the strongest winds of terrestrial storms. In some cases, the dark zones that separate the granules appear to be isolated points known as pores; they are very noticeable in the visible light against a light background. These too, usually, disperse very quickly: sometimes they grow very rapidly until becoming transformed into spots. Light granules can become joined, giving birth to brighter and larger spots: this is what is known as super-granulation. This phenomenon involves even the deepest layers of the chromosphere. Super-granules have an average diameter of about 30,000 km, last about one day and involve much larger gaseous masses.

Solar activity
In a moment of particularly intense activity an enormous prominence develops, shaped by the lines of force of the solar magnetic field. Other phenomena can also be observed: spots and filaments (dark), granulations, faculae (small, light spots), flares (white).

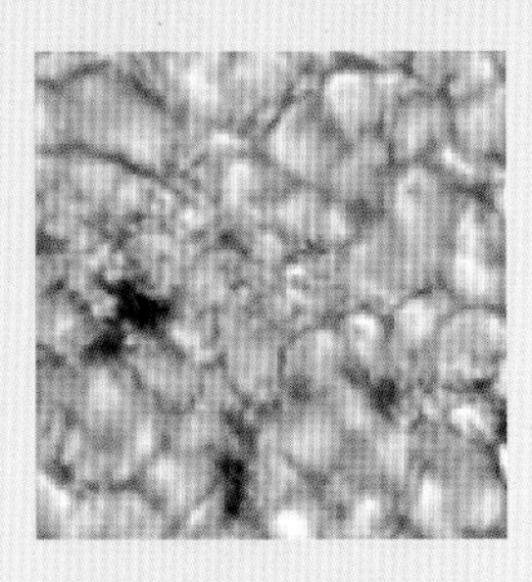

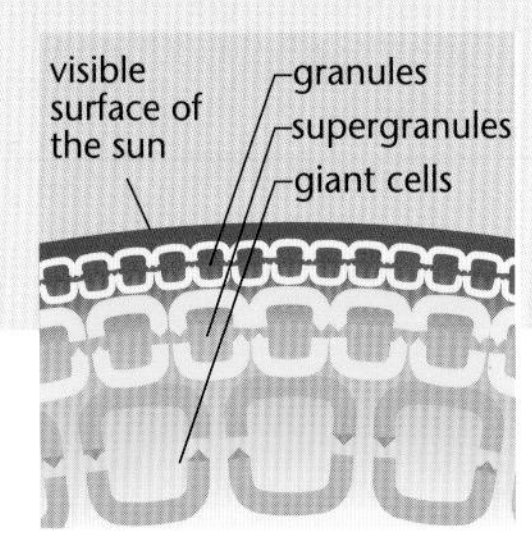

Surface phenomena
Granulation and diagram of the convection cells.

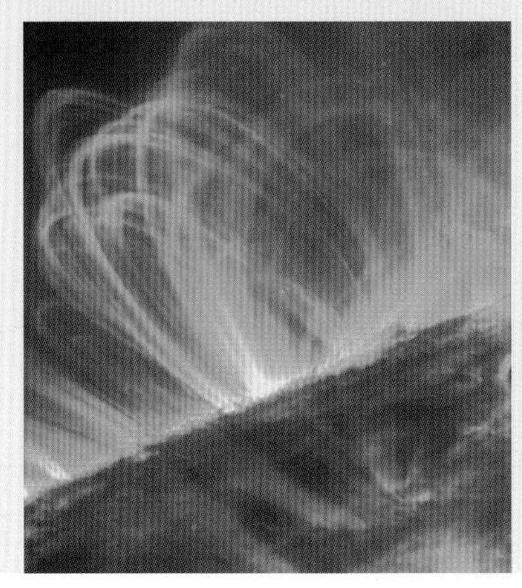

Surface phenomena
Photograph of the Sun from an angle in which can be seen: filaments, spots, faculae and, very clearly, the limb darkening phenomenon.

One of the ring prominences seen from the SOHO.

FACULAE

Literally 'small lights', they are phenomena of the upper photosphere and chromosphere associated with spots. Even appearing in areas close to the solar poles (up to a latitude of about 80°), in fact, they often appear in correlation to those places of the photosphere where the spots will be formed and remain even after the disappearance of the underlying spots. The cycle of activity that gives rise to the faculae is analogous to the cycle which gives birth the spots.

Faculae are masses of gas that are hotter than the surrounding gas (and therefore brighter). Since they have a brilliance that is almost equal to that of the centre of the solar disk, they can be observed better when they reach the limb of the solar disk where the photosphere is less luminous (limb darkening phenomenon). If however the Sun is observed at the wavelength corresponding to the centre of a strong spectral line (for example the hydrogen H-alpha line), the faculae can be seen even at the centre of the solar disk: they appear more extended and take up to 20% of the surface.

LIGHT BURSTS, OR FLARES

These phenomena are also associated with spots and therefore they occur very frequently during periods of greater solar activity. Their number is in proportion to the relative Wolf ➢153 number. Satellite observations have shown how the plasma involved in a flare is heated along arched structures 3,000-4,000 km in diameter. Very rapidly, and for short duration, small regions of the chromosphere become extremely bright (hence the adoption of the term 'flare') and millions of tons of plasma are pushed at a speed of 500 km/s up to the corona, where they provoke the so-called transient phenomena (that is, swift modifications of the structure). These potent, gaseous jets are accompanied by strong emissions of radiation and high energy atomic particles which make up the solar wind ➢147. These develop an energy equal to 10^{20}-10^{27} J, equal to that produced by 50,000 - 500 million atomic bomb explosions such as that detonated at Hiroshima. The flares involve the lowest part of the chromosphere and are often accompanied by flocculi➢146: they are like rounded flashes, long and deformed, and develop with a precise succession of stages:

- preparatory phase: can last from several hours to a day. There is an accumulation of energy, especially magnetic, and a slow increase in the flux of X, ultraviolet and visible rays.
- flash phase: last a few minutes, about half of the accumulated energy is freed in the form of radiation (radio, infrared, visible, ultraviolet, but

Surface phenomena
A solar earthquake spreads out over a radius of 100,000 km starting from a flare. The second, third and fourth images were taken 12min, 21min and 35min respectively after the first image. These seismic waves therefore spread at a rate of some 48 km/s: more than 3 times the speed of sound.

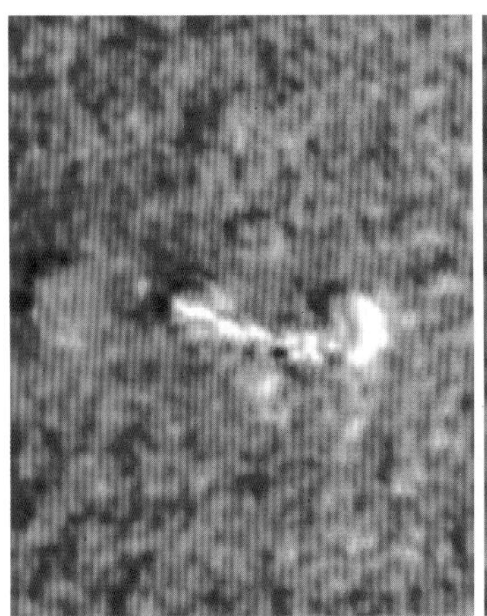

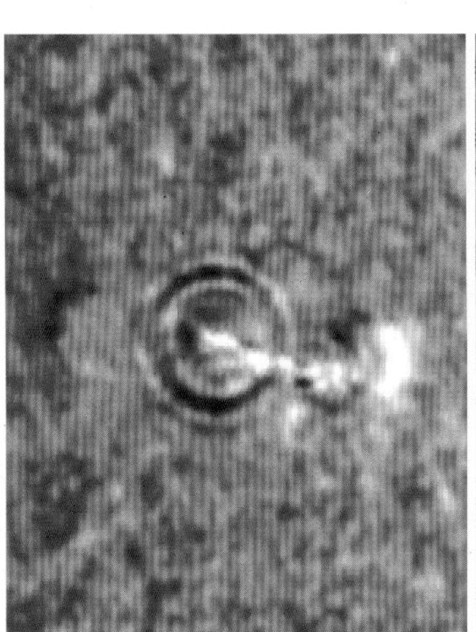

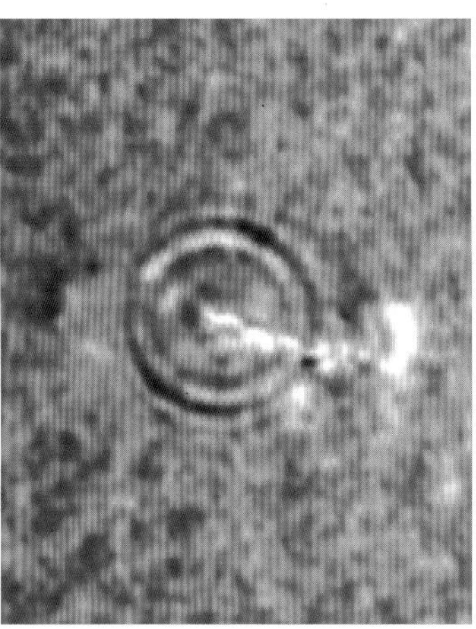

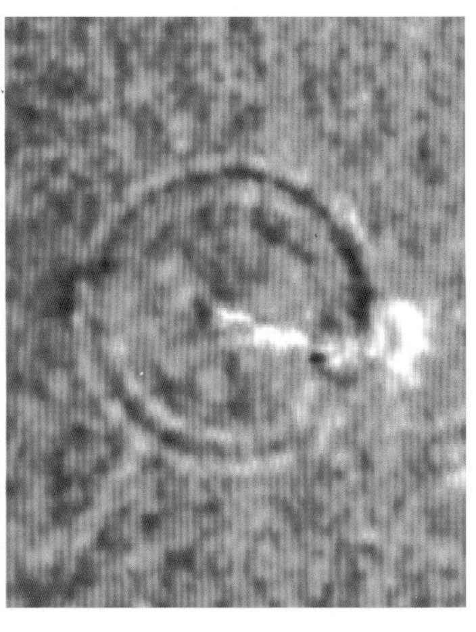

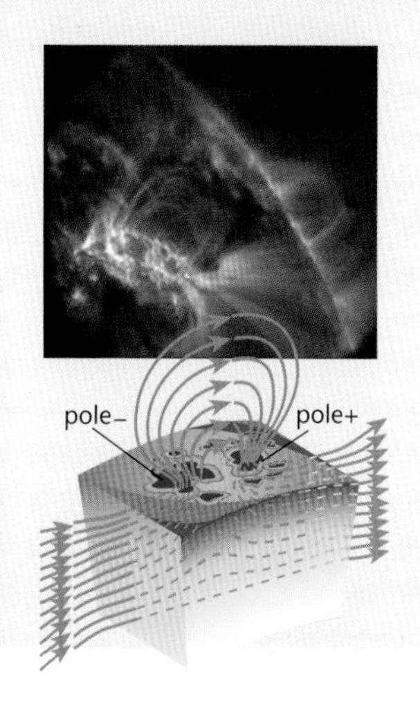

SPOTS
Lines of force of the magnetic field close to a group of sunspots highlighted by the ionized material. The image is a composite one, taken at different wavelengths. The diagram summarises the progression of the magnetic field in a bi-polar group of spots.

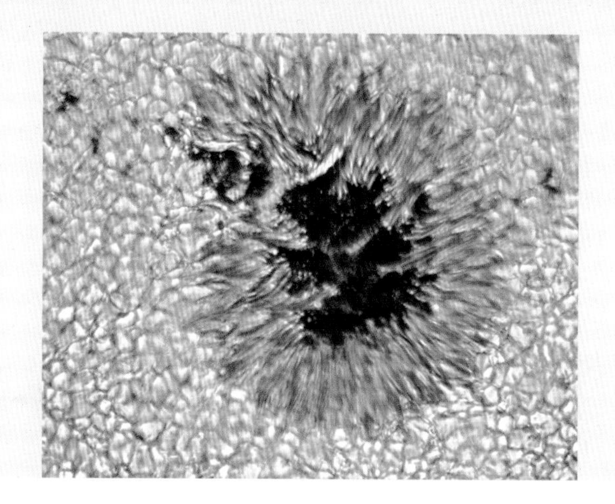

SUNSPOTS
A sunspot. The shadow and penumbra in the middle of the granulation are clearly visible.

above all X and, during more forceful events, even Ω rays) and in the form of kinetic energy of ions and electrons that abandon the Sun at a speed of 1,000-1,500 km/s (solar wind);

• decay phase: lasts from 20 minutes to a few hours. In this time the rest of the energy is freed, above all in the form of radiation.

SUNSPOTS

Already studied hundreds of years ago by Chinese astronomers, sunspots are cited repeatedly in the annals of scholars throughout the world from the Ancient Greeks to the Russians of the 14th century. In order for a sunspot to be visible to the naked eye, it is sufficient for its diameter to be no more than 40,000-45,000 km and this happens, on average, several times every 11 years.

Sunspots are photospheric phenomena, zones of the solar disk that appear dark because of their lower temperatures compared to that of the surrounding areas. The temperature of a sunspot's gas is usually in the region of 4,000 K as opposed to the almost 5,500 K of that by which it is surrounded. As in the case of the granules, the spot appears dark in contrast with the very bright surrounding photosphere. In fact even the shadow of a sunspot, its central and darker part, if isolated in the nocturnal sky, would blaze more brightly than the full Moon.

For a long time it was thought that the spots were of a funnel shape, because of the so-called Wilson effect. If a circular spot with a concentric penumbra around the shadow gets close to the solar disk limb, it can be seen that the part of the shadow closest to the centre of the disk thins out much more quickly compared to the opposite side. Spectroscopic analysis, however, and satellite images have shown that this is an optical effect: the spots have a two-dimensional structure.

Intense magnetic fields are found in the spots and the theories which explain the origins of these phenomena are still the subject of debate. The local alterations of the solar magnetic field certainly carry out an essential role in their formation, but how this comes about is still a controversial issue. Of the magnetic fields tied to the sunspots, in fact, only the direction, intensity and structure are known; regarding their origin, nothing is known for certain. The most credible theories regarding the formation of these 'cold' areas on the Sun's surface attribute the lower temperatures of the spots to the suppression of convection caused by the perturbations of the magnetic field or the intensification of the convection: the convection motions in the sunspots would be so efficient as to release more

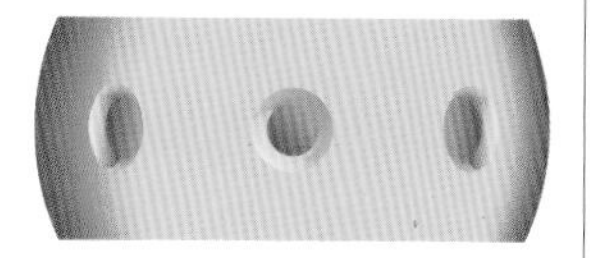

WILSON EFFECT
Different appearance of the sunspots on the solar surface.

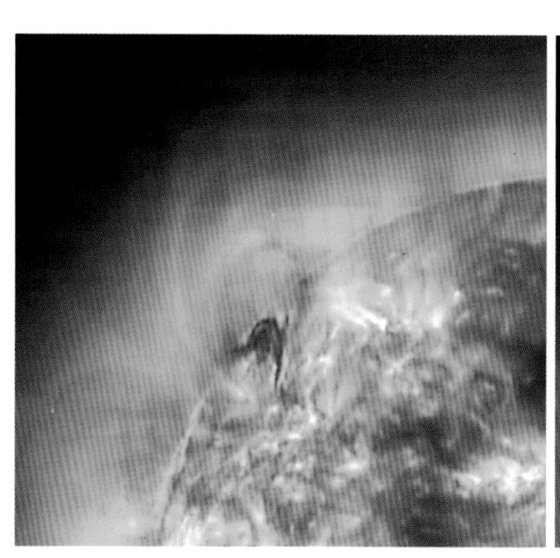

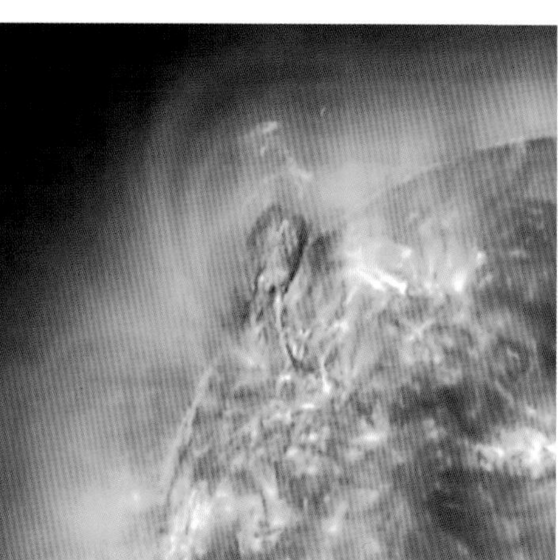

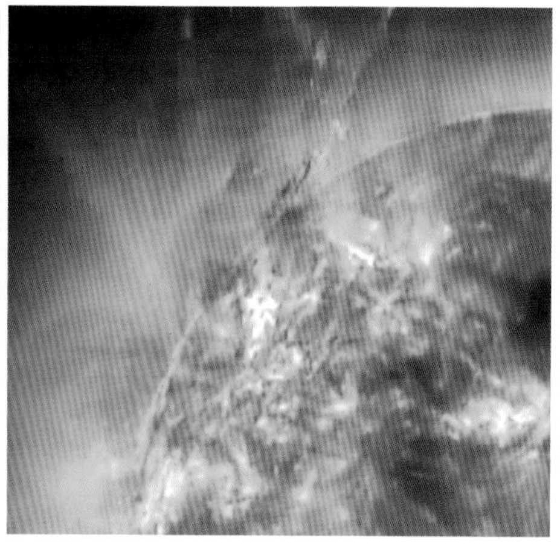

SUNSPOTS
Surface activity. The second and third image were taken 12min and 36min respectively after the first image.

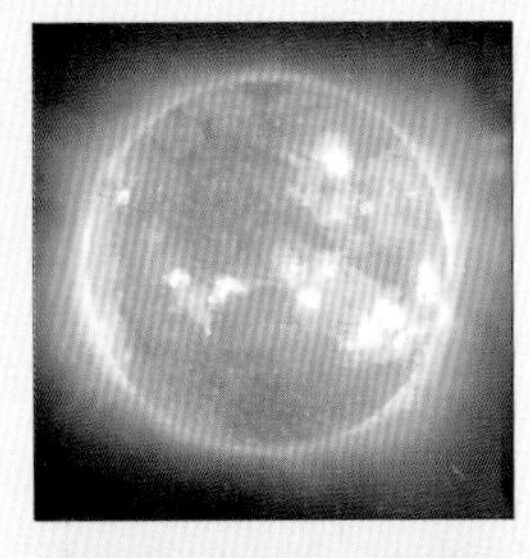

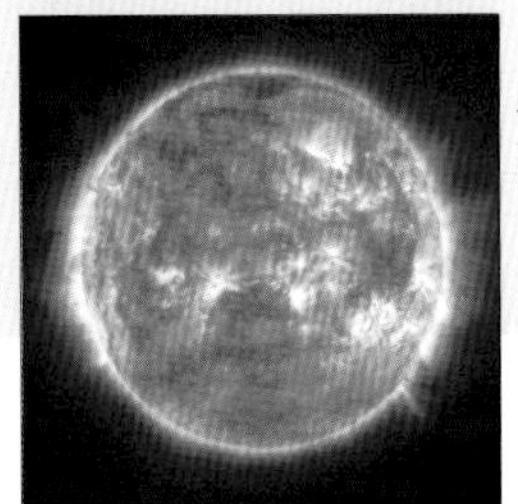

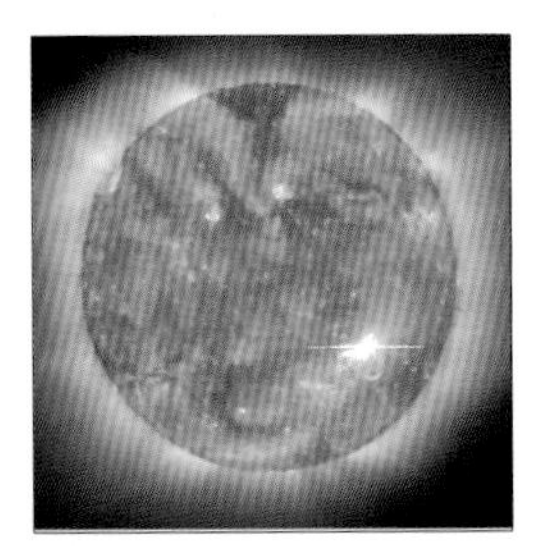

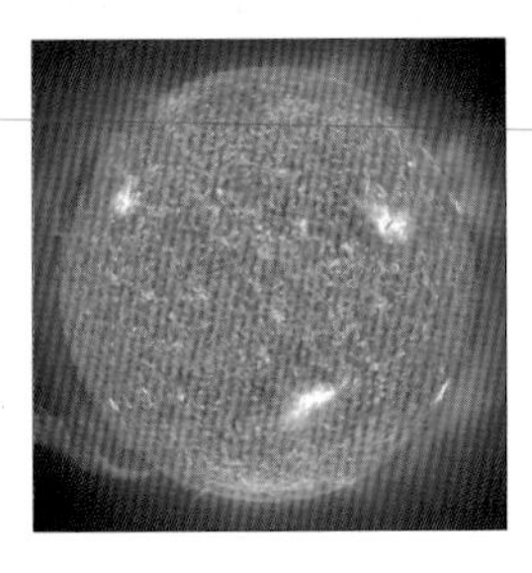

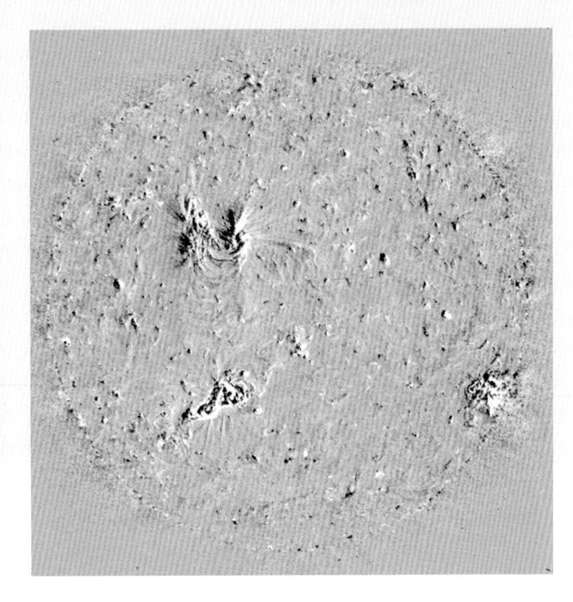

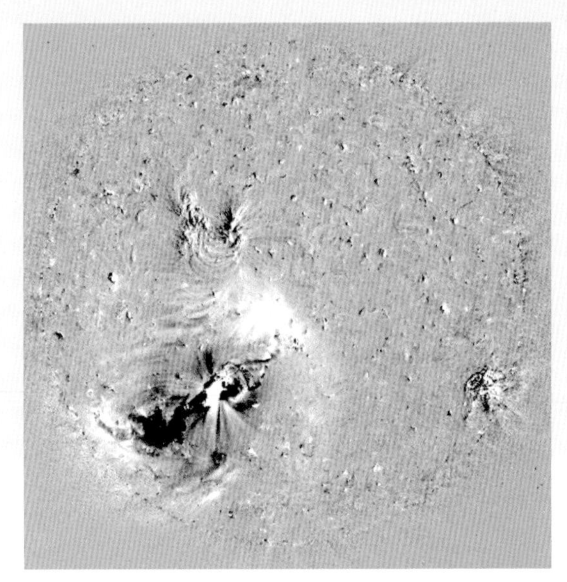

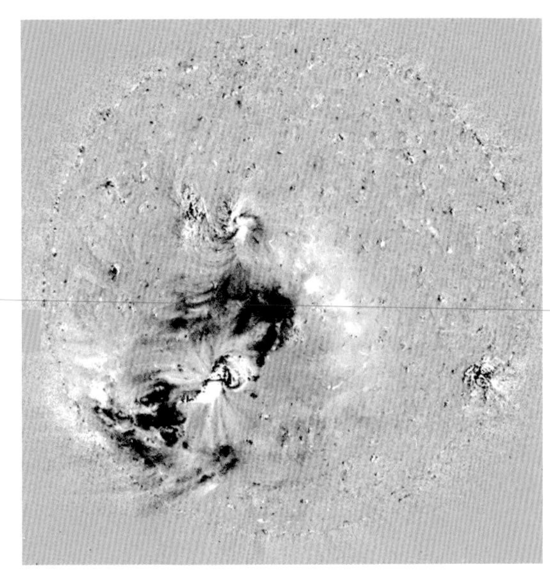

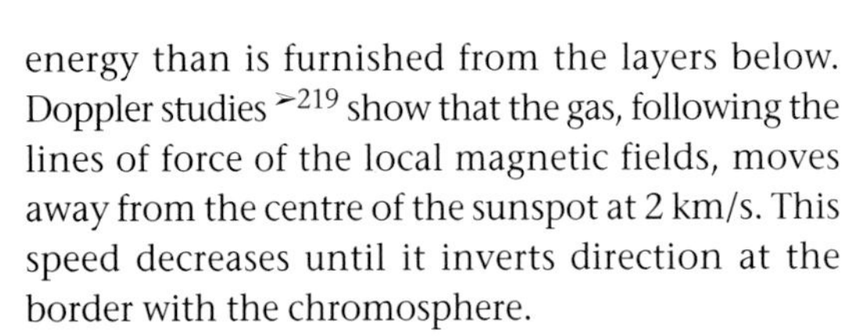

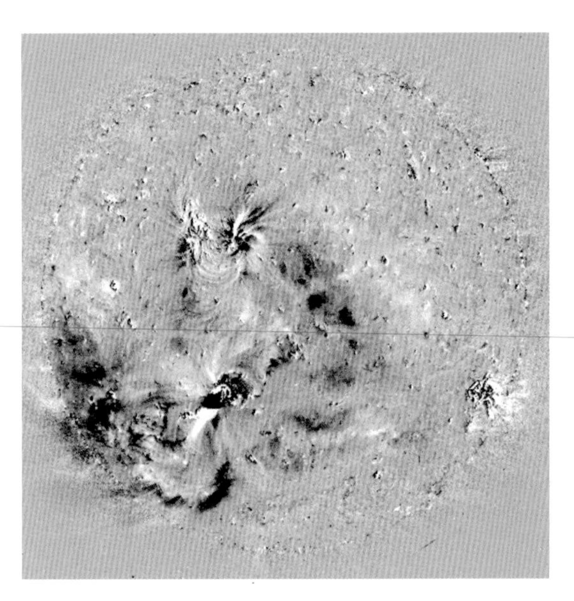

THE COLOURS OF THE SUN
These images of the Sun were taken from the S O H O using its extreme ultraviolet telescope. Each shows the Sun at different wavelengths, highlighting different aspects of the same surface phenomena.
FROM THE TOP:
- photograph taken in the XV ionised iron line (Fe XV, 195 Å);
- in Fe IX/X (171 Å);
- in Fe XII (195 Å);
- in He II (304 Å).

energy than is furnished from the layers below. Doppler studies ➢219 show that the gas, following the lines of force of the local magnetic fields, moves away from the centre of the sunspot at 2 km/s. This speed decreases until it inverts direction at the border with the chromosphere.
Sunspots vary greatly in size, but, generally speaking, they are very large. An average spot has a diameter that is equal to Earth's (about 13,000 km). Spots that are over 90,000 km wide are rare. Even isolated spots are fairly uncommon. Generally, they develop in pairs and even more frequently in groups born by the aggregation of pores. Their development is confined to a single area that lies between 40°N and 40°S. As a rule, they appear below 30° and those appearing between 40° and 30° are fairly uncommon. Exceptionally rare, small and lasting a very short time are those that are formed over 40°.

FLOCCULI
These phenomena of the lower chromosphere can only be observed with special filters or with spectroscopes, of particular wavelengths. They are relatively small zones where the gas is warmer than the surrounding one, probably brought about by abrupt variations in the electromagnetic field. Their development is not necessarily tied to that of sunspots.

SPICULE
These form in the chromosphere close to the borders of the super-granules ➢144, where the magnetic field becomes more intense and the temperatures higher. They are jets of gas with a mean diameter of 800 km that move to the outer part of the Sun at a speed of about 30 km/s and reach heights of around 6,000 km before disappearing within a few minutes. The highest spicules can be up to 50,000 km long and project beyond the chromosphere.

Moreton wave
A wave produced by the expulsion of coronal material, often linked to flare and sunspot activity. The wave front propagates on the solar surface at speeds of 300 km/s. Taken in the FE XII line, an ion that forms at about a million and a half degrees. The images were taken 12min, 21min and 35min respectively after the first one. Usually, the spots of a group are disposed along a solar parallel. The spot that precedes the others in the direction of the Sun's rotation is often also the most compact and the fastest, moving in an Easterly direction. At the two extremes each group of sunspots shows roughly opposite magnetic poles.

Magnetic storms
In this treated colour image, made by superimposing three solar images and taken at different wavelengths (171 Å, 195 Å and 284 Å), the eruptive activity next to a group of sunspots can be observed.

PROMINENCES OR FILAMENTS

Prominences are the most spectacular solar phenomena that can be seen at the wavelength of the H line, the dynamics and origins of which are still being debated, especially because of their unusual characteristics. In most cases, they are made up of coronal material with temperatures that are lower compared to that of their surrounding environment. The corona material is also 100 times denser than the corona in which it finds itself and thus tends to fall to lower levels.

Prominences are extremely rarefied objects: their mean volume is equal to about 1,000 times that of the Earth, yet their mass is a little less than 20 km^3 of water. They are also known as filaments, since if viewed at a visibility angle 'from above', projected onto the solar disk, they appear like long, dark filaments. At most, they remain on the solar surface for a few months. They change form rapidly and exchange material with other prominences of the same type relatively close by.

They are invariably incredibly large phenomena. Achieving heights of 150,000 km, they can reach lengths of 300,000 km and widths of 40,000 km and, for reasons that are still unknown, they can explode, projecting into space a mass of solar material of the order of 10^{13}-10^{15} kg at speeds that can reach 400,000 km/h. Various types of prominences can be discerned:

- active prominences: under the influx of magnetic fields they are usually tied to a group of sunspots, and have a relatively rapid evolution.
- quiescent prominences: these have a very slow evolution and, almost as though imprisoned in the magnetic field, they preserve their appearance for days and sometimes even for as long as a solar rotation. They are prevalent at high altitudes;
- eruptive prominences: they have a rapid evolution; they develop in a direction that is almost perpendicular to the photosphere and they fall towards the chromosphere in the space of a few hours. Tightly linked to flares, they leave trails of gas at 10^4 K that for a long time can remain confined to the corona. (100 times less dense at temperatures of 2-3 · 106 K). They are prevalent at low latitudes;
- loop prominences: typically they form in a closed ring, an arch, a jet, a fountain or a fan, since the material of which they are composed follow the lines of force of the magnetic field that exit and re-enter the photosphere;
- vortex filaments: typically have a spiral form resembling twisted rope;
- surge prominences: these are imposing displays of solar activity, closely related to both the sunspots and to the most energetic flares. They are jets of coronal material that are over 50,000 km high and travel at speeds of hundreds of km/s. They follow the cycle of solar activity ➢150.

Prominence
This immense cloud made up of relatively dense and cold plasma remains suspended in the middle of the corona. The hottest parts appear light, the darker ones are relatively colder. The information gathered by SOHO show the upper chromosphere to have a temperature of 60,000 K. Prominences can reach exceptional sizes, comparable to those of the Sun itself.

Diagram
Variations of the terrestrial ionosphere structure as a result of exposure to radiation and to the solar wind.

Magnetic field
Extension of the solar magnetic field in space to envelop Earth.

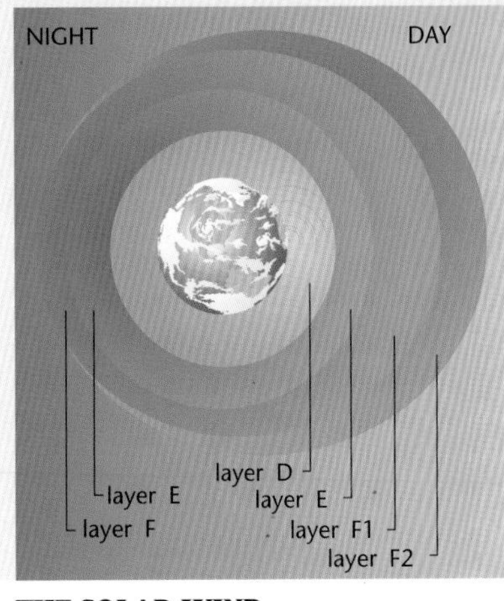

Solar wind
Radiation and currents of solar plasma collide and influence the entire solar system. The table shows the variations in concentration of particles at various distances from the Sun (number of protons/cm3).

THE SOLAR WIND

Like many other celestial bodies, the Sun is also the source of a magnetic field. It is much higher than the Earth's, which fluctuates between 0.2 and 0.7 G. In the Sun, it is 1 G at the poles, in the chromosphere it is 25-200 G. In prominences it varies between 10 and 100 G and in proximity to a sunspot it can reach 3,000 G and fluctuate as high as 5,000 G. The solar magnetic axis is inclined about 6° compared to the axis of rotation, and the magnetic poles are not found on the same solar diameter. Due to the differential rotation of the inner layers and the surface gasses at different latitudes -according to the most credible theories - the solar magnetic field controls the surface phenomena and expands in interplanetary space. The density of the gasses that move away from the solar surface reaches values that are comparable to those of the density of interplanetary space only at very great distances from the Sun. In proximity to Earth, it is still about 100 times higher. It is like saying that the solar corona extends as far as Earth and beyond, perhaps as far as 7.5×10^9-1.5×10^{10} km from the Sun, well beyond the orbit of Pluto. Travelling along the lines of force of the solar magnetic field at speeds of 450 km/s and with a density of about 5 protons/cm3, these solar particles at very high temperatures escape from the Sun's gravitational force creating the solar wind. In this way, the Sun loses, on average, about 10^{17}kg of material every day. The concentration of solar wind varies periodically, provoking disturbances in the terrestrial magnetosphere: this is due to the coronal holes, areas of the corona where all the lines of force of the solar magnetic field converge (coronal loop). Here the coronal particles are accelerated and

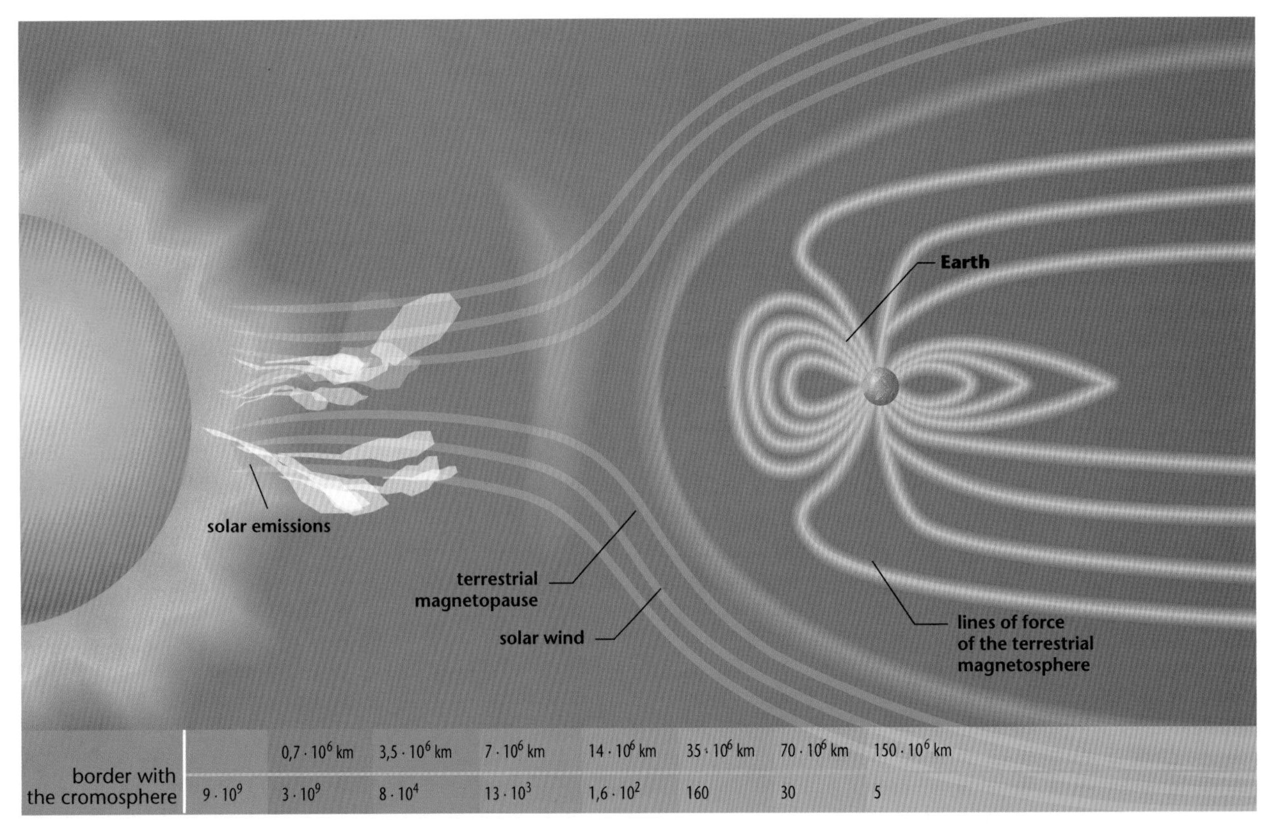

	$0,7 \cdot 10^6$ km	$3,5 \cdot 10^6$ km	$7 \cdot 10^6$ km	$14 \cdot 10^6$ km	$35 \cdot 10^6$ km	$70 \cdot 10^6$ km	$150 \cdot 10^6$ km
border with the cromosphere: $9 \cdot 10^9$	$3 \cdot 10^9$	$8 \cdot 10^4$	$13 \cdot 10^3$	$1,6 \cdot 10^2$	160	30	5

Magnetic field
The shape of the solar magnetic field is continually changing as can be verified by observing the shape of the corona during either minimum or maximum solar activity. The coronal material conforms to the coronal loop during minimum activity, whilst remaining uniform during maximum activity. This is obvious in the two images of a total eclipse: on the left during solar minimum activity, on the right during solar maximum activity.

Opposite: the corona highlighted by a coronograph from the SOHO. Some spiralling plasma can be seen around the lines of force. Below: a diagram of the solar magnetic field.

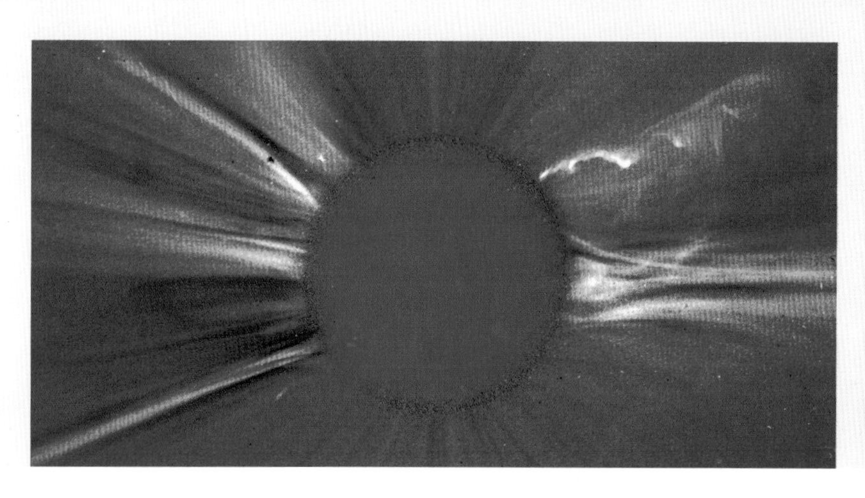

projected into space. The solar magnetic field also has an equatorial zone within which it is annulled. In consequence of the Sun's rotation, this neutral surface extends into interplanetary space undulating like the skirt of a ballerina, a little above and a little below the equatorial plain, as a result of the Sun's axis of rotation not coinciding with that of the magnetic field. The observations gleaned from the probes have shown that even the lines of force around the Sun are wound in a spiral, delimiting sectors which have a different polarity (positive and negative). As a result of the Sun's rotation, the lines of force of the magnetic field form an angle with the radial direction, which increases progressively. At about 1.50 x 10^8 km from the Sun (i.e. at the distance of the Earth) it is about 15° and at the distance of Jupiter it approaches 90°.

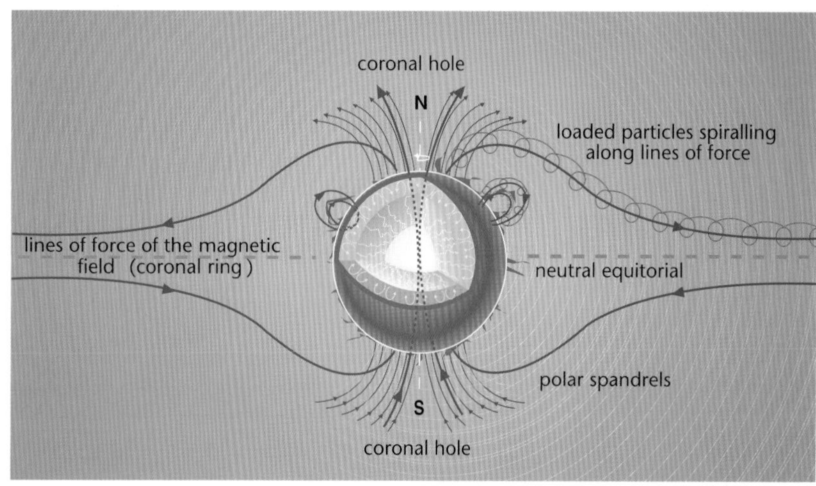

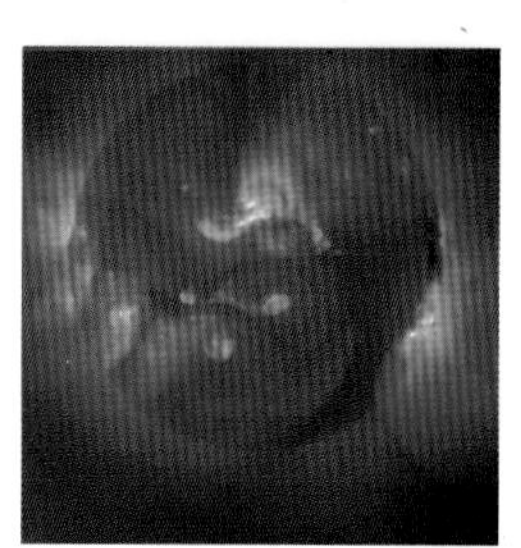

Coronal hole
An X-ray image of the Sun highlights a relatively 'dark' area. Here the lines of force of the magnetic field are almost radial, and the emission of solar wind is at its height.

WE KNOW NOTHING THAT CAN BEGIN TO GIVE US AN INKLING OF WHAT GOES ON INSIDE THE SUN. NO ATOMIC EXPLOSION, NO NATURAL PHENOMENON IS COMPARABLE TO THE NUCLEAR FURNACE HIDDEN INSIDE THE HEART OF OUR STAR. BUT WE KNOW THE ATOM AND ITS LAWS AND THIS ENABLES US TO DESCRIBE THE PROCESSES THAT ARE AT THE ROOT OF AN IMMENSE OUTPUT OF ENERGY AND IT ALLOWS US TO SPECULATE AS TO THE ORIGINS OF SOLAR CYCLES.

THE ACTIVITY OF THE SUN

DIAGRAM
The energy that ties subatomic particles to form the atomic nucleus of each element differs according to the number of particles involved. It does not grow in a linear manner with the sum of its components, but has, roughly speaking, a maximum ranging between 50 and 100. The diagram highlights the areas in which nuclear reactions release energy.

Reliable geological data indicate that many millions of years ago the Earth's temperatures were similar to those of today. Therefore, even the solar energy reached Earth in quantities comparable to those known to us today – i.e. about 1.4 kW/m^2. Every day, for thousand of million of years, a total of some 200 million million kilowatts of energy have rained down on our planet. Since the Sun gives out the same amount of energy in all other directions in space, on every surface that is comparable to the Earth, placed on a sphere that has the Sun at its centre and with a radius equal to the distance Earth-Sun, there therefore falls an equal amount of energy. With a simple calculation, the total quantity of energy the Sun releases into space every second can be determined.
It is an almost unintelligible value: 2 thousand million times the energy that reaches Earth. That is about 400,000 billion billion kilowatts. This has happened all day, every day, year after year for thousands of millions of years.

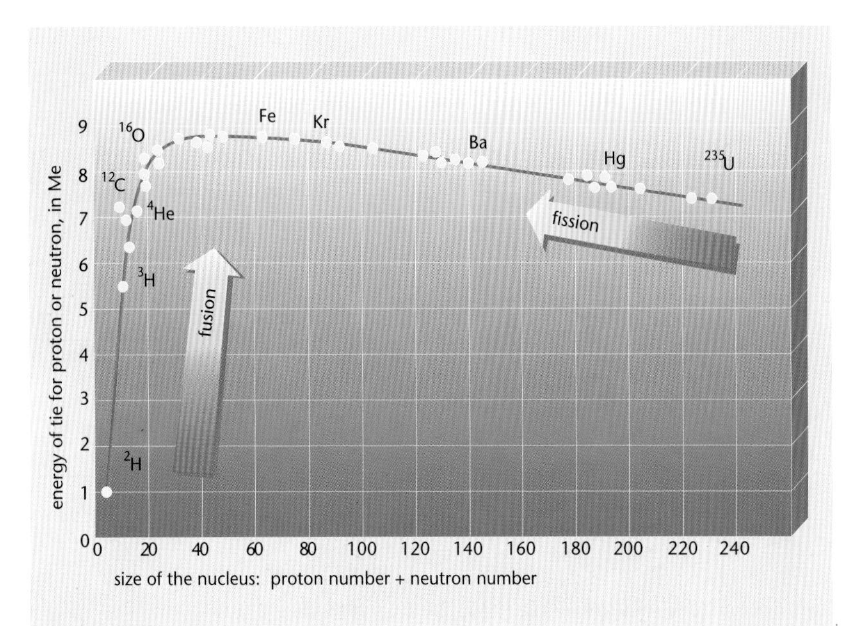

NUCLEAR REACTIONS

This enormous quantity of energy is produced by nuclear reactions that take place in the nucleus of the Sun. Under the conditions of temperature and pressure that prevail there, the fusion of the protons (hydrogen atom nuclei that in the solar nucleus have a concentration of around 5 x10^{25}/cm^3, for a total of 1.5 x 10^{58} protons) is spontaneous. This reaction leads to the formation of helium nuclei: for every gram of hydrogen used, 7 mg of mass become energy: 6.3 x 10^{11} J. The total mass of each atomic nucleus is not exactly equal to the sum mass of the particles which form it, but slightly less (mass defect). If one builds the nucleus of an element starting with its particles, or if one splits a nucleus into its components, the total mass of what is obtained is somewhat less than the original mass. The 'lost' mass is transformed into energy. The hydrogen present in the core of the Sun is sufficient to last at least another 4 or 5 thousand million years. Once it is all transformed into helium, it will be the helium that will fuse, producing carbon. If all the Sun's mass could be converted into energy, the Sun would shine for a further 15,000 thousand million years, one thousand times the age currently estimated for the universe. The nuclear reactions of a star continue only for as long as pressure and temperatures of the core remain high. In the Sun's case, the processes will presumably cease when all the carbon is transformed into oxygen.
Currently, each second, $5.7 \cdot 10^{11}$ kg of hydrogen are transformed in $5.6 \cdot 10^{11}$ kg of helium, freeing

FUSION REACTIONS IN THE SUN

THE CARBON CYCLE

Carbon-12 isotope (^{12}C) catalyses the reaction that is perhaps more frequent in stars with mass greater than the Sun's, which releases 25.03 MeV (about $4 \cdot 10^{-6}$ J) and occurs generally speaking as follows:

- a ^{12}C nucleus absorbs 1 proton transforming into nitrogen-13 (^{13}N) and releasing γ photons;
- ^{13}N rapidly decays in ^{13}C emitting 1 positron and 1 neutrino;
- ^{13}C absorbs 1 proton and emits a new Ωray transforming into stable nitrogen-14 (^{14}N);
- 1 proton fuses with ^{14}N and is transformed into unstable oxygen-15 (^{15}O);
- ^{15}O decays in nitrogen-15 (^{15}N) emitting 1 positron and _ rays;
- ^{15}N absorbs 1 proton and splits in two: a nucleus of helium-4 (^{4}He) and a nucleus of ^{12}C.

Once in 2,000 events, the reaction concludes differently, successively transforming itself into fluorine-17 and oxygen-16 and 17. The total energy released in this case is of 24.74 MeV (about 3.9·10-6 J).

THE PROTON-PROTON CYCLE

The most likely source of energy in the case of stars with the same mass as the Sun. After an initial commune succession, it runs different courses:

COMUNE SUCCESSION:

- 2 protons fuse producing 1 deuteron (nucleus of deuterium, or heavy hydrogen: 1 proton and 1 neutron), 1 positron and 1 neutrino, releasing 1.74 MeV of energy. In 0.25% of cases, (PEP reaction) 1 electron also intervenes: 1 neutrino is produced and 1.44 MeV of energy is released;
- deuterium fuses with another proton forming unstable helium-3 (^{3}He);

FIRST REACTION (91% of cases):

- 2^{3}He nuclei fuse producing ^{4}He and regenerating the 2 initial protons; 5.49 MeV are freed.

SECOND REACTION (9% of cases):

- 1^{3}He and 1^{4}He nucleus fuse in 1 nucleus of beryllium-7 (^{7}Be) releasing Ω rays;
- ^{7}Be absorbs an electron and transmutes in lithium-7 (^{7}Li) releasing 1 neutrino: the process can occur in 2 different ways which respectively release 0.86 MeV (90% of cases) or 0.38 MeV (10% of cases);
- ^{7}Li absorbs 1 proton and splits into 2 nuclei of ^{4}He.

THIRD REACTION (0,1% of cases):

- 1^{3}He reacts with 1 ^{4}He fusing with 1 nucleus of ^{7}Be and freeing Ω rays;
- ^{7}Be absorbs 1 proton and transmutes in boron-8 (^{8}B) and Ω rays;
- ^{8}B, unstable, is transformed in beryllium-8 (^{8}Be) releasing 1 positron, 1 neutrino and 14.06 MeV of energy;

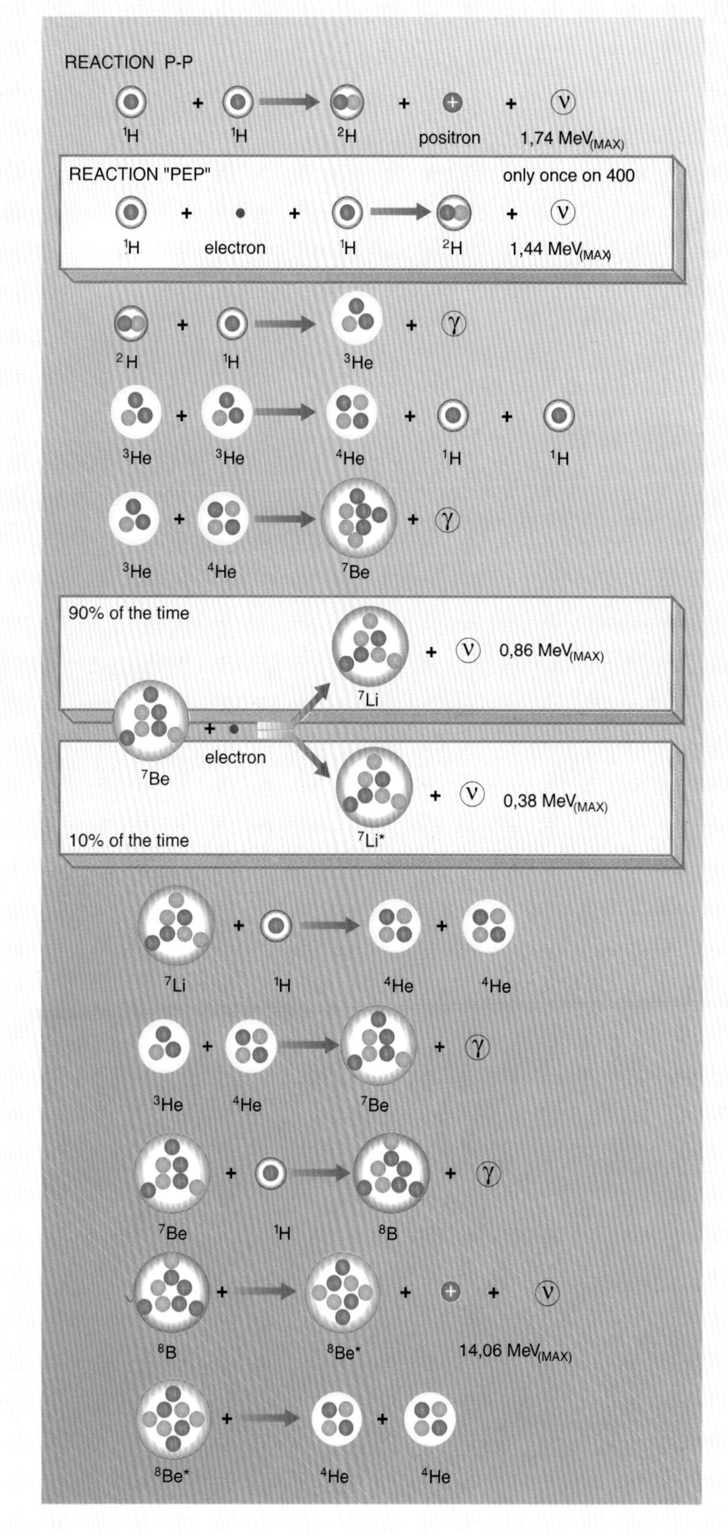

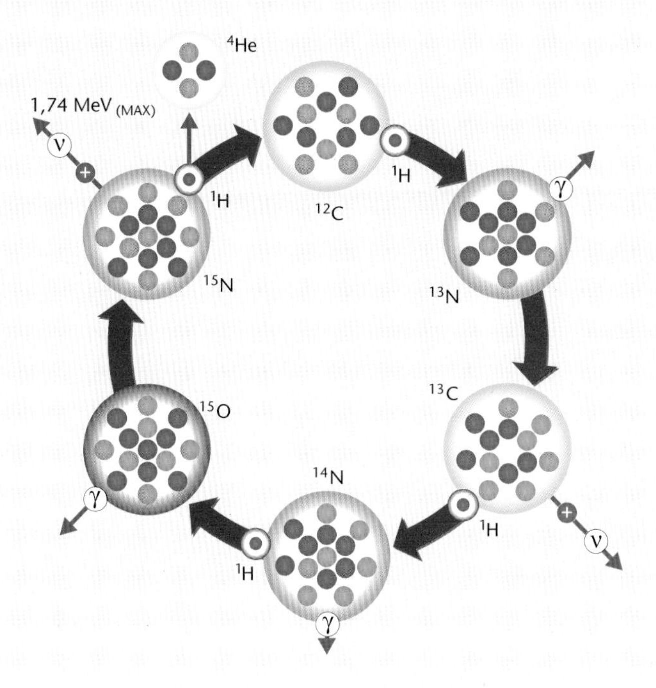

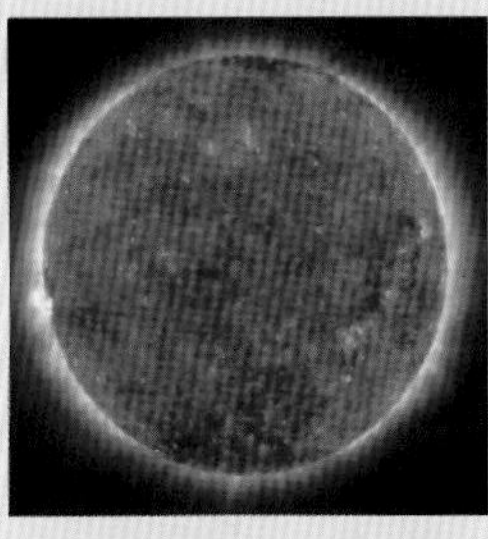

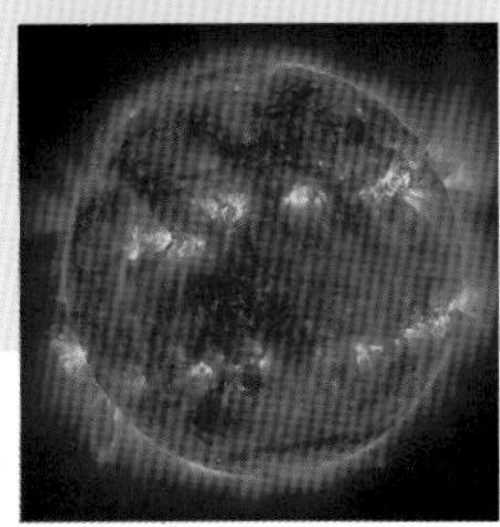

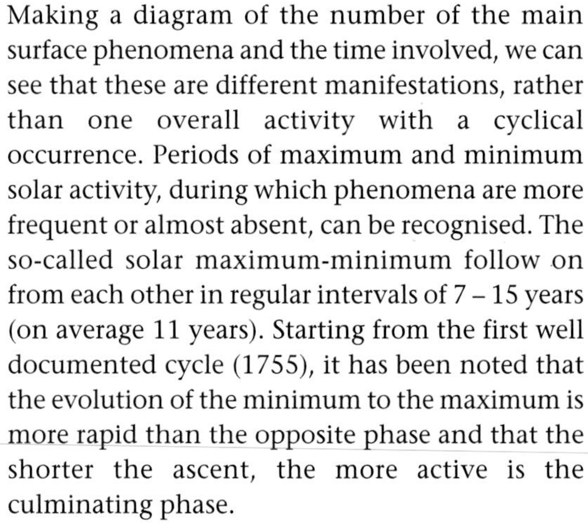

Magnetic field and solar cycle
The diagrams summarise how from a solar minimum one arrives at a maximum: because of the Sun's differential rotation, the lines of force of the magnetic field would 'become agitated' to the point of provoking a period of great instability (solar maximum). This would be followed by a progressive return to conditions of minimum.

Above: almost two years lapse between these two images of the Sun, taken in ultraviolet during a period of a minimum activity (above) and during a period of a progressive increase in activity (below).

energy, 97% of which is made up of Ω rays. But the process which leads 4 protons to fuse into a helium nucleus (2 protons and 2 neutrons) is exceedingly slow and can take two different paths, described by the German physicist Hans Albrecht Bethe.
Fusions in the Sun's nucleus are a self-regulating process. The Sun is in constant balance between gravitational pressures, which tend to make the gas fall towards the centre, and the internal pressure produced by the temperatures. If the production of energy increases, the temperature rises, the gas expands and the internal pressure increases making the Sun expand. But the expansion provokes a reduction in the pressure and therefore of the temperature. The internal reactions, that are highly sensitive to temperatures, slow down, the gas contracts again and everything returns to its previous state.
Conversely, if the production of energy diminishes, the internal pressure is lowered and the pressure of the gravity has the upper hand. With the concentration of gas towards the centre, the internal pressure is increased, therefore the temperature also, and also therefore the speed of the reaction. And everything returns to its previous state again.

THE SOLAR CYCLE

Making a diagram of the number of the main surface phenomena and the time involved, we can see that these are different manifestations, rather than one overall activity with a cyclical occurrence. Periods of maximum and minimum solar activity, during which phenomena are more frequent or almost absent, can be recognised. The so-called solar maximum-minimum follow on from each other in regular intervals of 7 – 15 years (on average 11 years). Starting from the first well documented cycle (1755), it has been noted that the evolution of the minimum to the maximum is more rapid than the opposite phase and that the shorter the ascent, the more active is the culminating phase.
This, however, is not always evident. The terminating cycle in fact always overlaps the initial one. But it is easy to tell one cycle from the other because of the magnetic properties of the sunspots: at each new cycle these invert polarity. Observing the Sun, starting during a minimum period, the following will be seen:
• very few sunspots near the equator (deriving from the old cycle) accompanied by some spots at 40° (new cycle);
• an increase of the total number of spots and the new appearance of spots at ever lower latitudes;
• when the bands of sunspots are at 15°latitude,

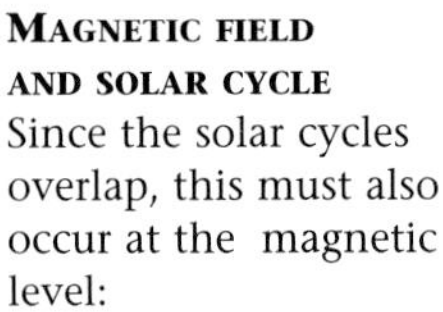

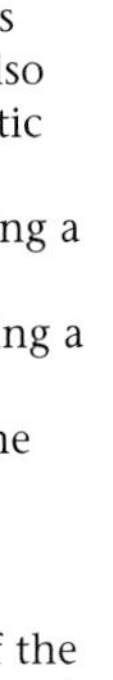

Magnetic field and solar cycle
Since the solar cycles overlap, this must also occur at the magnetic level:
a. lines of force during a solar minimum;
b. lines of force during a solar maximum;
c. lines of force at the beginning of a new cycle: those of the previous cycle are underlying; those of the new cycle with inverted polarity appear on the surface.

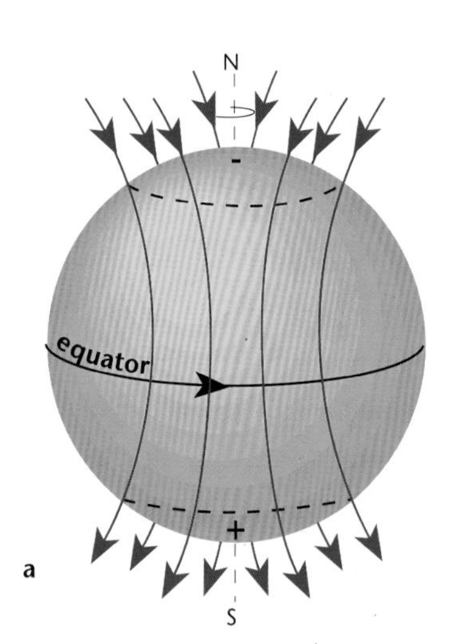

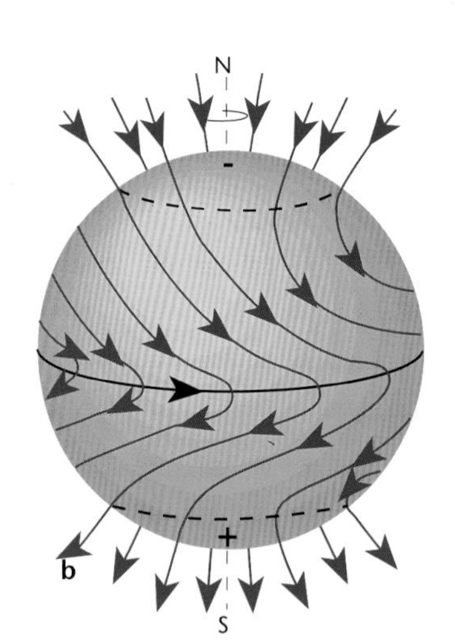

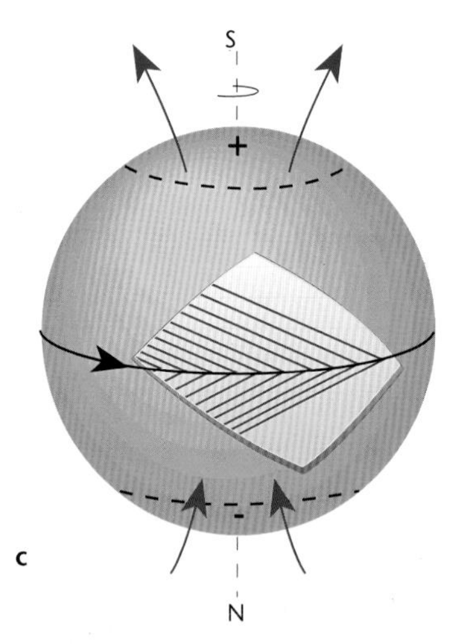

the total number of sunspots begin to decrease as they begin to migrate towards the equator;

• the cycle is concluded when the last few sunspots form between 5° and 10° latitude and the sunspots of the new series, with opposite polarity, begin forming at around 40°.

During each cycle of activity, all the groups of sunspots in the western extremity of the northern hemisphere have the same polarity, whilst all the sunspots in the southern hemisphere have opposite polarity at the western extremity.

To evaluate the level of solar activity based on the development of the spots, an index known as relative Wolf number is used. It is a system created in 1848 by the Swiss astronomer Rudolf Wolf, expressed by the formula $R = k (10 g + f)$, where R is a relative number, k is an instrumental constant, g is the number of the groups of sunspots and f the number of single sunspots. The solar sunspot cycle proves the existence of a magnetic cycle lasting about 22 years, the interpretation of which is still highly complex. If at the poles the lines of force follow a North-South direction, on approaching the equator the different velocity of rotation of the gas (which grows as the latitude decreases) deforms them. In 3 years, in fact, at latitude 0°, 5 more full rotations are completed than are completed at latitude 50°. During a solar minimum all the lines of force run along meridians and merge at the two magnetic poles. With the passing of time, the lines become ever more deformed until they follow the track of the parallels. Beyond a certain point, they become entangled, knotted in loops, twisted and imprisoned within the magnetic field. The plasma gives rise to phenomena that can be observed on the Sun's surface- but the mechanism is more complex.

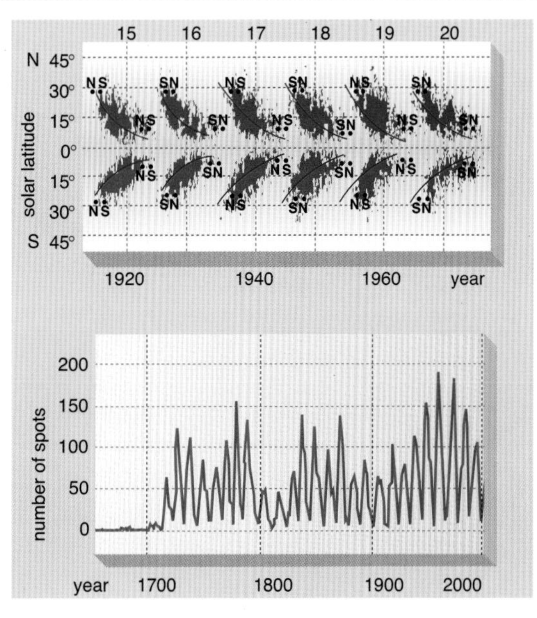

Solar cycle
The diagram shows the band within which lies Sun's equator and also shows the areas in which, with the passing of time, the spots are formed during each cycle. Shown also is both the drift towards lower latitudes and the change in polarity.

Number of spots
Each peak corresponding to a high number of spots corresponds to a solar maximum.

Increasing activity
Three images of the Sun taken in the FE XII line (195 Å) about 3 years apart, show a dramatic increase in activity. The corona's temperature is around 10^6K, and sunspots, eruptions and flares are found especially in areas of high magnetic instability.

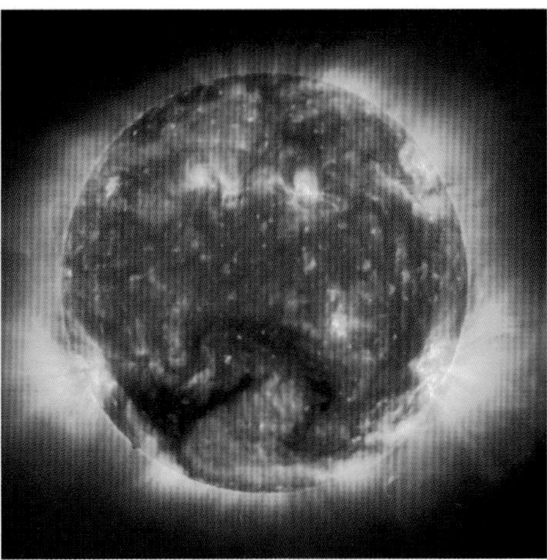

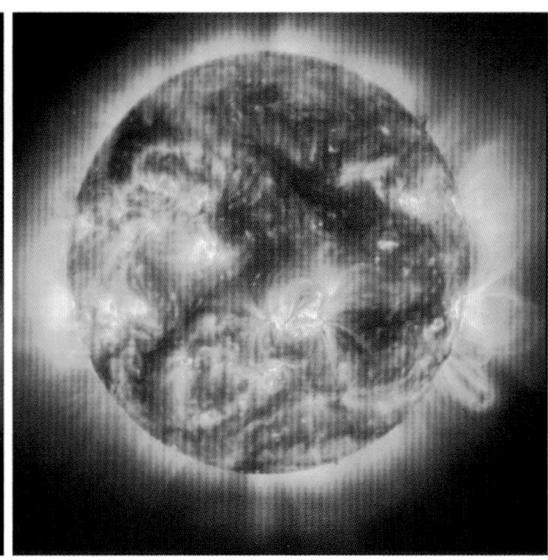

FROM MERCURY TO PLUTO

Errant lights, wanderers of the sky, the planets are the most visible objects of the solar system. Around the Sun, however, orbit millions of bodies more or less small and visible: they are satellites of the planets, asteroids, comets, but also masses of powder, molecular and atomic ions, particles and photons that fill up the "emptiness" of imperceptible matter and magnetic currents. It is clearly seen in the passage of the comets that, like pennants, evidence the solar wind that also deforms the planetary magnetic fields. Thanks to the space probes and to the great telescopes, today we know a lot about our system: all the planets except for Pluto, the comet 'Halley', and some asteroids have been observed close up, sampled, and analyzed.

Around the Sun move numerous bodies, many smaller and colder than our star: they are the planets and their satellites, but also the asteroids, the comets, the meteorites, the powders and ions of the interstellar medium legacy of solar gravity. A planet is a more or less spherical celestial body, in orbit around a star, that we can only see because it reflects the light. When talking about the Sun's various stars, however, the planets remain invisible: the light that they reflect from those distances is much too weak to be distinct from that transmitted by the star. In these cases it can be thought that in orbit around the star is a planet because gravitational effects are noticed on the motion of the star. With much difficulty, only large planets, of the dimensions of Jupiter and Saturn can be traced. The solar system is composed of nine planets (eight according to a group of scholars), with various structural and astronomical characteristics. Like the Earth, all revolve around the Sun rotating around its own axis more or less inclined with respect to the ecliptic and with respect to its plane of revolution. Moving away from the Sun encounter in the order: Mercury, Venus, Earth, Mars, Jupiter, Saturn, Uranus, Neptune and Pluto.

THE ROCKY, OR TERRESTRIAL, OR INNER PLANETS

They are those closest to the Sun (inner solar system): Mercury, Venus, Earth and Mars. They are relatively small, constituted

Solar System
The image summarizes the essential differences between inner planets and external planets.

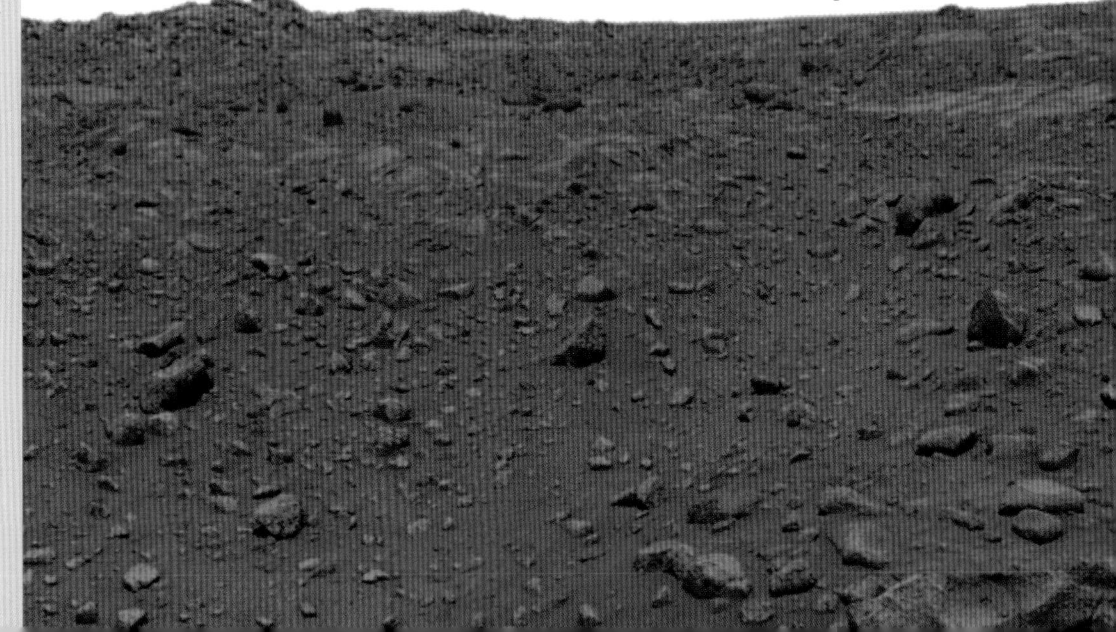

A rocky planet
Martian soil summary from the Pathfinder that removes every doubt on the physical or chemical characteristics of the planet.

The characteristics of rocky, or terrestrial, or internal planets

	Mercury	Venus	Earth	Mars
Equatorial radius (10^3 km)	2.439	6.052	6.378	3.396
Mass compared to Earth	0.06	0.8	1	0.1
Volume compared to Earth	0.06	0.88	1	0.15
Mean density (g/cm^3)	5.4	5.2	5.5	3.9
Polar flattening	0	0	0.003	0.005
Mean distance from Sun (10^6 km)	57.9	108.2	149.6	227.9
Number of satellites	0	0	1	2
Presence of rings	NO	NO	NO	NO
Principal atmospheric components	-	CO_2	N_2, O_2	CO_2

The characteristics of Jovian, or external planets

	Jupiter	Saturn	Uranus	Neptune
Equatorial radius (10^3 km)	70.85	60	25.4	24.3
Mass compared to Earth	317.8	95.1	14.6	17.2
Volume compared to Earth	1.316	755	67	87
Mean density (g/cm^3)	1.4	0.7	1.3	1.8
Polar flattening	0.061	0.109	0.03	0.03
Mean distance from Sun (10^6 km)	778.3	1.427	2.870	4.497
Number of satellites	16	23	15	8
Presence of rings	YES	YES	YES	YES
Principal atmospheric components	H, He	H_2, He	H_2, He, CH_4	H_2, He, CH_4

mostly from solid compounds like rocks (hence rocky), sands, powders. They are lacking in atmosphere or they have less extensive gaseous envelopes formed from relatively heavy elements: just like on the Earth (hence terrestrial). They do not have rings nor many satellites.

THE JOVIAN, OR EXTERNAL PLANETS

They are the four planets that follow orbits beyond (external) the asteroid belt, border of rocky bodies: Jupiter, Saturn, Uranus and Neptune. Characterized like Jupiter (hence Jovian) by very extensive atmospheres, small rocky cores surrounded by lighter elements (hydrogen, helium, methane, ammonia) that, subject to the low temperatures and high pressure inside the planets, are liquids or solids. Numerous rings of disparate materials orbit around every planet, together with a host of satellites of greater or lesser size.

OTHER BODIES

There are many smaller bodies of the solar system, even if their total mass is small:

• the ***satellites***: small and often spherical, sparkling from reflected light revolving around a planet;

• the ***asteroids*** or ***small planets***: with variable diameter from a few kilometers to nearly 1,000 km (on average less than 50 km) and irregular shape, they have orbits mostly between the Mars orbit and that of Jupiter;

• the ***comets***: objects rich in ice and little tens of kilometers in diameter, usually have very elongated orbits. Approaching the Sun, they become very visible;

• the ***meteorites***: smaller, they travel in space and at times enter the earth's atmosphere becoming incandescent from friction with the air.

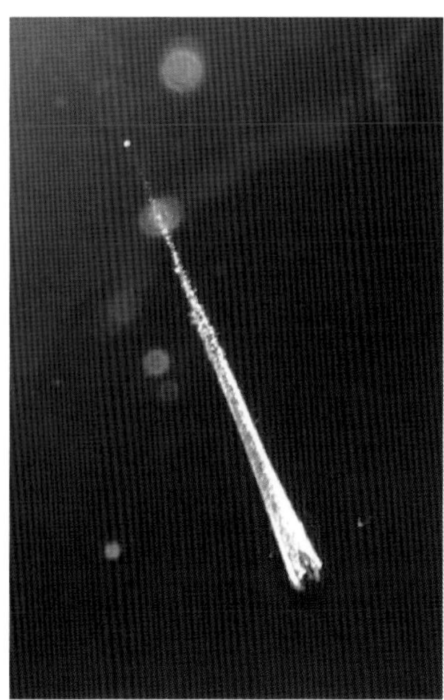

Planets and satellites
View of Jupiter and some of its larger satellites in a photo montage.

A meteor
The meteorites reach the Earth at a speed between 40,000 and 250,000 km/h.

THE MOTION OF PLANETS SEEN FROM EARTH IS NOT UNIFORM: IT SEEMS THAT THEY MOVE UP AND DOWN IN A CELESTIAL VAULT AND SOME EVEN TURN AROUND, THEN RESUME MOTION IN THE RIGHT DIRECTION. THAT IS DUE TO THE FACT THAT THE EARTH IS IN MOTION TOO. AND LIKE ALL THE OTHER PLANETS FOLLOWS KEPLER'S LAWS.

MOTIONS AND CONFIGURATIONS

KEPLER
The illustrations summarize the three empirical laws discovered by Kepler, that describe the planetary motions.
a. *First law: the orbits of planets are ellipses, and the Sun occupies one of two foci.*
Nearly all the planetary ellipses are a bit eccentric, but the distance from the Sun always varies between a minimum (**perihelion**) and a maximum (**aphelion**).
b. *Second law: the areas described by the Sun-planet vector are proportional to the times taken to describe them*
The velocity varies throughout the orbit: it is maximum at the perihelion, is minimal at the aphelion.
c. *Third law: the squares of the periods of revolution of planets are directly proportional to the cubes of the semi-major axes of the respective orbits.*

SOLAR SYSTEM
There are unavoidable errors in the proportions between bodies and distances, the image, summarizes the order of the main bodies of the solar system and the inclination of their orbits compared to the ecliptic.

The planets of the solar system (and like them also the satellites, the rings, the asteroids and the comets) are characterized by very complicated motions. They can be broken down like in the case of the Earth into simple motions that, put together, can more closely describe the true motion observed. In this way, they can be studied with greater ease using physics:

• all the bodies of the solar system, including the Sun, revolve around their own *axes of rotation;*
• all the bodies of the solar system revolve around the Sun along an *orbit;*
• all follow roughly *elliptic* trajectories;
• all the satellites revolve around planets following roughly elliptic trajectories;
• the axis of rotation of planets is inclined compared to the plane of their orbit around the Sun.

The physical laws that describe these celestial motions are the three Kepler's laws that find their

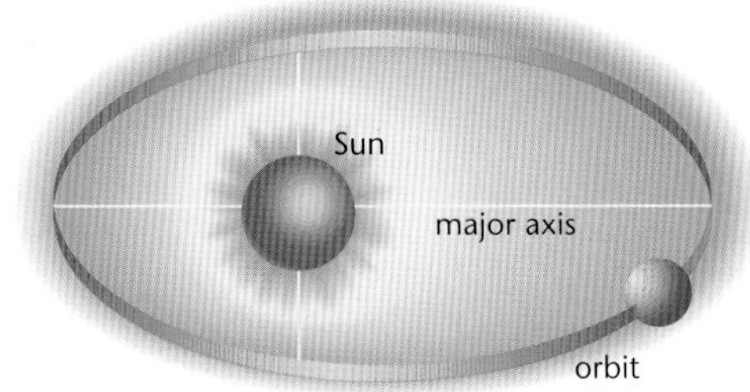

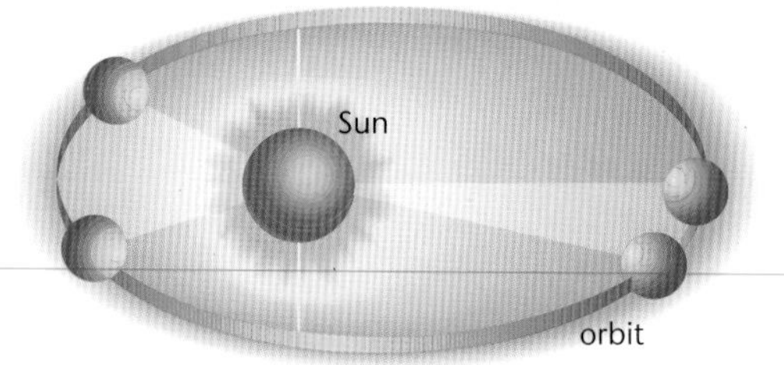

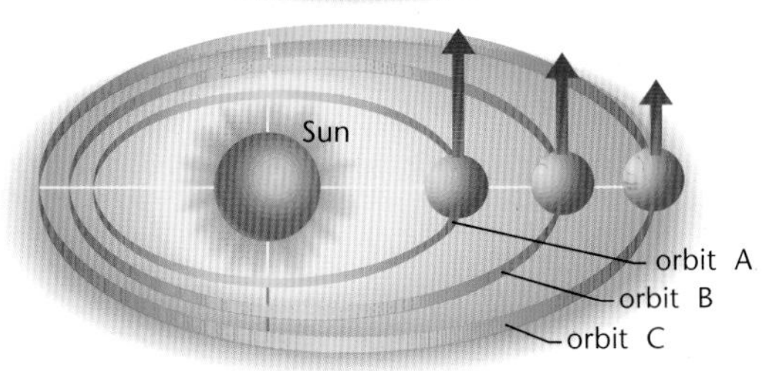

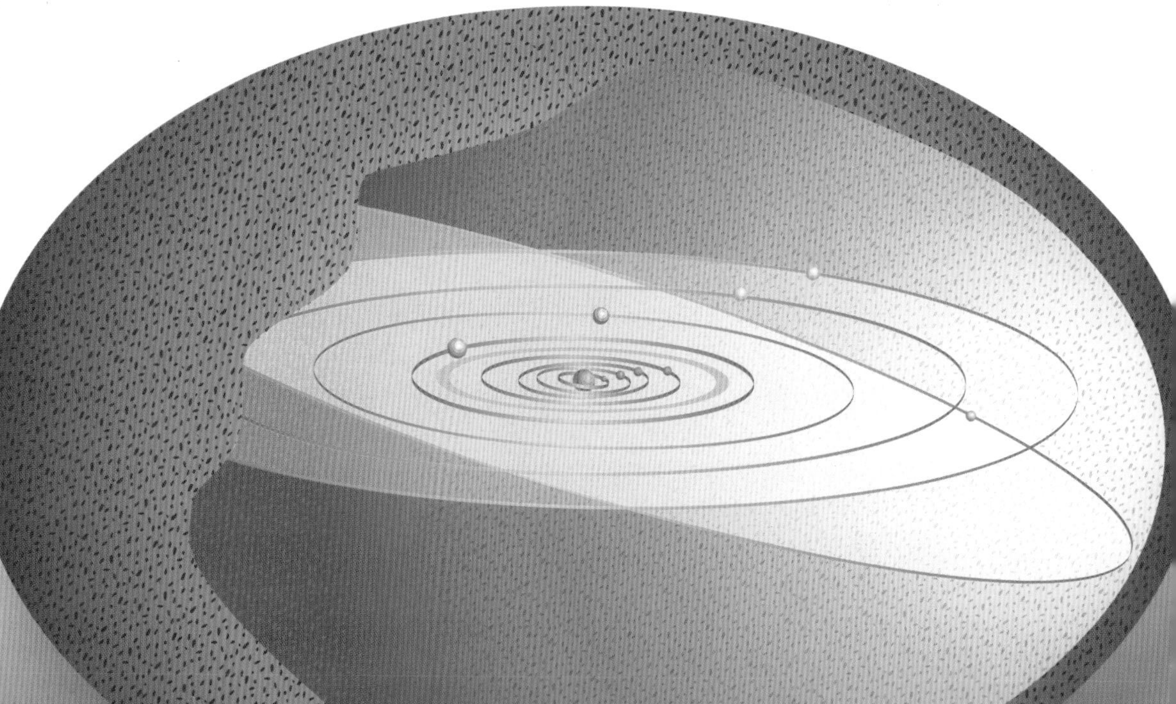

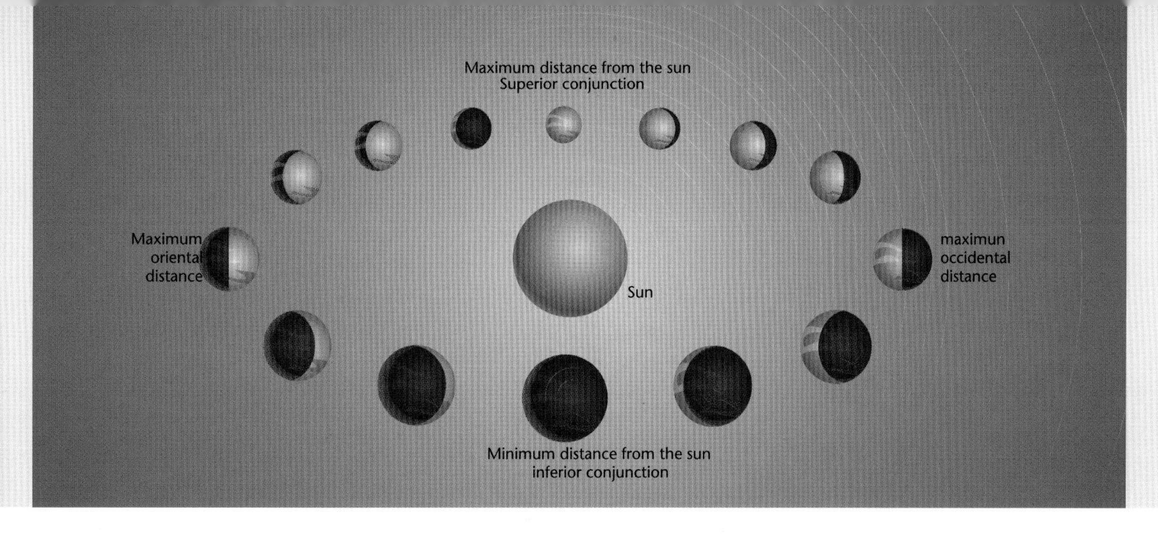

Variations
The schema summarizes the differences in magnitude which an internal planet shows in various positions compared to the Sun. In this case, it is Venus, which also shows a succession of phases like that of the Moon.

complete justification in Newton's universal law of gravitation. These laws apply both to planets in orbit around the Sun as well as satellites in orbit around planets and recurring comets or swarms of meteorites derived from the disintegration of old comets, or the asteroids which occupy the space between Mars and Jupiter.

DIRECTION OF MOTION

The sense in which all planets revolve around the Sun is "direct", that is to say counterclockwise for an observer positioned facing the Sun, with head towards the north pole of the ecliptic. The counterclockwise sense by which "having the head" towards the north pole of the ecliptic is also the sense in which rotation occurs around the own axes of almost all planets, and also in which rotation of almost all satellites occurs around the respective planets.

THE ORBIT AND APPARENT MOTIONS

Contrary to that which one would expect, the orbit of planets is not found all on the same plane: like the earth's is ecliptic, each planet has its own orbital plane defined by its respective orbit.
It is true, however, that the various orbital planes are little inclined between them: compared to earth's orbital plane, for example, the inclination of various orbits is contained within 5° (with the exception of the orbit of Mercury and Pluto). For this reason, often, in the night sky, can be seen some planets which, aligned with the Moon, "visualize" the ecliptic of the celestial sphere.
The various velocity with which the planets proceed along respective orbits (for planets found farther away from the sun, the speed is slower).

ALIGNMENT

The angular distance of a planet from Sun-Earth alignment is called elongation if the planet is external to the Earth, at 0° referred to as conjunction, at 90° referred to as quadrature, at 180° referred to as superior conjunction if the planet is internal, quadrature is not achieved and an elongation of 0° is referred to as superior conjunction, of 180° of inferior conjunction. the maximum elongation (East or West) is 28° for Mercury and 48° for Venus.

"Zig-zag" movement
The schema shows how it is possible to see a planet move "zig-zag" in the sky. In order to simplify things, we hypothesized that the planet is motionless but things do not change much if it moves very slowly from the Earth, as occurs in the case, noted, of the external planets.

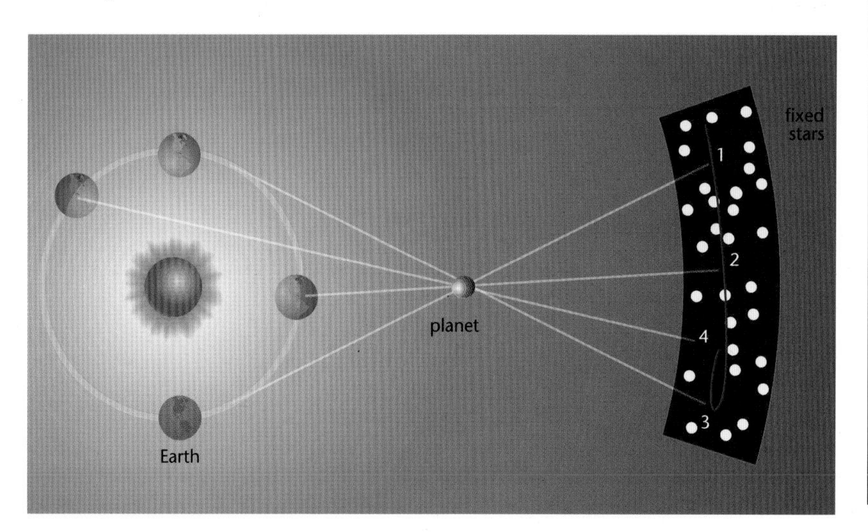

Distance and movement

Planet	Maximum distance from the sun (10^6 km)	Revolution in days (d) or mean solar years (a)	Rotation (h)	Orbital velocity (km/s)
Mercury	69.7	88 d	1,407	47.9
Venus	109.0	224.7 d	24.62	35.0
Earth	152.1	365.26 d	23.93	29.8
Mars	249.1	687 d	24.62	24.1
Jupiter	815.7	11.86 a	9.84	13.1
Saturn	1.507	29.46 a	10.24	9.6
Uranus	3.004	84.01 a	15.6	6.8
Neptune	4.537	164.8 a	18.5	5.4
Pluto	7.375	247.7 a	153	4.7

IT IS THE PLANET CLOSEST TO THE SUN, DENSEST AFTER THE EARTH, SMALLEST (WITHOUT COUNTING PLUTO), AND IS ALSO THE PLANET WHERE THE THERMAL JOLT BETWEEN NIGHT AND DAY IS MOST DRAMATIC: ON AVERAGE 550 °C! FROM A SCORCHING MID-DAY INDEED (427 °C) TO A FREEZING MIDNIGHT (-123 °C) IN LITTLE MORE THAN 500 HOURS: THE LONGEST DAY OF THE SOLAR SYSTEM.

MERCURY

Mercury is indeed a special planet and, for many aspects, all yet to be discovered: because of the remarkable vicinity to the Sun, in fact, is very difficult to observe it from earth. On an average as far away as "only" 43 million kilometers from the visible surface of our star, straying to a maximum of 28° from the solar edge. Therefore, from earth Mercury is visible only at daybreak or sunset. It also has a very rapid orbital motion (rushing in haste towards the Sun): it finishes a complete voyage in only 88 days. Therefore it can only be observed for a few successive days, and then disappears. From earth, moreover, like in the case of the Moon and of Venus, also in the case of Mercury, phases can be observed: use of binoculars or a telescope, however, can be dangerous for the eyes, and it is necessary to pay particular attention not to aim the instrument directly at the Sun. Beyond the little information gathered with enormous patience by the astronomers of the past - Mercury is known since antiquity – we have a firm mass of data thanks to the explorations carried out using space probes. In particular, in 1974 the planet was explored by Mariner 10, an American probe which approached within some hundreds of kilometers from the

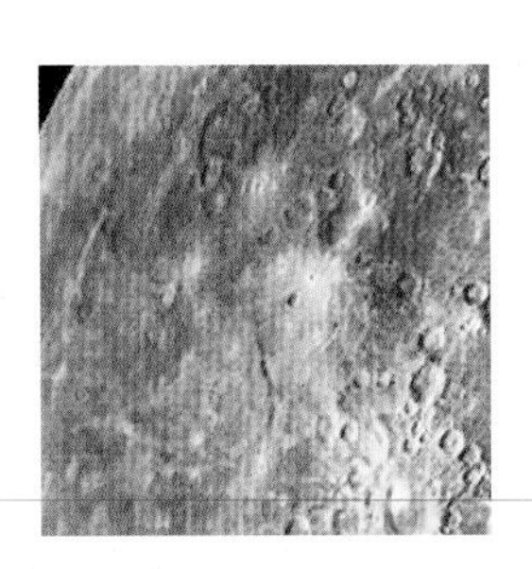

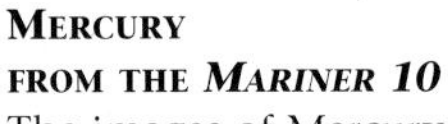

MERCURY FROM THE *MARINER 10*
The images of Mercury on these pages show three collage of photographs released from Mariner 10. Knowledge of this planet is not very advanced, even though solar probes have gathered new data. Recent surveys carried out by the researchers of Arecibo have evidenced some polar zones using high power radio wave reflection, explained with the existence of a layer of ice a couple of meters thick. New European, Japanese and American expeditions are previewed for 2004 and 2009.

ABOVE: False color processing shows the diversity of chemicals making up the soil.

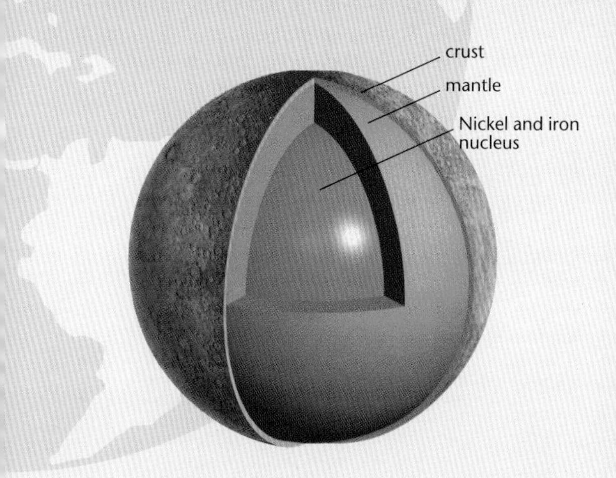

THE CHARACTERISTICS OF MERCURY

EQUATORIAL RADIUS (IN km)	2,439.7
MASS (IN 10^{23} kg)	3.3
• COMPARED TO EARTH	0.055
VOLUME (IN 10^{23} cm^3)	0.07
• COMPARED TO EARTH	0.054
MEAN DENSITY (IN g/cm^3)	5.43
MEAN TEMPERATURE (K = °C)	
• OF DAY	700.15 = 427
• OF NIGHT	100.15 = -173
SUPERFICIAL GRAVITY (IN m/s^2)	2.78
POLAR FLATTENING	0
MEAN DISTANCE (IN 10^6 km)	
• FROM SUN	57.9
• FROM EARTH (INFERIOR ELONGATION)	91.7
PERIHELION (IN 10^6 km)	45.9
SATELLITES: NONE	
APHELION (IN 10^6 km)	69.7
APPARENT DIAMETER OF SUN	1°22.7'
PERIOD	
• OF ROTATION (IN MEAN SOLAR H)	1407
• OF REVOLUTION (IN MEAN SOLAR D)	87.969
ECCENTRICITY OF ORBIT	0.2056
MEAN ORBITAL VELOCITY (IN KM/S)	47.9
INCLINATIONE	
• OF ORBIT COMPARED TO ECLIPTIC	7°
• OF EQUATOR TO ORBIT	<28°
ALBEDO	0.10
ATMOSPHERIC PRESSURE (IN 10^{-7} Pa)	1
PRINCIPAL ATMOSPHERIC COMPONENTS (PERCENTAGE)	HE (42%), NA (42%), O (15%)

surface transmitting approximately 6,000 photographs, covering approximately 40% of the surface. Therefore, for the first time, they have been able to observe from "close up" the arid panoramas of Mercury, subjected to the strongest solar radiation. Mercury turns around on its axis (sidereal day) in 58.65 Earth days, but because the Sun turns on the same point of the surface, it requires 176 Earth days: in this period, the planet turns twice around to the Sun. This slow spin exposes to the Sun for long periods the same hemisphere: made red-hot at more than 400 °C, Mercury is cooled off in the long night to 180 °C below zero (if it is at aphelion, even to -430 °C). This thermal excursion is the greatest of the entire solar system: on average 500 °C between night and day. That is also due to the absence of atmosphere: blasted by the solar wind, with a decidedly low gravity - it is the smallest planet of the system, Pluto excluded - Mercury is surrounded by a gaseous veil made up predominantly of helium. The particles α (helium nuclei) that constitute the more massive part of the solar wind would be captured continuously and undergo loss to the planet. Mercury has a lot in common with the Moon: due to its position, it has been bombarded by meteorites captured by the Sun, and is literally covered with craters. The lowlands are also very similar to the lunar river basins, presumably also produced by lava flow. The more important formation on Mercury is the Caloris Basin, an approximately 1,400 km wide, 9 km deep crater, encircled by mountain ridges approximately 2 km high: it is believed to have had meteoritic origins, approximately 3.54 billions of years ago. The frightful collision threatened to destroy the planet: at the antipodal to the Caloris Sea, in fact, a complex network of fractures is observed, in all probability, formed due to the repercussion of the impact. On the inner structure hypotheses can only be made.

VENUS IS A PLANET VERY SIMILAR, IN DIMENSIONS, TO THE EARTH. LIKE THE EARTH, IT HAS SOME VOLCANOS THAT ARE NO LONGER ACTIVE. BUT THE LIKENESS ENDS HERE: FLAT FOR 65% OF THE SURFACE, VENUS IS WRAPPED IN A THICK CARBON DIOXIDE ATMOSPHERE, SWEPT BY VIOLENT WINDS LIKE HURRICANES, HEATED FROM THE GREENHOUSE EFFECT TO AN AVERAGE TEMPERATURE OF NEARLY 500 °C

VENUS

Like Mercury, Venus is an inner planet and orbits the space between the Sun and the Earth. For this reason, Venus can also be better observed at sunset or dawn. If it is visible at sunset, it is the first star to appear in the sky, and for this the ancients called it Vesper (from the Latin name of the sunset), or the evening star. If it is visible at dawn, it precedes the Sun slightly: for this the ancients called it Lucifer (that in Latin means bearer of light), or the morning star. To observe Venus with the naked eye, or with binoculars or a telescope is, indeed, very easy: it can be seen disappearing (or rising) 3 hours after (or before) the Sun (sunset or dawn). For this reason, even in antiquity, its motions and phases were well known. In fact, because its motion is more close to the Sun than that of the Earth, as happens with the Moon also Venus is illuminated in various ways and, seen from earth, passes through a decreasing phase and an increasing phase, until disappearing in order then to resume the cycle.

Unlike the Moon, however, during the phases, Venus changes in a very remarkable way also in the dimensions of its apparent diameters: because of the strong difference in distance from the Earth when it is in full phase (2.60×10^8 km) or in crescent phase (only 0.4×10^8 km), in fact, passes from an apparent diameter of 10" of arc to an

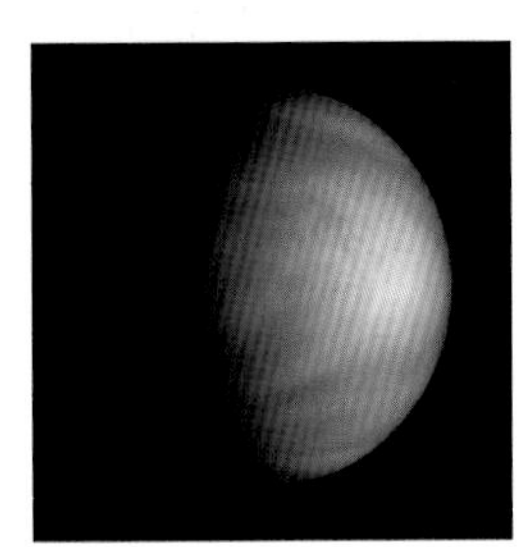

VENUS FROM SPACE
Three images of Venus: ABOVE, IN VISIBLE LIGHT AND ULTRAVIOLET ACQUIRED BY GALILEO; NEXT, A COLLAGE

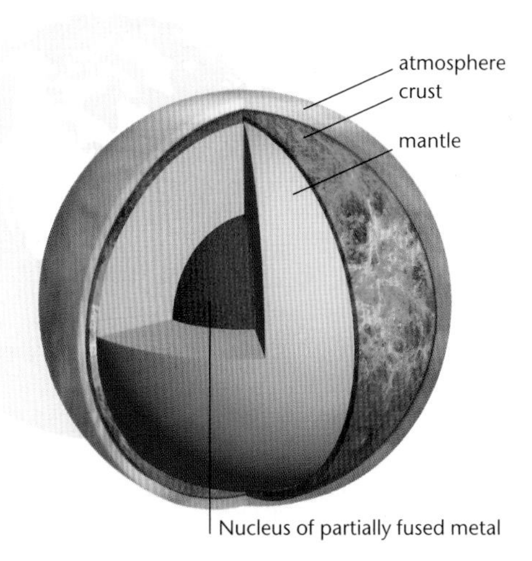

THE CHARACTERISTICS OF VENUS

EQUATORIAL RADIUS (IN km)	6,052
MASS (IN 10^{24} kg)	4.90
• COMPARED TO EARTH	0.8149
VOLUME (IN 10^{27} cm^3)	0.95
• COMPARED TO EARTH	0.84
MEAN DENSITY (g/cm^3)	5.25
MEAN TEMPERATURE (K = °C)	
• OF DAY	243.15 = - 30
• OF NIGHT	753.15 = 480
SUPERFICIAL GRAVITY (IN m/s^2)	8.87
POLAR FLATTENING	0
MEAN DISTANCE (IN 10^6 km)	
• FROM SUN	108.2
• FROM EARTH (INFERIOR ELONGATION)	41.4
PERIHELION (IN 10^6 km)	107.4
SATELLITES: NONE	
APHELION (IN 10^6 km)	109.0
APPARENT DIAMETER OF SUN	44.2'
PERIOD	
• OF ROTATION (IN MEAN SOLAR H)	5,832.23
• OF REVOLUTION (IN MEAN SOLAR D)	224.7
ECCENTRICITY OF ORBIT	0.0068
MEAN ORBITAL VELOCITY (IN KM/S)	35.03
INCLINATION	
• OF ORBIT COMPARED TO ECLIPTIC	3.394°
• OF EQUATOR TO ORBIT	177°3'
ALBEDO	0.65
ATMOSPHERIC PRESSURE (IN 10^6 Pa)	9
PRINCIPAL ATMOSPHERIC COMPONENTS	CO_2 (96%) N (3%)

absolute and apparent diameter 66" of arc. The observations of this phenomenon allowed Galileo to propose one of the most concrete tests of validation of the Copernican theory. Unlike Mercury's orbit, that it is very inclined compared to the ecliptic, Venus' orbit is inclined only 3.4°. Accordingly, the planet is in some periods completely hidden by the Sun (occultation) and in others, it crosses the disc from one side to another (transit): in this case, it appears as a black disk projected on the luminous background of the Sun. Differently from the majority of other planets, moreover, Venus revolves around its own axis of rotation in a retrograde sense: whoever lived on the planet, would see the Sun rise in the West and set in East.

A PLANET LIKE THE EARTH, BUT MORE DIVERSE

Venus is very similar to the Earth: the dimensions, the mass, the density, the gravity, the presence of a thick gaseous atmosphere - responsible, on the other hand, being able reflecting, of its brightness - are the characteristics that this planet shares with ours. For this reason it has been long believed to accommodate forms of life similar to those on Earth.

And in search of these life forms, numerous space expeditions have been sent by the United States and Soviet Union. But all the types of probes that have explored the atmosphere of the planet and its surface, have collapsed every hope of finding even the smallest living being: Venus is similar the Earth, but not greatly. The data collected on the atmospheric constitution, the characteristics of the surface and the soil composition, in fact, have sufficiently allowed knowledge of the "mean" conditions of this inhospitable planet.

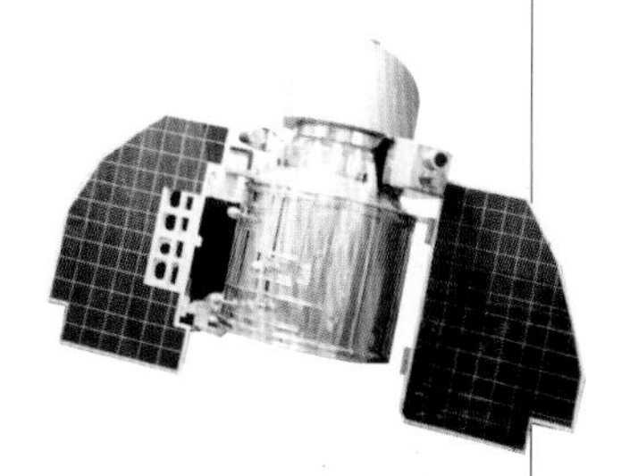

VENERA 1
was the first probe to travel towards Venus, planned in order to float in the hypothetical venusian ocean after the descent to the planet. 7 days from the launch, however, contact was lost. In 1961 it passed within 100,000 km of Venus, entering a heliocentric orbit.

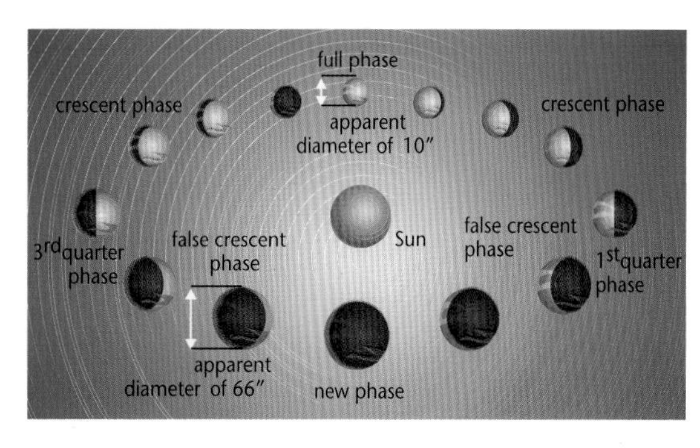

PHASES
Phases and apparent diameters of Venus.

TOPOGRAPHY
This topographical map of Venus, realized on the base of the data collected from the Magellan expeditions, show in one sinusoidal projection the various geologic characteristics of the venusian soil. False color processing evidences the inferior zones and altitudes.

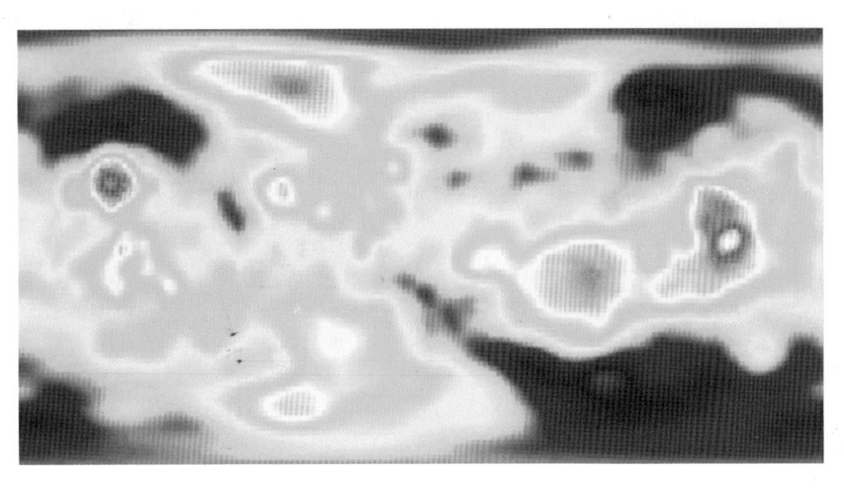

Gravity
The Venusian planisphere shows the distribution of gravimetric data gathered by Magellan.

Computer
The black-blue colors signal the presence of thicker cloud

Calendar of explorations

1961 *Sputnik 7* and *Venera 1*: Attempts to reach Venus
1962 *Mariner 1* and *2*; *Sputnik 19*, *20* and *21* attempt to fly by the planet: : only *Mariner 2* succeeded (at 35.000 km)
1963 *Cosmos 21*: attempt failed
1964 *Venera 64A*, *64B*; *Cosmos 27*, *Zond 1*: attempt failed
1965 *Venera 2* and *3*, *Venera 65A*: attempt failed
1967 *Venera 4* atmospheric data; *Mariner 5* data collection, *Cosmos 167* failed
1969 *Venera 5* and *6* atmospheric data
1970 *Venera 7* landed; *Cosmos 359* failed
1972 *Venera 8* landed; *Cosmos 482* failed
1973 *Mariner 10* flew past Venus and Mercury
1975 *Venera 9* and *10*: orbiter was placed in orbit and released lander with probe
1978 *Pioneer-Venus 1* and *2*, *Venera 11* and *12* placed in orbit and samples collected
1981 *Venera 13* and *14* placed in orbit and samples collected
1983 *Venera 15* and *16* placed in orbit
1984 *Vegas 1* and *2* placed in orbit and released balloon-probe
1989 *Magellano* placed in orbit

Missions URSS: *Cosmos, Sputnik, Vegas, Venera, Zond*

Missions USA: *Mariner, Magellan, Pioneer-Venus*

A "HEAVY" ATMOSPHERE AND VOLCANIC SOIL

The thick Venusian atmospheric blanket, approximately 85 km high, is formed 96% of carbon dioxide and on the surface pressure develops approximately 92 times greater than that that which is measured on the Earth at sea level. There is decidedly little water in the atmosphere of Venus and, therefore, on the entire planet and the clouds, suspended to a suspended to a mean 50 km, are constituted mostly from sulfuric acid. In order to have an idea of the dimensions of the Venusian atmosphere, we can remember that, on the Earth, nearly all the air is accumulated in the first 5.6 km and that all the meteorological phenomena take place exclusively in the first 10 km above the surface. It is not the case that, from space, the Earth's surface can be easily seen while the surface of Venus remains invisible also to the more powerful instruments: the refraction is too strong. That which is observed from the telescope, therefore, is not the surface of Venus, but the top of its atmosphere: carbon dioxide, in fact, is a gas that reflects visible and infrared radiation in a more efficient way. For this same reason, the mean temperature of the planet is extremely elevated: in the neighborhood of 500 °C, quite higher than on Mercury. This value - decidedly too elevated to be produced only from solar radiation, even if this is, however, greater compared to that found on the Earth - explained as the result of millennia of heaviest greenhouse effect due indeed to the high percentage of atmospheric carbon dioxide. Moreover in the high atmosphere, winds blow at a velocity of more than 400 km/h: the speed of rotation of the

A mountain on Venus
Computer generated three-dimensional image on the basis of the Magellan surveys that reproduces a panorama of Mount Maat.

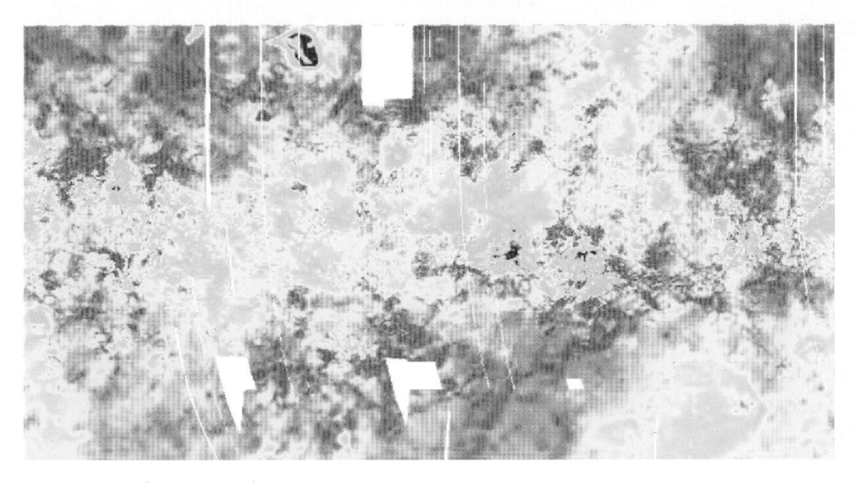

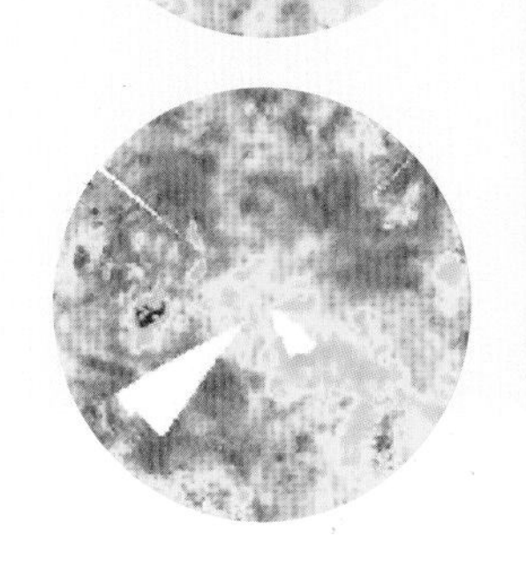

Venus microwaves
Analysis of the microwave emissions of the planet: polar planisphere and zones (on the left) North and South. The white parts (not colored with false color processing inserted by computer) indicate zones in which emissions have not been detected.

planet, in fact, it varies considerably whether it is the clouds or the surface which are observed. While the clouds complete a rotation in little more than 4 days, the surface takes approximately 243 days to complete a rotation. Since a Venusian year (that is the time it takes to complete an entire revolution around the Sun) is 224 days, the day on the surface of Venus is longer than a year. For this reason, on the surface, the wind never exceeds a few kilometers per hour. But not enough. The equipment released from Venera 10 that succeeded in landing undamaged, was destroyed after few hours of operation above all because of the strong pressure. However, in those hours they sended unequivocable data: under the planet's enormous gas cloak, the sunlight scarcely penetrates and the brightness of the Venusian sky is equal to that of a day of intense fog on Earth. In this warmest penumbra, moreover, there is a constant rain of sulfuric acid droplets mixed with water. Surveyings of the probes have also allowed us to know a lot about the surface of Venus. In large part, it is flat with deep depressions, such as the Venus Rift Valley that is 4 km in depth and 1,400 km long, ridges that exceed 1,000 m. Moreover, it is suspected that Venus has had an intense phase of volcanism, supported by tectonic plates similar to that of Earth, even though no volcanic emission is currently recorded. Venus does not have satellites, nor a magnetic field: it is probable that this last characteristic is due to the slow motion of the planet rotation around its axis.

Magellan
The probe, here still anchored in Earth orbit, waiting to be launched towards Venus.

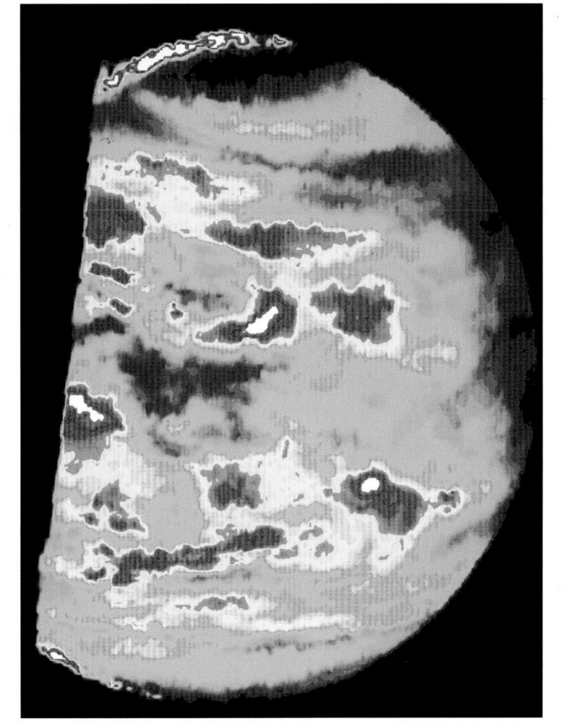

Venus by infrared
The Venusian night investigated by the spectrometer of Galileo, in travel towards Jupiter (1990). From a distance of approximately 100,000 km, the Near Infrared Mapping Spectrometer (NIMS) analyzes the emission of the planet to a wavelength of 2.3 μm. The false color processing evidences the zones in which the heat irradiated from the Venusian surface succeeds in crossing the blanket of clouds, 10 times darker. The red-white colors indicate light clouds, while black-blue colors signal the presence of thicker cloud.

THE RED PLANET, CALLED THUS FOR THE COLOR OF THE SURFACE, WELL VISIBLE ALSO BY NAKED EYE, IS THE FIRST "EXTERNAL" PLANET IN TERRESTRIAL ORBIT. VERY SMALL, VERY COLD, FOR MANY REMINISCENT OF THE EARTH: THE ICECAPS CHANGE APPEARANCE VARYING WITH THE SEASONS, THE SURFACE SHOWS TRACES OF RIVERS AND ANCIENT VULCANISM. AND SINCE ANTIQUITY IT IS SUSPECTED THAT IT HOSTS LIFE FORMS.

MARS

Seen with a telescope or good binoculars, appears like a small reddish disc: it is by reason of its manifest coloration that it comes to be called the " red planet". It is much smaller than the Earth: it has a diameter that is approximately half that of our planet, and perhaps also for this it is not too adapted to observation and: in fact, although it is a planet relatively close - to the perigee it is little more than 56 x 10^6 km - also in the best conditions, it is not possible to distinguish the details of the surface. What can in truth be seen, is the thin network of dark and regular structures that Schiaparelli described in 1877: not referring to channels constructed by living beings, as he interpreted it, but of optical illusions produced from the atmospheric turbulence. On Mars, in fact, there has been no evidence of life found.

Mars always has been considered very similar to the Earth: the duration of the day is nearly equal to that of Earth since the Martian period of rotation is only slightly longer than that of Earth (approximately 24.5 hours). But likenesses ends here: as has happened in the case of Venus, all the data collected from the probes emphasize only the enormous difference between Mars and our planet. Its mass, as an example, is only a tenth of that of the Earth, and therefore, the gravity is also much less; the temperature fluctuates between many tens of subzero degrees and few degrees above zero, with great thermal excursions every day (about 50 °C of difference between night and day). The cause of these strong thermal differences, is contributed by the extremely

C.

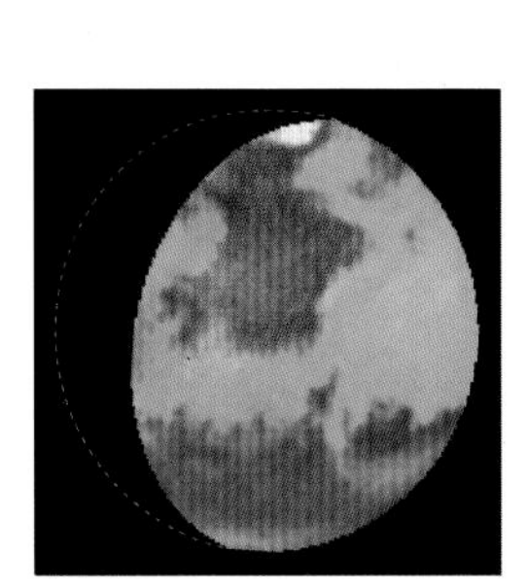

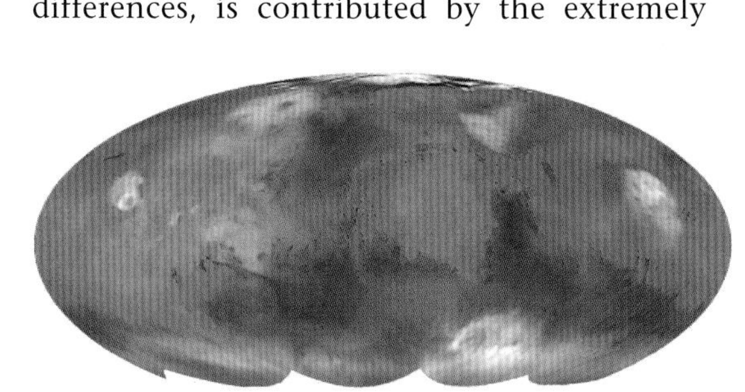

VIEW OF MARS
CENTER: Mars. Valles Marineris visible.

ABOVE: Mars in infrared. Red zones could well be water deposits

BELOW: Mars taken by Hubble in a photograph made up of three images using different filters (410nm, blue; 502nm, green; and 673 nm, red).

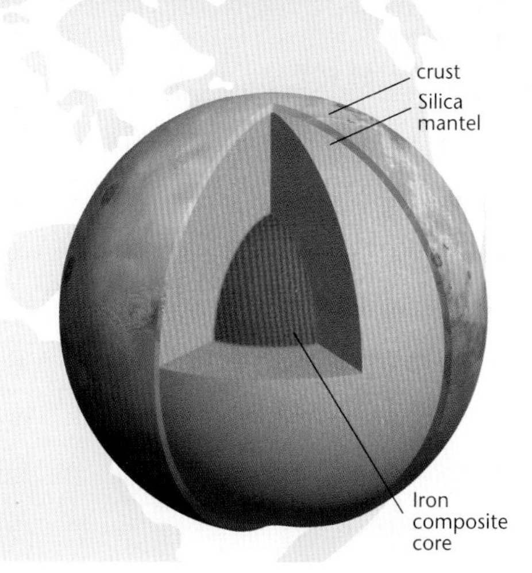

THE CHARACTERISTICS OF MARS

EQUATORIAL RADIUS (IN km)	3,397
MASS (IN 10^{23} kg)	6.42
• COMPARED TO EARTH	0.108
VOLUME (IN 10^{27} cm^3)	0.16
• COMPARED TO EARTH	0.15
DENSITÀ MEDIA (IN g/cm^3)	3.94
MEAN TEMPERATURE (IN K = °C)	
• OF DAY	293.15 = 20
• OF NIGHT	133.15 = -140
SUPERFICIAL GRAVITY (IN m/s^2)	3.8
POLAR FLATTENING	0.005
MEAN DISTANCE (IN 10^6 km)	
• FROM SUN	227.940
• FROM EARTH (INFERIOR ELONGATION)	78.4
PERIHELION (IN 10^6 km)	206.7

SATELLITES: 2	
APHELION (IN 10^6 km)	249.1
APPARENT DIAMETER OF SUN	21'
PERIOD	
• OF ROTATION (IN MEAN SOLAR h)	24.62
• OF REVOLUTION (IN MEAN SOLAR d)	687
ECCENTRICITY OF ORBIT	0.0934
MEAN ORBITAL VELOCITY (IN km/s)	24.1
INCLINATION	
• OF ORBIT COMPARED TO ECLIPTIC	1.85°
• OF EQUATOR TO ORBIT	24°11'
ALBEDO	0.37
ATMOSPHERIC PRESSURE (IN 10^2 Pa)	7
PRINCIPAL ATMOSPHERIC COMPONENTS	CO_2 (95%). N_2 (2.7%). AR (1.6%)

rarefied atmosphere: approximately 100 times less dense than Earth, it is not in a position to develop a sensitive greenhouse effect, despite being constituted mostly from carbon dioxide. For the low temperatures reached, especially at the poles, this gas becomes solid during Winter and diffuses during Summer. Summer can be torrid: on average the temperature fluctuates -68 °C, but it is sufficient to decrease the icecaps. The dramatic thermal excursions generate a veritable "atmospheric sea", with currents much stronger than those found on Earth. Even in the rarefied Martian atmosphere, agitation provokes immense sandstorms that, periodically, wrap the entire planet. The Mars surface, in fact, is desert, covered by iron oxide powder, of red rust color. Raised by wind, this powder also gives a rosy coloration to the Martian atmosphere: it is seen very well in the data sent back to earth by the American probes that have succeeded in landing on the planet.

GEOLOGY

But it is very probable that, at one time, Mars had a very different appearance.
Wide canyons and plains of alluvial type

MGS IN MARTIAN ORBIT
The Mars Global Surveyor is currently in orbit around Mars and collects data fundamental for the next missions. The picture is a reproduction of it while photographing Mount Olympus.

MARTIAN SOIL
Landing point of Viking 1 module

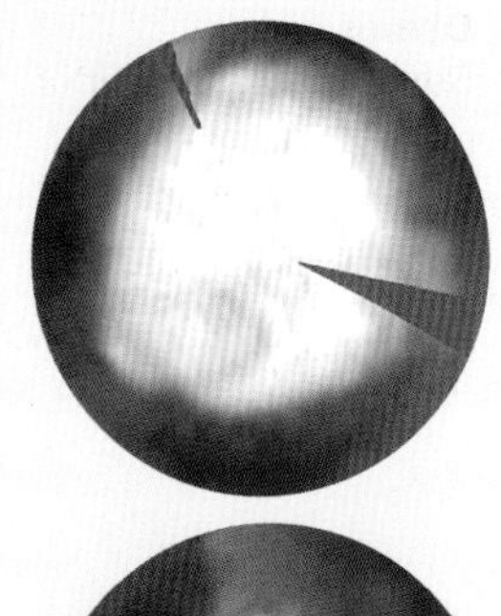

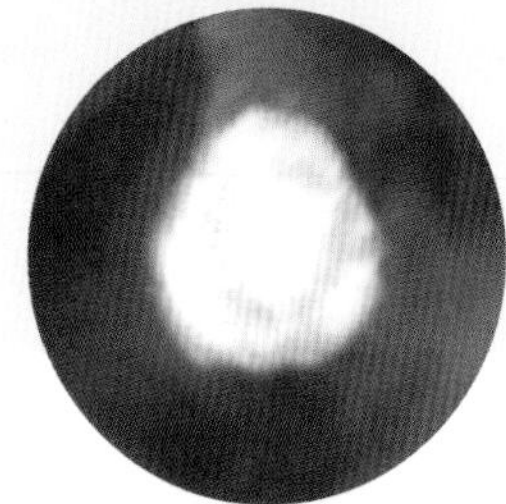

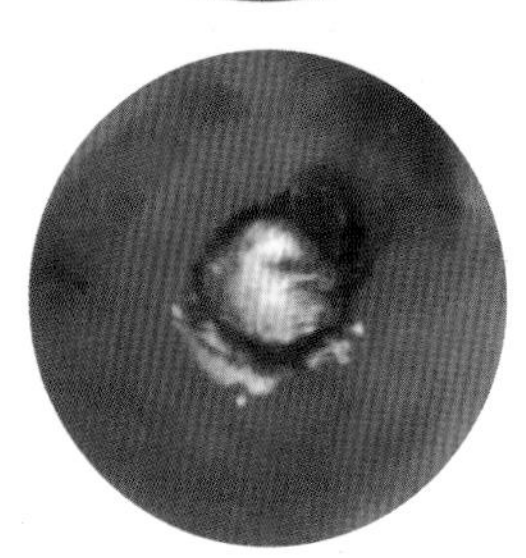

CAP AT NORTH POLE
ABOVE: in Winter, in Autumn and Spring, in Summer

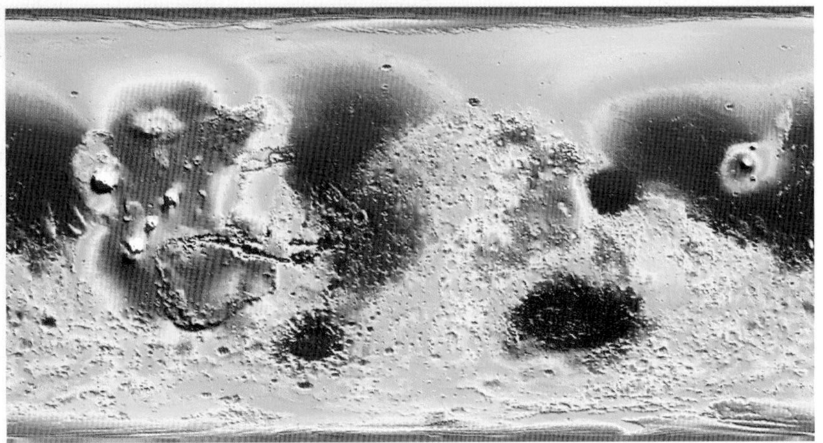

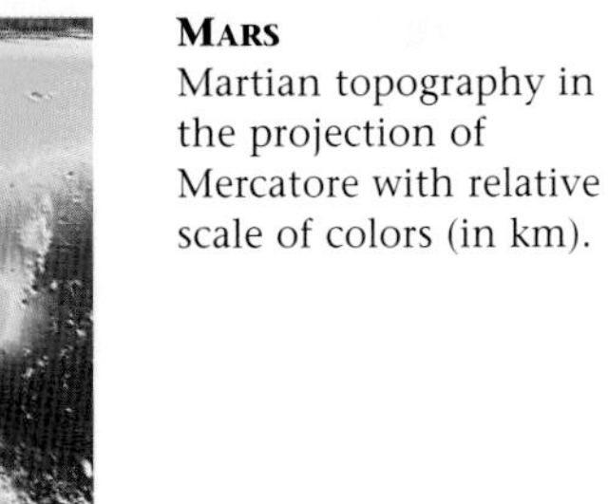

MARS
Martian topography in the projection of Mercatore with relative scale of colors (in km).

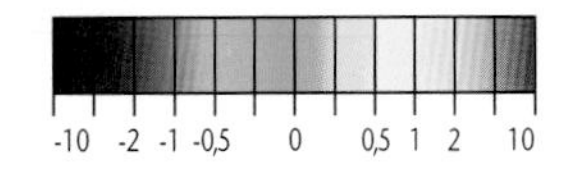

demonstrate that the surface has been shaped by fluid flow. More probable is the hypothesis that this was water: some billions of years ago, Mars had rivers, presumablly of torrential character, that has left the design of their voyage upon the rocky surface of the planet.
Of this water, however, there is no longer any trace: it is found neither in the liquid state, nor in the gaseous state, but it could be present in the subsoil in the form of ice. The evidence of an intense volcanic activity is numerous and indisputable: in the equatorial belt, the two immense regions of Tharsis and of Elysium present a high concentration of extinguished volcanoes of remarkable dimensions: Mount Olympus, largest, exceeds 26 km in height, higher than three Mt. Everests put together! The immense lava streams that encircle them (the base of Mount Olympus alone would cover the entire state of Arizona) demonstrate that, in the past, the endogenous activity of the planet has been considerable. Perhaps, as has happened on the Moon, lesser gravity has favored the spread of the lava streams; however it is certain that the mass of materials ejected has often been important in erasing, in many zones, every trace of meteoritic craters.

THE MARTIAN LANDSCAPE

On Mars the landscape is more varied than how much they show the images collections from the

VOLCANOES OF THE SOLAR SYSTEM

VOLCANOES of the SOLAR SYSTEM magma wells up uninterrupted from the planetary depths, just like the volcanoes of Hawaii. This fact, united to the lack of superficial tectonic movements,
The volcanoes are apertures in the solid crust of a planet or a satellite, from which fused material, gas and vapors escape.
The volcano presence is closely tied to the inner dynamics of the celestial body: there must, in fact, some such inner power source sufficient to fuse the materials or to heat them until rendering them gaseous. On the Earth, this energy derives, most probably, from the nuclear decay of a series of elements present in depth beyond that ofthe diffusion of tectonic mechanisms also in the other rocky planets, is found on Mercury, Venus, the Moon and Mars: enormous plains of solidified lava, funnel shaped buildings, craters, stratified cones, rock "flood" channels, to calderas (that is the collapsed volcanic structure) testify to geologic activity by now extinguished on every one of these celestial bodies. On Mars, in particular, rises the highest volcano of the solar system: Mount Olympus has geologic characteristics similar to those of shield volcanoes would explain the great height which these structures reach. Very different is the discussion that regards volcanoes of Io, satellite of Jupiter. Without considering Earth, certainly the planet geologically most active of the entire system, is the body richest in still active volcanoes. Numerous images have been acquired by the Voyager probes. But the differences are many: the "warm" zones

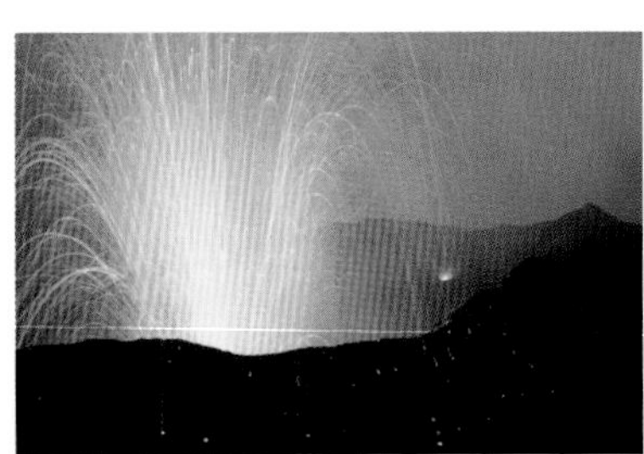

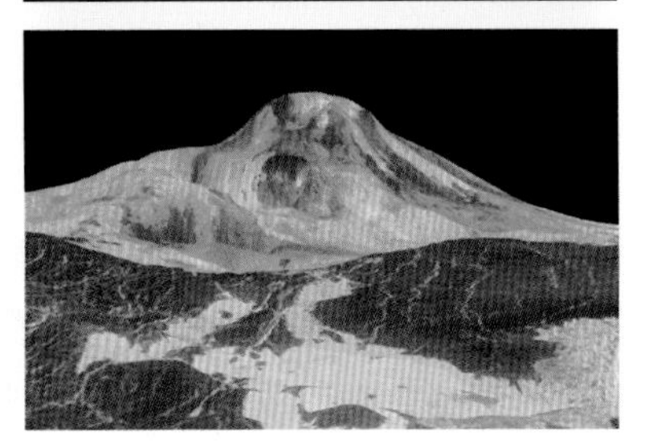

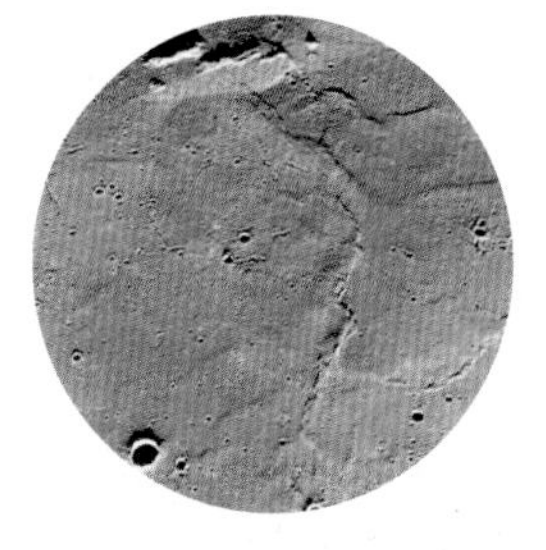

TYPES OF VOLCANOES
Hawaiian Volcanoes, computer image of the Venusian volcano Mount Maat; traces of lava streams on the Moon.

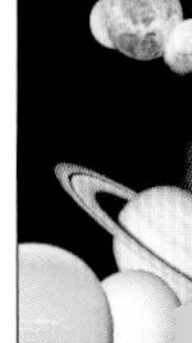

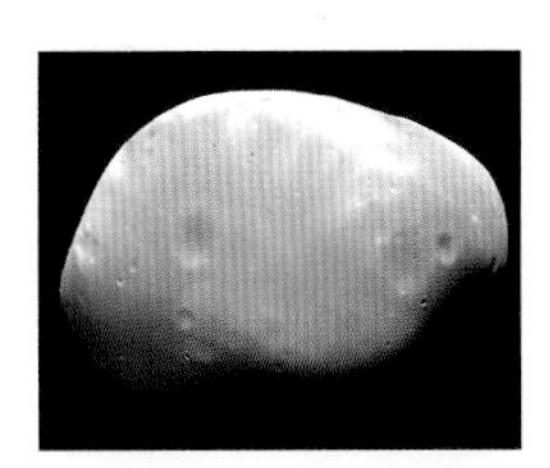

Phobos
Since its orbital motion decreases nearly 2 m each century, it is thought that the tidal forces which cause precipitation to fall on Mars or, more probably, will produce fracture. Phobos, then, will be transformed into a ring.

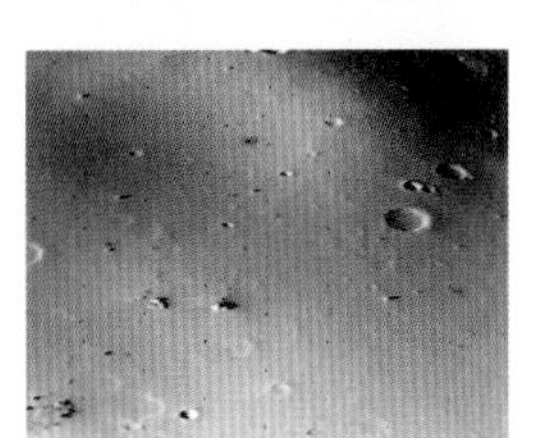

Below: surface of Phobos

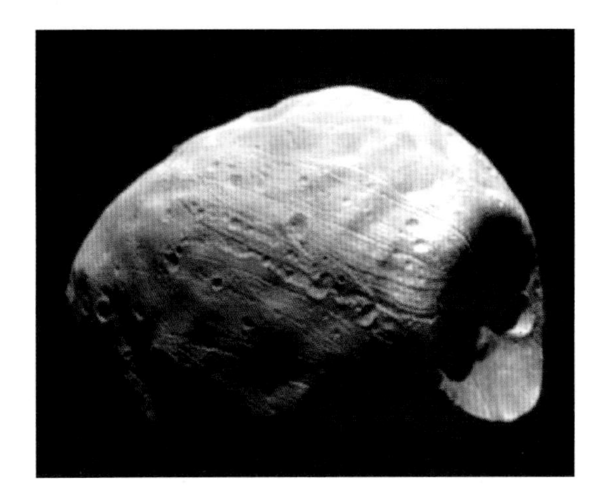

Deimos
Famous for the crater of an immense meteoritic impact.

Sassi
Below center: a detail of the Martian surface. Rust red dust and stones.

probes: in the hemisphere North, in fact, pianeggianti and desert regions, covered predominate of cliff rossastra; in the hemisphere South, instead, the territory is deeply uneven, interrupted continuously from numerous craters of meteoritica origin. Hellas, one of the largest, has one of 3km deep and a diameter of 1,800 km. The Tharsis region, then, is crossed from an enormous fracture that extends for approximately 5.000 km, par nearly sixth of the circumference of the entire planet. Referred to is the Valles Marineris, about 100 km wide. The poles, with their resplendent ice caps of solid carbon dioxide, are the most changeable geologic formation of the Martian surface. They, in fact, vary remarkably in size, changing with the season.

PHOBOS AND DEIMOS

Are the two Mars satellites: Phobos (Fear) and Deimos (Terror) remember respectively the son and the companion of Mars, Greek god of war. They are not visible using an amateur telescope: their respective diameters of 25 and 13 km are too small. From what we can tell they seem to be meteorites captured by Mars' gravity.

scarcely reach 17 °C (everything is relative: the surrounding atmosphere is -176 °C!), and only locally active the volcanic orifices reach approximately 500 °C. On the Earth, geysers also reach higher temperatures. At similar temperatures, only some sulfur materials liquefy; however it is probable that also some silicas are involved in eruptions. There are also various causes at the origin of this increase in temperature: rather than tied to an insufficient radioactive decay, they would be due to the energy of the tidal forces, to which the satellite is subordinate.

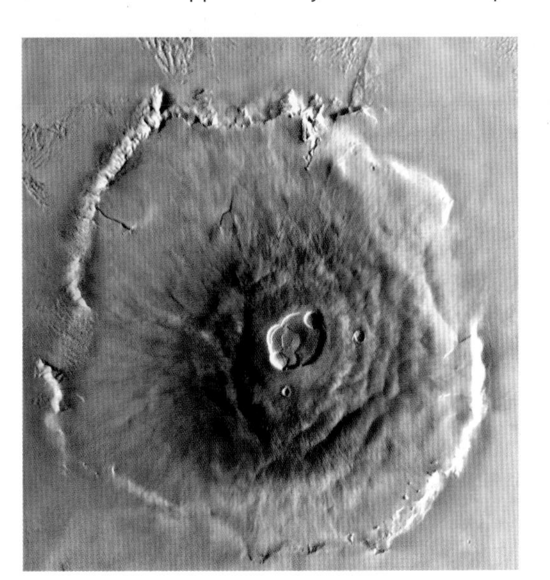

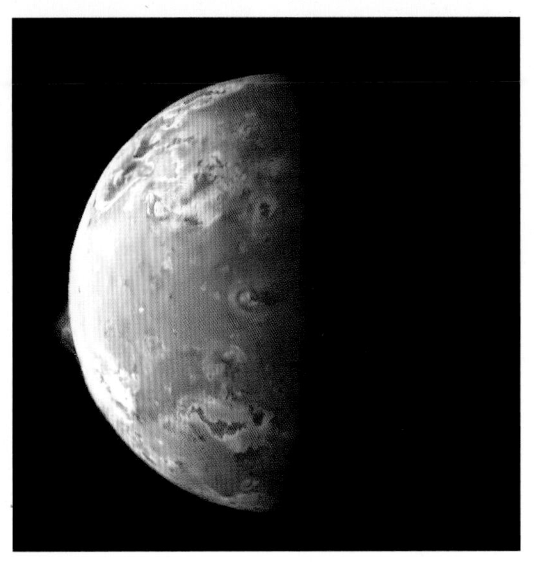

Types of Volcanoes
Mount Olympus eruption: the plume reaches 280 km in height.

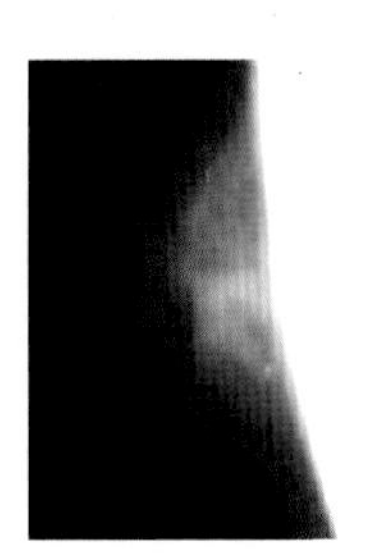

BLOCKS OF ROCK MOVE IN THE IMMENSE SPACE COMPRISED BETWEEN THE ORBIT OF MARS AND THAT OF JUPITER: THEY ARE THE ASTEROIDS, OR SMALL PLANETS.
THE LARGEST HAVE A DIAMETER OF THOUSANDS OF KILOMETERS, THE SMALLEST ARE PEBBLES.
THEY ARE SUBJECT TO DESTRUCTION AND THEIR FRAGMENTS CAN FALL TOWARDS THE INNER PLANETS.

THE ASTEROID BELT

ASTEROIDS
a. Gaspra;
b. Ida and Dattilo, its small "satellite";
c. Giano;
d. Matilde;
e. Eros.

In the region of wide space more than double the Earth-Sun distance (on average 2.8 UA) between the orbit of Mars and that of Jupiter, there are billions of prevalently rocky celestial objects called asteroids or small planets. They have reduced dimensions and irregular shape: their greater diameter, on average, 2- 3 km, and rarely exceeds 50 km. Largest is Cerere, with a diameter of approximately 1,000 km: discovered in 1801 by Giuseppe Piazzi, at the Observatory of Palermo. Since the beginning of the past century, there have been many discoveries: initially were chosen for them names taken from Greek mythology, Cerere, goddess of the harvest, followed Pallas Athena, Juno, Vesta... But very soon there were not enough names: the asteroids today catalogue to approximately 35,000, and have names of objects, queens, sweethearts and betrothed of astronomers, or also of famous personages, from those of the Beatles to Hemingway, Rembrandt and Clapton. But more than half are identified alone with a number. The fascia - or belt - of asteroids has mysterious origins: since according to the law of Titius-Bode in this zone of the solar system, a planet should be found, some students think that the asteroids are that which remains of a rocky planet mysteriously exploded or crushed. Others think, instead, that they are that the remains of small planets formed in the primitiva nebula, in the primordial solar system: they would be objects that "did not succeed" in accreting and forming a new planet. Whichever of these is their origin, the astronomers agree on the fact that the asteroids have the same origin as planets. The second hypothesis, however, rouses greater consensus: is more probable, in fact, that the contrary and equally strong gravitational attractions of Jupiter and the Sun have prevented the primitive matter that was found in this zone of the solar system to aggregate in a single planetary body. Losing heat, it would consequently have "condensed" into a myriad of different bodies.

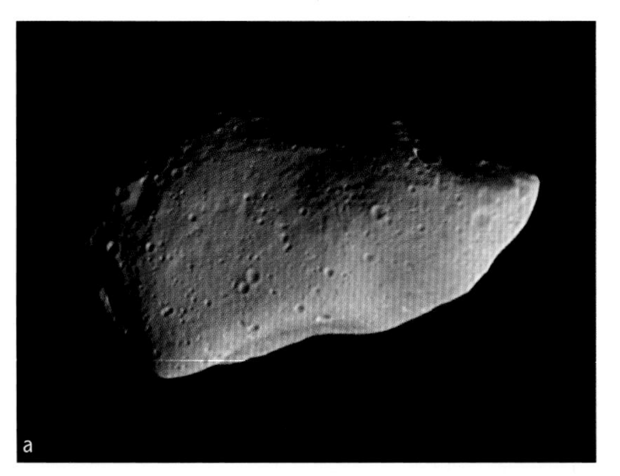
a

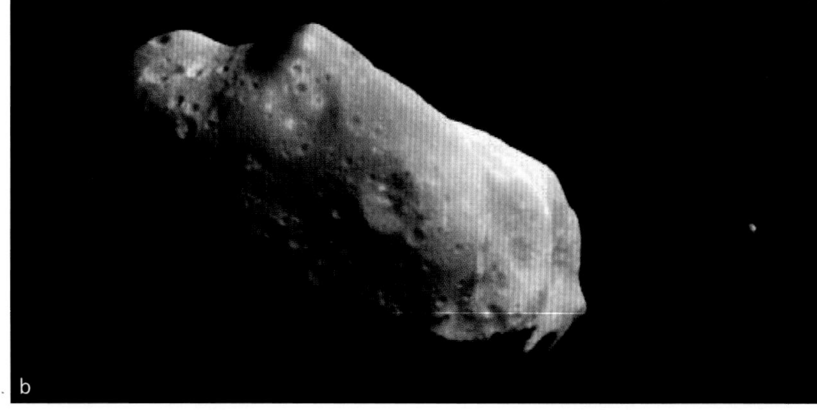
b

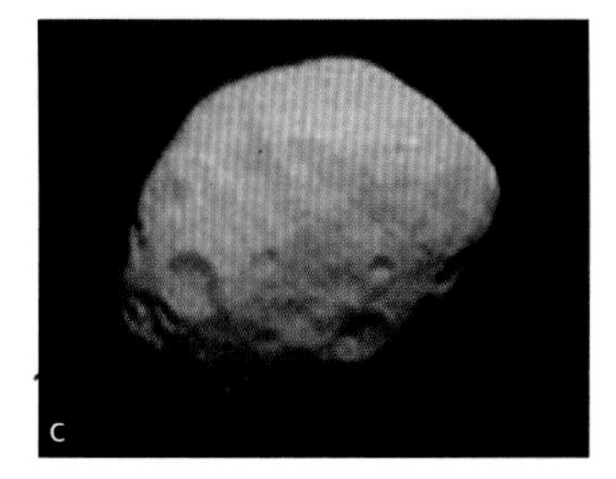
c

d

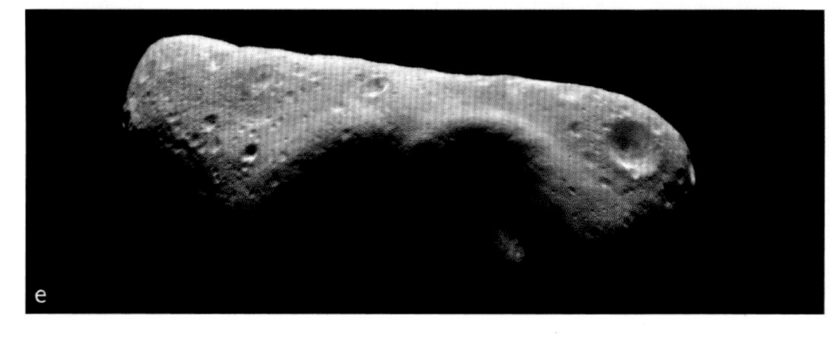
e

THE CHARACTERISTICS OF THE PRINCIPAL ASTEROIDS

NAME	MEAN DISTANCE FROM SUN (IN 10^6 km)	DIMENSIONS OR RADIUS(IN km)	ALBEDO
GASPRA	205.0	17 x 10	0.20
IDA	270.0	58 x 23	?
EUNOMIA	395.5	136	0.19
CERERE	413.9	457	0.10
PALLADE	414.5	261	0.14
PSICHE	437.1	132	0.10
INTERAMNIA	458.1	167	0.06
EUROPA	463.3	156	0.06
IGEA	470.3	215	0.08
DAVIDA	475.4	168	0.05
SILVIA	521.5	136	0.04
VESTA	353.4	262.5	0.38

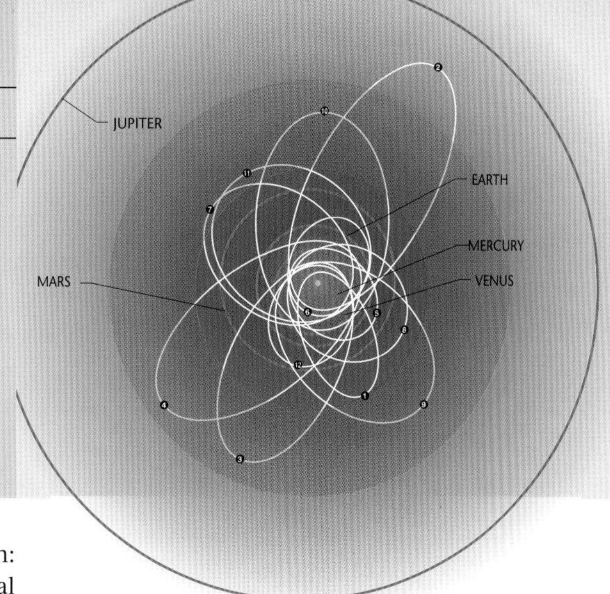

ORBIT
The orbit of NEO in relation to Jupiter's orbit and the other planets. The colored ring shows where the majority of asteroids' orbits. The smaller figures indicate the Sun's distance.

The surface of asteroids reminds one of the Moon: craters, striature, tormented peaks, these natural masses, predominantly rocky and mostly metallic, mark the crash with meteorites or other celestial bodies.

DANGER OF IMPACTS

Many asteroids "travel in pairs" revolving around a center of common gravity; moreover often have a very elongated orbit: some of them, like Eros, Icaro or Apollo, in their motion of revolution around the Sun intersect the orbits of the Earth, Venus, Mercury. Since they have a very small mass, they are particularly sensitive to the gravitational interferences of the greater bodies, and may possibly, for collisions or particular planetary configurations, change their distance without warning.
The study of their motions, therefore, is very important: it is certain that a part of the meteorites which reach the Earth come from this zone of the cosmos, and an impact with an asteroid, however remote, can never be excluded. The EGA, Earth Grazing Asteroids (asteroids that graze the Earth), or NEO, Near Earth Objects (objects near the Earth) are all asteroids with a diameter under 50 km that have an orbit such that it renders them possible dangers for our planet. They pass "only " within several hundreds of thousands of kilometers from our planet, and thus they are continuously monitored so as to be able to preview each possible impact with the Earth with a sure margin of intervention, in case it should change trajectory. It is indeed a difficult job if it is believed that the NEO are very many, and with complex orbits: there are 150 million with a diameter beyond 10 m, 300,000 of approximately 100 m, 1,000 of 500 m, more than 2,000 of 1 km, approximately 400 of 2 km and ten of 10 km.

VESTA
This is perhaps the most studied asteroid.
From left: photographic reproduction of the Hubble; computer reconstruction; planisphere and computerized three-dimensional image that summarize the data relative to the altitude.

IS THE LARGEST PLANET OF THE SOLAR SYSTEM AND THE SECOND OF EXTERNAL PLANETS. ITS THICK ATMOSPHERE PREVENTS SEEING THE PLANET PROPERLY, BUT GIVES A BRIGHTNESS THAT AT TIMES EQUALS THAT OF VENUS.
IN MANY ASPECTS, IT IS SIMILAR TO THE SUN: IT EXHIBITS A DIFFERENTIAL ROTATION, HAS SUPERFICIAL PHENOMENA OF GRAND DIMENSIONS AND SHORT DURATION BUT, ABOVE ALL, EMITS ENERGY. MORE THAN THAT RECEIVED FROM THE SUN.

JUPITER

IMAGES FROM JUPITER
BELOW: Jupiter in the visible acquisition from the Hubble.
BELOW NEXT: the same frame, infrared data allows visibility, beyond to clouds, also the thin ring that is shown on the page opposite in a more profound analysis.

TOP: computer elaboration of Jupiter's rings.

The largest planet is not alone: after Venus, when it is in opposition, it is also the brightest planet. Jupiter lends itself very well to observation and with binoculars it is already possible to see the planet surrounded by its four main satellites. Through a telescope or good binoculars, it can be easily observed: it has the appearance of a floppy disk flattened at the poles criss-crossed by light and dark streaks parallel to the Equator. In order to distinguish other details of the surface it is necessary to use a telescope: we will be able to then observe a rapid alternation of changing formations that in the rotation of which some day or quite few hours arise and disappear, but better still, we will be able to see also the famous red spot, a hurricane of immense proportions that has churned the atmosphere of Jupiter and has done so for at least three centuries. The maximum measures of this superficial phenomenon have been 39,000 km by 14,000 km: it was sighted for the first time more than 300 years ago, and the planetologists are inclined to think that it is a phenomenon

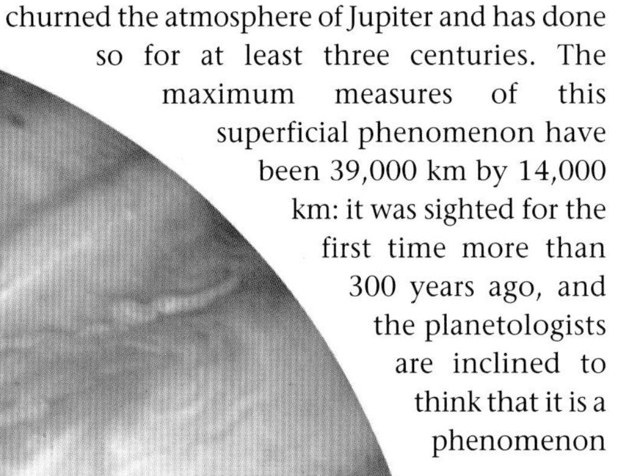

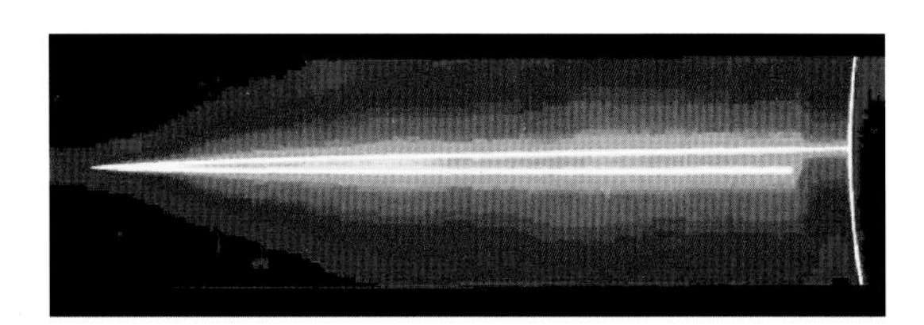

of long but unstable duration. Also the coloration of horizontal colorful bands is produced by strong winds and convective movements that churn gases of the atmosphere: in the clear zones the warmer gases rise quickly to high elevation where they are crystallized, returning to descend down to the darker zones. The dominant colors – yellow, orange, ochre - are produced from ammonia and ammonium hydrosulfide that is condensed in thick clouds, while the more intense colors are produced from compounds of fluorine.

NOT A TRUE PLANET NOT A TRUE STAR

After the Sun, Jupiter is the largest body of the solar system: compared to the Sun it has a smaller diameter 10 times smaller, a mass 1,000 times smaller and a nearly equal density. Its volume is far greater than that of Earth: more than 1,300 times, and its mass is such to produce sensitive

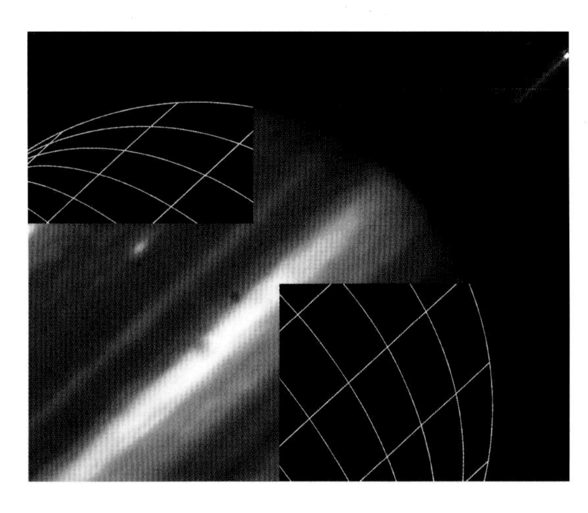

atmosphere
liquid hydrogen and helium
metallic hydrogen and helium
rocky core

THE CHARACTERISTICS OF JUPITER

EQUATORIAL RADIUS (IN km)	71.492
MASS (IN 10^{27} kg)	1,9
• COMPARED TO EARTH	317,938
VOLUME (IN 10^{27} cm^3)	1.425,23
• COMPARED TO EARTH	1.316
MEAN DENSITY (g/cm^3)	1,33
MEAN TEMPERATURE (IN K = °C)	
• OF DAY	152,15 = -121
• OF NIGHT	303,15 = 30
SUPERFICIAL GRAVITY (IN m/s^2)	22,87
POLAR FLATTENING	0,061
MEAN DISTANCE (IN 10^6 km)	
• FROM SUN	778,4
• FROM EARTH (INFERIOR ELONGATION)	628,8
PERIHELION (IN 10^6 km)	740,9
SATELLITES: 17	
APHELION (IN 10^6 km)	815,7
APPARENT DIAMETER OF SUN	6,2'
PERIOD	
• OF ROTATION (IN MEAN SOLAR H)	9,84
• OF REVOLUTION (IN MEAN SOLAR D)	11,86
ECCENTRICITY OF ORBIT	0,0483
MEAN ORBITAL VELOCITY (IN km/s)	13,1
INCLINATION	
• OF ORBIT COMPARED TO ECLIPTIC	1,308°
• OF EQUATOR TO ORBIT	25°11'
ALBEDO	0,52
ATMOSPHERIC PRESSURE (IN 10^4 Pa)	7
PRINCIPAL ATMOSPHERIC COMPONENTS	H (90%), HE (10%)

perturbations in the orbits of asteroids and comets.

The motion of rotation around the axis is very fast: in scarcely 9h50min it finishes a complete rotation. It follows that the linear speed at the Equator reaches 12.6 km/s; that is to say provokes considerable flattening at the poles and the obvious rotation differential similar to that which is verified on the Sun: Jupiter rotates not like a rigid body, but more quickly at the Equator than at the poles.

With the Sun, Jupiter has in common also the principal constituents: they are gas like hydrogen (90%) and helium (9%), and in moderation clearly inferior methane and ammonia. Moreover, the data collected with the probes Pioneer, Galileo and Voyager indicate that the planet is contracting: every year its diameter would be reduced by approximately 1 millimeter. This consideration and numerous others, confirm the hypothesis that Jupiter is a mostly gaseous body, and that the energy that it emits is in greater measure compared with the quantity received from the Sun as well as is produced only from the "gravitational concentration" demonstrated by the probes. If Jupiter had been slightly larger,these processes could have flowed in a nuclear fusion reaction, just like the Sun. And our system would have been a double system. Instead the mass of Jupiter is not sufficient: for this reason it is referred to as a "missing star", and its temperature remains considerably low. The fast movement of rotation provokes other consequences as well. In the first place it has determined a geostrophic equilibrium between the atmospheric pressure and the Coriolis force, hindering the establishment of atmospheric currents in meridian direction. Moreover, the high speed of rotation would be the cause of the intense magnetic field that has origin on the planet: the magnetosphere of Jupiter extends in space for millions of kilometers, eliciting beautiful polar auroras similar to those that are observed on the Earth. On the base of the data collected from the probes, moreover, it is believed that the interior of Jupiter is made up of a small rocky core of iron-sillicate surrounded by an ocean of liquid metallic

JUPITER HEAT
Image of Jupiter. The total amount of the energy emitted from the planet exceeds that which it received from the Sun.

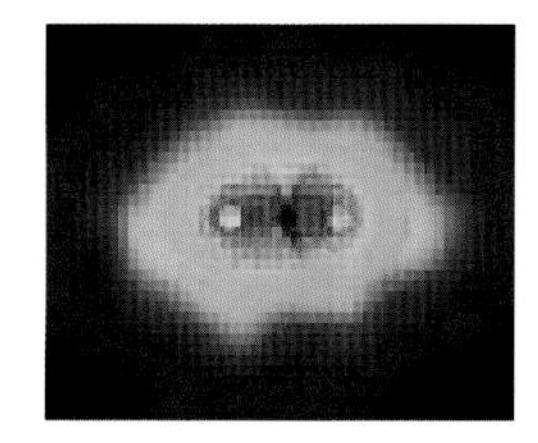

MAGNETISM
The magnetosphere of Jupiter and, under, a collage with the polar auroras produced from the magnetic field and the solar wind.

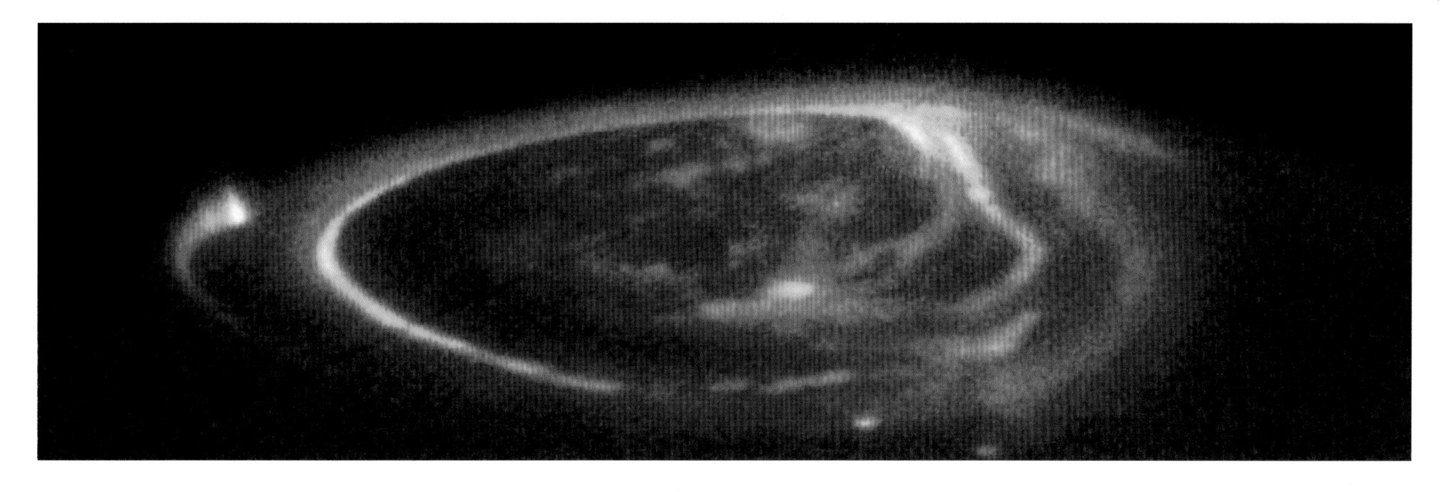

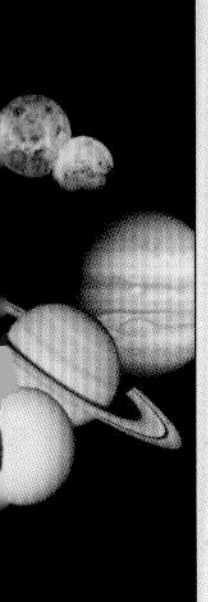

Jupiter and its entourage
Right: Io while it journeys along its orbit, data returned by the Hubble. In the background, the clouds of Jupiter.

Below: Io in infrared: the clear colors highlight the warmer zones.

hydrogen. This structure would contribute to explain the magnetic field , but it is only a working hypothesis. Above, the atmosphere extends nearly 1,000 km.

RINGS AND SATELLITES

Beyond an entourage of numerous satellites (currently 17 are known, but the number is certainly destined to grow as space exploration provides new data), Jupiter has a ring of incoherent materials similar to the much more spectacular one which orbits around Saturn. Invisibile to the telescope, it is revealed for the first time from the Voyager probe: often only 4 km, orbit around the planet to a distance of approximately 60,000 km from the top of clouds. The largest of Jupiter's satellites are noted: Io, Europa, Ganimede and Callisto. They were discoveries by Galileo in 1610 and called "Medician satellites" in homage to the Florentine lineage. Easily observable with good binoculars, they appear like very small stars around the planet but, since the plan of their orbits is slightly inclined compared to Jupiter, eclipses are very frequent. It was studying their motions that Ole Rømer, in 1675, succeeded in measuring for the first times the speed of the light. Thanks to the

Principal rings and satellites

Name	Distance from center of Jupiter (10^3 km)	Dimensions width x depth or radius (km)	Mass (kg)
Alone	92	$3 \cdot 10^4 \times 2 \cdot 10^4$	?
Principal rings	122.5	$6.4 \cdot 10^3 \times <30$	1.0×10^{13}
Metis	127.97	20	9.56×10^{16}
Adrastea	128.98	12.5 x 10 x 7.5	1.91×10^{16}
Amalthea	181.30	135 x 84 x 75	7.17×10^{18}
Thebe	221.90	55 x 45	7.77×10^{17}
Io	421.60	1.8015	8.94×10^{22}
Europa	670.90	1.569	4.80×10^{22}
Ganimede	1,070.00	2.631	1.48×10^{23}
Callisto	1,883.00	2.400	1.08×10^{23}
Leda	11,094.00	8	5.68×10^{15}
Himalia	11,480.00	93	9.56×10^{18}
Lysithea	11,720.00	18	7.77×10^{16}
Elara	11,737.00	38	7.77×10^{17}
Ananke	21,200.00	15	3.82×10^{16}
Carme	22,600.00	20	9.56×10^{16}
Pasiphae	23,500.00	25	1.91×10^{17}
Sinope	23,700.00	18	7.77×10^{16}

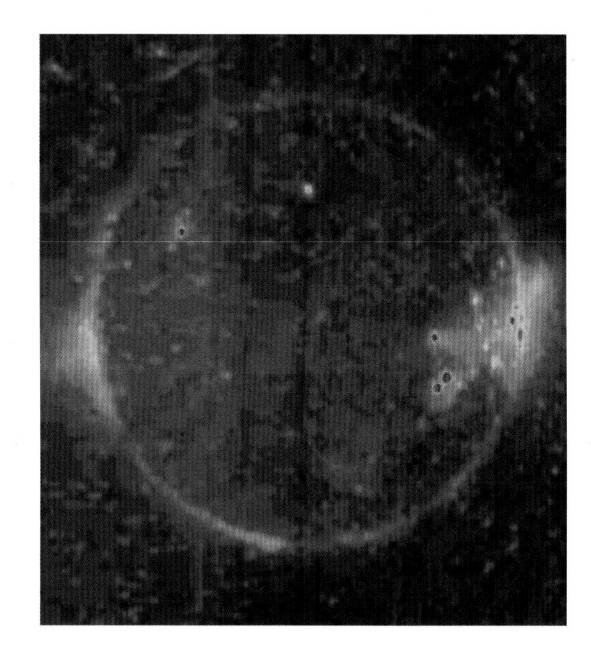

explorations of American probes Pioneer 10e, 11, Voyager 1 and 2 and Galileo, we have a great deal of information on the satellites of Jupiter. While the smallest are similar to the moons of other planets, and at times it is obvious that they have origins in asteroids, the Medician satellites are spherical, and have very different characteristics.

Io: relatively hot (2,000 K in the eruptive points, but on average around 130 K), rich in active volcanoes and lacking in meteoritic craters. Its surface, therefore, is quite recent: it is the single body of the solar system on which an eruption has been photographed. The eruptive material is probably liquid sulfur, or fused silicas rich in sodium. Some recent data returned by the Galileo probe suggest that it has an iron (or iron sulfate) core at least 900 km in diameter and its own magnetic field. The origin of Io's energy would be in the tidal interactions that tie it to Europa, Ganimede and Jupiter: while Io turns the same face to Jupiter (as happens with the Moon and the Earth), the gravity of Europa and Ganimede deform the crust with unevenness ranging to 100 m in height. These continuous deformations are translated in thermal energy, just as happens with iron threads folded repeatedly.

Europa: a little smaller than the Moon, completely wrapped in ice under which the probe Galileo has uncovered an immense salt sea, still liquid due to the heat developed from tidal interactions. Europa is the smoother body of the solar system: the most elevated superficial formations, in fact, do not exceed a few hundred of meters. Moreover, Io, Ganimede, Titan and Triton are small satellites endowed with an atmosphere: while Io is wrapped in sulfuric vapors, Europa has a light oxygen layer. It also has a weak magnetic field that, for its rotation around Jupiter, periodically varies under the influence of the strong magnetic field of the planet.

Ganimede: it is the largest satellite of the solar system, and also exceeds Mercurio and Pluto. Below a layer of thick ice between 100 and 200km deep, large volumes of water exist, although from the outside the crust resembles the Moon with an array of craters and fissures. It has a small magnetic field.

Satellites
Dimensional relationship between the Galilean satellites.
From left: Io, Europe, Ganimede, Callisto.

Details
The surface of the same four satellites in color (above) and in black and white (below).
From left: Io and its volcanic surface; Europa cracked and its rich ice surface full of crevaces, under which it is suspected exists the liquid ocean; the surface dust of Ganimede and that of Callisto, bombarded by meteorites. All the images are collages of Voyager 2 photographs or recent spectrometer data from Galileo.

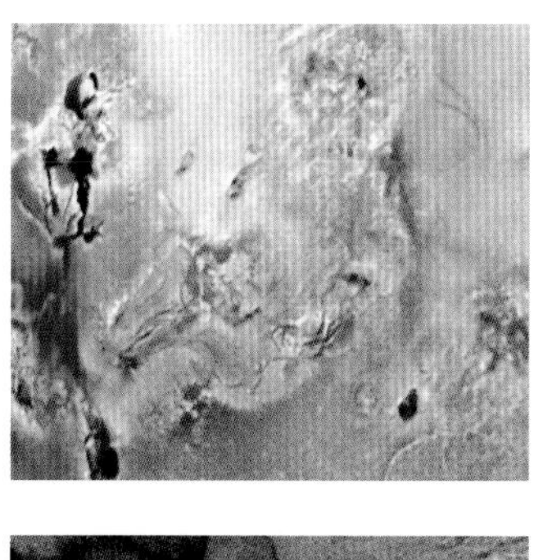
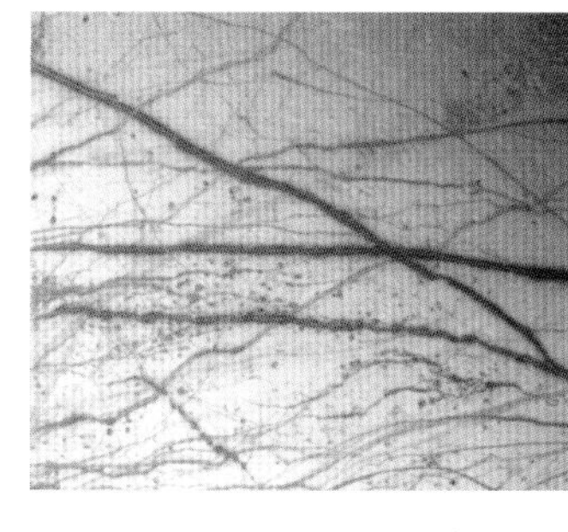
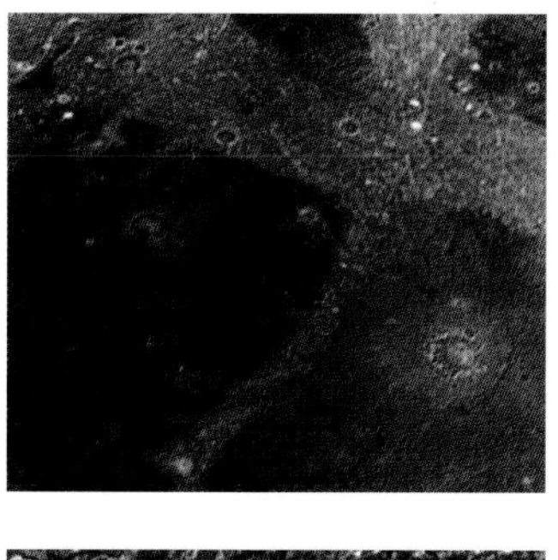

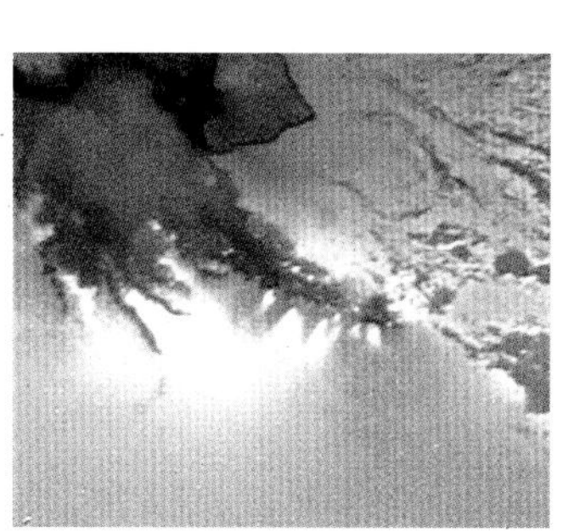
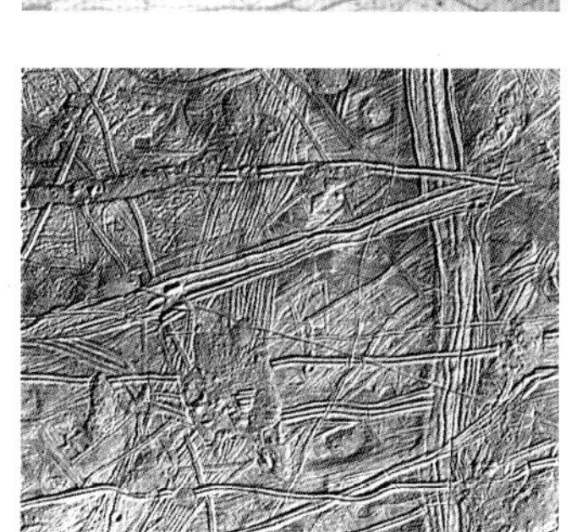
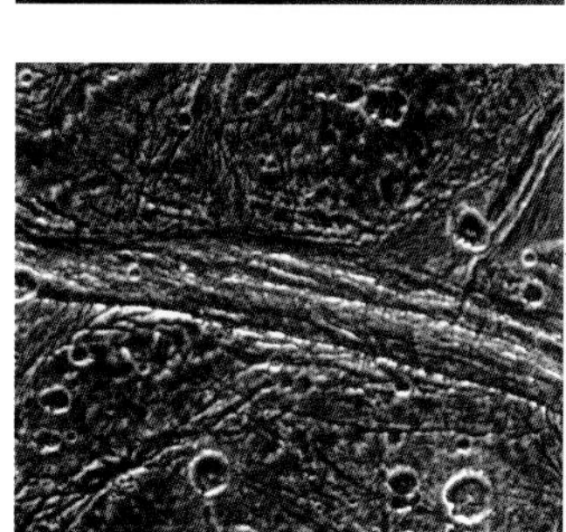
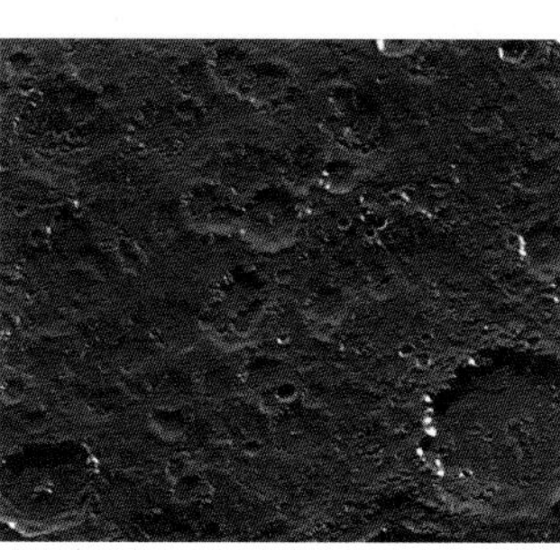

SATURN OCCUPIES THE SIXTH POSITION FROM THE SUN: ALMOST BUT NOT QUITE AS LARGE AS JUPITER, DISTANT NEARLY A BILLION AND A HALF KILOMETERS FROM OUR STAR, THANKS TO ITS LUMINOUS RINGS, THE MOST FAMOUS PLANET, STUDIED AND OBSERVED SINCE THE TIMES OF GALILEO

SATURN

SATURN
Like Jupiter, Saturn also has a strong magnetic field that generates immense polar auroras. BELOW: a complete image of the planet returned by the Hubble.

Although in relation to Jupiter, it appears smaller due to the objective dimensions and the greater distance, good obervations of Saturn can be made using decent binoculars. Pearly white in color, it has a beautiful ring system that, but, it is not always visible.
Saturn is very similar to Jupiter: it has a thick atmosphere of hydrogen (75%) and helium (25%) and it is believed that it has a rocky core covered by an ocean of liquid metallic hydrogen and a blanket of molecular hydrogen and ice with varied characteristics.
Like Jupiter, it has a differential rotation: its visible surface, in fact, is not solid. This is confirmed by its density (below that of water), the lowest in the solar system.
Since. Saturn also has a high speed of rotation which in that it is less than that of Jupiter's rotation (10h40min instead that 9 h 50 min) influences both the form and atmospheric stratification: polar flattening and the coloration in light and dark bands parallel to the Equator are the consequence, like on Jupiter. Indeed, on Saturn, the form is flatter while the parallel bands have fainter and wider profiles at the equator
Saturn also shows long term atmospheric formations, similar to the great red spot of Jupiter, and the various tonalities of color of the bands are due to differences in the chemical composition (compounds of sulfur and phosphorus) and in the thickness of clouds. Like Jupiter, moreover, Saturn also has an intense magnetic field, and irradiates

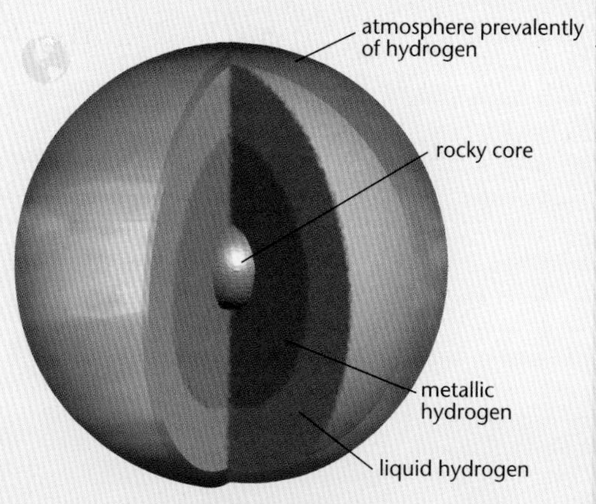

THE CHARACTERISTICS OF SATURN

EQUATORIAL RADIUS (IN km)	60,268
MASS (IN 10^{26} kg)	5,688
• COMPARED TO EARTH	95.1
VOLUME (IN 10^{27} cm^3)	817,67
• COMPARED TO EARTH	755
MEAN DENSITY (g/cm^3)	0.69
MEAN TEMPERATURE (IN K = °C)	
• OF CLOUDS	148,15 = -125
SUPERFICIAL GRAVITY (IN m/s^2)	9.05
POLAR FLATTENING	0.109
MEAN DISTANCE (IN 10^6 km)	
• FROM SUN	1,429.4
• FROM EARTH (INFERIOR ELONGATION)	1,277.4
PERIHELION (IN 10^6 km)	1,347
SATELLITES: 23	
APHELION (IN 10^6 km)	1,507
APPARENT DIAMETER OF SUN	3.4'
PERIOD	
• OF ROTATION (IN MEAN SOLAR H)	10,233
• OF REVOLUTION (IN EARTH YEARS)	29.458
ECCENTRICITY OF ORBIT	0.056
MEAN ORBITAL VELOCITY (IN km/s)	9.6
INCLINATION	
• OF ORBIT COMPARED TO ECLIPTIC	2.488°
• OF EQUATOR TO ORBIT	26.7°
ALBEDO	0.47
ATMOSPHERIC PRESSURE (IN 10^5 Pa)	1.4
PRINCIPAL ATMOSPHERIC COMPONENTS	H_2 (97%), HE (3%)

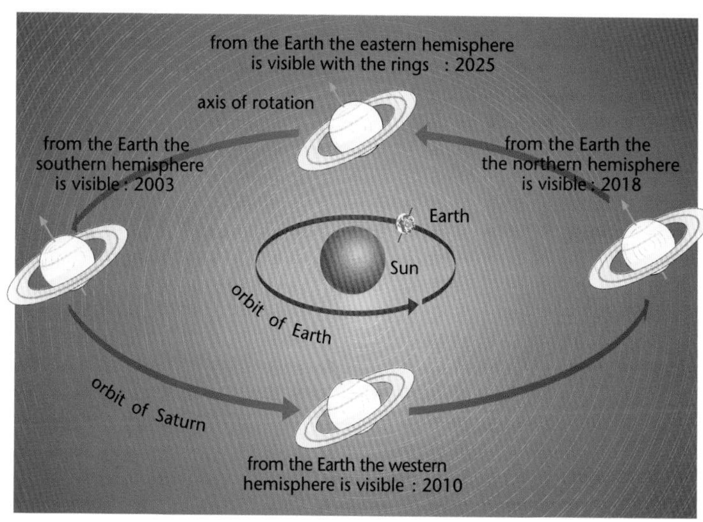

more energy than it receives some from the Sun: the interior reaches 12,000 K, probably by the same process of contraction that is observed on Jupiter. But Saturn fascinates above all by the wonder of its rings. Observed for the first time by Galileo – who, however, did not succeed in grasping what they were - the ring of Saturn is actually formed of a numerous succession of concentric rings of somewhat diverse diameters. They are constituted from powders, ice corpuscles and dry ice (carbon dioxide), frozen in orbit around the planet that show a wide spectrum in mass, dimensions (between 1 cm and some meters) and shape.

This ring system has a very complex structure: by telescope, two or three separate rings are seen and empty space between. In reality, it extends with no break in continuity for approximately 65,000 km: the most inner ring orbiting at the boundaries of the atmosphere, the most external reaches beyond the visible edge of the planet by approximately 250,000 km.

Its thickness is comparatively very small: on average a few kilometers. Since all the rings lie

ASPECTS OF SATURN
According to its position compared to the Earth, Saturn's rings are shown variously tilted, as exemplified in the collages of photographs by distance, time and structure.

RINGS
FROM LEFT:
a. Saturn's rings are seen edge-on: the rings disappear. Some satellites are visible.
b. Encke division (smaller) and that of Cassini in contrast of colors.
c. Ring C in a photograph taken from about 3 x 10^6 km (ultraviolet, visible and with a green filter)
d. Voyager 2 is nearly 9,106 km away, showing the different chemical make up of the rings.

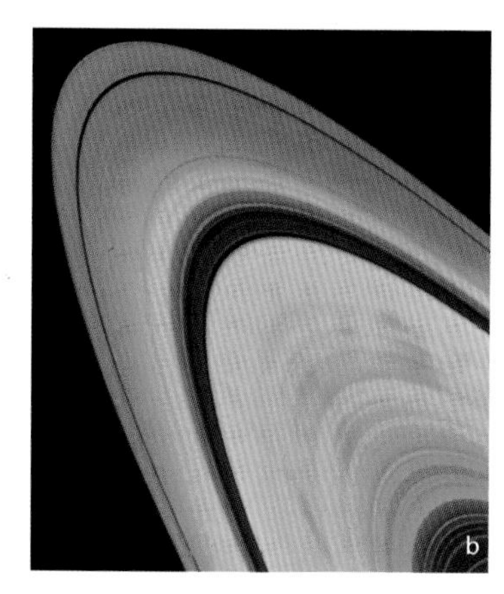

SATELLITES AND RINGS Relationship between the dimensions of the main satellites of Saturn and those of the planet. BELOW: structure of rings. The positioning of the main divisions and the orbits of the inner satellites can be seen clearly.

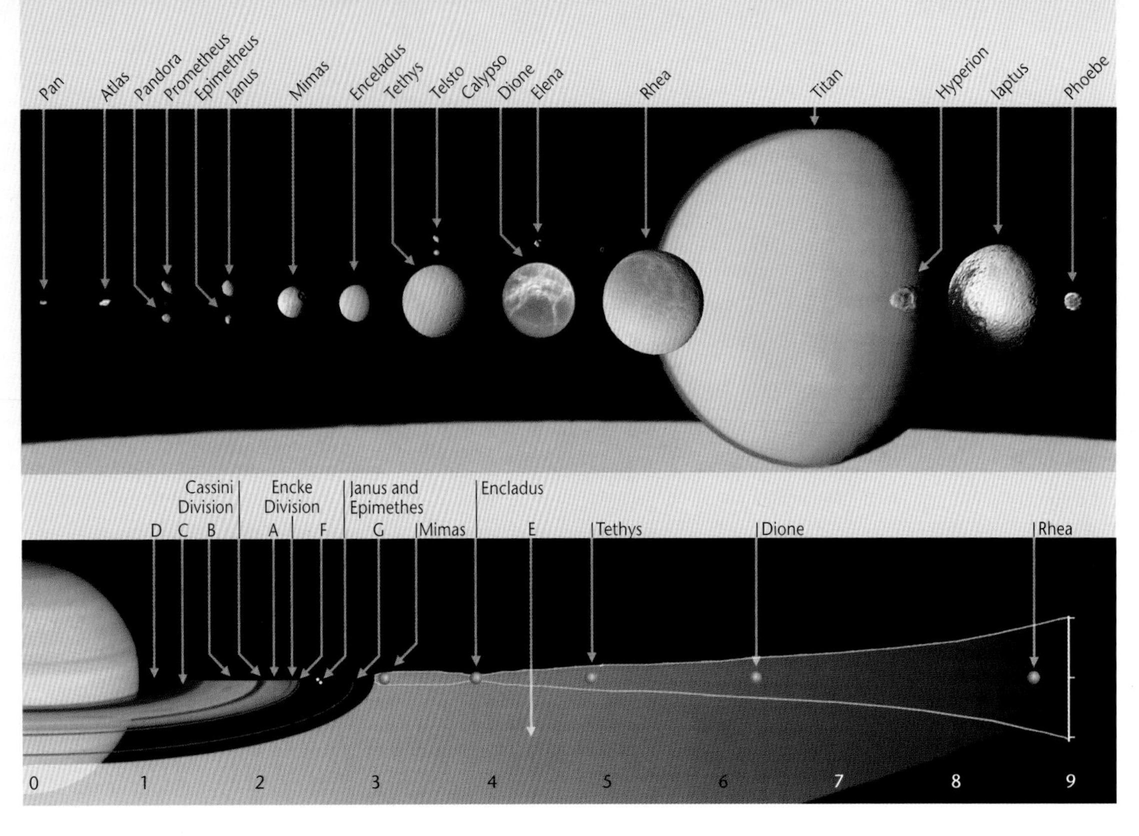

along the equatorial plane of the planet, and since the axis of rotation is inclined with respect to the plane of orbit, observed from Earth they changes appearance with the passing of time and, at approximately 14 and a half year intervals, "disappear" from view, found to be "edge-on" from the vantage point of the observer. The origin of the rings of Saturn (as of rings of other external planets) are not known: they could have their origin in one or more satellites shattered by tidal forces, but it is more probable that they are the remains of primordial material that, because of the close proximity of great planets did not aggregate to form a satellite. On the other hand, no planet has captured as much "space debris" as Saturn.: Voyager 1 has localized asteroids, cosmic rocks, powder masses, small moons, blocks of ice in, indeed, enormous amount. Saturn, with its 23 satellites, 18 of them named, is surrounded by a miniature solar system. Only Titan, however, known of long before the sending of probes, is large enough: of the dimensions of Ganimede, it is bigger than Mercury, and particularly interesting because it has an atmosphere composed of nitrogen, methane, ammonia, hydrocarbons and acetylene. Since it is believed that this gas mixture is similar to the atmosphere of the primitive Earth, thinks that Titan is one of the extraterrestrial bodies most adapted to the formation of the life. Unfortunately the atmospheric blanket does not allow us to see any details of the surface, and one only presumes that it is similar to the surface of the primordial Earth: the probes have revealed the presence of water and ice.

PRINCIPAL RINGS AND SATELLITES

NAME	DISTANCE FROM CENTER OF SATURN (10^3 km)	DIMENSIONS WIDTH X DEPTH OF RADIUS (km)	MASS (kg)
RING D	67	$7.5 \times 10^2 \times$?	?
RING C	74.5	$1.8 \times 10^4 \times$?	1.1×10^{18}
RING B	92.0	$2.6 \times 10^4 \times 0.1 \div 1$	2.8×10^{19}
RING A	122.2	$1.5 \times 10^4 \times 0.1 \div 1$	6.2×10^{18}
PAN	133.58	9.66×10^3	?
ATLAS	137.64	20 x 15	?
PROMETHEUS	139.35	72.5 x 42.5 x 32.5	2.7×10^{17}
RING F	140.2	$30 \div 500 \times$?	?
PANDORA	141.70	57 x 42 x 31	2.2×10^{17}
EPIMETHEUS	151.42	72 x 54 x 49	5.6×10^{17}
JANUS	151.47	98 x 96 x 75	2.01×10^{18}
RING G	165.8	$8 \cdot 10^3 \times 1 \times 10^2 \div 10^3$	$6 \div 23 \times 10^6$
RING E	180.0	$3.0 \times 10^5 \times 1 \times 10^3$	?
MIMAS	185.52	196	3.80×10^{19}
ENCLADUS	238.02	250	8.40×10^{19}
TETHYS	294.66	530	7.55×10^{20}
TELESTO	294.66	17 x 14 x 13	?
CALYPSO	294.66	17 x 11 x 11	?
DIONE	377.40	560	1.05×10^{21}
ELENA	377.40	18 x 16 x 15	?
REA	527.04	756	2.49×10^{21}
TITAN	1,221.85	2,575	1.35×10^{23}
HYPERION	1,481.00	205 x 130 x 110	1.77×10^{19}
IAPETO	3,561.30	730	1.88×10^{21}
PHOEBE	12,952.00	110	4.0×10^{18}

On moonless nights, Uranus is nearly invisible to those who watch the sky without the aid of an instrument. It is a modern planet that has claimed its definitive name only since 1850. But to the Voyager 2 probe that has approached and photographed it, Uranus has shown its most beautiful face and some of its more important characteristics.

URANUS

Through binoculars, this planet appears as a luminous point, while by telescope it assumes a green-blue coloration, remaining a disc too small to distinguish details. The planet was uncovered by William Herschel who, in 1781, was observing the sky in a systematic way with his new telescope. Due to the small magnitude (only 5.6), in fact, although it had been seen by the ancient astronomers, it had been considered a fixed star.

Uranus is a unque planet in the solar system: beyond having a retrograde rotatory motion, its axis of rotation is tilted 82° compared to the plane of orbit: in other words, the Equator is nearly perpendicular to the direction of the motion of revolution. One supposes that the violent impact with a body of the dimensions of the Earth "jolted" it from a position similar to that of other planets. A collision with this entity would also explain the retrograde motion: if the planet had been "jolted" 98°, the North Pole would be that under the plane of orbit, and the planet would still be turning "the right way". The consequences of this characteristic are numerous: the pole exposed to the Sun has a mean temperature of -208 °C (62.4 K) while the pole hidden to the Sun has a mean temperature of -215 °C (64.5 K); in contrast to that which happens on other planets, the poles are observable, therefore as the system of 11 concentric

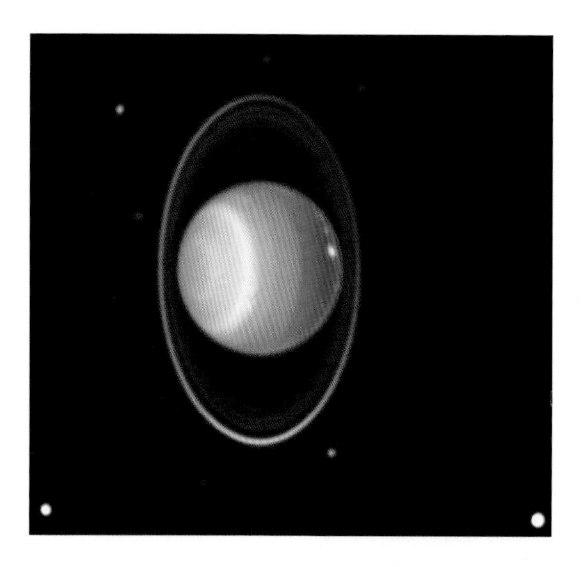

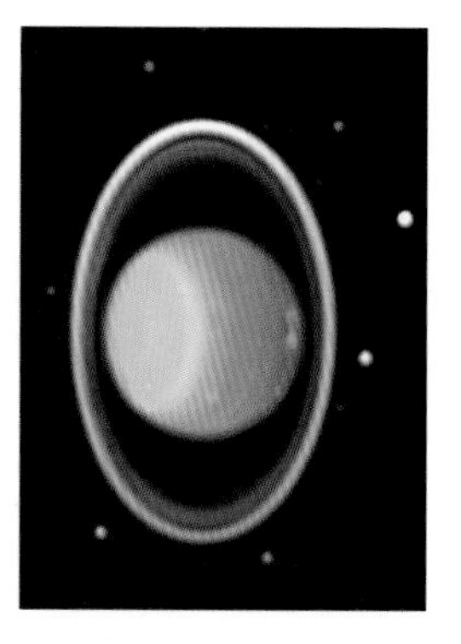

Uranus
Two images of Uranus and some of its satellites returned by the Hubble. The rings are also visible.

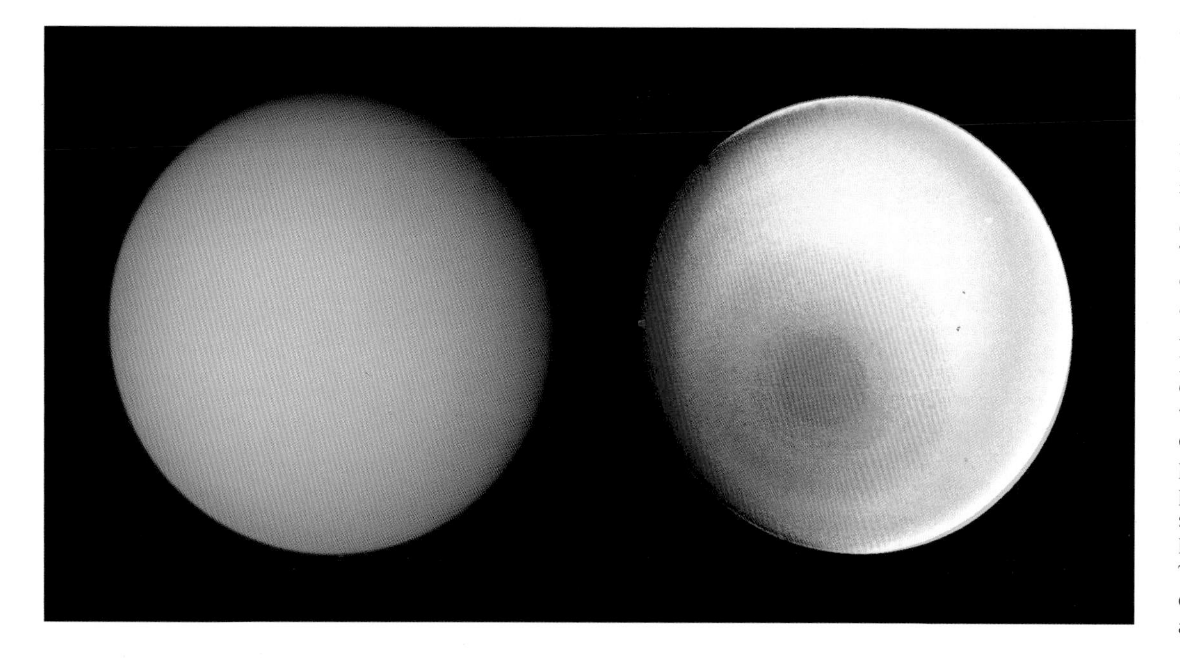

Uranus
Image in the visible (left) and in false colour (right) of the pole of rotation of the planet photographed by Voyager 2 from a distance of 18 x 10^6 km. The grey-green visible colour is due to atmospheric methane. After electronic processing of images captured at different wavelengths, the image on the right highlights a polar zone of smoke-like particles surrounded by a series of increasingly light convection bands. The brown at the equator is perhaps due to atmospheric acetylene.

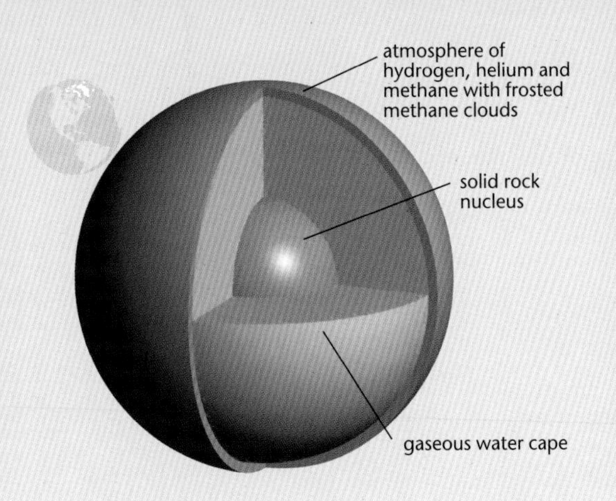

THE CHARACTERISTICS OF URANUS

EQUATORIAL RADIUS (IN km)	25,560
MASS (IN 10^{25} kg)	8.686
• COMPARED TO EARTH	14.6
VOLUME (IN 10^{27} cm^3)	72.56
• COMPARED TO EARTH	67
MEAN DENSITY (g/cm^3)	1.29
MEAN TEMPERATURE (IN K = °C) OF CLOUDS	80.15 = -193
SUPERFICIAL GRAVITY (IN m/s^2)	7.77
POLAR FLATTENING	0.03
MEAN DISTANCE (IN 10^6 km)	
• FROM SUN	2,871
• FROM EARTH (INFERIOR ELONGATION)	2,719.7
PERIHELION (IN 10^6 km)	2,735
APHELION (IN 10^6 km)	3,004
SATELLITES: 15	
APPARENT DIAMETER OF SUN	1.7'
PERIOD	
• OF ROTATION (IN mean solar h)	17.9
• OF REVOLUTION (IN EARTH YEARS)	84.01
ECCENTRICITY OF ORBIT	0.046
MEAN ORBITAL VELOCITY (IN km/s)	6.8
INCLINATION	
• OF ORBIT COMPARED TO ECLIPTIC	0.774°
• OF EQUATOR TO ORBIT	82.1°
ALBEDO	0.51
ATMOSPHERIC PRESSURE (IN 10^5 Pa)	1.2
PRINCIPAL ATMOSPHERIC COMPONENTS	H_2 (82,5%), HE (15,2%),CH_2 (2,4%)

THE 11 RINGS OF URANUS
The composition is similar to that of the rings of Jupiter and Saturn. The major difference lies in the structure of the system: the rings of Uranus are darker, tightened and wide it separates those of the other two planets.

rings, formed of powders and minuscule solid fragments, it remains nearly always completely visible. Moreover, the magnetic field, has an intensity of approximately 1/3 of that Earth (0.25 G), has an axis inclined 35° compared to the axis of rotation. Like the great external planets, Uranus has a very developed atmosphere composed of hydrogen and helium but also of methane, which gives an orange-blue color to it. The Voyager 2 probe has shown clouds in the high atmosphere, where the temperature, close to absolute zero, oscillates around 200° C.

Like Saturn, moreover, the Uranus rings have a wealth of numerous satellites: from Earth, it it is possible to see 5 (Miranda, Ariel, Umbriel, Titania and Oberon), but the Voyager 2 probe has discovered another 10 and it is probable that there are more still, also 9 of the new satellites are simple cosmic rocks captured from the planet in remote times. The satellites of Uranus orbit with retrograde motion along the equatorial plane. One thinks that, as the other systems of satellites of great planets, also their evolution has remained.

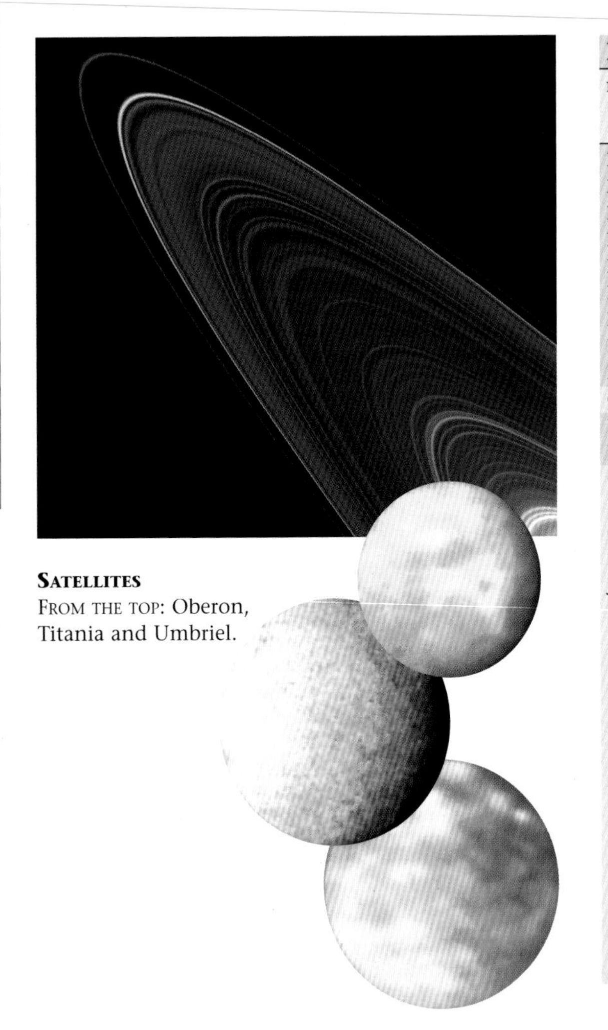

SATELLITES
FROM THE TOP: Oberon, Titania and Umbriel.

PRINCIPAL RINGS AND SATELLITES

NAME	DISTANCE FROM CENTER OF URANUS (10^3 km)	DIMENSIONS WIDTH X DEPTH OR RADIUS (km)	MASS (kg)
RING 1968U2R	38.00	2.500 x 0.1	
RING 6	41.84	1÷3 x 0.1	?
RING 5	42.23	2÷3 x 0.1	?
RING 4	42.58	2÷3 x 0.1	?
RING ALPHA	44.72	7÷12 x 0.1	?
RING BETA	45.67	7÷12 x 0.1	?
RING ETA	47.19	0÷2 x 0.1	?
RING GAMMA	47.63	1÷4 x 0.1	?
RING DELTA	48.29	3÷9 x 0.1	?
CORDELIA	49.75	13	?
RING 1986U1R	50.02	1÷2 x 0.1	?
RING EPSILON	51.14	20÷100 x < 0.15	?
OPHELIA	53.76	16	?
BIANCA	59.16	22	?
CRESSIDA	61.77	33	?
DESDEMONA	62.66	29	?
JULIET	64.36	42	?
PORTIA	66.10	55	?
ROSALINDA	69.93	27	?
BELINDA	75.26	34	?
MOON 1986U10	75.00	20	?
PUCK	86.01	77	?
MIRANDA	129.78	235	6.33 x 10^{19}
ARIEL	191.24	578	1.27 x 10^{21}
UMBRIEL	265.97	584	1.27 x 10^{21}
TITANIA	435.84	788	3.49 x 10^{21}
OBERON	582.60	761	3.03 x 10^{21}
CALIBANO	7,100.00	30	?
MOON 1999U1	10,000.00	20	?
SYCORAX	12,200.00	60	?
MOON 1999U2	25,000.00	15	?

MANY STUDENTS WHO "DEMOTE" PLUTO TO THE STATUS OF A MISSING COMET, CONSIDER NEPTUNE THE LAST PLANET OF THE SOLAR SYSTEM. INVISIBLE TO THE NAKED EYE, IT IS SIMILAR TO URANUS ALSO IN THE BEAUTIFUL COLORS THAT CHARACTERIZE ITS CLOUDY ATMOSPHERE, DUE TO THE PRESENCE OF METHANE

NEPTUNE

The eighth planet of the solar system in distance from the Sun and the last Jovian planet. Pluto, in fact, has radically different characteristics from those of other external planets. For the geometry of the orbits of Neptune and Pluto, moreover, comes to pass that Neptune finds itself much further away from the Sun than Pluto: this has happened, as an example, in years 1979-1999. However, generally, Pluto is more far away: it reaches the maximum distance from the Sun that is approximately one and a half the maximum distance reached by Neptune.
The presence of an eighth planet already had been suspected by Adams and Le Verrier in that some irregularity in the orbital motion of Uranus is evidenced, justifiable with the gravitational action of an external body of great dimensions. But Neptune was at the boundaries of the possibilities for observation of the instruments of the age, and its discovery, realized by the astronomers of the observatory of Berlin December 23, 1846 after a series of patient investigations. While to astronomers of that time, it appeared to be a small blue-green disc, we know that, in reality, it is certainly not small: in terms of dimensions, it follows only Jupiter, Saturn and Uranus. This remained almost all that that was known on Neptune, until in 1989 it was reached by the Voyager 2 probe, that has changed the way of thinking about this planet. In the photographs, in fact, Neptune appears as a beautiful blue planet, with shadings, spots and tenuous white streaks. On its surface was observed a dark spot as large as the Earth: as in the case of the great red spot of Jupiter, it was a matter of a storm in stationary conditions that, subsequently, beside the planet, in fact. More recent information on Neptune has been collected taking advantage of the most powerful telescopes on Earth

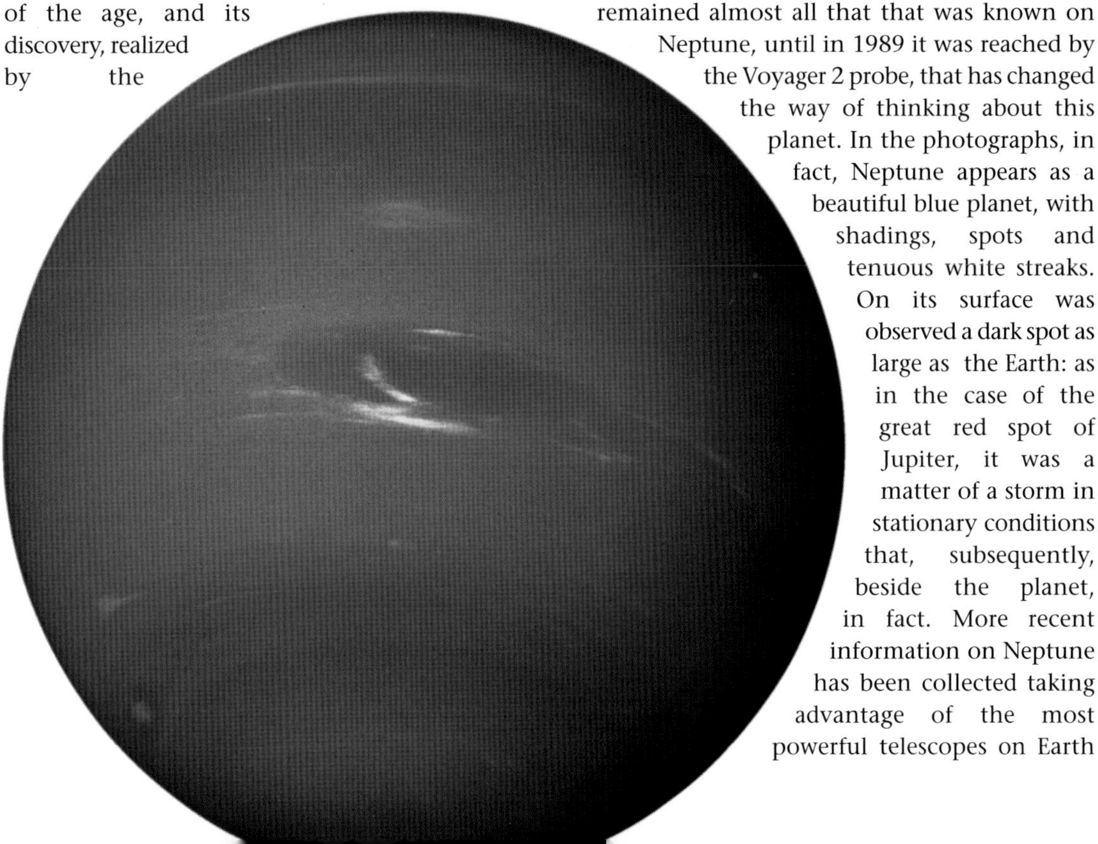

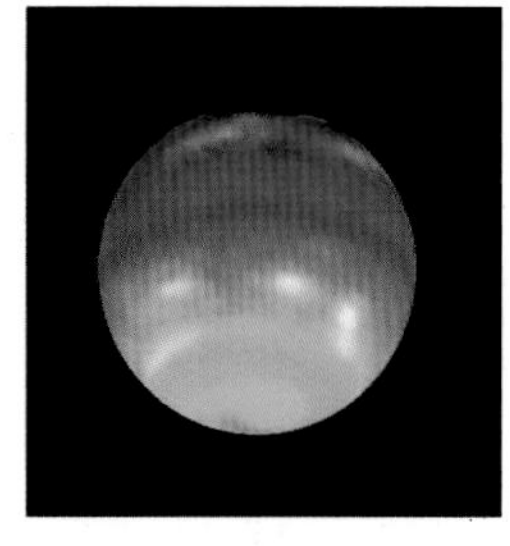

IMAGES OF NEPTUNE
The great dark spot, a hydrogen sulphide cloud returned from the Voyager 2 probe, disappearing in the images returned recently from the Hubble.

ABOVE: in this image returned from the Hubble the methane clouds (white), the powerful equatorial jetstream (dark blue) with a wind which reaches speeds close to 1400 km/h and a region where the atmosphere absorbs the ultraviolet rays near the South Pole (green steak).

LEFT: view of the planet.

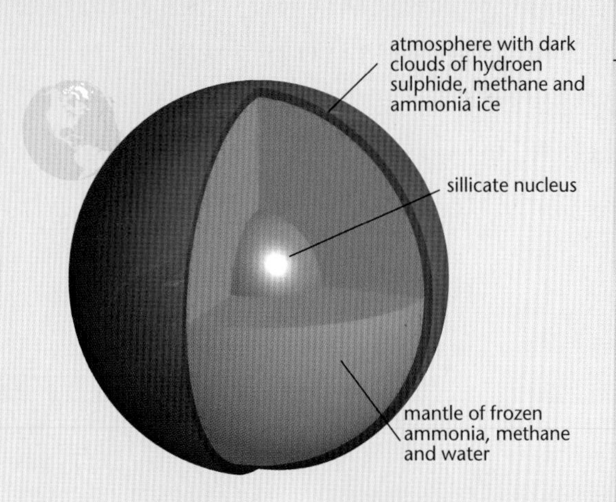

THE CHARACTERISTICS OF NEPTUNE

EQUATORIAL RADIUS (IN km)	24.746
MASS (IN 10^{26} kg)	1,024
• COMPARED TO EARTH	17,2
VOLUME (IN 10^{27} cm^3)	61,73
• COMPARED TO EARTH	57
MEAN DENSITY (g/cm^3)	1,64
MEAN TEMPERATURE OF CLOUDS (IN K = °C)	
• MAXIMUM	120,15 = -153
• MINIMUM	80,15 = -193
SUPERFICIAL GRAVITY (IN m/s^2)	11,0
POLAR FLATTENING	0,003
MEAN DISTANCE (IN 10^6 km)	
• FROM SUN	4.504,3
• FROM EARTH (INFERIOR ELONGATION)	4.347,4
PERIHELION (IN 10^6 km)	4.456
SATELLITES: 8	
APHELION (IN 10^6 km)	4.537
APPARENT DIAMETER OF SUN	1,1'
PERIOD	
• OF ROTATION (IN mean solar h)	16,11
• OF REVOLUTION (IN EARTH YEARS)	164,79
ECCENTRICITY OF ORBIT	0,0097
MEAN ORBITAL VELOCITY (IN km/s)	5,45
INCLINATION	
• OF ORBIT COMPARED TO ECLIPTIC	1,774°
• OF EQUATOR OF ORBIT	28,8°
ALBEDO	0,41
ATMOSPHERIC PRESSURE (IN 10^4 Pa)	1÷3
PRINCIPAL ATMOSPHERIC COMPONENTS	H_2 (85%), HE (13%), CH_4 (2%)

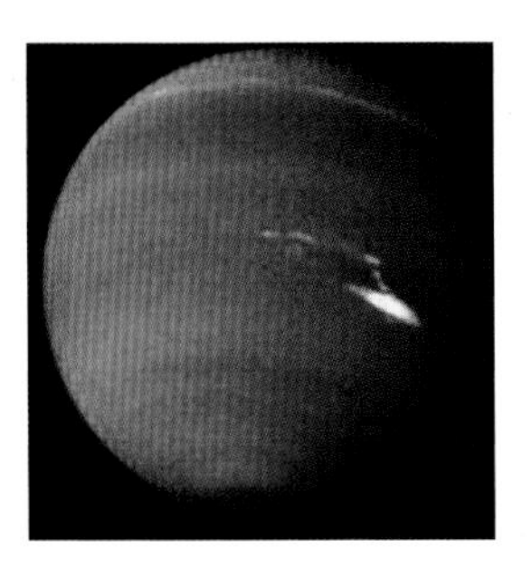

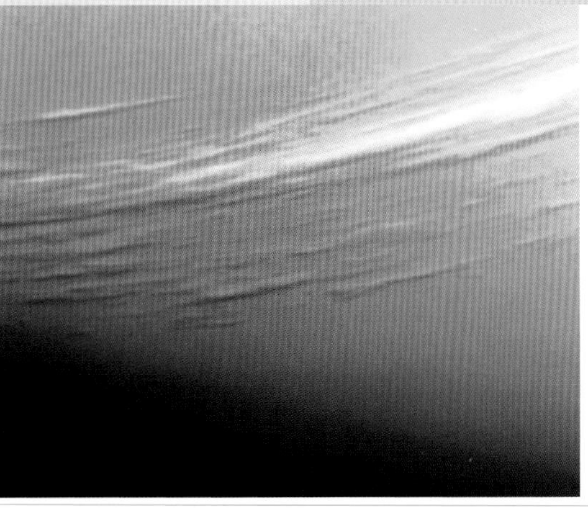

CLOUDS
Neptune by false color processing returned from the Voyager 2 probe, at 5,000 km from the surface.
The image shows various levels of clouds in the atmosphere.

TRITON
Collage of photographs released from the Voyager 2.

and the Hubble. If the Neptune name reminds one of the God of the sea and its blue-green appearance reminds one of the color of the water, that does not imply that this planet is covered with oceans: like in the case of Uranus, the color of the atmosphere - formed above all of hydrogen and helium - has a small percentage of methane. As in other external planets, the atmosphere prevents from seeing the frosted surface of the planet: perturbed by vortices and storms, it is crossed by winds that blow parallel to the Equator reaching up to 2,000 km/h. The winds of Neptune are the fastest of the solar system. The mean temperature is – 200 °C, and like the other external planets Neptune also emits more energy than that which it receives. Like Uranus, it has a magnetic field more weak than that of Earth, a system of 4 rings, very dark and tenuous, made above all of water-ice particles with dimensions that range from millionth of millimeter to a few meters. To distinguish it from the other famous rings, the rings of Neptune do not have one uniform density: in some it is high, while elsewhere the material is very rarefied. At the moment, we know of 8 satellites of Neptune, all of dimensions less than those of our Moon. The largest are Proteous, the darker body of the solar system that reflects only 6% of the solar light; Nereid, that follows a very elliptic trajectory and very inclined compared to the equatorial plane of Neptune; and Triton, the most massive, rotating in a retrograde fashion. In particular, Triton has dimensions similar to the Moon, and provokes the interest of the planetologists with some geysers of gaseous nitrogen that have been photographed and that reach to many kilometers of height. Moreover, it has its own atmosphere, a probable content of 25% water and rocky structure. The surface, at the lowest temperature similar to that of Pluto (approximately 34.5 K) presumably is covered by an atmosphere with dark clouds of hydrogen sulfide, methane and ammoniac ice The surface, with very low temperatures similar to those of Pluto (about 34.5 K) is presumably covered with frozen methane, carbon dioxide and nitrogen.

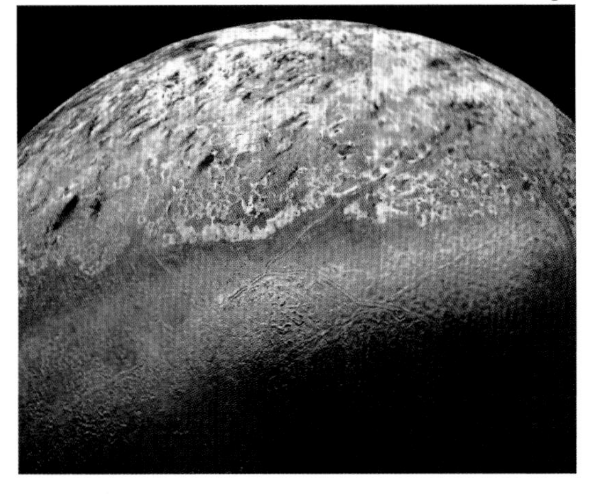

PRINCIPAL RINGS AND SATELLITES

NAME	DISTANCE FROM CENTER OF NEPTUNE (10^3 km)	DIMENSIONS WIDTH X DEPTH OR RADIUS (km)	MASS (kg)
RING 1989N3R	41.90	15	
NAIADE	48.00	29	?
THALASSA	50.00	40	?
DESPINA	52.50	74	?
RING 1989N2R	53.20	15	?
RING 1989N4R	53.20	5.80	?
GALATEA	62.00	79	?
RING 1989N1R	62.93	< 50	?
LARISSA	73.60	104 x 89	?
PROTEOS	117.60	200	?
TRITON	354.80	1,350	2.14 x 10^{22}
NEREID	5.513.40	170	?

Pluto and Charon: are they a double planet? Are they lost satellites? Are they captive comets of the solar gravitational field?
The hypotheses advanced in these last decades must still find definitive confirmation. But an increasing number of scholars thinks that these two celestial bodies are "intruders".

PLUTO AND CHARON

The smaller planet of the solar system: it is also smaller than the Moon. Uncovered in 1930, very little is known of it. It covers its orbit in 248 years, and the perihelion is slightly inside the orbit of Neptune. Charon, with dimensions approximately half those of Pluto, is found at only 17,000 km and forms with Pluto a veritable double planet: Charon and Pluto rotate around a gravitational center in such a way always showing one another the same hemisphere. The origin of these two bodies is still controversial. Pluto, according to the more recent hypotheses, could be a large comet, or one of the largest "Kuiper bodies" orbiting relatively small distances from the Sun. Its characteristics, in fact, are therefore different from those of the planets that precede it making it very difficult to believe that it has the same origin.
First of all, the orbit is eccentric, while the orbit of other planets is nearly circular. Moreover, the plane is very inclined compared to the ecliptic plan, while the orbit of other planets is nearly coincident. Unique among external planets, finally, it is nearly lacking in atmosphere (perhaps a small amount of nitrogen, carbon dioxide and methane), and probably with a constitution similar to Triton: 70% rock and 30% ice. The high amounts of water could assimilate Pluto to a comet. However, these differences could be

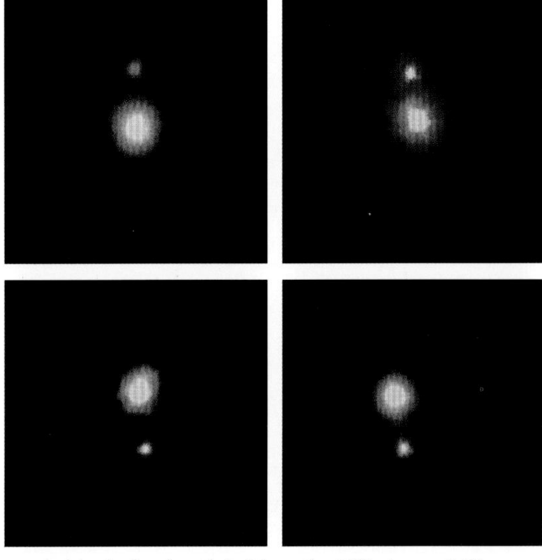

explained also by thinking that Pluto and Charon may have been satellites of one of the great planets, strongly torn from their orbit by a collision or gravitational interference. Pluto has been searched after for 24 years, in the attempt to justify the irregularity of the orbit of Uranus that the discovery of Neptune did not explain completely. Since its discovery it has only travelled along one fifth of its orbit, and since 2000 it has started along the sector that will take it further away from the Earth.

Couple of planets, satellites or comets?
Pluto and Charon returned by the powerful instruments of the Hubble.
Below: orbital dynamics of the system.

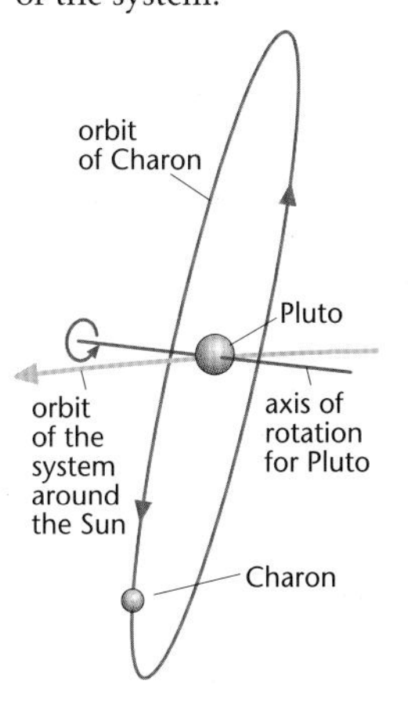

Origins
The debate about the true nature of the Pluto-Charon system is still ongoing, and its highly eccentric orbit is one of the subjects debated.

The characteristics of Pluto and Charon

Equatorial radius (in km)	
• Pluto	1,137
• Charon	586
Mass (in 10^{22} kg)	
• Pluto	1.27
• compared to Earth	0.002
• Charon	0.19
Volume of Pluto(in 10^{27} cm^3)	0.11
• comapared to Earth	0.1
Mean density (g/cm^3)	1
Mean surface temperature of Pluto (in K = °C)	
• maximum	63 = -210
• minimum	38 = -235
Medium distance of Charon from Pluto (in 10^4 km)	1.964
Mean distance from Pluto (in 10^6 km)	
• from the Sun	5,909.200
• from Earth (in opposition)	5,665.6
Perihelion of Pluto (in 10^6 km)	4,425
Aphelion of Pluto (in 10^6 km)	7,375
Apparent diameter of the Sun	0.8'
Period	
• of rotation (in mean solar h)	153
• of revolution (in Earth years)	247.7
Eccentricity of orbit	0.25
Mean orbital velocity (in km/s)	4.7
Inclination of orbit in respect to the ecliptic	17.2°
Albedo of Pluto	0.41÷0.50

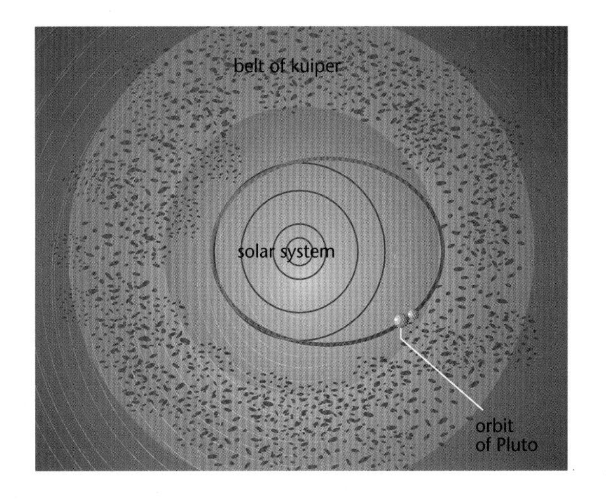

"Shovels of dirty snow", the comets are described with majestic orbits traveling invisibly through the darkness. It is near the sun that they are transformed into the most beautiful objects of the solar system. The meteorites, on the contrary, catch fire on contact with the atmosphere. Furthermore, space may appear void but in the seeming emptiness are powders, ions, molecules and particles.

COMETS, METEORITES AND INTERPLANETARY MEANS

Wind and radiation
The tail of a comet is the work of the sun: the heat sublimates the icy nucleus, the pressure of radiation (photons) and solar wind (ions and magnetic field) "blows away" the tail. A comet has at least two tails. The straight tail is made up of molecular and atomic ions and of electrons which shine with flourescence like a polar aurora, and the curved tail is made of dust that moves along the orbit of the comet pushed by force generated when molecules are expelled from the nucleus, from gravitational effects of the sun, and pressure of the light.

West is Hale Bopp
Two comets that have been observed from Earth.

The comets are celestial bodies of small dimensions characterized by strongly elongated and eccentric orbits. Some comets orbit around the long Sun trajectories that, also are very elongated, bringing them in proximity of the Sun in a relatively short time. Others, instead, have rather long periods that for us are visible one single time. Their origin is uncertain but consider that a large spherical cloud exists named the Oort Cloud with a radius of 0.5-1 x 10^3 km formed from 100-1,000 billion comets: gravitational perturbation due to other starts or external planets, the comets can rush towards the sun or remain in a very low orbit, or lose themselves in space. Other can follow the Kuiper belt: an equally abundant ring, of small, bodies, comets or asteroids, that orbit at 4.5-7.2 x 10^6 km from the Sun.
When traveling through space, the comet is a body like many others, of the dimensions of a small asteroid (at maximum 60 km), formed of dusts, rocky meteoritic materials and ices. But when reaching approximately at 3 x 10^8 km from the Sun, the ice of the nucleus sublimes, that is to say passes from a gaseous state, giving origin to a gas cloud that constitutes the tail (the "atmosphere" of the comet). The solar radiation pushes gases of the tail that go to form the tail: for this reason, always pointed in a direction opposite to the Sun. The head of the comet (nucleus and chioma) can

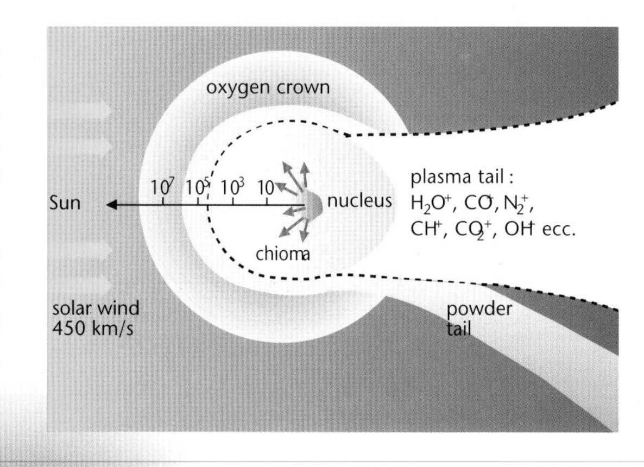

plasma tail
perihelion
Sun
orbit of Earth
orbit of comet

Comet shower

Name	Date of maximum	Mean duration (days)	Mean number (hours)	Name* of shower
Thatcher (1861 I)	21 April	3	5	Lirid
Halley	4 May	4	5	η Acquarid
1862 III	12 August	10	40	Perseid
Giacobini-Zinner	9 October	1	irregular	Draconid
Halley	20 October	8	12	Orionid
Temple-Tuttle (1866 I)	6 November	5	irregular	Leonid
Biela	22 November	10	1	Andromedid
?	13 December	6	50	Geminid

*Chosen from the constellation in which the radiating point is found, from which the falling stars seem to come.

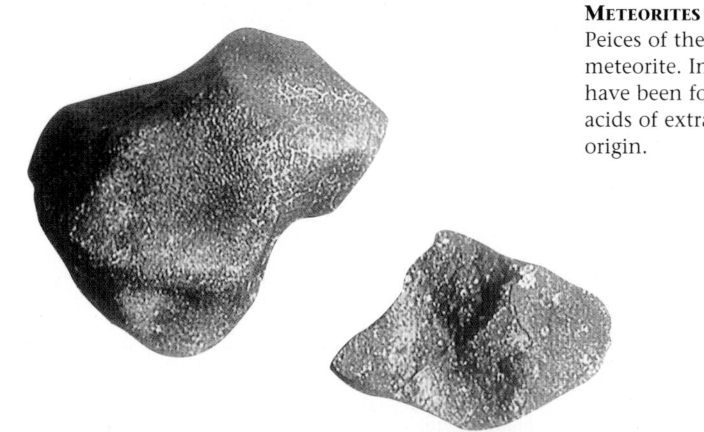

Meteorites
Peices of the Murchison meteorite. In which have been found amino acids of extra terrestrial origin.

reach 200,000 km of diameter, while the tail can reach hundreds of million of kilometers in length.

FALLING METEORITES AND STARS

Innumerable, small bodies cover interplanetary space: if they fall to Earth, they are called meteorites or shooting stars. Every year, the Earth reaps several million kilograms of them. The smallest (dusts), fall slowly, like snow; the largest can reach Earth with catastrophic effects burning, due to friction, in the atmosphere, illuminated by the ionization and recombination of its heated up atoms. They are the falling stars that, in particular periods of the year, fall with much more elevated frequency. They are that the remains of ancient comets: every time it comes close to the Sun, the cometary nucleus loses great amounts of matter and becomes subordinate to gravitational stresses that can shatter it in many pieces. The comet, as such, disappear, but its pieces continue to travel along the old trajectory of the comet. When the Earth or an other body crosses this fragment ring, they are captured.

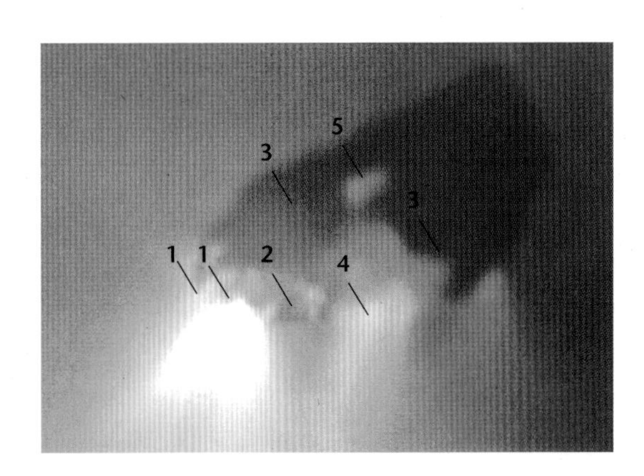

Nucleus of Halley
Photographed from the Giotto probe at 600 km, the nucleus of the comet shows, from the illuminated part, powerful jets similar to a geyser: the gases form the tail and, bend from the solar wind and the pressure of the light, the plasma tail
1. luminous spots (origin of jets);
2. crater;
3. terminator (border between night and day);
4. luminous region;
5. bulk.

ZODIACAL LIGHT

The great amount of gas, powders and particles that "fill up" interplanetary space produced from all the objects of the solar system, is not distributed in a uniform way but is found in abundance above all on the plane of the system that nearly coincides with the ecliptic. Like the terrestrial fine dust, that in space reflects and diffuses the light of the Sun.

Burger and linear from the Hubble
The second comet has been shattered apart in numerous blocks of which one can here be observed.

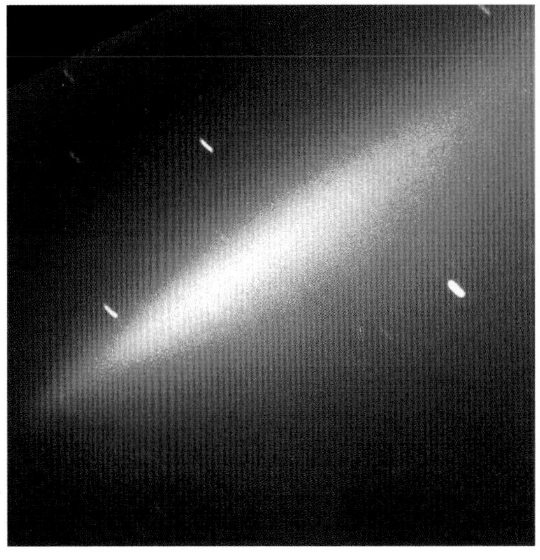

On the base of the data collected by the planetary probes and, above all, on the analysis of meteorites which have fallen to Earth, evolutionary theories are elaborated that take current observations into account. But no hypothesis advanced at provides a completely satisfactory answer and the theoretical models undergo continuous revaluations and modernizations

HYPOTHESIS ON THE ORIGIN OF THE SOLAR SYSTEM

Supernova
Schematization of the effect that the shock wave produced by the explosion of a star has on the nebular mass: a solar nebula originates from concentration of gas and dust.

New star
A series of very young stars observed along the expansion border of energy flow generated by the formation of a large star NCG2264IRS.

Since all planets orbit around the Sun in the same direction and according to orbits that are nearly on the same plane, already Kant and Leplace thought that all the bodies of the solar system had a common origin, and derive from the same gas and dust whirlwind storm, even the Sun. The majority of astronomers today are convinced of this given the considerable amount of evidence and number of observations. We know that the star born in a region in which gas and dust are sufficiently dense to prime a movement of gravitational concentration. According to this hypothesis, a supernova can produce these concentrations: the shock wave generated by an explosion triggers off the processes that through accumulation of material, produces the formation of a solar nebula that rotates around its own axis, while collapsing towards the center.
Here, when the pressure and temperature exceed

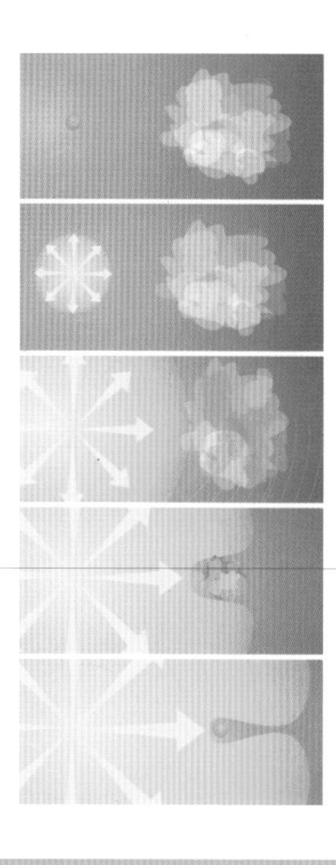

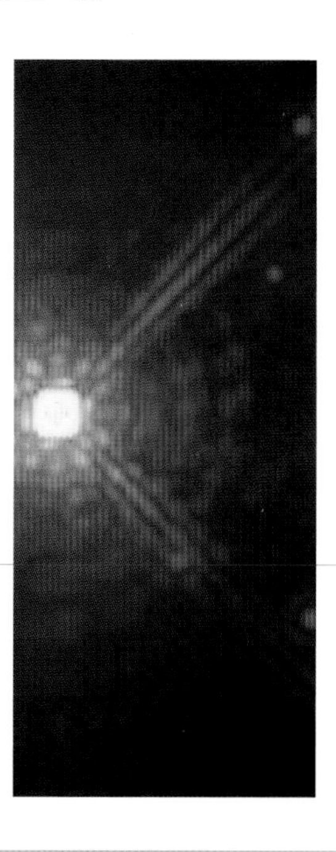

thousands of years ago	5,000,000	4,999,950	4,999,925
	initiates the contraction of clouds; the combined action of gravitation and rotation produce a flat disc like that which Hubble photographed in the Orion nebula.	in only 50,000 years, the cloud differentiates: in the center the Sun is being formed, still invisible because it is enveloped by clouds; the very heavy dust condenses on a flat disk while the light and turbulent gas envelops all.	in another 25,000 years, at the interior of the central zone, the necessary temperature to trigger the prime nuclear reaction is reached, meanwhile, in the dust disc the small planets begin to form.

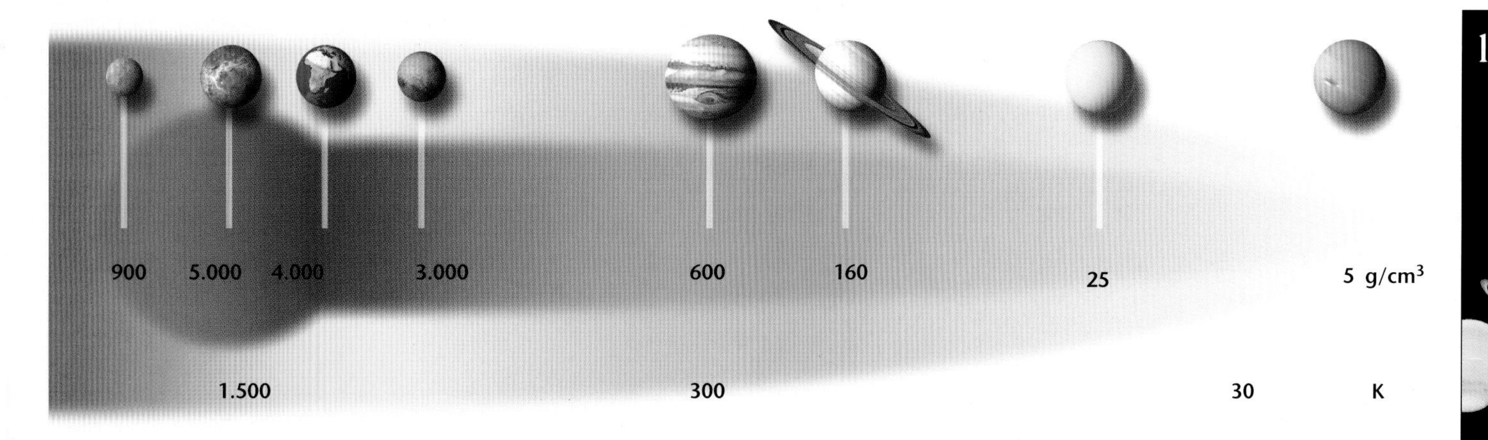

the threshhold level, the nuclear reaction is triggered: a proto-Sun is born that develops in various ways according to the available mass ➢198-203.

The rest of the nebula would be concentrated, at the same time, on the equatorial plane: the planets and other bodies of our solar system originate from this material.

Although the evolutive model of the solar system is continually modified in accordance with the new data gathered, in principle, the astronomers agree on this so-called "planetesimal" model: beyond the Sun, where the major part of the nebular mass should concentrate, relatively small objects should form from condensation of the equatorial disc (in order of kilometers) called small planets. In turn, by successive aggregation, they would have provided the origin to planets and satellites: a few thousand years would have been sufficient for this. That would happen around 4.6 · 10^9 years ago. this, in fact, was the estimated age (radioactive decay) of meteorites and older lunar fields. According to some scholars, the asteroids between Mars and Jupiter would be those that remain of the small primordial planets.

The large part of the mass of originating nebula, however, would be lost: the solar wind would have swept away the lighter elements, only minimally captured in the gravitation of the external planets already formed bulkier planets.

The strong emissions of solar wind happened in remote periods that would also justify the fact that although containing 99.9% of the mass of the entire system, the Sun conserves only 2% angular moment. In a certain sense, the solar wind would have 'transported' the greater part of the weight of the waves to the outer reaches of the galaxy.

Primitive nebula
Indicative values of temperature (in K) and density (in g/cm3) in the equatorial ring of primitive nebula at a distance from the center corresponding to that of various planets. This data is still very controversial.

Evolution of the system
Phases of formation of the solar system and principal events. The time scale is only indicative and not true to scale.

4,999,900	4,975,000	4,920,000
initiates the fusion of hydrogen into helium: the Sun ignites and the intense radiation emitted in this phase - known with the name of T Tauri, the name of the star in which this phenomenon was observed for the first time - swept away the cloud near the star	the action of solar wind has a diverse impact on lighter gas and dust: close to the Sun remains a small amount of heavy materials, farther away remains lighter but more abundant materials.	the small planets that they form move in all the possible orbits, with all the orientation compatible with the structure of the cloud: a very hidden but chaotic system; colliding, they accrete in bodies with ever greater dimensions until the planet is left, with circular orbit and not much intersecting.

THE INVENTION OF EVER MORE SENSITIVE INSTRUMENTS CONTINUES TO OPEN UP NEW FRONTIERS, EXPAND THE HORIZONS OF THE KNOWN UNIVERSE AND PROMOTE NEW IDEAS, THEORIES AND INTERPRETATIONS ABOUT NATURE. THE SOLAR SYSTEM THAT SEEMED SO VAST AND ALL-ENCOMPASSING TO GALILEO AND NEWTON HAS, WITHIN THE SPACE OF LITTLE MORE THAN A CENTURY, BECOME A SMALL GROUP OF ALMOST COMMONPLACE CELESTIAL BODIES AND OUR SUN IS A "MEDIUM" STAR, LIKE BILLIONS OF OTHERS, ORBITED BY COLD BODIES AND WITH A "VOID" THAT THE SOLAR WIND FILLS WITH PARTICLES. EVEN FURTHER INTO SPACE, AWAY FROM THE SUN'S INFLUENCE, THERE ARE INCREDIBLE OBJECTS AND PHENOMENA THAT WE CAN OBSERVE: HUGE STARS WHOSE DIMENSIONS TEST THE LIMITS OF OUR IMAGINATION AS WELL AS VERY DENSE, TINY STARS, DOUBLE AND TRIPLE STARS THAT ORBIT EACH OTHER, COSMIC GAS AND DUST CLOUDS THAT ARE SLOWLY COMPACTING AND THAT MOVE THROUGH SPACE AT HIGH SPEED, FORMING THE LARGE CELESTIAL "VORTEX" THAT WE CALL THE MILKY WAY. TECHNOLOGY MOVES ON AND NEW INSTRUMENTS HELP US TO EXTEND THE DISTANCES OVER WHICH WE EXTEND OUR GAZE. WE DISCOVER THAT THE MILKY WAY IS NOT A UNIVERSE, JUST A SMALL PART OF ONE. OUTSIDE ITS CONFINES, MILLIONS OF GALAXIES ARE SPEEDING AWAY AT A VELOCITY THAT IS CLOSE TO THE SPEED OF LIGHT. THESE ARE HUGE OBJECTS, EACH ONE MADE UP OF HUNDREDS OF BILLIONS OF STARS AND THE LIGHT WE OBSERVE LEFT THE STARS BILLIONS OF YEARS AGO. OBSERVING THESE OBJECTS MEANS LOOKING INTO THE PAST AND A MATHEMATICAL INTERPRETATION OF THE FACTS TAKES ON THE ASPECT OF PHILOSOPHICAL THEORISING.

Stars, galaxies and other celestial bodies

A UNIVERSE OF SUNS

THE SUN IS OUR BEST-KNOWN STAR: WE KNOW A GREAT DEAL ABOUT THE NUCLEAR PROCESSES FROM WHICH IT DERIVES ITS ENERGY AND WE ALSO KNOW A LOT ABOUT ITS SURFACE PHENOMENA, ITS ORIGINS AND ITS EVOLUTION. WE DON'T, HOWEVER, REALLY KNOW A LOT ABOUT THE SUN AND SOLAR RESEARCH IS A WIDE-REACHING FIELD OF STUDY.
HOW CAN WE THEN TALK ABOUT OTHER STARS?
HOW CAN WE OBTAIN INFORMATION ON THOSE POINTS OF LIGHT THAT ARE SO FAR AWAY THAT THEY STILL APPEAR AS PINPRICKS, EVEN TO THE MOST POWERFUL TELESCOPES?
THE INGENUITY AND FANTASY OF SOME SCIENTISTS HAVE ENABLED THEM TO FIND A WAY OF ANALYSING THE FINE RAYS OF LIGHT THAT REACH US FROM LIGHT-YEARS AWAY AND TO GATHER MASSES OF DATA FROM THEM.

If the Earth's "expulsion" from the centre of the universe set a revolution in motion, the parallel ejection of the solar system did not cause a murmur. It did, however, radically change the perspective of astronomical research: saying that the stars are bodies that are essentially similar to the sun and that there are billions of them throughout the universe is a leap into the unknown that has had and continues to have numerous implications. We can, for instance theorise that stars like our Sun also have planetary systems and that billions of planets could therefore support life. This is an area of research that has always provoked a lot of interest and controversy, above all amongst non-specialists. We, however, will confine our discussions to the stars.

PLANETARY RINGS
The Hubble Telescope gathered these images of dust rings orbiting Pictor b - an example of a planetary system being formed. The central (clearer) area is the same size as the solar system and this is where the planets are taking shape.

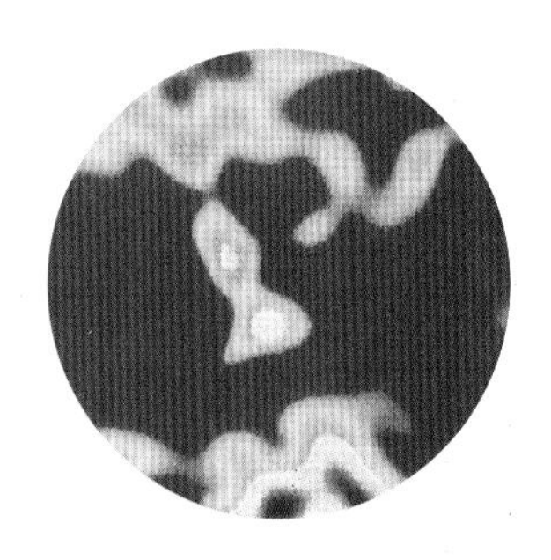

BETELGEUSE
This huge red star (above left of the photograph of the constellation of Orion) was the first star whose photospheric emissions were studied in detail. To the right is a computer-enhanced image illustrating the different emission characteristics recorded.

THE NEAREST STARS

SUN	8 LIGHT MINUTES
PROXIMA	4.2 LIGHT YEARS
α CENTAURI	4.3 LIGHT YEARS
BARNARD'S STAR	5.9 LIGHT YEARS
WOLF 359	7.6 LIGHT YEARS
BD +36° 2147	8.1 LIGHT YEARS
LALANDE 21185	8.3 LIGHT YEARS
SIRIUS	8.6 LIGHT YEARS
LUYTEN 726-8	8.9 LIGHT YEARS
ROSS 154	9.4 LIGHT YEARS
ROSS 248	10.3 LIGHT YEARS
ε ERIDANI	10.7 LIGHT YEARS

YERKES, OR MKK, SYSTEM

LUMINOSITY	DESIGNATION
IA - 0	SUPER SUPERGIANT
IA, IB	SUPERGIANT
IIA, IIB	WHITE GIANT
IIIA, IIIB	GIANT
IVA, IVB	SUB-GIANT
VA, VB	DWARF (MAIN SEQUENCE)
VI	SUB-DWARF

STARS, COLOURS, CLASSIFICATION

MINTAKA	VIOLET, O
RIGEL	INDIGO, B
SIRIO	BLUE-WHITE A
PROCIONE	GREEN, F
CAPELLA	YELLOW, G
ALDEBARAN	ORANGE, K
BETELGEUSE	RED, M

THE NEAREST LARGE STARS

	COLOR	CLASS	SOLAR DIAMETER
BETELGEUSE	●	M2	750
ANTARES	●	M1	640
αHERCULES	●	M2	500
MIRA	●	M6	420
DENEB	○	A2	110
RIGEL	○	B8	70
ALDEBARAN	○	K2	45
BOOTES	◍	F0	45
ARCTURUS	○	K2	23

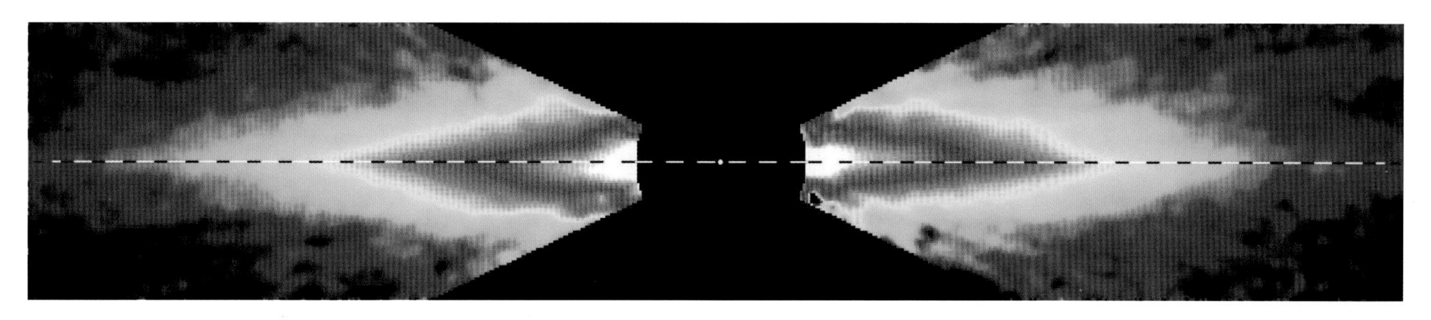

WHAT IS A STAR?

Like our Sun, stars are large gaseous bodies predominantly made up of hydrogen. They radiate energy at a variety of wavelengths due to the nuclear reactions that take place deep below the surface. Their origin lies in the gravitational contraction of large clouds of interstellar material and the stars continue to produce energy until the nuclear fusion reactions develop more energy than they need.

DISTANCES AND DIMENSIONS

The first question asked when observing the sky is usually "How far away are those points of light?"
We can give more and more precise answers as instruments become more and more accurate. The nearest star to us is Proxima in Alpha Centauri which is 4.3 light years away – i.e. 10^{13} Km – a truly "astronomical" number. If the Sun died this instant, we would be in darkness within 8 minutes. If Proxima died, it would be 4.2 years before its final rays reached the Earth. The distance of our neighbouring stars is measured in terms of parallax (the angle subtended by the earth's orbit with the star's vertex). Indirect physical methods are used to measure the distance of distant stars or extra-galactic bodies.
Although chemical composition and mass have relatively uniform values, a star's dimensions and distances can vary greatly. Some stars, such as Antares, have a diameter equal to 640 times that of the Sun (like the orbit of Mars); some are even larger, but the majority are similar to our Sun, and many are even smaller, the same size as a planet.

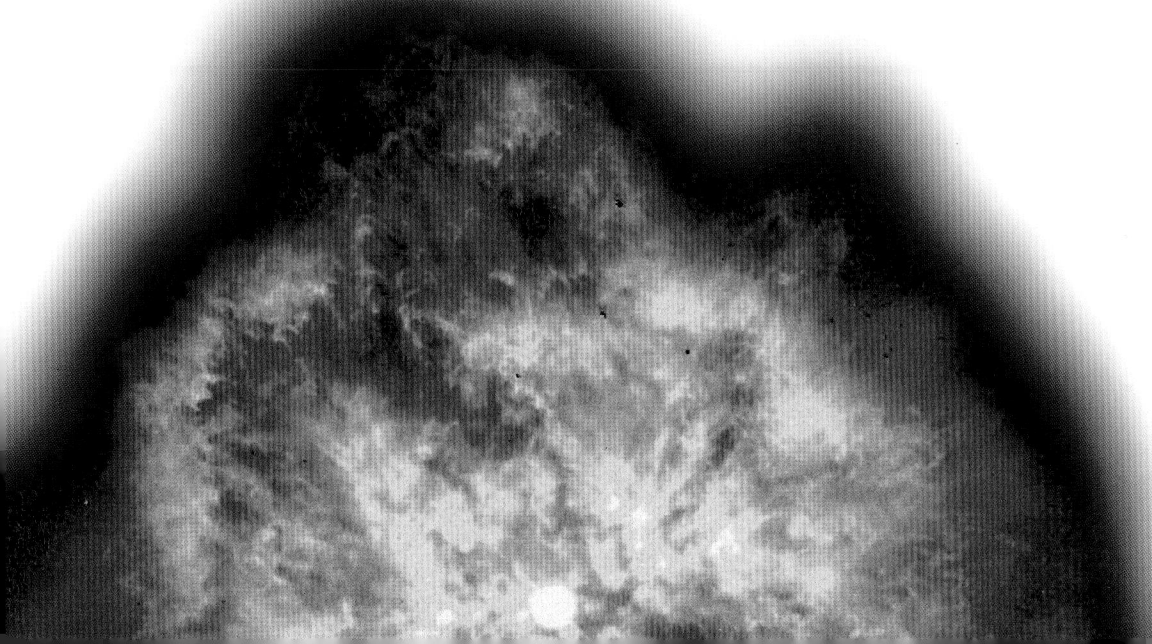

DIMENSIONS
The huge star WR 124 taken from Hubble is enveloped in a layer of super-heated plasma hurled into space at a rate of 200,000 km/h. Like its mass, the size of a star is linked to its internal activity. The quicker the nuclear fusion processes, the more radiation produced and the more the star expands.

A LARGE FAR-OFF OBJECT SEEMS SMALLER THAN A SMALL OBJECT THAT IS NEARER. HOW DO YOU MAKE PEOPLE UNDERSTAND THAT A STAR IS ONLY LARGER AND MORE LUMINOUS THAN ANOTHER BECAUSE IT IS CLOSER? THE BEST CLASSIFICATION OF STARS IS BASED ON THEIR LUMINOSITY.

LUMINOSITY AND DISTANCE

LUMINOSITY AND DISTANCE
The radiation produced by a source radiates uniformly in all directions like a sphere with an increasing radius. The further away the observer, the less radiation and, therefore, the less apparent luminosity that reaches him.

Anyone can see that the stars are not the same – some shine more brightly than others. The apparent luminosity of a star is called its apparent magnitude (abbreviated to "m") and depends on the star's intrinsic magnitude i.e. the amount of light emitted every second per surface unit and its distance from the Earth.
Apparent magnitude was used by Hipparch➢14 to classify hundreds of stars that could be observed with the naked eye. Hipparch believed that all the stars were located in a celestial vault and that they were all the same distance from the Earth. Any difference in luminosity therefore corresponded to a difference in visibility. He divided them into 6 groups: the brightest were of the "first size" and the faintest belonged to the "sixth size". The brighter of two stars was therefore smaller in size (or magnitude) and vice versa.

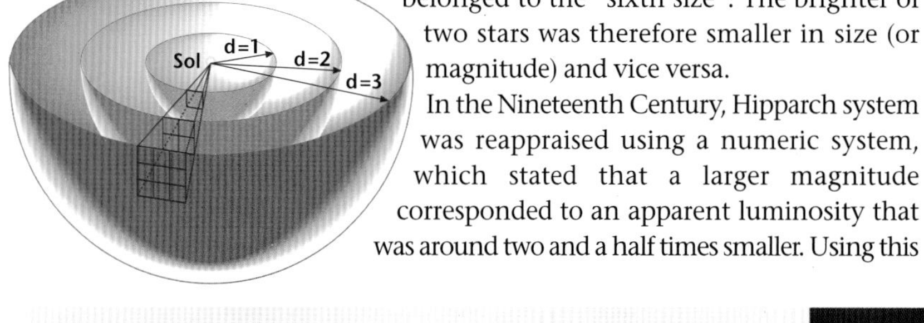

In the Nineteenth Century, Hipparch system was reappraised using a numeric system, which stated that a larger magnitude corresponded to an apparent luminosity that was around two and a half times smaller. Using this scale, a star of m=1 has a luminosity of 1, roughly speaking, and one with m=2 has a luminosity of 0.4. M=3 represents a luminosity of 0.16, m=4 that of 0.064, m=5 0.026 and m=6 - a luminosity of 0.01.
A difference of 5 apparent magnitudes corresponds to a 100-fold difference in apparent luminosity and a star with a magnitude of 3 is 10,000 times brighter than a star with a magnitude of 13. the zero, of course, is arbitrary.
Subsequently, in order to retain the traditional and widely used way of classifying stars and taking into account new ways of cataloguing observable celestial objects, this scale was slightly modified to give brighter bodies a negative magnitude value. Sirius, for example, has a magnitude of -1.5. Venus, depending on its phase, can achieve a maximum magnitude of -4 and the Sun is -27.
The highest magnitude visible to the naked eye is still 6 (under excellent conditions for observation), whilst the largest telescopes can distinguish objects with a magnitude of 25. In addition, since a star's apparent magnitude also depends on its distance from the Earth, this measurement, with no physical

THE SEARCH FOR PLANETARY SYSTEMS

Although we know billions of stars, we have only observed 9 planets, all of which are in our solar system. Man has always been on the look-out for other planets and it is more than probable that the pattern of our solar system is duplicated throughout the universe. However, it is very difficult to see an object that is not bright and that is also usually a lot smaller than a star as well as being several light years away.
Hubble has captured several very detailed images of dust rings that orbit Auriga AB. To the right is a composite image of the field of observation (9 times more extended than our solar system) and to the left can be seen the bands.
spiral-form structures and groups that are interpreted as

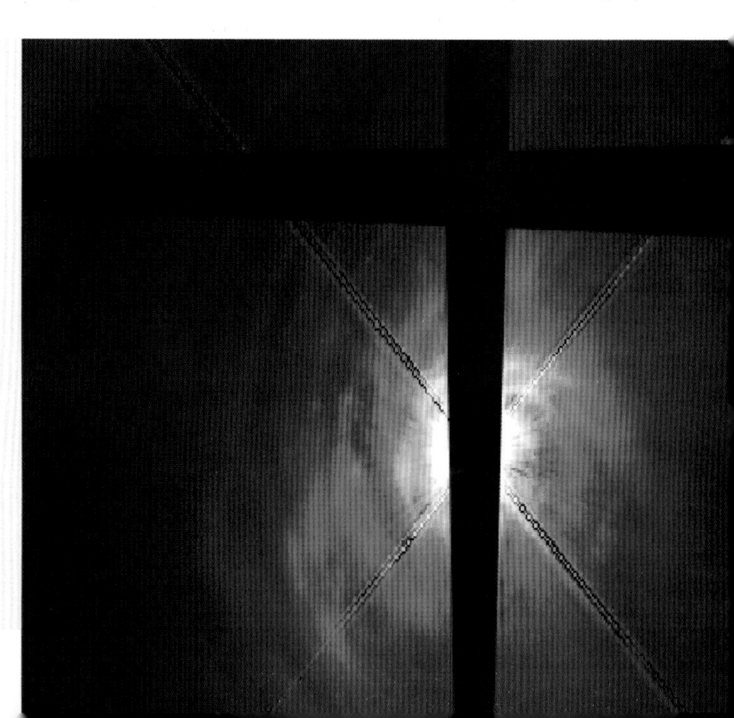

THE BRIGHTEST STARS
Recently discovered by Hubble, this star, which emits energy 10 million times that of our Sun ,has an intrinsic luminosity greater than any other star in the Galaxy. Before it was captured by HUBBLE'S powerful NIMCOS camera using infrared and spectrometry, the star was invisible, hidden in the clouds at the centre of the galaxy.

DIMENSIONAL RATIOS
The difference in diameter between the four main star types. A red giant, a star similar to our Sun, a white dwarf and a black hole. Their density increases in inverse proportion to their dimensions. The opposite is the case with luminosity.

value, is little used by astronomers who prefer the term absolute magnitude "M". Celestial objects are classified according to a scale in which the apparent magnitude is modified in terms of distance and where M indicates the luminosity that the star would have if it was at a set distance of 10 pc (equal to 32.6 light years). The absolute magnitude of a celestial object can be calculated if we know either the actual distance or the apparent magnitude. The two magnitudes are connected by means of a simple formula:

$$M = (m+5) -5 \log d$$

where d is the distance of the observed object from the Earth. The absolute magnitude of a star is a physical unit that allows direct comparisons of stellar luminosity. Using this scale, the most luminous star is also the largest energy source.

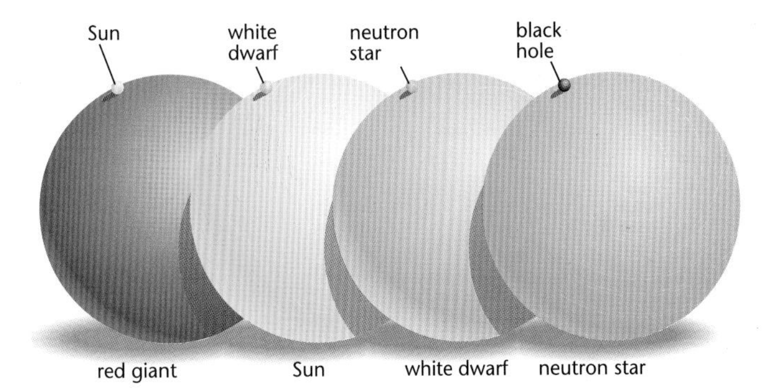

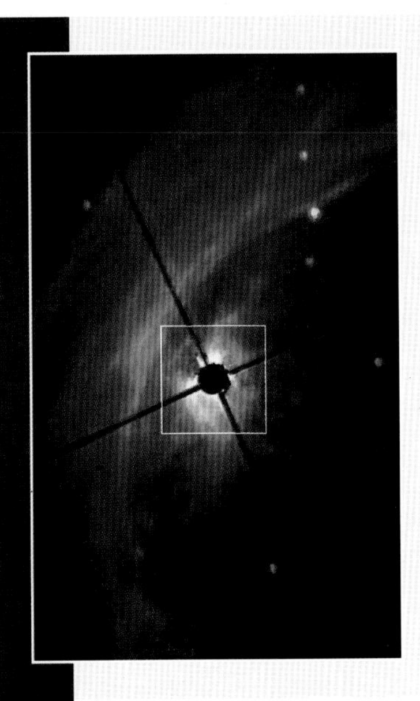

planets being formed. Another image captured by Hubble shows a detail of the dust ring around Pictor b. The nodules are concentric rings of dust produced when the star passed close to a small star around 100,000 years ago and which caused major gravitational disturbances. This model of evolution of a planetary system from dust and gas in orbit around a star, by means of gravitational disturbances, had already been proposed as an explanation for the origin of the solar system and, in particular, the arrangement of the terrestrial and young planets.
These are the best images taken to date. Any other evidence of large planets close to known stars is limited to the observation of regular oscillation of stellar luminosity that can be explain with the passage of a planet which partially eclipses the stellar disk.

THE KNOWLEDGE ACQUIRED ABOUT DIMENSIONS, DISTANCE CHEMICAL CONSTITUTION, TEMPERATURE AND SPEED OF STARS AND OTHER CELESTIAL BODIES IS CLOSELY LINKED TO OUR KNOWLEDGE OF ATOMIC DYNAMICS. THE GATES TO THE INFINITELY HUGE WERE ONLY OPENED AFTER THOSE OF THE INFINITELY SMALL HAD BEEN FLUNG APART.

STELLAR SPECTRA

STARS AND COLOURS
This image of a stellar field taken by Hubble contains stars with a variety of colours ranging from blue to orange.

Stars usually appear to be completely white. At these levels of luminosity, the eye loses its capacity to distinguish different colours. At times, however, when observing brighter stars (or when looking at a photograph), you will notice that some are blue others yellow and yet more are red. Looking at their spectra will also cause these differences to be observed. Researchers have used this diversity and its reasons to amass data about the rays of light that reach us from every celestial body.

SPECTRA, ATOMS, CHEMICAL ELEMENTS

LUMINOSITY SPECTRUM
A prism is used to split visible light into its component colours. The manuscript on which the prism rests is the one where Newton described the phenomenon – a corpuscular explanation countering the wave theory proposed by Hugens and Hooke.

The term radiation source spectrum is given to all the radiation emitted by a given source. Instruments such as the spectroscope and spectrograph ➢121 use refraction to disperse the light ray into its various components of different wavelengths that, projected onto a photosensitive plate, only make an impression where they "hit". The radiation then passes through a crack and the spectrum appears to be made up of a series of parallel lines, one for each wavelength present in the analysed radiation.
In order to understand how radiation is formed and how it is linked to a star's temperature and chemical constitution, we must know how energy is emitted and the links between atoms and energy and which atomic processes give off energy and which ones absorb it. Röntgen, the Curies, Maxwell, Einstein and Patrick were the first to discover the link between atoms, chemical elements and radiation, the origins of energy and the chemical constitution of stars, although their knowledge was limited.

RADIATION AND PHOTONS

Electromagnetic radiation is an electrical field, whose intensity varies over a period of time. This variation occurs in the same direction (radially from the source) as the speed of light ($c=3.10^8$m/s). We therefore use the terms electromagnetic waves and wavelength, according to the interpretation given by Maxwell in 1865. Solar radiation, however, is a much more complex phenomenon. Fifty years later, Einstein expounded the theory that light can be considered as a corporeal flow deprived of mass, i.e. photons, "packets" or "amounts" of energy that move at the speed of light each "carrying" energy that is inversely proportional to its own wavelength. The greater the wavelength, the less energy is "transported". In other words, "red" photons "transport" less energy than "violet" photons.
Photons, however, have strange patterns of

ATOMIC STRUCTURE AND ENERGY LEVELS

The link between radiation and the structure of matter is very complex. It can be summed up using only a few of the most essential facts to draw up an approximate and stylised image. Each of the ideas must, however, be clearly understood, in order to grasp how stellar spectra can be used to calculate the chemical composition and the temperature of celestial bodies.

Matter is made up of atoms each of which is made up of different component parts, i.e.:

Protons – each with an equal mass and single positive electrical charge.

Neutrons – with a mass only slightly different from a proton's, but with no electrical charge.

Electrons – negligible mass (-2000 times less than a proton) and with a single negative charge.

Of the atomic models proposed so far, by far the simplest is the planetary model, in which the centre of the atom is occupied by the nucleus comprising neutrons and protons, orbited by electrons. The number of protons in the nucleus is called the atomic number and, when added to the number of neutrons is called the atomic mass. The nucleus contains almost all the atom's mass and its positive charge (the total charge of all the protons). Under normal conditions, there are as many electrons circling the nucleus as there are protons in it. The atom is called "neutral" because the positive charge is equal to the negative charge.

This model of atom is not in fact similar to the solar system. Planets have an orbital path that is almost constant in terms of time, but which can have a variety of shapes (circular, elliptical, parabolic) and can be at varying distances from the sun (as shown by satellites). Electrons, however, are always changing their trajectory, leaping from one orbit to another and only occupying an orbit that is a specific distance from the nucleus (see Plank's quantitative theory).

In addition, electrostatic attraction means that each electron is more closely tied to the nucleus if it is internal orbit. To move from an internal to an external orbit, it must "absorb" enough energy to overcome the attraction of the nucleus – in other words, the electron (or atom) must "be excited". This happens when the atom absorbs energy -if it is hit or absorbs radiation. The energy makes the electron "leap" to an outer orbit, but, immediately after, it loses its energy again and falls back into its internal orbit. The energy freed by the electron when it "leaps" back inside is dispersed "all at once". The atom releases a "body of energy" called a "quantum light photon". The emitted photon's energy is "quantified" and is as large as the "leap" made by the electron.

The energy photon emitted then travels in space. It is a radiation with a defined wavelength λ that can be calculated using the following formula:

$$\lambda = \frac{hc}{\varepsilon}$$

Using the formula, the greater the electron's "leap" (i.e. the more energy absorbed) the shorter the wavelength of the radiation emitted.

The photons with the most energy are g (gamma) rays, followed by X rays and ultraviolet rays (all three invisible to the human eye) and the lowest, visible photons. The more energetic ones seem to give off the colour violet, followed by various colours ending with red. Photons with less energy are invisible to our eyes. We see the most energetic in terms of heat – infrared, whilst those with the very least energy – microwaves and radio waves- are only detectable using specialist instruments.

Atoms can only absorb a fixed amount of energy – enough to enable an electron to "leap" between two allowed orbits. It is as if each atom were an automatic petrol pump. If you don't insert the correct banknote (absorption) to distribute (emit) the petrol (energy) the pump will not work. Since the different locations of the electron in terms of the nucleus correspond to the atom's different potential energies, it is better (and more correct) to talk in terms of energy levels or excitation levels rather than orbits and of leaps between energy levels instead of leaps from one orbit to another.

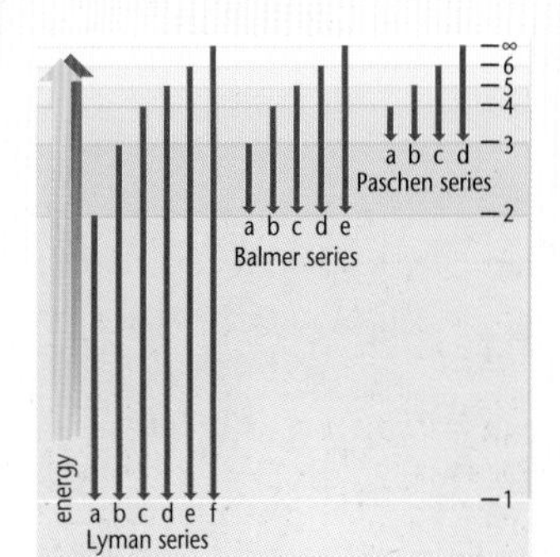

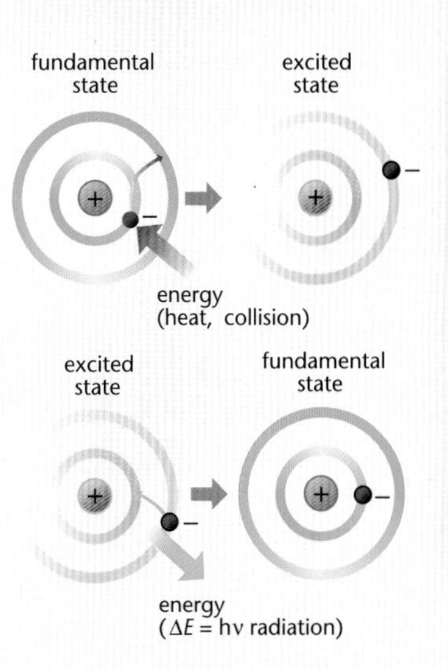

Therefore, using hydrogen, (with 1 electron the simplest atom) we can draw the diagram above called an energy level diagram. The horizontal lines (those at the bottom corresponding to the fundamental or neutral state) of the atom, are spaced in proportion to the differences in energy between each level. Each level represents the energy needed to carry an electron from a lower level to that considered. The top line corresponds to the ionisation energy, the typical excitation energy value for each atomic configuration where the connection between nucleus and electron is broken and the electron moves away from the rest of the atom (called the ion). The ionised atom still belongs to the same chemical element. A carbon atom that has lost one electron is a carbon atom ionised once, with two lost it is a twice ionised carbon atom and so on. In addition, atoms from the same element ionised in different ways have a different configuration of energy levels. We can therefore consider the different elements. When an ion captures a free electron, this will occupy a free space emitting an o quantity of energy.

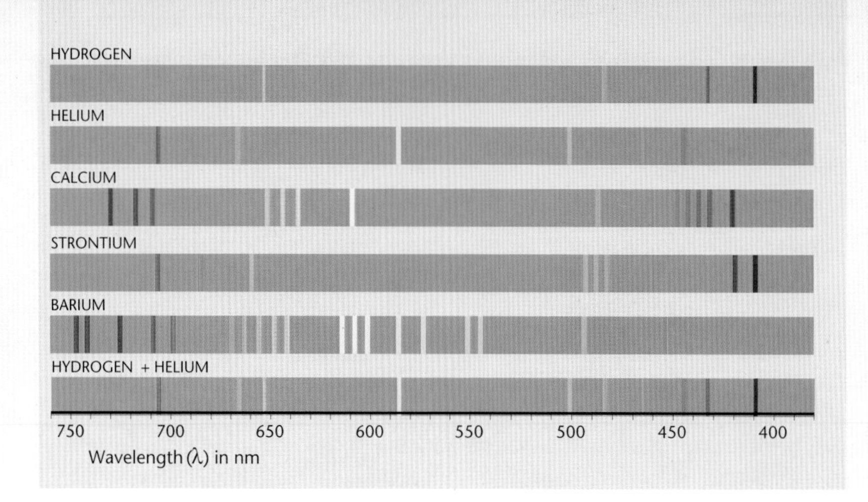

Emission's spectra
The lines that correspond to these spectra of various elements are linked to the different types of "leaps" of the electrons according to each atom's structure. An emission spectrum produced by a gas mixture is equivalent to the sum of the lines produced by each chemical element.

behaviour. They interact with matter as if they had bodies. They are absorbed by atoms and molecules, deviated from their paths, they rebound, pass undisturbed through matter... and they can also behave like "packages of waves" producing phenomena of interference. The radiation of celestial bodies, therefore, sometimes behaves like an electromagnetic wave and sometimes like a photon flow. For this, when a spectrum of visible radiation is created, we can either say that the wavelengths divide or that the photon groups that constitute them are selected.

AN EMISSION SPECTRUM

Let's suppose that to make the spectrum of radiation emitted by atoms of a single chemical element heats a gas sample in a lamp. We see a succession of brilliant, coloured, luminous lights. These are the emission lines – typical and indicative of a chemical element. Each line is linked to an "energy leap" that the electrons from each atom excited by the heat can use to return to their fundamental state. The wavelength of radiations emitted is specific to each type of "leap" (linked to the energy difference between the end and initial levels) and determined by the atom's structure. The emission lines of our element have specific colours. In "leaps", only photons of a particular colour can be produced (i.e. with a specific wavelength).

All of the emission lines of one source taken together are called an emission spectrum. Most of the lines cannot usually be seen by our eyes because their wavelengths lie outside the visible range.

If a source is made up of one single type of chemical element, the number and type of "leaps" that the electrons can make are always the same and the radiation wavelength emitted by the source corresponds - qualitatively – to that emitted by a single atom. Vice versa, if the source is made up of several elements, the number and type of "leaps" that the electrons can carry out are different and the radiation wavelength emitted by the source corresponds – qualitatively – to the sum of the wavelengths emitted by each atom. This source's spectrum includes all the typical emission lines for each element mixed and in increasing order of wavelength.

THE CONTINUOUS SPECTRUM

Under specific pressure and temperature conditions, any source is also made up of atomic ions that can capture free electrons. An electron that combines with an ion, however, can have different energies. Each electron can be "inserted" into a different atomic levels. Each electron that combines with an ion, however, emits a radiation with a different wavelength according to the "leap" that it is allowed by its energy and by the structure of the ion. A source of this type emits all sorts of radiation: the lines "cling" to each other and the spectrum is a continuous spectrum. The capacity to emit a spectrum thus depends on the source's physical state.

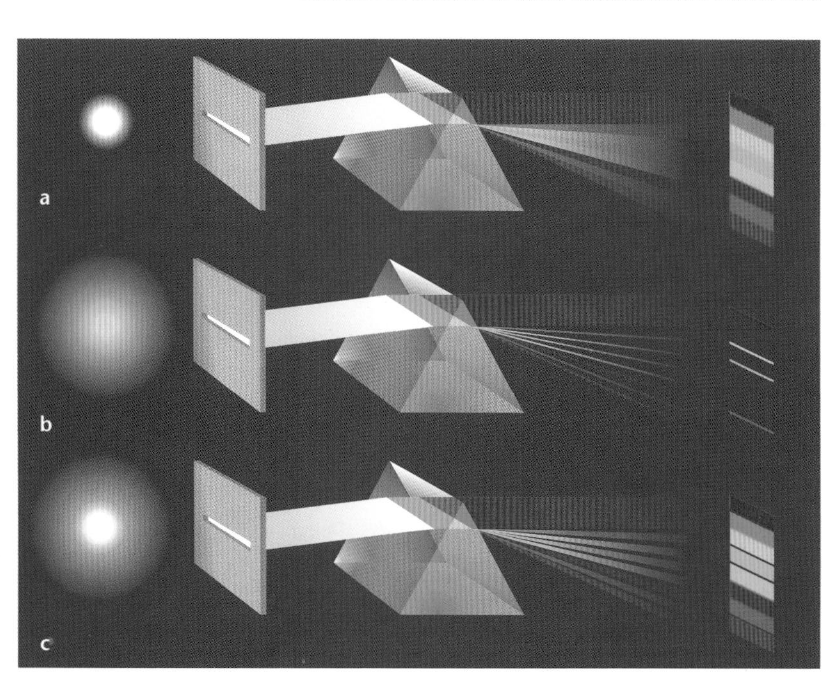

Prisms and spectra
Production of various spectrum.
a. ***Continuous spectrum*** – the radiation that reaches the prism is produced by a solid body, a liquid or high-pressure gas (or by a black body)
b. ***Emission spectrum*** - the radiation emitted by a gas reaches the prism in an unaltered state
c. ***Absorption spectrum*** – the radiation produced by a source arrives at the prism having been absorbed. The black lines correspond in negative to the coloured lines that the chemical elements that have absorbed it would emit if they were the source.

Visible solar spectrum
First of all radiation passes through the atmosphere of the Sun and then that of the Earth, that is to say, through the strata of atoms of diverse chemical elements that absorb certain photons. Other photons, of differing wavelengths carry on their journey, others are absorbed.

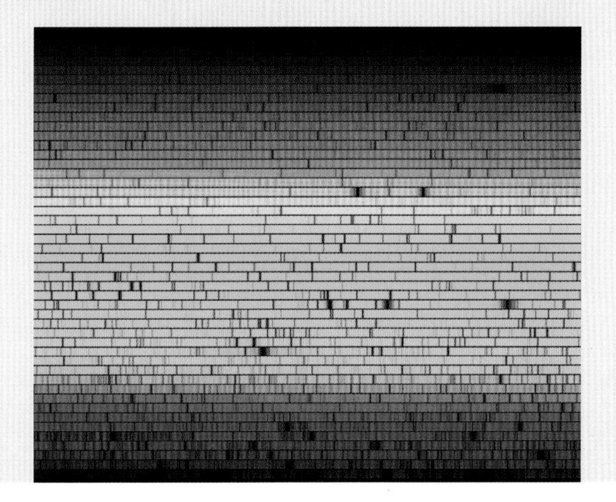

THE BLACK BODY

One example of a continuous spectrum is a black body – i.e. a body that can absorb all the radiation that it receives. Such a body does not exist in reality, but this abstraction is very useful in the study of relations between temperature and radiation. Like any other body, a heated black body would emit radiation, but, unlike real bodies, the radiation absorbed by each square centimetre of a black body at a certain temperature is that emitted by its surface. They are the same and their relationship only depends on the radiation wavelength and the temperature attained.
Thanks to the artificial concept of the black body, we arrive at the mathematical rule that links the radiation emitted (the body's energy) and temperature. As the graph shows, as the temperature increases, the more radiation is emitted in absolute. The more radiation emitted, the greater the part of radiation at shorter wavelengths. This is also the reason why a black body, like any other body, changes colour when the temperature rises. Since most if the photons emitted must have an even shorter wavelength, at around 3,000 K it would become red, at 6,000 K yellow and resplendent like the sun and at 10,000 K it would be blue.
Let's return to the stars. If some are red, some yellow and others blue – this means that the radiation emitted by them is more "abundant" in red, yellow and blue or, with the same emitted energy and the same distance, each has an emission peak in different spectral zones. We now know that this corresponds to a surface temperature difference. These, then, will be, respectively, 3,000K, 6,000 K and 10,000 K. If it were even higher the emission peak would be in the ultraviolet, just as if it were lower than 3,000 K the peak would be in the infrared. With the right instruments, spectra can be made produced from these signals.

THE SPECTRUM OF THE SUN AND THE STARS

Like spectra from natural sources, the Sun's spectrum is continuous and demonstrates several typical dark lines called absorption lines. Let's see what they are made of through the Sun's and the Earth's atmosphere, i.e. through layers of atoms of various chemical elements that can absorb photons with specific and typical wavelengths, exciting them as they pass. Whilst photons with other wavelengths continue their journey, those with energy corresponding to the excitation energy of the atoms in the atmosphere are absorbed and immediately re-emitted in all directions. The original radiation is therefore "deprived" of part of its photons and, as photons arrive with different wavelengths, relatively few photons arrive with wavelengths that are typical of elements found in the atmosphere. So, in the spectrum corresponding to these wavelengths where the photons are "missing", absorption lines are formed called Fraunhofer lines.

DATA FROM SPECTRUM

We know that a certain element can only emit a certain type of radiation. If we find that radiation

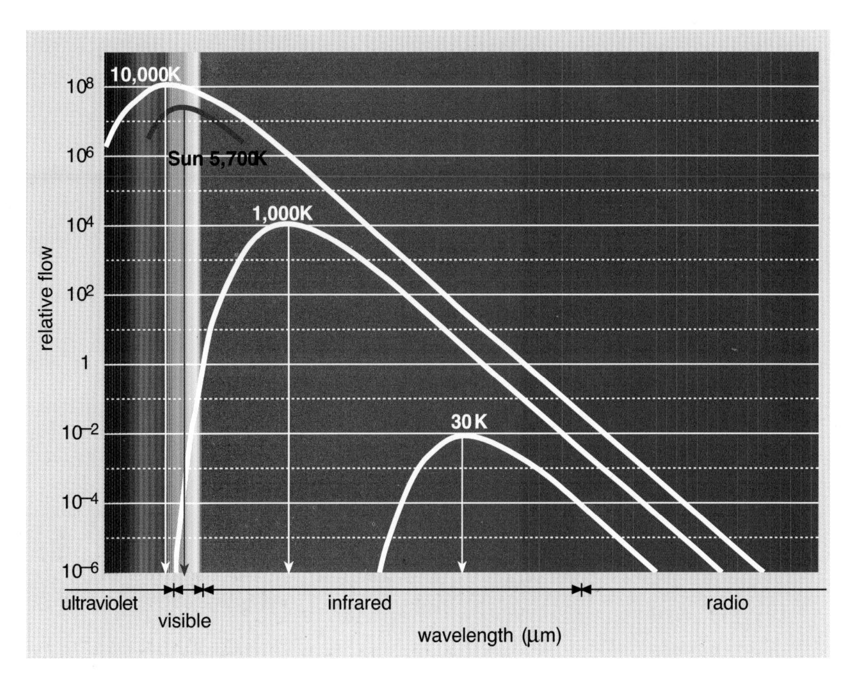

Diagram of black body
The radiation of a black body at three different temperatures compared with the nearest part of solar radiation to that of a black body. The radiation peak produced by the sun is in the visible range.

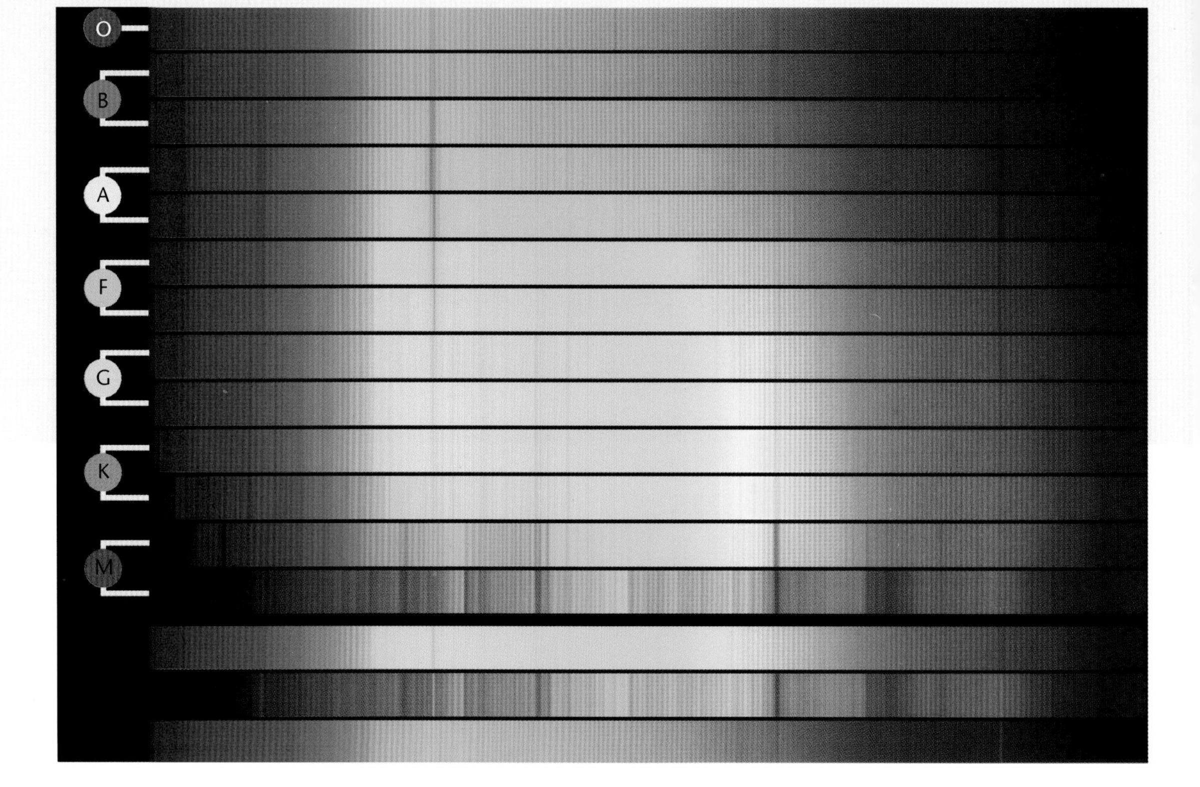

STELLAR SPECTRA
The visible spectra of stars from different stellar groups, in order, are:
• HD 12993 TYPE O6.5
• HD 158659, TYPE B0;
• HD 30584, TYPE B6;
• HD 116608, TYPE A1;
• HD 9547, TYPE A5;
• HD 10032, TYPE F0;
• BD 610367, TYPE F5;
• HD 28099, TYPE G0;
• HD 70178, TYPE G5;
• HD 23524, TYPE K0;
• SAO 76803, TYPE K5;
• HD 260655, TYPE M0;
• YALE 1755, TYPE M5.
The last three spectra belong to stars with specific characteristics:
• HD 94028, type F4 metal dusts;
• SAO 81292, type M4,5 with brilliant emission lines;
• HD 13256, type B1 with brilliant emission lines.

in the light from a star, we can say that the star contains that element. In other words, we can carry out a chemical analysis using the light emitted.
It seems simple, but it isn't. in a gas with the typical temperature, pressure and density of a star, the atoms move at high speeds and collisions are always taking place. Each collision brings about an energy change, excites the atoms and causes further radiation emissions. In addition, the atoms belong to more than one chemical element (even if some only exist as trace elements) and each atom has its own diagram which becomes more complicated the higher its atomic number. The number of possible emissions is, therefore, very complex. Finally, each atom can be ionised many times and this also alters the atomic characteristics that determine the emissions.
A star's radiation characteristics would vary even if it was made up of only one atom, either because of the different temperature conditions (that limit the possibilities for "leaps" and determine whether ions are formed or not and what type they are) or through the different density conditions (that influence the number of collisions).
By comparing the wavelength measurements using gases from single chemical elements and taking into account the relative intensity of the single emission lines, astrophysicists have identified all the elements present in the atmospheres of most stars. With additional evaluations, the chemical analysis has been made quantitative in order to extrapolate the proportions of the elements that make up these sources. We can definitively say "what" the stars are made of and "how much" of each element exists.

SPECTRAL CLASSIFICATION

Stars can be put into one of seven spectral classifications. Each of these is, in turn, divided into ten sub-classes. Starting with the highest temperatures, we have stars of
• **Class O:** blue, very luminous and massive, with surface temperatures between 40,000 and 20,000 K. We can see the helium lines and those of other atoms ionised several times. Those of almost completely ionised hydrogen are weak.
• **Class B:** blue-white, with surface temperatures between 20,000 and 10,000 K. There are several very ionised metallic atom lines. Those from helium diminish whereas the hydrogen ones increase. To this group belong b Crucis (B0),Spica and b Centauri (B1), g Orionis (B2), Alkaid (B3), Archernar (B5), Regulo and Alcione (B7), Rigel (B8), Alpheratz (B9).
• **Class A:** Green-white, with surface temperatures between 10,000 and 7,000 K. The Balmer series lines for hydrogen reach their maximum intensity. We begin to see lines of ionised calcium. Vega a Coronae, Alioth (A0), Sirius, b CUrsa Maioris (A1), Deneb, b Aurigae (A2), Formalehaut (A3), Rasalhague (A5) and Bellatrix belong to this group

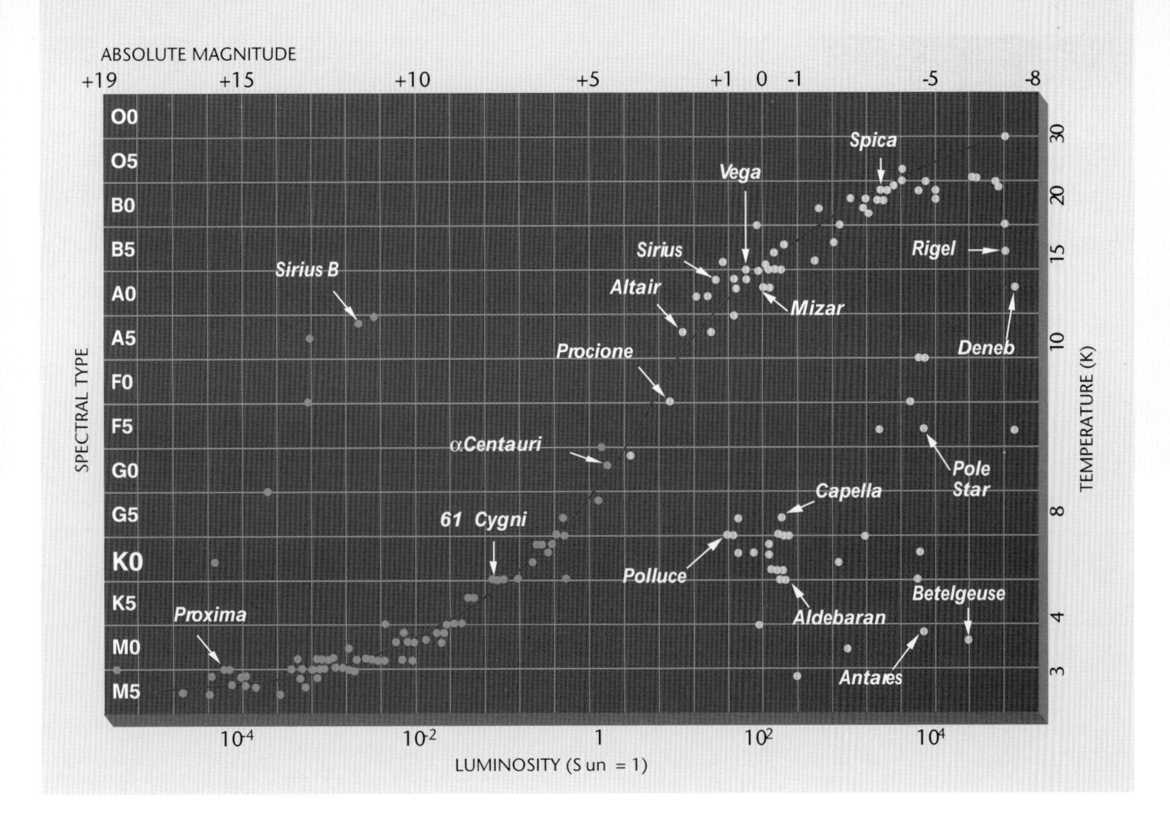

and some of them have huge magnetic fields.

• **Class F:** green, with surface temperatures between 7,000 and 6,000 K. The hydrogen lines are intense and neutral calcium lines appear in addition to those of ionised calcium. Canapo, gScorpionis, aHydri (F0), bCassiopeae (F2), Procione, aPersei, Mirfak (F8), rPuppis, the Polar Star (F8) belong to this group;

• **Class G:** like our sun (G2) are yellow, with a surface temperature between 6,000 K (G0 dwarf) and 4,800 K (G0 giant). The Blamer lines for hydrogen decrease and those of metal appear. To this classification belong Capella, zHerculis, hBootis (G0), bHydri (G1), bCarvi (G5) bGeminorum, bPersei (G8) and the main component of aCentauri.

• Type K: yellow-orange, with surface temperatures between 4,800 K (K0 dwarf) and 3,100 K (K0 giant). The neutral calcium line shows the most intensity and the metallic lines are more in relief. The hydrogen lines disappear. Aldebran and Schedir belong to this group.

• **Class M:** more or less dark red with surface temperatures between 3,400 K (dwarf) and 2,000 K (giant). Titanium oxide lines are very intense (molecular composition). This group comprises bAndromeda (M0), Antares (M1), Betelgeuse, bPegasi, Rasalgehti (M2), gCrucis, mGeminorum (M3), Mira (M6), Menkar.

There are huge variations within the classifications. There are very luminous whites and others that are barely brilliant. There are red dwarf and giant stars and so on. We also use absolute magnitude to classify them.

THE HERTZPRUNG-RUSSELL DIAGRAM

By constructing a diagram with two axes that has on one side the spectrum classification dependent on temperature and on the other absolute magnitude, we can obtain the celebrated Hertzprung-Russell diagram, for the friends of HR diagrams. It is obvious that the distribution of stars is not fortuitous: the main grouping is denser, above are the blue stars (O,B), in the centre the white and the yellow (A,F,G) and below the red stars (K,M).

Outside the main series there are two groups: one lower and on the right comprised only of large, bright red stars; the other, on the diagonal are white and blue stars that are not very bright- though this diagram leaves quite a bit out.

Hertzprung-Russell Diagram
Along the diagonal are to be found dwarf stars. Below are the group of giant and Supergiant stars (the most luminous of the group). Above the diagonal are the white dwarfs. Constructed using values of stars close to the Earth of which were known the distance by means of trigonometrical parallax measurements. The diagram enables us to calculate the distance or to define the absolute magnitude of stars for which we can define the spectral classification. In this way, we can determine stellar distances if we can photograph the stars. The areas of the diagram where there are the largest clusters are those that indicate the stable phase in which the stars remain for long periods of time.

IF WE LOOK AT THE SKY, WE ARE SEEING STARS BEING FORMED, STARS THAT ARE FULL OF ACTIVITY, STARS THAT EXPLODE, STARS IN TRANSFORMATION...
THE HERTZPRUNG-RUSSELL DIAGRAM DESCRIBES THE JOURNEY THAT EACH STAR MAKES, FROM THE BEGINNING TO THE END AND THAT IT ONLY DEPENDS ON ITS MASS.

THE EVOLUTION OF A STAR

The HR diagram shows us that most stars have some common characteristics. They are celestial bodies formed from hydrogen and other gases, with very high internal temperatures, density and pressure. Since a star's position in the diagram depends only on its mass (larger = more luminous, absolute magnitudes are compared), the HR diagram can be used as an instrument to establish a star's age and to work out its evolution, once the departure mass is known.

A STAR IS BORN

A star cannot live for ever. It radiates energy at the expense of its mass and, since its mass is not infinite, sooner or later, it is used up. And, sooner or later, it must be formed from scattered matter. Space is teeming with clouds of gas and dust and they are immense. We can therefore assume that a star is being formed at any given time. However, the atoms of interstellar matter have a very low density. To form a nucleus around which enough matter can eventually be collected to form a star, more atoms must be in each others vicinity that cannot move away from each other. The gravitational interference of other bodies or other phenomena such as the explosion of a supernova ➢203 that occurs close by can start this process of stellar formation. The proto-star is an object that is almost always invisible. It is also "cold" and can only be seen "in negative" (dark on a light background) or illuminated by an adjacent source. As we would expect, we can observe many dark

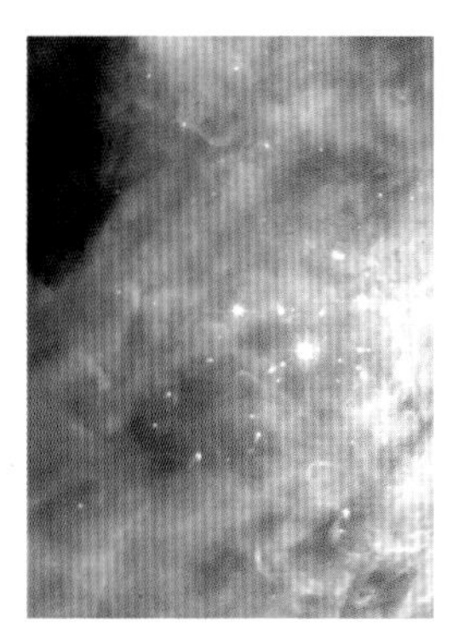

INTERSTELLAR CLOUDS
Two images captured by Hubble from the constellation of Orion
a. In the mass of the Trapezium, new stars are being formed.
b. These recently formed stars are still enveloped by thick nebulosity.

HR Diagram
The relationship between the position of a star in the HR diagram, its spectral classification, its diameter and its surface temperature (colour) can be seen.

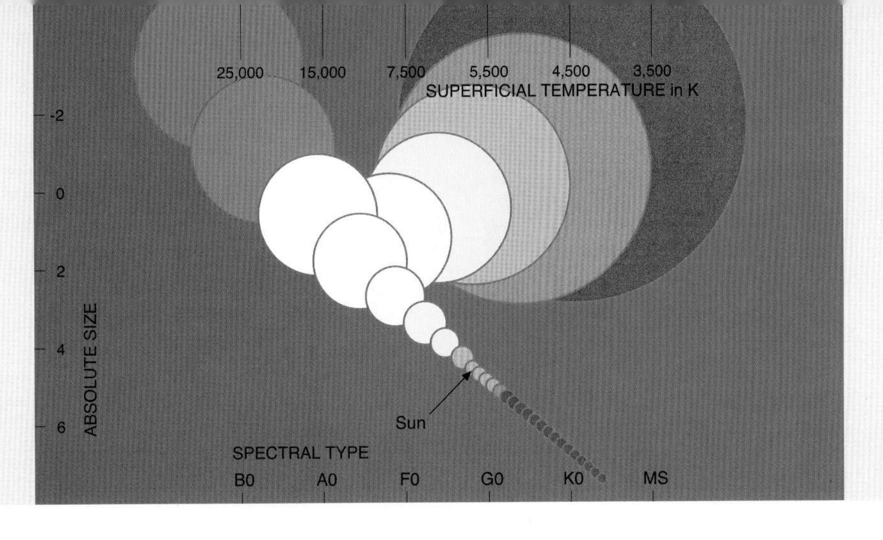

globular masses (Bok globules) of large dimensions (in the order of light-years).

THE NEW STAR

Gravity causes atoms to "precipitate" towards the centre of the globule mass. As the speed increases, the pressure increases and the gas gets hotter. The proto-star, now infrared, gets smaller and denser and its temperature rises from 100K to 50,000K. Weak and fading, it still has a place in the HR Diagram and, over a period of time, its transformations cause it to change position from one location to another in the diagram. We cannot see it since the times are too great. We can, however, imagine the journey and find confirmation in theories and experiments. The gases, for example, will continue to precipitate towards the centre of the mass, the diameter continues to decrease and the temperature to rise.

A STAR LIKE THE SUN, ALIKE MANY OTHERS

The diameter of the proto-star has reached 10^8km and its surface temperature is 3,500K. Its radiation flow is still, however, low but it is still very large, above 4000 time the sun's surface. It is an enormous red object and it is represented in the HR diagram in the area of red giants.
It will only be here for a short period of time, since changes will take place quickly.
The radius – and the irradiating surface – decrease gradually in size and the luminosity will decrease, until the temperature at the core drops to 10^7K. At

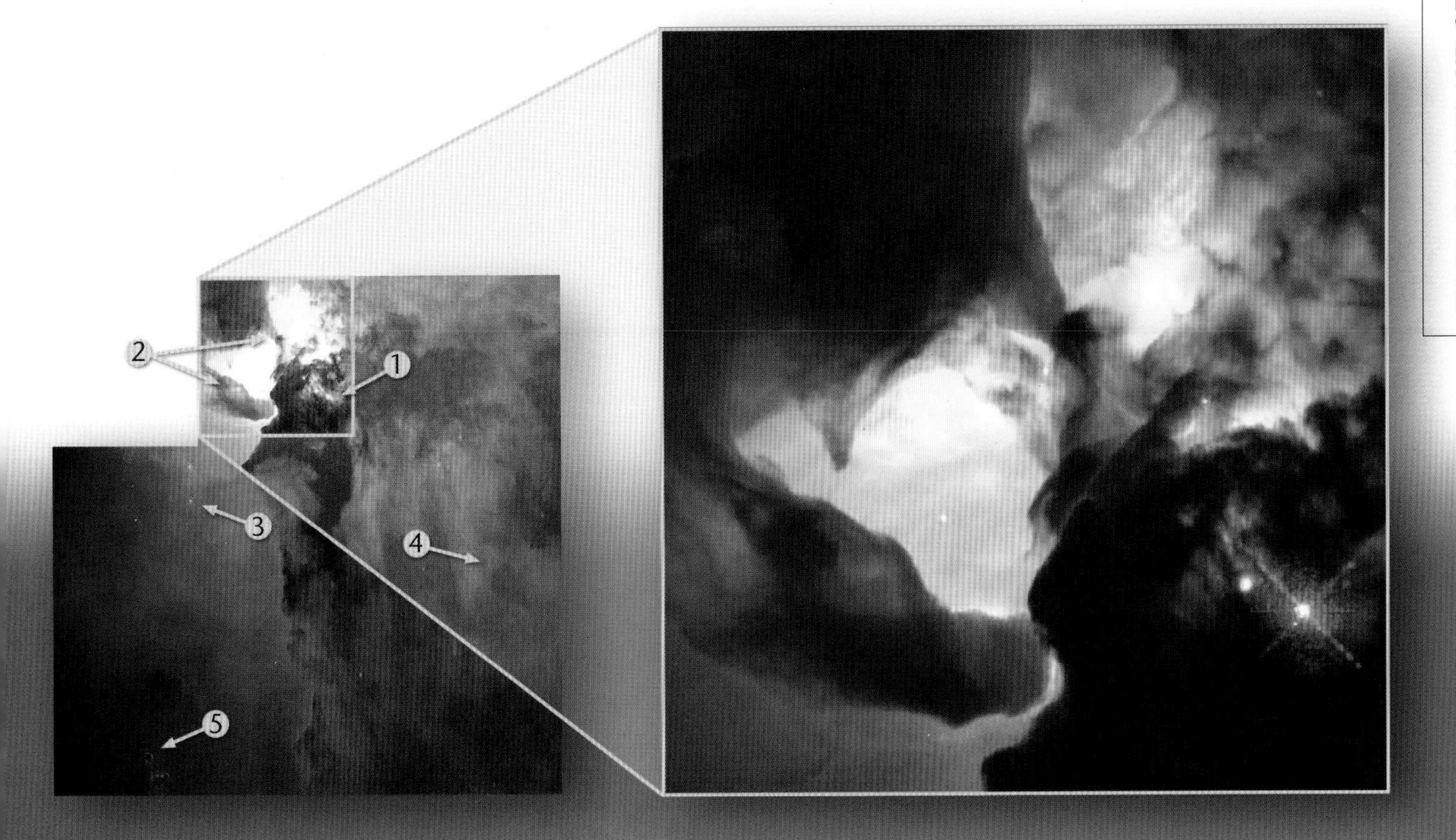

Lake Nebula
Hubble allows us to distinguish several dynamic elements:
1. central star Herschel 36; **2.** vortices;
3. shockwave; **4.** ring and jet; **5.** Bok globule.

Stars in formation
The image taken by VLT shows a cloud of gas illuminated by a group of nascent stars in the RCW38 region.

this temperature, fusion processes will start up and the star will "light up" and radiate energy. After a settling-in period, temperature, radiation and luminosity will stabilise. The diameter is slightly smaller than our sun's and the surface temperature is around 6,000K. After 27 million years of formation, the star has its own stable position in the main sequence. In 10 billion years, it will be hydrogen. As it changes into hydrogen, its size and luminosity will increase in proportion. After 4 and a half billion years' activity, the star is similar to our Sun.
Up to this point, the process would be the same, but the time scales would vary, even though the proto-star's mass is much larger. The main sequence is not in the most populated area of the diagram, but things change, when fuel begins to get scarce.

THE END OF THE SUN

The interruption in the fusion chain reactionof the hydrogen breaks the hydrostatic equilibrium that held the external areas of the star "suspended" around its nucleus: The energy developed in the area in which the fusions take place (with a diameter around one tenth that of the star), can no longer develop sufficient pressure to counteract the gravity. The time of the star's permanence in the main sequence comes to an end here. The luminosity is reduced by one and a half and the diameter contracts to around 2 million kilometres

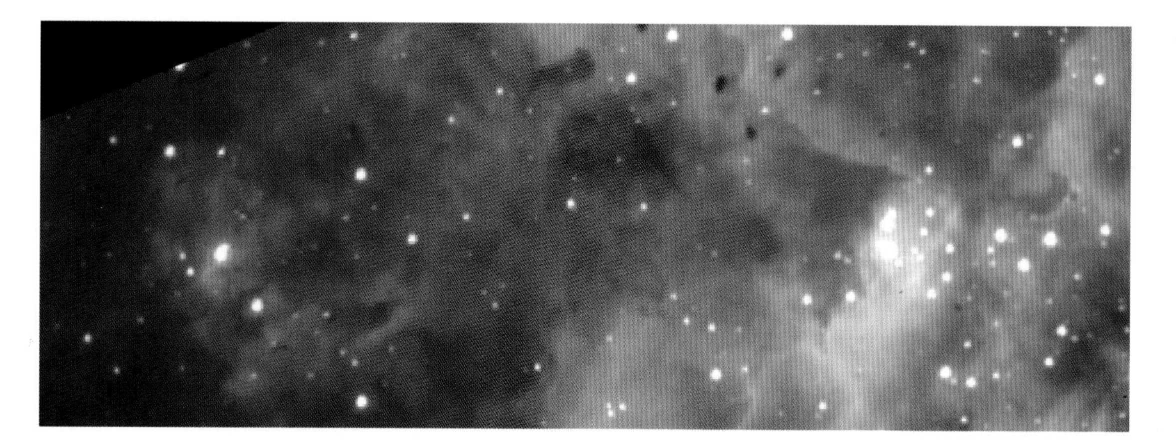

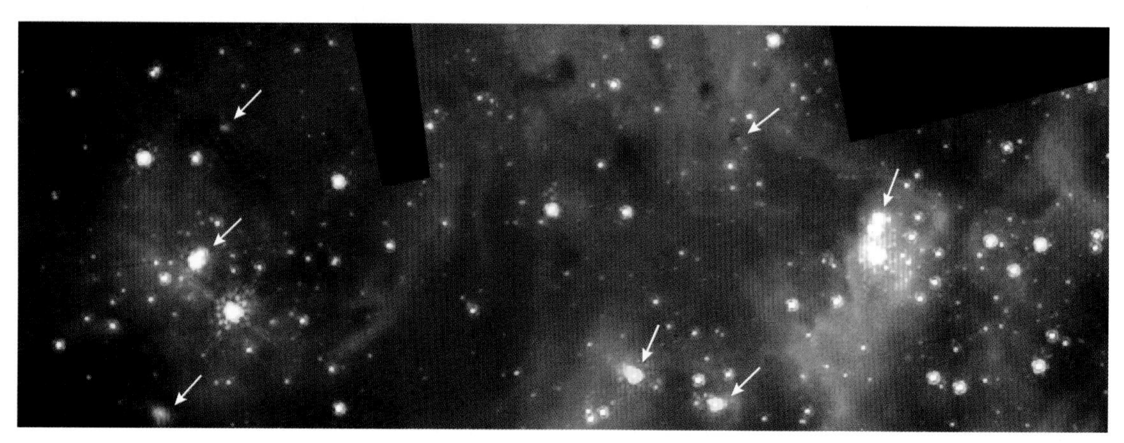

Dorado Nebula
Two images taken by Hubble in visible light (above) and infrared (below) show numerous stars in formation (indicated by the arrows). It can't be seen in the visible, but the infrared shows that the temperature is a lot higher in these areas of the nebula.

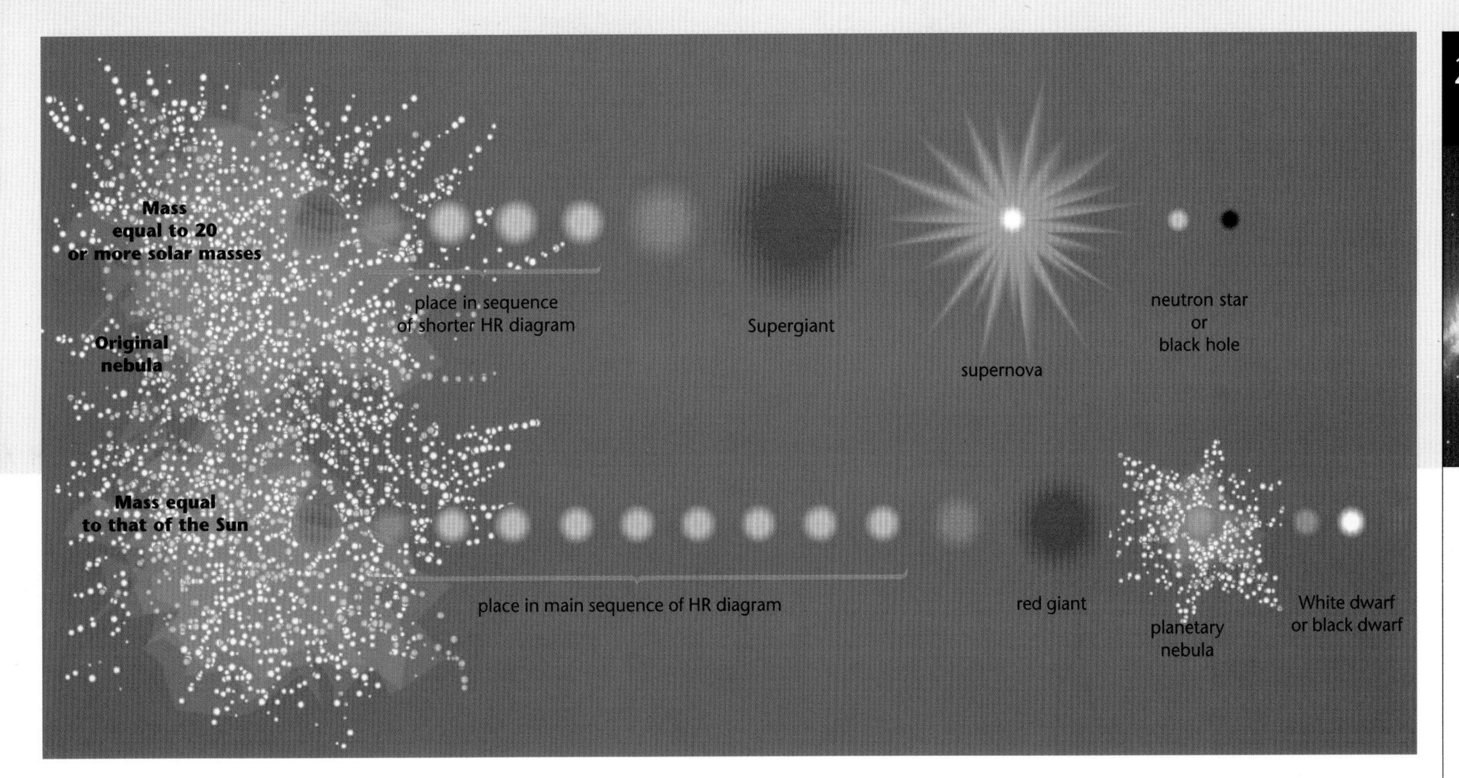

whilst the rapid contraction of the gas causes the temperature to increase again. The hydrogen in the external part of the star's helium core reheats and the fusion processes start again in the surface layers. The amount of energy emitted, at a certain point in time, becomes even greater than that produced in the first phase.
The star, however, does not become more luminous, just larger. It expands and its representative point in the HR diagram is in the area of the red giants. The expansion brings about a decrease in the surface temperature that decreases to 4,000K in little more than 1×10^9 years. The star becomes redder until it enters into a phase of constant luminosity.
However, it continues to produce energy, until it is a hundred times that produced in the main sequence phase and the surface expansion and cooling cannot balance it. The star increases in size, as well as luminosity and in a hundred million years and it becomes a red giant with an atmosphere of rarefied hydrogen (average density approx.2×10^8g/cm^3, equal to around 10^{16} atoms/cm^3), a small nucleus of helium that, whilst containing _ of the total mass has a diameter barely equal to 10^{-3} – 10^{-2} that of the star. The first envelope of hydrogen around the nucleus - where the fusions take place - is several thousand kilometres thick.
When the core temperature reaches 10^8 K, fusion begins in the helium clouds. With a very high density, the nucleus of a star with a solar mass expands very little despite the energy released by the fusions and does not cool relatively quickly.
The fusion of the helium nuclei accelerates with a further temperature increase that causes an increase in the amount of fusions and this goes on, until in a few hours, i.e. practically instantaneously, the nucleus explodes. This is the so-called helium lamp phase.
The shattered nucleus expands at a great velocity and the star's structure changes rapidly. The internal temperature collapses and the carbon synthesis stops. These precipitate the emission of energy and luminosity.
No longer "sustained", the external gas collapses towards the centre of gravity. The star's diameter begins to diminish, the core temperature increases and, at a certain point in time, the carbon production processes start up again. 10,000 years after the flash at the star's core, the temperature is

Evolution of stars
According to the mass present in the original nebulosity, each star evolves differently according to different time scales. The larger they are, the quicker their evolution. All of them, however, spend most of their existence in a stationary phase that corresponds to the main sequence of the HR diagram.

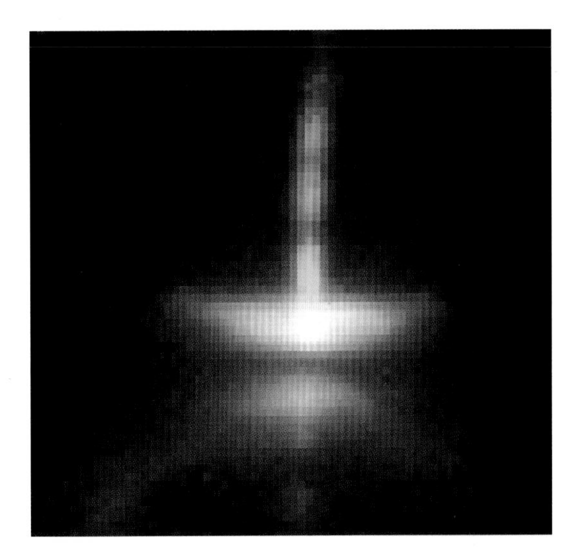

Herbig-Haro 30
This example of a reformed star photographed by Hubble illuminate from the inside the nebula that envelopes it, cut into two halves by a thin disk of dark matter. Two strong jets of gas are produced at the poles.

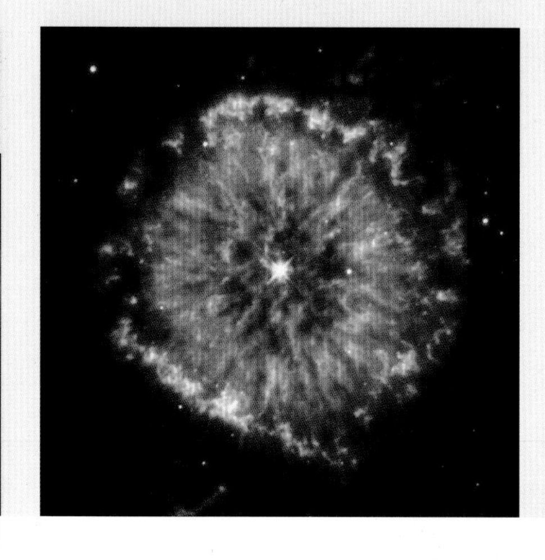

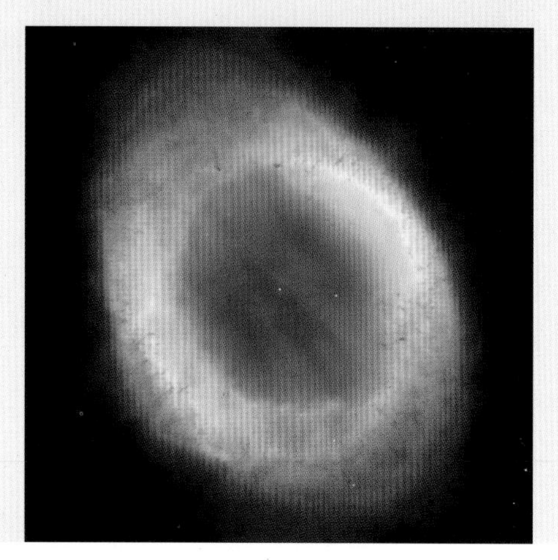

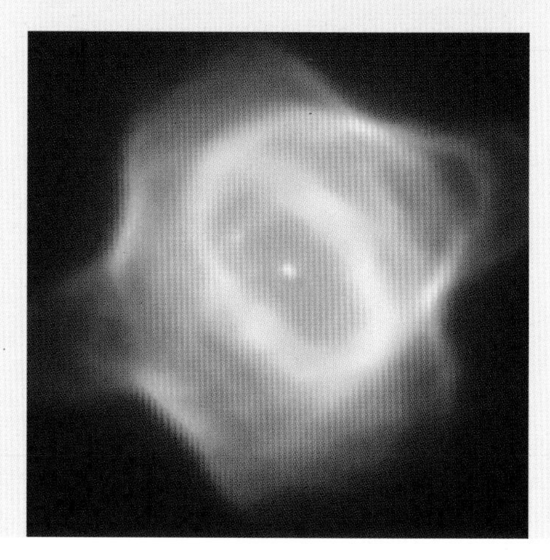

Planetary Nebulae
From left to right, the images taken by Hubble of NGC 6751 in Aquila, M57 or Ring 2000 light years away in the Lyre, Hen-1357 the "youngest" planetary nebula ever observed.

To the right: a stellar field with a number of white dwarfs taken by Hubble in M4.

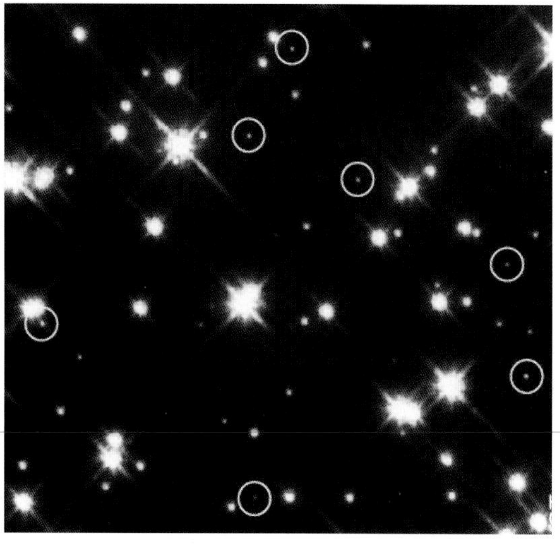

2×10^8K.
There is then a new period of constant luminosity in which a carbon nucleus is formed at the core of the helium nucleus, just as previously a helium nucleus had been formed inside a hydrogen one - history repeating itself. At a certain point, a carbon nucleus will appear that begins to contract, raising the temperature of the helium that envelopes it and setting off nuclear reactions. Nuclear reactions also take place in the outer hydrogen layers whilst a huge external envelope of hydrogen expands.
The surface temperature then decreases and the luminosity increases. The star appears in the red giant area of the diagram.
The process is more accelerated now. In a few million years, the star will probably undergo several flashes.
After this period of instability, the star enters the final phase of its existence. Since it does not have a sufficiently large mass, the temperature increase in the central areas produced by gravity is never enough to cause carbon fusion nuclear reactions that only take place at 6×10^8K or higher.
The carbon is compressed and the temperature rise increases energy production by the helium around the nucleus. The star is a red giant again, but this time, because it has expanded, the temperature in the outer layers lessens to promote the formation of neutral atoms that absorb and re-emit photons. There is a chain of events: heating, expansion, cooling, new neutral atoms, heating, expansion, cooling...
The envelope expands at a quicker rate until it abandons the star.
The nucleus remains at the core. It is a small white object with a central core and a gaseous atmosphere of neutral atoms. A planetary nebula is then formed.
Whilst the overall luminosity remains unaltered, the diameter goes through drastic variations. The star's surface rapidly reaches the temperature of the external part of the nucleus (1×10^4K).
When the star's diameter shrinks to 3×10^4K, the density is $10^6 - 10^7$ g/cm^3 and the nucleus stops contracting. The star has become a white dwarf, 10^2 times less luminous than the Sun. The radiation emitted by the core parts is emitted almost undisturbed and the surface temperature, at its maximum, is 10 times higher that of the nucleus – around 1×10^4K.
We have reached the end now. In time, the star will radiate its last energy, like a body with a certain temperature abandoned in the void. Its colour turns from white to yellow, red, dark red, until it disappears - it has become a black dwarf.

A STAR MORE MASSIVE THAN OUR SUN

If, at the moment that the helium fusion processes stop, the sun has a mass approximately equal to eight times that of the Sun, contraction proceeds until the atoms "compact" into a diameter of only ten kilometres and a neutron star or pulsar is born. Characterised by an extremely high density (10^{13} g/cm_), it orbits at a very high speed. It orbits every 20'''s approx and emits hundreds of radio pulses every second, with perfect regularity. The arrival of a signal from a pulsar can be predicted weeks in advance with an error of no more than 10'''s.
Of all the pulsars identified by the radio telescope, only a few are also visible to the human eye.

Planetary Nebula
Planetary nebula Abell 39 taken by one of the powerful telescopes at Kitt Peak.

"Butterfly" Nebula
Hubble captured the M2-9 nebula, an example of a bi-polar or "twin jet" nebula where a dying star emits matter at supersonic speeds.

If, at the moment the star is formed, it has a mass equal to 20 solar masses, when the helium fusion processes stop it continues, charged with carbon atoms and is transformed into a cloud of neon, magnesium and oxygen. When all the carbon has been transformed, the processes continue until, in the nucleus, all the silicon is transformed into iron and the synthesis comes to an end. Making iron nuclei requires more energy than the fusion process can supply.
In a few seconds, the star collapses and from dimensions equivalent to those of our sun, it passes to a diameter of around ten kilometres. The gravitational energy freed by this process produces a flow of photons that are so rich in energy that they cause the atomic nuclei to fragment into elementary particles. Just like the star's nucleus, it is transformed into a huge atomic nucleus 200 km in diameter with a mass approximately equal to one and a half that of the Sun. This compression produces a rapidly expanding shockwave at the core. The wave collides with the rest of the star and a mass almost 15 times that of the sun "hurtles" towards the nucleus. The collision gives rise to nuclear fusions that generate heavier elements such as iron and the star breaks up completely. A Supernova can be observed in the sky that, in the space of a day, reaches a luminosity 15-20 magnitudes greater - only to disappear within the space of a year.

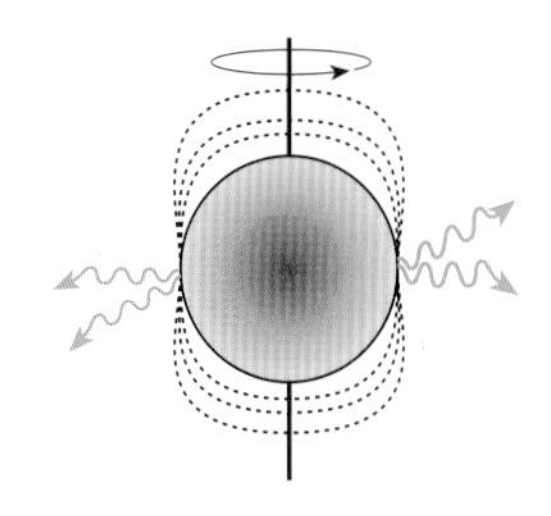

Pulsar
Standard model of a pulsar, where the magnetic fields (broken lines) concentrate radiation exiting from the magnetic poles.

Hubble again
The NGC2440 planetary nebula that is 4,000 light years away in the direction of Puppis has a structure which is more chaotic than normal. It has a core white dwarf that represents one of the hottest known stars. It surface temperature will be in the region of 2×10^5K.
Black clouds and clouds made fluorescent by ultraviolet radiation orbit the star.

Above: an image of the brown dwarf Gliese 229B next to its more active companion.

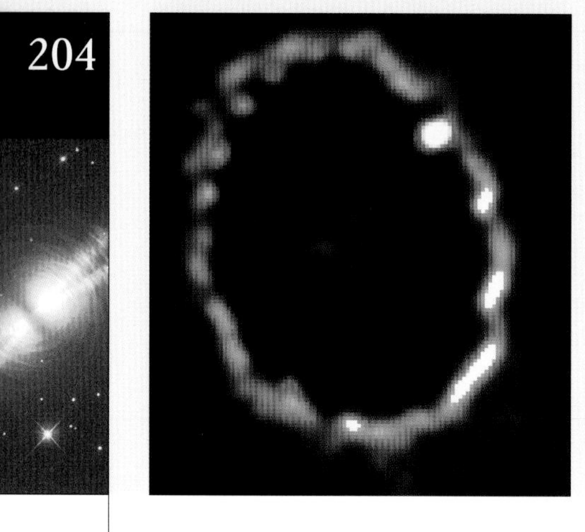

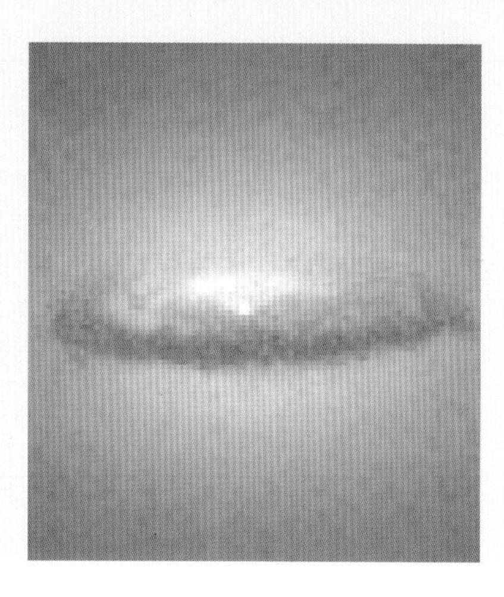

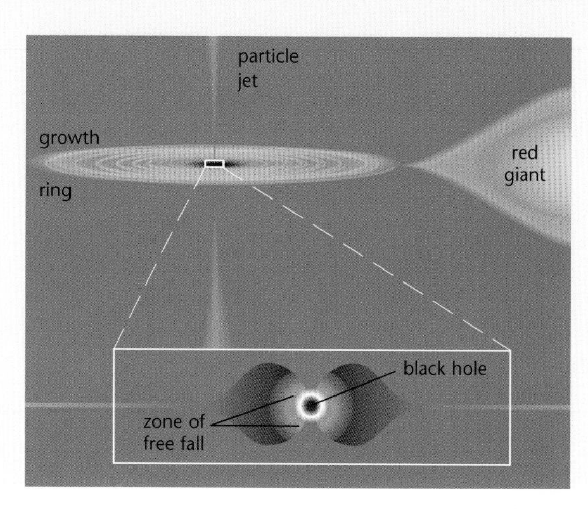

SUPERNOVA *1987 A*
Hubble shows us the ring of gas encircling the area occupied by the star that exploded in 2000. Some gas areas are visible, probably heated by a residual shockwave.

CRAB NEBULA
Hubble shows us the visible remains of the supernova from the constellation of Taurus and their X-ray emissions (below).

BLACK HOLES

When a star's mass is equal to at least 20 solar masses, what remains after the explosion is still sufficient to promote other processes. The force of gravity takes the absolute upper hand, causing an unstoppable concentration of the entire mass. Since, for every spherical mass, there exists a radial limit (called the Schwarzschild radius for the scholar who calculated it in 1916) that provokes an extreme distortion in the space-time continuum around the mass, separating it from the rest of the universe. Nothing emitted by the star has enough energy to escape from the surface, not even the most energetic radiation.
The star then becomes a black hole and is completely invisible, unless it is next to another star with a large mass. In this instance, matter from the twin star can be attracted by the black hole, orbit it and be drawn into it, thereby helping the mass and gravitational pull to grow.
The matter of this "growth ring" due to the speed that it attains, becomes extremely hot and starts to emit radiation – predominantly X rays. This indirectly reveals the presence of a black hole. This, like all the other arguments proposed, is over-simplified. There are numerous theories and scientists still sceptical of the existence of black holes, despite the results. The research of X sources such as those predicted in the theories has enabled probable growth rings to be traced. This is the case in Cygnus X-1 - a supergiant that emits strong, but irregular x-ray pulses with orbital dynamics characteristic of a binary system ➢206 in which an invisible companion had dimensions equal to that of the Earth and a mass greater than the predicted theoretical limit. After this first possible black hole,

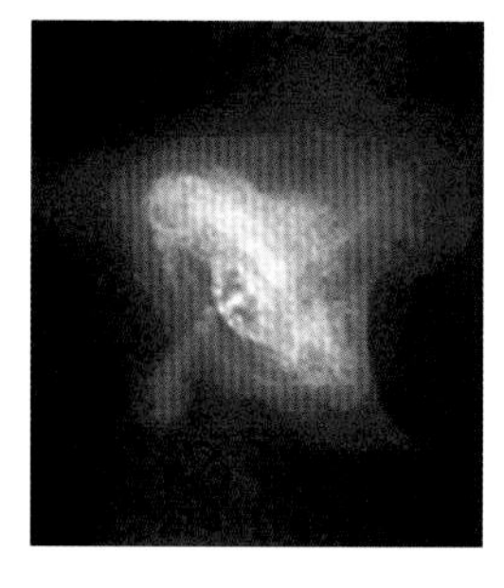

It was once thought that the pulsar's origins lay in explosions from supernovae, but to date, only the Crab and Ship Nebulae have a radiating source at the core. The Crab pulsar has a rotation period of 33 milliseconds – it completes 30 rotations per second.

Black Holes
Facing page, centre: A ring of residual dust and gas from a supernova with a diameter of 3,700 light years around a possible black hole with a mass equal to 300 million Suns. The existence of black holes, even if that was not the term used, was theorised in 1783 by John Michel. Next to the photo is a diagram that illustrates a black hole growing in a binary system. BELOW, RIGHT: an image of the centre of the M84 galaxy taken by Hubble. Astronomers suspect that a black hole is to be found here. As the "S" diagram of gas velocities shows in colours, a central ring would exist in which the matter would rotate at very high speeds, as predicted by the theory in a growth ring.

Possible Black Holes

	Companion	Mass (Sol = 1)	Magnitude	Distance (light years)
A0620-200	K dwarf	8-15	18	3,000
Cygnus X-1	O supergiant	10-15	9	8,000
GS2000+25	K dwarf	5.3-8.2	22	8,000
J0422+32	M dwarf	4,5	22	8,000
Nova Muscae 1991	K dwarf	4-6	20	10,000
V404 Gygni	K dwarf	4-6	18	11,000
LMCX-1	O giant	4-10	14	175,000
LMCX-3	B dwarf	4-11	17	175,000

many others were found using the same method, amongst which was the V404 Cygni System. However it is more than likely that much larger black holes will be discovered in the centre of active galaxies.

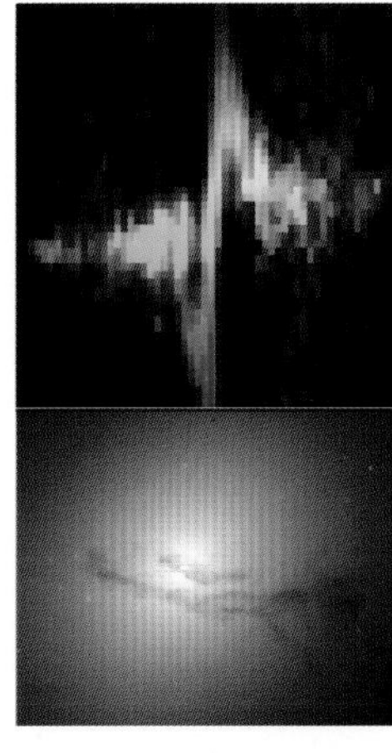

Next to a Black Hole
a. This image of the M87 galaxy taken by the VLA telescope shows a gigantic structure with radiation emissions that are increased by jets of sub-atomic particles produced by the black hole at the core of the galaxy itself.
b. This photograph of the same object was taken by Hubble and reveals a brilliant jet of electrons emitted at high speed by the nucleus produced by a black hole with a mass equal to 3 billion Suns.
c. Radio image taken by VLBA (very Long Baseline Array) that demonstrates the jet produced by the black hole.

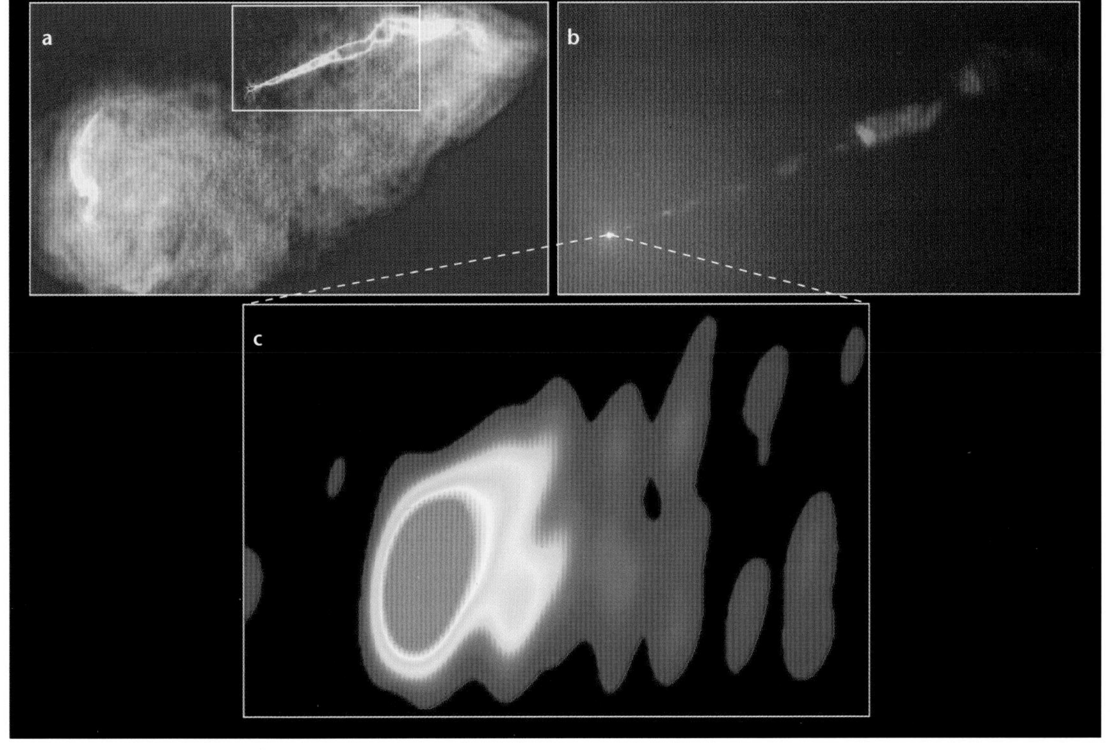

SOME STARS TRAVEL IN PAIRS AND SOME IN CLOSE GROUPS. IF WE DON'T LOOK CLOSELY, THEY SEEM TO US TO BE A SINGLE POINT OF LIGHT. WE OFTEN SPOT THAT THERE IS MORE THAN ONE STAR BECAUSE THE LUMINOSITY VARIES. ON OCCASIONS, THIS CHANGES INTO A NOVA AND LIGHTS UP THE SKY. OTHER STARS CHANGE MAGNITUDE RAPIDLY – THESE ARE VARIABLE STARS, OF WHICH THERE ARE MANY TYPES.

BINARY AND VARIABLE STARS

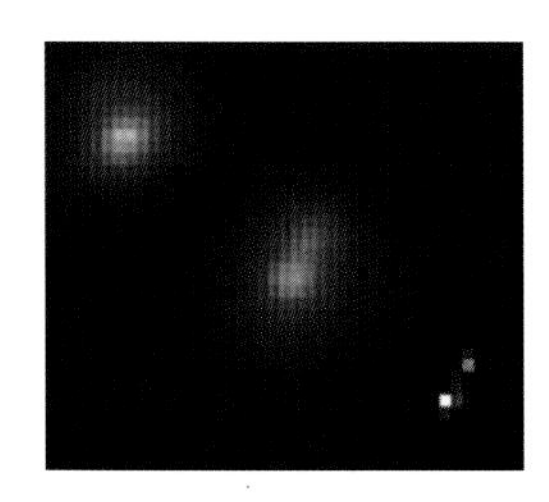

BINARY SYSTEM
The two adjacent stars that make up a binary system only become visible when a top quality, sensitive instrument is used.

RECURRENT NOVA
What remains of the nova tPyxidis (t Compass Box) as an ultraviolet image taken by Hubble

These stars formed in pairs or groups (which then split up) remain linked by gravity and are called binary. When seen with the naked eye, as with Mizar –a group of 7 stars of which we can only see two (visual binary) – we can make out the distinct, but adjacent points of light. It could be the perspective, but studying the motions of each star allows us to see that each orbits around a common mass core in accordance with Kepler's laws. We can use this method to find "invisible companions". If one of the two stars is a white dwarf, for example, and is very difficult to see, its mass remains and affects the movement of its companion. Studying the movement of binaries is the only direct method of determining the mass of these stars and of constructing a mass-luminosity diagram that allows us to calculate a star's mass when we only know its luminosity.

If the stars are very close to each other (close on a stellar scale, although the could be millions of kilometres apart), the tidal pull provokes the movement of matter from one star to the other. This method is also used to identify black holes, around which growth rings are formed. This is also the way in which novae are formed.

NOVAE

These are stars that suddenly increase in visibility, due to an explosive event and seem to spring "brand new" out of nowhere, which is how they got their name from the Ancients. in reality, only a few are

THE DYNAMICS OF TWIN STARS

The drawing shows a probable evolution of a binary system, using a calculated simulation, from a nebula with a mass equal to 30m times that of the sun. one of the two stars will have a greater mass than the other. Let's suppose that the respective masses are 6 and 20 solar masses (1). Linked by gravitational pull, the two stars remain active for at least 107 years, until one begins to expand (2) developing normally into a red giant. The outer layers are captured by the gravitational field of the second star, which increases in mass (3). The nucleus of the first star becomes exposed and the star precipitates towards its supernova phase, given the initial mass, like neutron stars or black holes with around 3 solar masses (4).

But the increased mass of the second star causes it to "age" rapidly (5). Transformed into a massive star, with high temperatures and rapid rotation, it will also become a red giant in a few million years and will transfer its mass to its twin (6). The matter, captured by the gravitational field, will form a growth ring that, as it heats up, will produce X or g radiation (7).

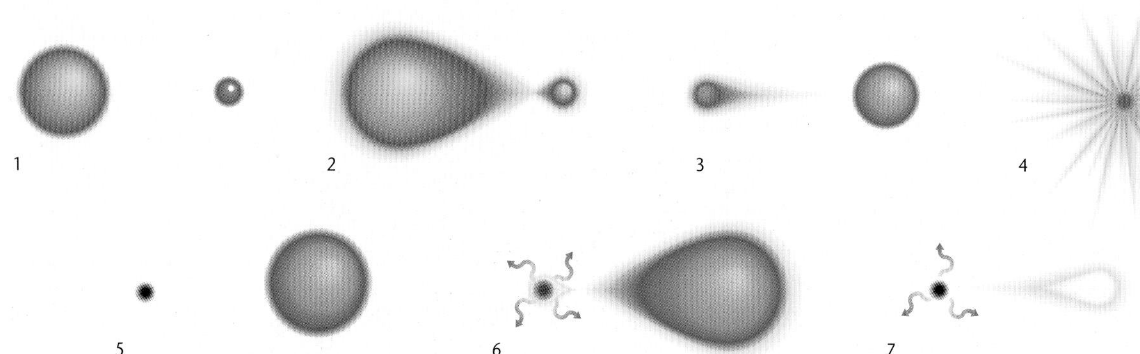

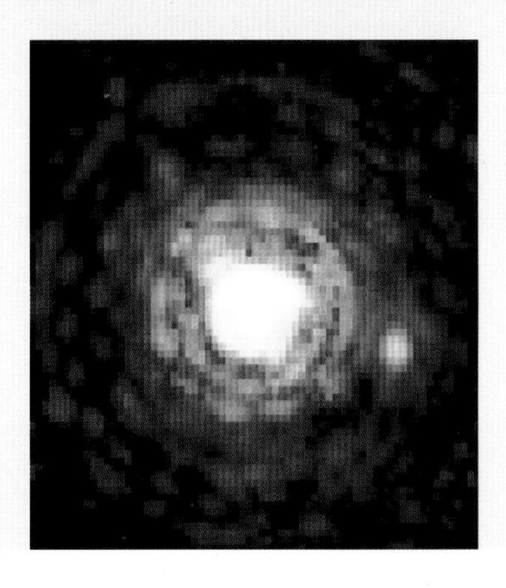

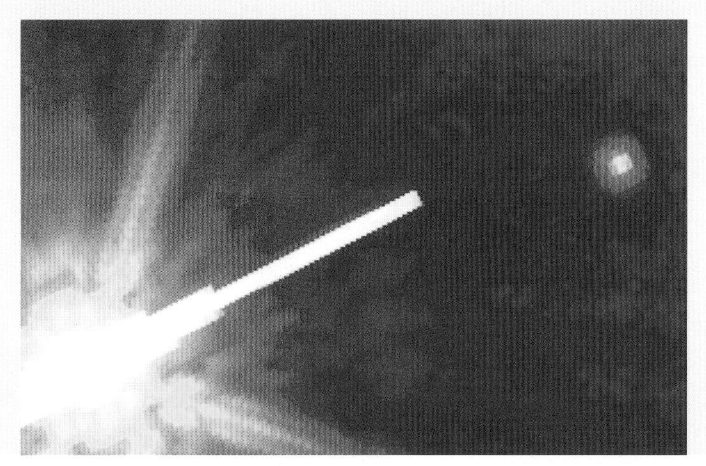

GLIESE 623B AND 105A
Twin stars taken by Hubble in visible mode (LEFT) and at various wavelengths (RIGHT) Gliese 105A, the smallest star is a brown dwarf.

luminous before the explosion. We were not aware of this until photography allowed us to compare the sky before and after the event,.
These stars are almost all "contact binaries" a white dwarf and a red giant – very close to each other. The matter expanding from the red giant forms a growth ring around the white dwarf and, when the mass of the ring exceeds a certain limit, the conditions in the layers nearest the dwarf start nuclear fusion processes. The large explosion makes the star 100,000 times more luminous within the space of a few days. The external shell shatters and forms a gas cloud that expands at thousands of km/s. The amount of matter expelled is not great, but it does contain a lot of heavy elements. The explosion does not signal the end of the star, which, within a few months, will return to its previous state.
The process will repeat itself several times.

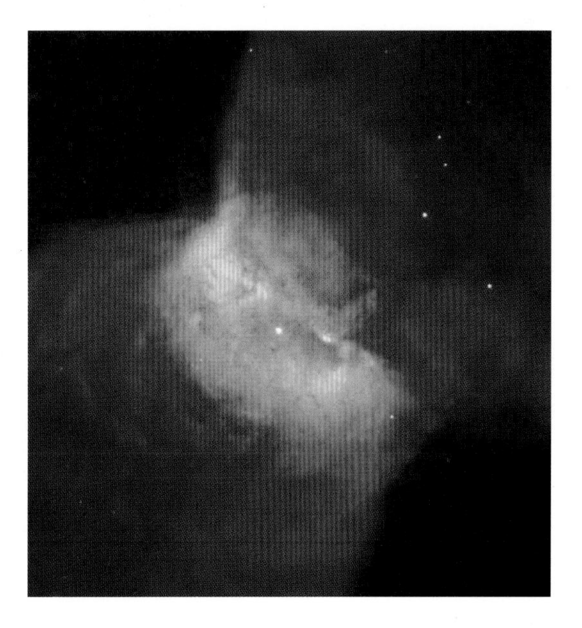

BINARY SYSTEM
The NGC 2346 planetary nebula was the first such observed with a binary system at its centre. The two stars are very close together. They complete an orbit every 16 days.

VARIABLE

Many novae have "re-ignited" and more or less all stars do it. They are classified as follows: cataclysmic or eruptive variables (novae and supernovae), eclipse variables, in which the variation is due to the interposition between us and those of another star, rotation variables that show us over a period of time areas with different luminosity and pulsing variables with regular variations. These are stars with different masses that have specific phases of evolution and can be found some distance from the main sequence in the HR diagram. As the Doppler studies on gas velocity have shown these stars vary in diameter over periods of between 1 and 50 days (Delta Cepheid) or 9 to 17 hours (RR Lyrae). They are at their most luminous when the gas is contracting (when it is further from us) and less bright when it extends.

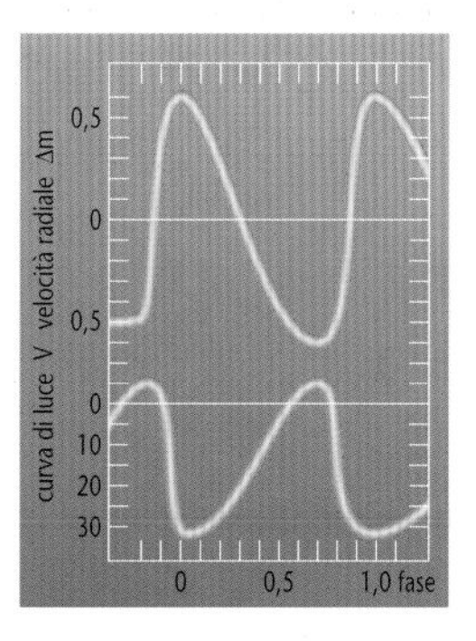

DELTA CEPHEID
The diagrams illustrate the relationship between radial speed and luminosity.

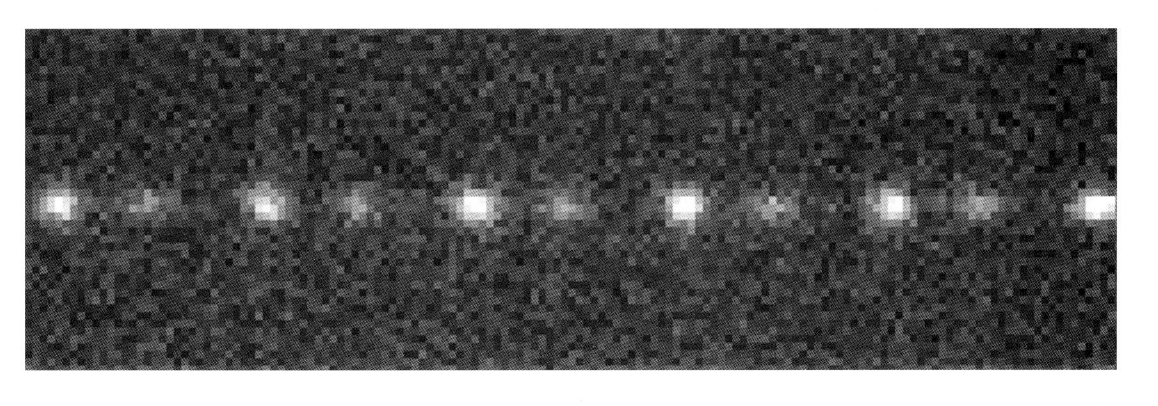

PULSAR
Photographic sequence of the Crab pulsar taken by VLT. Even though the name pulsar comes from pulsating radio sources these stars do not pulse – their rotation is variable.

THE MILKY WAY AND BEYOND

It seems impossible, but even at light year distance stars are in contact with one another. As a matter of fact, they are kept together by gravitational forces which drag them in a very rapid waltz, causing them to cluster into huge systems called galaxies. Even the Sun, along with other stars, forms a galaxy.
The ancients named our galaxy the Milky Way without knowing what it was, by simply observing that band of light in the sky that today we recognize as the plane where most stars, nebulae, and dusts revolve around the galactic center. Just like millions of other galaxies, even ours has a structure that we come to know more and more every day by observing other galaxies and by coming up with new methods of analysis.

The ancient people made up many theories about what that band of light diffused in the sky and dotted with stars is all about: for the Greeks it was the milk spread by Hera, mother of all gods. Galaxy, or better galàxias kùklos (milky circle), was the name given to her and that we still use today which translates as milky way. The Chinese, instead, thought that it was a Celestial River - thus, they named it so - where thousands of fish-stars swam and where, by mistaking the Moon's crescent for a hook, they would swim away and disappear as the Moon advanced. The inhabitants of Siberia, convinced a while ago that the sky was split in two halves, thought that the band of light was the welding that held the two hemispheres together. Although visible from any place on Earth, it wasn't until Galileo pointed his telescope to it that anyone noticed that, in reality, the diffused luminosity was the sum of billions of tiny stars. It was William Herschel who, for the first time, gave a "body" to the

Milky Way
In this summery sky the Galactic Center glows about 30,000 light years across. Numerous clusters of stars and nebulae can be observed.
Below, left to the brightest side, we can see the constellation of Sagittarius.

The origins of the Milky Way
In this 1582 painting, kept at London's National Gallery, Jacopo Tintoretto recalls the Greek myth about the origin of the galaxy by portraying Hera breastfeeding and spraying stars from her breast.

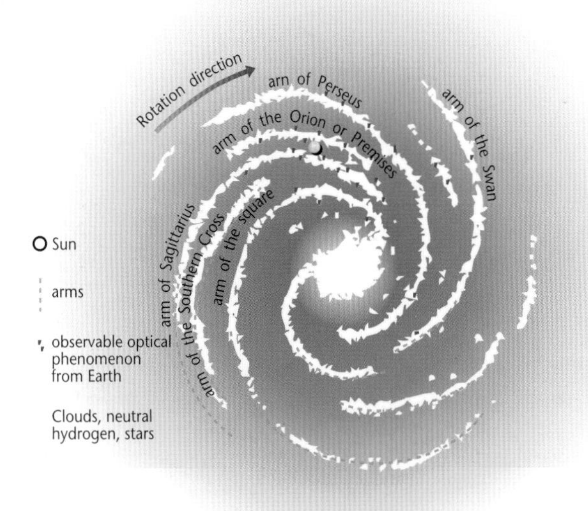

Numbers for the milky way

Component stars	2-3 x 10^{11}	Spectral type stars O e B	11%
Disk diameter	10^5 l/y. = 30 kpc	Spectral type stars A e F	41%
Thickness of disk center	10^3 l/y. = 300 pc	Spectral type stars G, K e M	48%
Total age	1.5 x 10^{10} year	Planetary nebulae	~700
Mass(in 10^{33} kg)	1.99	Globular clusters	~500
Average gas density	10^5 atom/cm^3	Pulsar	~400
Composition:	H 73%, He 25%, others 2%	Open clusters	~18,000
Percentage of double stars	70%	Novae/year (average)	~2
White dwarfs	15%	Supernovae/century (average)	2

globular mass
alone
disc plane
Sun
nucleus
15.000 pc

Galaxy: indeed, wanting to know what place the Sun had among the stars, he created a bi-dimensional mapping of the sky indicating clearly how stars were distributed in a flat shape whose inside would even contain the Sun. It was only at the end of the last century when we learned precisely that there are billions of stars in our system (between 200 and 300), that the center of our Galaxy is about 30,000 light years across from the Sun in the direction of Sagittarius, and that it exhibits a spiral structure with regions less populated by stars and others that are mostly populated by them. Thus, we started talking about spiral arms; about the Galactic Center, which is located in the region where all arms extend from; and about the galactic halo, shaped as a sphere, which envelopes the Galactic disk containing very old stars and globular clusters. The Sun is situated in the outer regions, and the Galaxy can be observed from its inner region projected onto the celestial sphere.
The observations of other galaxies beyond our stellar system confirm what has already been learned about the Milky Way and validate those observations which suggested that it might have motion.

THE GALAXY MOTION

Stellar kinetics studies galaxy motion, and the spectroscopy's applications allow us to have very reliable evaluations. The Galaxy spins around itself at a radial speed of about 220 km/s and makes a complete revolution about every 2.4 x 10^8 years. As in other galaxies, the rotation in ours is differential: the near stars at the center move as a solid body and rotate at a greater speed than the further ones which, presumably, following Kepler's laws. But, there are many more motions: for instance, the Sun moves at a motion of its own in the direction of Hercules m, and this motion overlaps with the general galactic one. And just like the Sun, even all other objects in the galaxy, stars, nebulae, open clusters and globular clusters, move through independent motions.

Galaxy
This term, used by the ancients to refer to the milky way, came to be used later as a general name (written in lower case) to refer to other stellar systems. Galaxy, in upper case, as well as Milky Way, remains the name of our stellar system: even Dante refers to it by that name in Convivio (II, 14):
«That dawn, that Galaxy we call »
And in the Comedy (Par. XIV, 96-98):
«As, graced with lesser and with larger lights between the poles of the world, the Galaxy gleams so that even sages are perplexed».
In this image, a galaxy very similar to ours, even though 108 million light years across from us: the NGC 4603.

A GALAXY IS A COMPLEX SYSTEM MADE UP OF BILLIONS OF STARS, JUST LIKE THE SUN, BIGGER AND SMALLER STARS OF RECENT FORMATION AND NEAR THE END OF THEIR EVOLUTION. MOREOVER, QUANTITIES OF "DARK MATERIAL" ALSO CONTRIBUTE TO FORM THE GALAXY: GAS, DUSTS, INVISIBLE ATOMIC PARTICLES, WHICH ARE ALL A MUST IF WE WANT TO EXPLAIN THE GALACTIC DYNAMICS OBSERVED.

THE STRUCTURE

The spiral structure of the Galaxy has even been confirmed by studies on neutral hydrogen distribution on the galactic plane emitted in a wave length of 21 cm: the areas of major concentration correspond to the disk, the nucleus and the arms. On the outer region, the halo, we find the open clusters and globular clusters, which are stars that revolve around the galactic nucleus. Even if these are not found in the deepest part of the galactic vortex, they are, nevertheless, gravitationally bound to our system.

ZONES AND FEATURES

The Milky Way has a structure which is very similar to millions of other systems observed in space. And, like these, even by having a diameter of 10^5 l/y (=30 kpc), in its outer regions it has a thickness of 10^2 l/y (= 30 pc). At the center it has a bulge called the nucleus, around which the Disk extends, a layer of 17 x 10^2 l/y about (500pc) in thickness. This has some areas where the material is denser, where the nearest stars, called arms, are shaped as a half moon due do their differential motion.

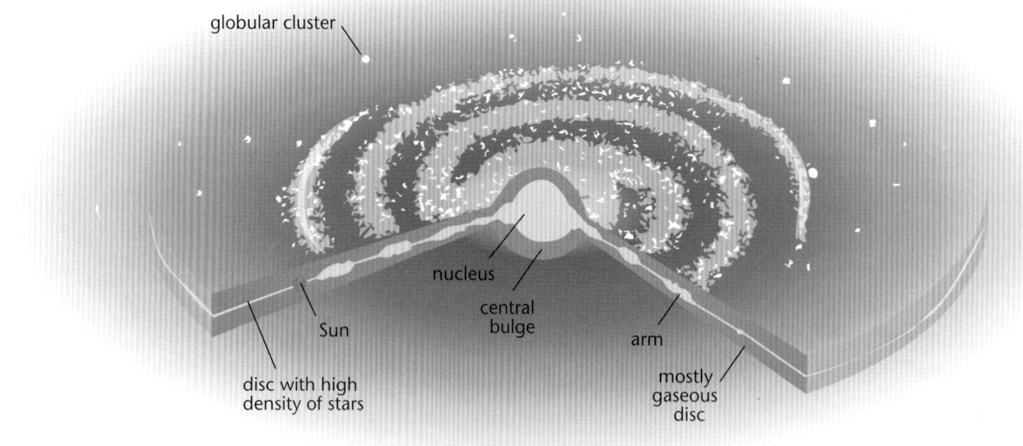

SCHEME
The Galaxy structure, as shown from optical and radio observations, and integrated by the latest infrared and ultraviolet data. In reality the side opposite to the Sun is just hypothetical: it has, indeed, been only a short while that, thanks to instruments such as the Hubble Telescope, we have been able to "look beyond" the galactic nucleus full of stars and luminous gas .

GALACTIC PROFILE
The NGC4 0 1 3 spiral shape of the Galaxy profile allows us to identify the various zones among which we can also distinguish our Galaxy . 1 arm, 2 halos, 3 nuclei, 4 galactic planes, 5 globular clusters.

Nucleus
A detail of the disk galaxy Sombrero taken with VLT which shows how great quantities of luminous stars mass up at the galactic center. In the periphery, instead, we see huge dark clouds, made of dust and gas which, not being lit by any near stars, they function as a screen to the central radiation.

Gas
Cloudiness of the Swan, taken from Hubble galactic gases lit by the stars re-emit the radiation absorbed in wavelength in different color.

THE NUCLEUS

Of a diameter of about 2 x 10^4 l/y (= 7 kpc) and about 3 x 10^3 l/y (= 1 kpc) thick, seen from earth, the galactic nucleus is located in the direction of Sagittarius. It is made of "old" stars near the end of their evolution that have been active for about 15 x 10^9 years: these are called Population II stars, and their features are similar to the stars that form the halo and, particularly, the globular clusters. As a matter of fact, these are red stars, relatively poor of heavy elements, which originated from primordial elements before hydrogen and helium could be transformed into other atoms. For the same reason it is believed that these stars are less likely to have developed planetary systems with rocky planets (made of, indeed, heavy elements). In the nucleus, gas and dust are present in limited quantities, compared to other galactic zones. Stars are all that is left from the remote period of the Galaxy formation. The orbital speed of the stars that form the nucleus is approximately equal: it is believed that the nucleus rotates around the galactic axis, just as if it were a solid body. It is further thought, as we suspect with other galaxies, that there is a black hole even in the Milky Way ➢206.

THE DISK

The disk contains mostly "young" stars at the beginning of their evolution, or on their way to formation: these, along with stars from the galactic halo's open clusters, make Population I stars. These are stars that are found mostly on the galactic planes and on the arms: they can be blue, very hot and even colder; they formed from heavy clouds relatively rich of heavy elements and, for this reason, it is believed that very likely they might have developed planetary systems.

The disk also contains different stars, like the Sun, which form the disk population stars. These stars can be also very old, up to 10^9 years. Those stars with age between 2 and 5 billions of years - like the Sun, which has a "defiladed" position - are more numerous especially closer to the galactic

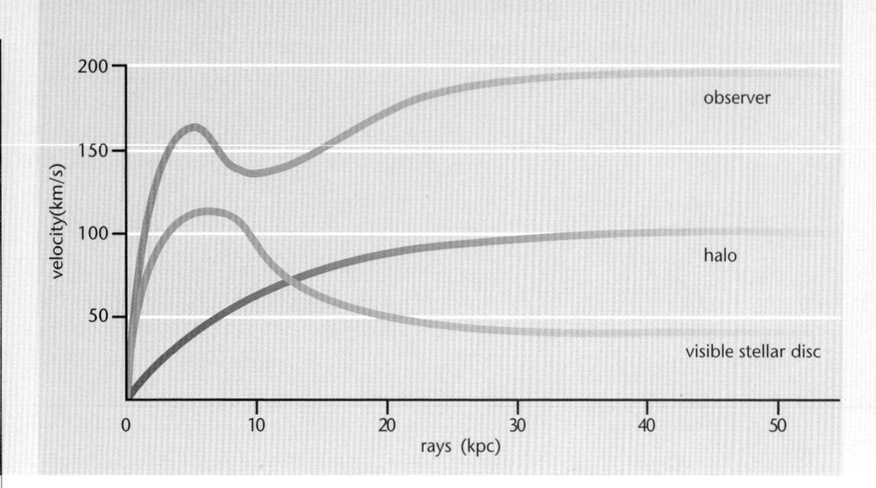

Rotation speed
The Galaxy moves in a non-uniform way: its speed increases starting from the center of the nucleus up to about3 x 10^4 l/y (= 10 kpc) then it slows down to once again increase till it stabilizes at about 7.5 x 10^4 l/y (= 22.5 kpc) from the center. If we observe the curves of speed of objects, we obtain different graphs: evidently, the speed distribution cannot be explained only on the basis of gravity distribution of visible stars. The Sun, located about 3 x 10^4 l/y (= 9 kpc) from the galactic center, has an orbital speed of about 250 km/s: the cosmic year, which is the time that it takes to make a galactic rotation, lasts approximately an average of 225 millions of solar years.

plane, while gas and dust, although considerably present in this zone of the Galaxy, are mostly present on the outer bands and in between the arms. The orbital speed of each star depends, at first, on its distance from the galaxy center: up to about 1.5 x10^4 l/y (= 4.5 kpc) from the center, the further a star is located, the slower its speed is. Then, contrary to what would be expected due to the dynamics that also move our solar system, the speed increases: this was especially emphasized by the observation made with radio telescopes that have allowed us to quantify the speed of carbon monoxide clouds, traceable even at a distance of over 4 x 10^4 l/y from the galactic center. This data suggests that the mass involved in the galactic motions might not be the only one observed: the Galaxy periphery should contain 5 times as much as the matter contained inside the galactic orbit of the Sun. As in other notorious galaxies, and therefore as in the whole universe, a great quantity of Dark Matter invisible to all instruments must exist: this is composed of: stars that are too weak to be seen; planets that have a certain mass but that remain invisible at great distances; atomic nuclei and elementary particles, such as neutrinos, spread throughout the interstellar medium. There are many assumptions, but the problem with the "missing" mass, both at a galactic and universal level, is still far from a solution. It is a fact, however, that the "dynamic" mass of the Galaxy

Witchead
Witch-Head Nebula taken with Hubble Space Telescope is an example of an emission nebula..

M51 Whirlpool Galaxy
The nucleus of this spiral galaxy taken from Hubble Space Telescope reveals an intense activity of stellar formation.

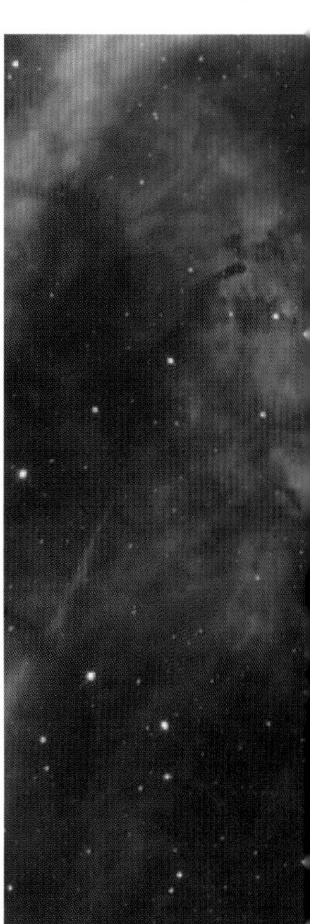

Triffid Nebula
Right: Cloudiness and new born stars taken from Hubble Space Telescope in the Sagittarius constellation: according to the zones, we can observe parts of dark nebulae and reflecting parts

Nebulae and new-born stars
Below: A detail of the great Magellan's cloud, taken from Hubble Space Telescope rich with stars.

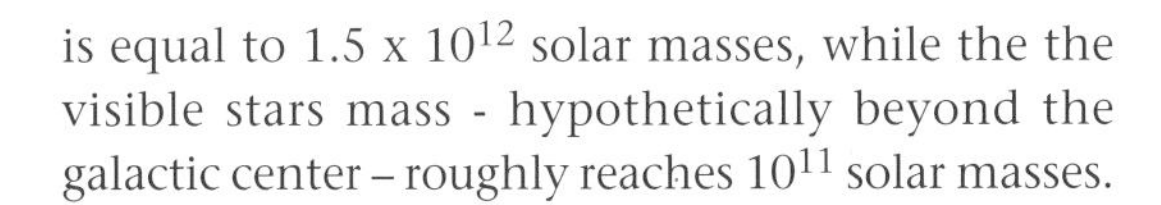

is equal to 1.5 x 10^{12} solar masses, while the the visible stars mass - hypothetically beyond the galactic center – roughly reaches 10^{11} solar masses.

THE ARMS

Just like speed, even the distribution of the matter observed in the Galaxy is not uniform: there are zones where the matter is less dense, and the stars are at a further distance, and there are more luminous parts. These last ones are called Arms, and they are produced, probably, from the differential rotation of the Galaxy . Nevertheless, if that were the only reason for their existence, they should disappear in a few rotations, in a matter of a billion of years.

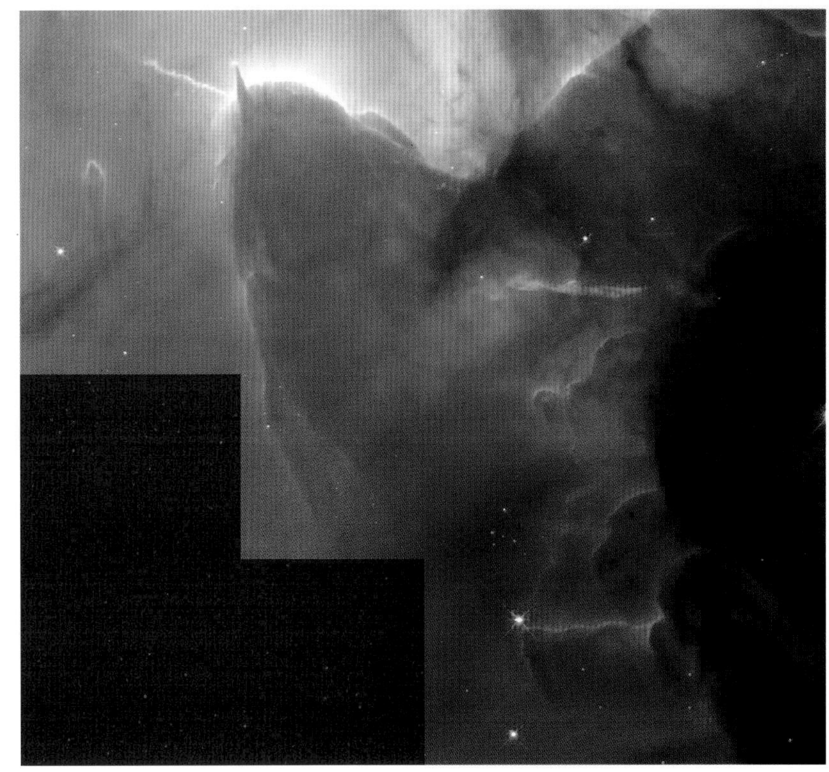

The Galaxy's spiraling, just like the other galaxies', is in reality the visible portion of a density wave that moves around the Galaxy in the same direction as the stars, but at a lesser speed (about 30 km/s, while stars move at an average of 200-

MOST VISIBLE OPEN CLUSTERS

NAME	CONSTELLATION	FEATURES
HYADES	TAURUS	OVER 100 STARS, OF WHICH MANY WITH MAGNITUDE BETWEEN 3 AND 4
M45	TAURUS	KNOWN AS PLEIADS; OVER 100 STARS, 6 WITH MAGNITUDE < 5
M47	PUPPIS	30 STARS WITH MAGNITUDE < 5
M44	CANCER	KNOWN AS BEEHIVE CLUSTER; ABOUT 50 STARS, AT LEAST 10 WITH MAGNITUDE < 7
M6	SCORPIUS	ABOUT 80 STARS WITH MAGNITUDE > 7
M7	SCORPIUS	ABOUT 80 STARS WITH MAGNITUDE > 7
M35	GEMINI	ABOUT 200 STARS WITH MAGNITUDE > 8

MOST VISIBLE NUBULAE

NAME	RIGHT ASCENSION	DECLINATION
M8, LAGOON	$18^h04',1$	-24°20′
M16, EAGLE	$18^h18',8$	-13°49′
M17, OMEGA	$18^h20',9$	-15°59′
M20, TRIFID	$18^h02'$	-22°60′
M42, ORION	$5^h35',5$	-5°28′
ROSETTA	$6^h33',7$	+4°58′

INTERSTELLAR MEDIUM

	TEMPERATURE IN K	DENSITY IN ATOMS/cm^3	MASS IN SOLAR MASSES
INTERNEBULAR MEDIUM	~10	10^7	5×10^7
DARK CLOUDS	10	10^9	10-30
MOLECULAR CLOUDS	30-100	10^{10}-10^{12}	10^5
EMISSION CLOUDS	10^4	10^7-2×10^3	10

300 km/s). Therefore, stars can reach and overpass the density wave producing a crash wave that also rotates along the spiral. The gas and dust clouds, very present in the disk, are therefore compressed right behind the arch of what we see as a spiral's arm, originating an intense stellar activity formation. What we see as the "edges of the arms" are, in reality, zones that are bright because of the intense production of stars: very hot, glowing "young stars" cluster here. Even though these stars have a large mass, they explode in a short period of time ("short" according to the astronomical time scale) causing crash waves that help trigger processes of stellar formation in the near clouds. If these stars are small, they follow the galactic course along with the remaining interstellar matter. In any event, sooner or later they enrich the interstellar medium with nebular matter which, on the other hand, recycles to new stars, and so on. The rhythm that transforms interstellar nebulae into stars has been estimated to be equal to the one that stars in terminal phase have when wasting all the matter that makes up their interstellar medium. On the basis of this theory and of the gathered data, some computer simulations confirm that the perturbation produced from the everlasting process of stellar formation in a rotating galactic body, indeed, creates a spiral picture. Furthermore, these mathematical models have led to a conclusion: unless a great quantity of dark matter really exists, as already assumed, the spiral is bound to destroy

THE PLEIADS
The open cluster of the Pleiades is visible to the naked eye. The "youth" of these stars is indicated by the presence of gas and dust halos that have been left since their formation. They are bright because they reflect starlight (reflection nebulae).

itself. On the other hand, the presence of dark matter would stabilize the spiral structure by transforming it, slowly once at a time, into a barred spiral ➢222. The open clusters, or galactic clusters, are also typical of the spiral arms in the galaxy. These are molecular clouds and cosmic dust that are not located outside the Galaxy, but that characterize even other galaxies .

OPEN CLUSTERS OR GALACTIC CLUSTERS
These are irregular and relatively small clusters of a few tenths or several thousands Population I stars, distributed irregularly and extremely clustered within a surface that can vary from about 6 (= 2 pc) to 30 l/y (= 10 pc; in average 12 l/y = 4 pc.) Of recent formation, these stars are often very bright and can be easily observed: for example, the Pleiades in the constellation of Taurus; the stars in the Beehive in the constellation of Cancer; the double cluster of Perseus; the M6 cluster, in the constellation of Scorpio and known as "the Butterfly";the Hyads, in Taurus, about 140 light years across from Earth, which were very helpful in establishing a galactic "scale of distance". All stars of an open cluster originated from the same gas cloud: this is shown by the relative uniformity of their chemical composition. Moreover, almost all of them have the same "age", and the differences noticed from one another can only be justified by their evolutionary course (different original mass). These stars are kept together by the gravitational force which, very often, can't win over the disintegrating forces in a matter of a few hundreds of millions of years.

GALACTIC CLOUDS OR NEBULAE
Among the countless images taken with powerful instruments, such as the Hubble or the V LT, those of galactic clouds are surely the most suggestive ones. These are huge masses of relatively cold gas, so much that the gas is primarily found in a molecular state. Overall, these clouds constitute a mass of 3×10^9 solar masses, equal to 15% of the total mass of stars that are present in the disk. They can be distinguished according to their dimension:

- ***Small Clouds***, so to speak, they reach a few light years of diameter and a density of 10^3 - 10^4 molecules/cm3. They have a temperature of 10^- ^{20}K and are primarily composed of molecular hydrogen. Their temperature keeps this low because there are no stars nearby that can heat them up: some of these have even colder zones where the concentration of hydrogen goes up by even a factor of 10;
- ***Huge clouds***, each of them reaches a diameter of $1.5\text{-}2.5 \times 10^2$ l/y and a density of 10^5 molecules/c m^3, for a total mass that can reach the value of 10^6 solar masses. Identified thanks to the radio astronomical research, they are made of great quantities of hydrogen , carbon monoxide and- even if in lesser quantities - other molecules: Up to 60 different types of molecules have been recognized in some of the bigger clouds. In other words, those clouds are made of - and mostly made of - matter that was expelled from those stars that ended their evolutionary cycle and that can here combine with one another. Unlike smaller clouds, the huge clouds give origin to new stars and, therefore, they have a very elevated temperature, so much that they can be detected at infrared. Furthermore, these clouds contain several nuclei composed of a mass of up

Crescent Nebula
NGC6 8 8 8 taken from the Hubble. A big star near its end is tearing the cloud of matter emerged 250,000 years ago thanks to a strong stellar wind. The cloud appears in this image for the first time as a complex net of denser filaments and knots located around a blue gas film (blue because of the emission).

To observe nebulas
The galactic nebulae can be visible due to three different processes:

a. Emission of proper radiation due to the stimulation of nebular gas from a star that is located opposite to the Earth;

b. reflection of the radiation coming from a source situated between the clouds and Earth;

c. interception and darkening of the radiation coming from a diffused object situated opposite to Earth.

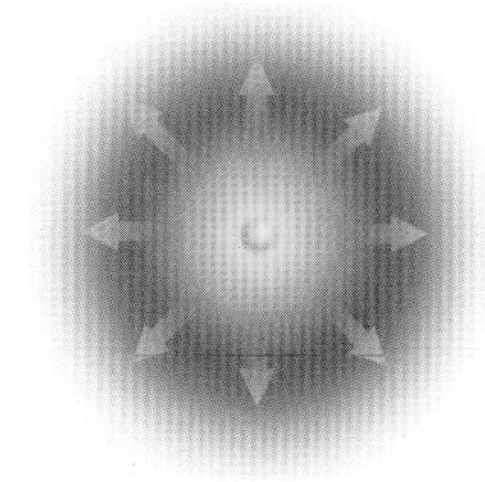

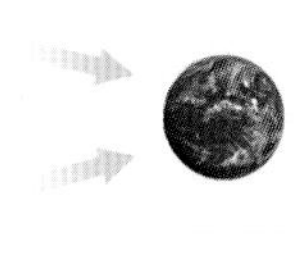

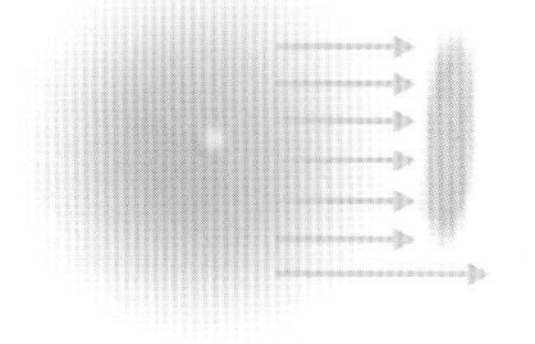

Nebula detail
In the Carina Nebula (NGC3372), Hubble Space Telescope shows us a detail of the "keyhole" nebula, so called by Sir John Herschel in the nineteen hundreds: a dark cloud, made of cold molecules and dusts along a filament of fluorescent gas.

to 10^3 solar masses and dense (up to 10^5 molec ules/cm^3), supposedly made of stars in a stage of formation, that are clearly visible in infrared.
There are also wells with a high MASER activity (Microwave Amplification by Stimulated Emission of Radiation): the molecules, stimulated by the radiations of the near stars, emit microwave radiations which subsequently stimulate other molecules to produce a chain reaction of microwaves that forms a deep band of radiation with a specific wavelength. As a matter of fact, the nebulae are visible because their gas emits radiations after being stimulated by a near source (Emission Nebulae), either because, being rich with dust, they reflect the light produced by a source (reflection nebula), or because they appear as dark masses on a dark background (dark nebulae).

COSMIC DUST
This is the name given to small particles of matter that "fill in" the interstellar space: dust grains, mostly of graphite and silicates, often coated with ice, with frozen ammonia or with dry ice (solid carbon dioxide), have a diameter of 1.01-10 mm. They are revealed by the fact that they absorb and diffuse star radiation , particularly along the blue band of the visible spectrum and the ultraviolet. Therefore, the presence of this dust makes stars appear more red than they really are, just like at sunset on Earth when the atmospheric dust

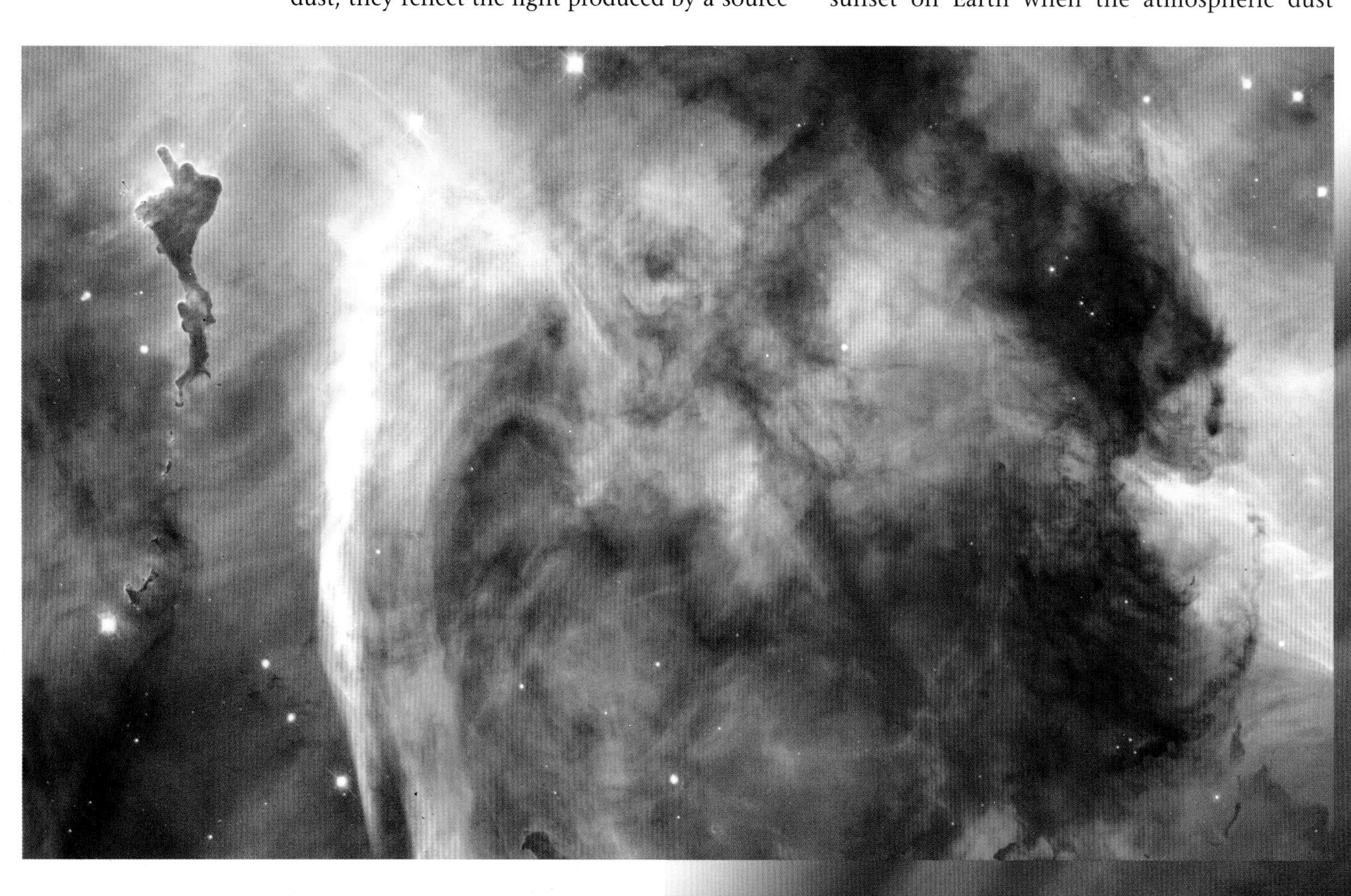

NEBULAE
Dark nebulae in the Magellan's Great Cloud shot by the Hubble Space Telescope and emission nebula shot by the X rays from Rosat in the small NGC2300, approximately 1.5 x 10^8 l/y from Earth in the direction of Cepheus.

"BUBBLE" NEBULA
There couldn't be a more fit name for the NGC7635 nebula in Cassiopeia, shown unbelievably clear by Hubble. The expanding spherical edge delineates the beginning of a calmer and colder region of the nebula, where a very strong stellar wind blows.

modifies the radiation.. The presence of interstellar dust has been calculated to reduce the luminosity of our Galaxy stars of one magnitude every approximately 10^3pc (= circa 3 x 10^4 l/y) traveled by light. These grains are likely to be made of matter that has formed in the atmosphere by "cold" stars spread in the sky : the estimated amount of dust is considerable and, just in our Galaxy, this constitutes a mass of 2 x 10^8 solar masses. It is thought that these icy grains spread throughout the huge clouds provide a surface fit to produce complex molecules such as, for example, amino acids. This theory is supported by the discovery of glycine (one of the simplest amino acids) in a stellar cloud.

NEAR THE GALAXY

The galaxy, as it looks from its dynamics, surely does not end where it seems to. On the other hand , it has been known for a while that other "exterior" bodies are linked to its gravitation.

THE GALACTIC HALO

This is the name given to a spherical area that surrounds the Galaxy, dominated by the gravitational influence of the dark galactic matter. It is formed by two parts:

- a luminous area, of the same diameter as the disk (30 kpc = 10^5 l/y approximately) and of very hot gas, Population II stars and stars caught in clusters;

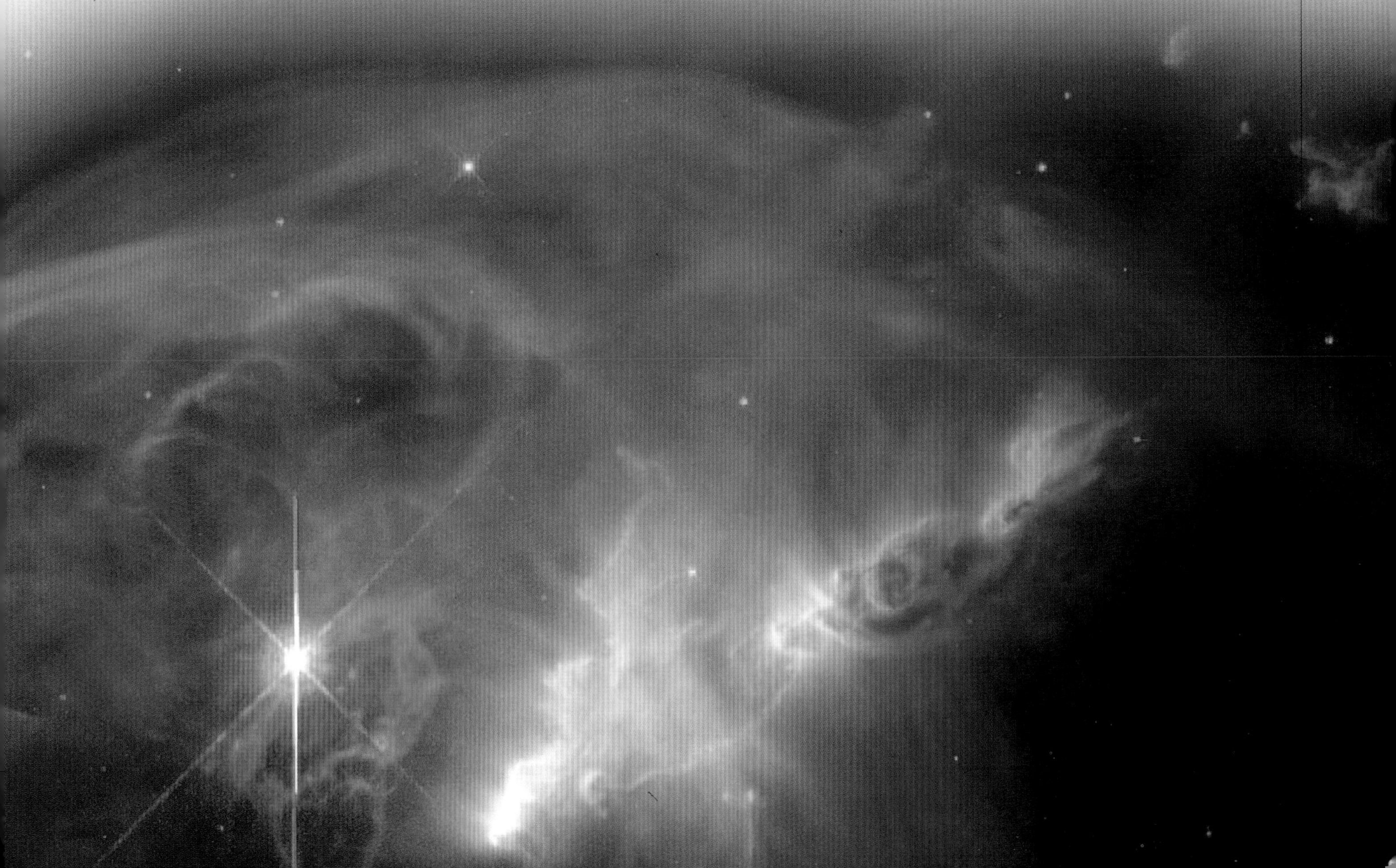

Gravitational Lens
This image, captured by Hubble in the 0024+1654 galaxies' cluster, shows the effects of a gravitational lens: an object's gravity deviates the light from a source, which can become lighter, or even, split in multiple images. There are those who think, though, that these observations can prove that the space is folded rather than being flat and endless: the light from the stars would be arriving to us from several directions.

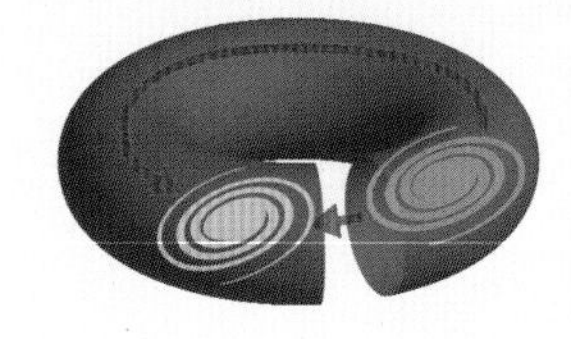

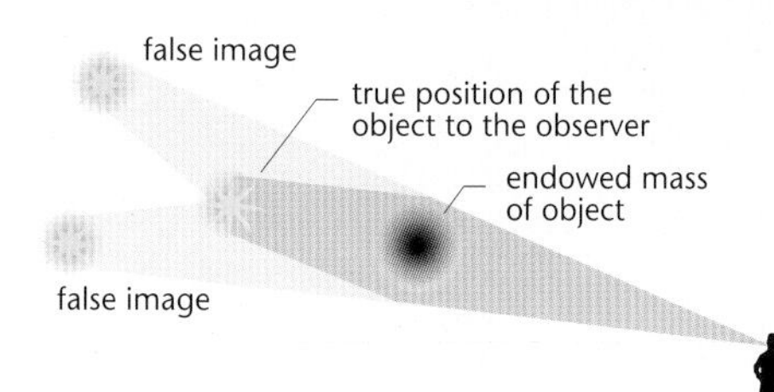

- a dark area, much more extended, probably rich with very weak stars(dark dwarfs). The existence of these objects is demonstrated by the gravitational lens effect that they produce, particularly over the light coming from objects located outside the Galaxy

According to the dynamic calculations, the "missing" matter, that is the dark matter, should be all located on the halo: about 90% of the Galaxy matter is located here. This theoretical model has been supported by a series of observations about other galaxies: in 1994, in particular, astronomers from the Kitt Peack National Observatory noticed all around a far galaxy a kind of luminosity with dimensions presumed to be that of a halo . But, in order to further confirm that, researches are still being conducted.

Cluster
Even other galaxies have globular clusters that, like ours, orbit around the galactic center along strongly elliptical orbits, crossing periodically the disk. In this image, shot by the Hubble, we see G 1 or Mayall II, a globular cluster of at least 300,000 stars located at approximately 1.3 x 10^4 l/y from the center of Andromeda.

GLOBULAR CLUSTERS

These are Population II star clusters, many of which have reached a red giant stage, containing hundreds of thousands, or even millions, of stars in very few parsec of diameter. In some cases the stars are so concentrated that one can find even up to 1,000 in a sphere of radius slightly over 2 l/y. To have an idea of what this means, just remember that the closest star to the Sun (Proxima) is 4.2 l/y across. Usually these clusters are almost perfectly spherical, and only a few of them appear flat: all of them orbit around the Galaxy by staying in the halo and, even if located in the galactic plain, only few of them are further from the center of the spiral than the Sun. According to their spatial distribution and to their chemical composition, poor of matter, it is thought that stars of globular clusters originated 10-20 billions of years ago, when the Galaxy was still forming and the big bang had just ended.

Clusters
Named N GC1 8 5 0, they are located in the Magellan's Great Cloud.

Globular Clusters
Shot by the Hubble Space Telescope, we can see the M80 (NGC6093) globular cluster; below is the center of Cluster 47 Tucanae. The core of globular clusters is often so dense with stars that you cannot distinguish one from the other even with the most powerful instruments.

DOPPLER EFFECT AND SPEED

The Doppler Effect, which originally described the compression of of sound waves as their source gets closer (with a consequent increase of sonic frequency and intensity) and their successive distension as their source gets further (with subsequent decrease of sonic frequency and intensity), was extended to light and other electromagnetic waves in the mid 1800's by the Frenchman Fizeay. By using this effect, which works the same whether we talk about sonic or electromagnetic waves, we can determine the speed that the source has compared to the observer. The radiation of a luminous object that moves toward us is compressed, moving the spectrum towards the blue; vice-versa, if the source gets further, the emitted radiation expands and the spectrum moves towards the red. The more the spectrum moves, the faster the object moves along the conjunction with Earth. If we can also measure a transversal speed, combining it to a Doppler speed, we can determine the real motion of a star. This technique is fundamental in order to determine the distances on a cosmic scale. The Doppler effect is also useful in determining the mass of stars that form binary or multiple systems, to quantify the rotational speed of the Galaxy or the relative speeds of galaxies that are located with the same groups.

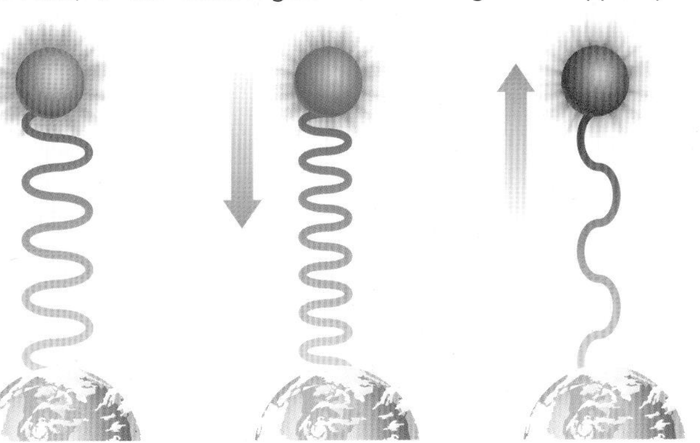

SINCE 1925, WHEN THE AMERICAN ASTRONOMER EDWIN POWELL HUBBLE WAS ABLE TO SWEEP AWAY ANY DOUBT ABOUT ANDROMEDA'S DISTANCE FROM THE GALAXY BY LOCATING IT AT MORE THAN 1 MILLION LIGHT YEARS AWAY FROM US, THE KNOWN PART OF THE UNIVERSE HAS STARTED TO EXPAND AT ASTRONOMICAL SPEED: INSTRUMENTS, OBSERVATIONS, DATA ABOUT UNTHINKABLE CELESTIAL SYSTEMS AND OBJECTS HAVE ALL TRIGGERED THE CREATION OF MORE AND MORE COMPLEX COSMOLOGICAL THEORIES AND SCENARIOS.

BEYOND THE MILKY WAY

Once beyond the Galaxy borders, we can see the universe. The Earth is no longer at the center of the solar system; the Sun is no longer at the center of a sphere of stars; and the Galaxy is not at the center of the universe, but just one among the billions of stellar systems that rotate in space at incredible speed. We can see different kinds: simple ones with stellar clusters as complex as the Milky way, long ones shaped as a barred spiral, as a butterfly. Hubble was the first one in classifying them and, by proposing the first theory about their evolution, he even gave an explanation about the "strange" shapes that characterizes a small group of galaxies.

THE SHAPES OF GALAXIES AND HUBBLE CLASSIFICATION

Hubble proposed to subdivide the galaxies in two big groups according to their shape.

Elliptical galaxies

They barely have any structure and, if it were not for the fact that they are bigger, they would resemble globular clusters. Whether spherical or ellipsoidal, they all have a brightness that decreases as they get further from the center. They have no arms, or nucleus, or disk, and they show no condensations or dust clouds, and they are only made of

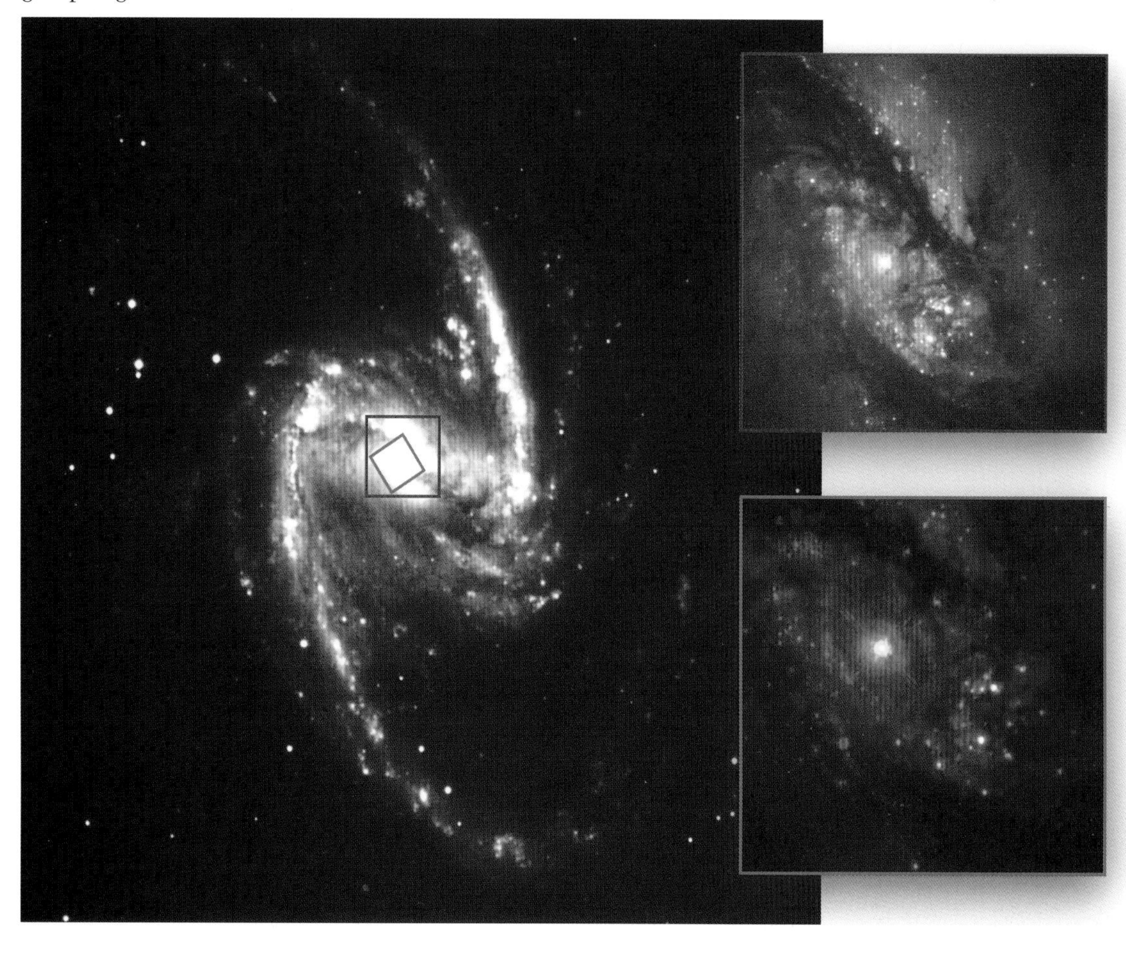

ANALISI DI UNA GALASSIA
The powerful Hubble Telescope, along with the V LT and other telescopes made in the last few years, have all made it possible for us to see details about distant galaxies that only a few years ago were unthinkable. Here we see the N GC1 3 6 5 barred spiral galaxy: around the photo captured from the eye of an earth telescope, we can observe:
ABOVE ON THE RIGHT, the central region as it appears to Hubble in visible light. Around the yellow sphere of the nucleus, the dark matter composed of gas and dusts is pushed from the bar toward the galactic center.
BELOW ON THE RIGHT We can observe the center area of the central regions as seen by Hubble at infrared with clusters of forming stars. It is thought that a black hole is at its center.

Famous galaxies
In these two images, taken by the Hubble Space Telescope in visible light, we can see:
LEFT, Andromeda's galaxy (M 3 1), a spiral galaxy similar to the Milky Way, of which we have been able to calculate for the first time with precision its distance from our solar system;
RIGHT, the NGC4650 A featured by a polar ring. While the first one is at 2.3×10^6 l/y from us, the second one is located at about 1.3×10^8 l/y from Earth. It is one of the 100 known galaxies that show a polar ring, whose nature, structure and dynamics we still know very little of. The theory is that it is what remains from a collision between two galaxies that happened at least a billion years ago.

Galaxies as seen from the Hubble
LEFT: a detail Stefan's Quintet, a groups of 5 galaxies which, due to the strong reciprocal gravitational strengths, have a distorted shape and long gas currents that keep them together.
RIGHT: the interaction between NGC6872 and IC4970 and the huge tidal-force derived from it modify the barred spiral shape in a prominent way.

Population II stars: without any gas no new stars can form. It is likely that galaxies of this kind, with a little bit of gas and dust, have incorporated more recent ones: the elliptical galaxies, indeed, contain some of the oldest stars that are known. The more the galaxy's inclination is toward the direction of Earth, the more pronounced is its flat shape; nevertheless, lacking superficial details, it is hard to say what their symmetry is: the flattening could have originated indeed from the greater or lesser speed of rotation. Some recent observations suggest, though, that the dynamic of the elliptical galaxies is more linked to the chaotic dynamic of each individual star rather than to an ordered collective motion, such as the one present in our Galaxy. They do not tend to collapse as expected, but rather maintain their shape because of the chaotic distribution of speed of every single star: kind of what happens to a gas clouds at an elevated temperature which only forms into a shape according to the speed of each molecule. Hubble classified elliptical galaxies by attributing to each of them a number from 0 to 7 that characterizes their "ellipticity": a E 0 galaxy (E is for elliptical) is almost spherical, an E7 one in very elongated.

Barred spiral galaxy
The barred spiral NGC4314 as seen with an optical telescope from the McDonald observatory, in Texas, and by Hubble: the detail shows nuclei with active stellar formation.

HCG87
Known as Hickson Compact Group 87 (HCG87), these four galaxies are so close to one another that their gravity causes an exchange of gas that distorts their shape and evolution. The HCG87a galaxy, as seen by its profile, and the elliptical HCG87b one, on the right, have both active nuclei , whose center would contain black holes. In the Rather close HCG87c spiral galaxy, we can observe a strong stars formation activity. From the spectral analysis it has not been ascertained yet whether the small spiral galaxy at the center has really been involved at a gravitational level by the other three galaxies, or whether it is just a galaxy connected to them just by its perspective.

Galaxy collision
Even though there are light-year distances between stars, the collision between galaxies can have devastating consequences on the shape and the dynamics of these immense celestial objects.
In these images, which compare the view of a group and a detail taken by the Hubble Space Telescope , we can observe how as spiral galaxies NGC4038 and NGC4039 clash with each other they change shape.

The spiral galaxies

These are distinguished as ordinary spiral galaxies, such as our Milky Way, and as barred spiral galaxies.

• Ordinary spiral galaxies: they have a curved, elongated and very bright structures (the spiral arms) which can vary in number and all branch out of the very luminous nucleus. The nucleus has usually features that are similar to those of elliptical galaxies, except that it is smaller: more or less flat, it is formed of great quantities of Population II stars and it is almost absent of gas and dust. A very thin disk composed of Population I stars surrounds it, as well as dust, gas and clouds that show a strong activity of star formation. The relative dimensions of the nucleus, the disk, and the arms are variable and are used to subdivide spiral galaxies into subgroups: we have , therefore, type Sa galaxies (where S is for spiral) with a very large nucleus compared to the disk and to the arms which are very dense and compact; type Sc galaxies with a small nucleus and arms that are very spread out and well separated; type Sb galaxies with features that are in between the former two. Type S0 galaxies, also called Lenticular Galaxies, are considered an element of transition between elliptical and spiral galaxies.

• Barred spiral galaxies: besides the nucleus and the disk, they have a third component shaped roughly like a cylinder, named indeed bar, which compared to the nucleus, is usually symmetrical. At the bar's extremities two main spiral arms branch out and more arms can develop from any point of a

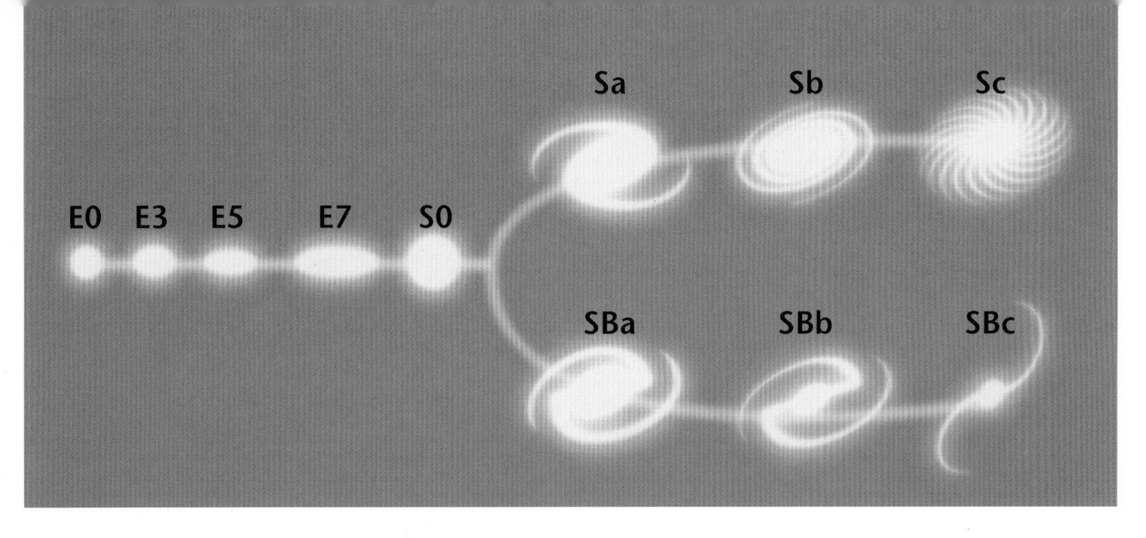

Hubble fork
Even though considered too simplified, the method proposed by Hubble about classifying galaxies and explaining their evolution is still thought to be a valid working model.

Galaxy Galleries
Above, the image in the visible of each galaxy allows to see what the detail observed by the Hubble Space Telescope is, by combining information captured in the visible and in the infrared. This way the most ancient star populations in the core of spiral galaxies are emphasized.
From the left:
NGC5838 (S0),
NGC5689 (Sa),
NGC5965 (Sb),
NGC7537 (Sbc).
This last one is the most similar to our Milky Way.

luminous ring centered in the nucleus and with a diameter that matches the bar's length. Even in this case, the appearance and the relative dimensions of the galactic elements (nucleus, bar, arms) are highly variable, and are used to classify galaxies into subgroups. We distinguish then type Sba galaxies (where SB is for barred spiral), with a very large nucleus, a relatively short bar and often a large number of compact arms, type SBb with a small nucleus, a very long bar and thin arms, often a single one; Type SBc with features in between the former two.

Irregular, dwarf and giant galaxies

These are galaxies that Hubble could not classify and that most likely are what is left from the collision of one or more galaxies. Irregular galaxies, where it is not possible to distinguish a particular symmetry, are rich with dust and interstellar matter, but on a regular basis they have a small mass and a relatively low luminosity. The same applies to the dwarf galaxies, despite their clear spiral shape, or even to the elliptical giants which often are located at the center of groups of galaxies.

THEORIES ON THE FORMATION OF GALAXIES

For some time researchers thought that the classification proposed by Hubble could reflect an evolutionary scheme: so type Sa and SBa galaxies were called "primitive", while the Sc and the SBc were called "evolved". Despite today's usage of this terminology, though, it is thought that such simplification is too premature in order to describe correctly the phenomena involved in the galactic evolution. As a matter of fact, despite the fact that much is already known about the stellar evolution, in the case of galaxy evolution, starting with their formation, there is nothing certain that can be said. The problem is mostly due to the fact that, while inside the same galaxy whole populations of stars at different stages of evolution can be observed, the galaxies that can be observed appear all to have the same age: only a few of them, located extremely far away, appear relatively younger. Without having a wide spectrum of cases, it is therefore hard to describe an evolutionary course: one must rely on the "Statistical" knowledge and on theoretical elaborations. On an average, out of 100 galaxies observed 13 are elliptic, 22 lenticular, 61 spiral and 4 irregular, and for the most part they concentrate

Lenticular galaxy
The shape of NGC6745 is clearly due to the interaction with the small galaxy passing by below to the right.

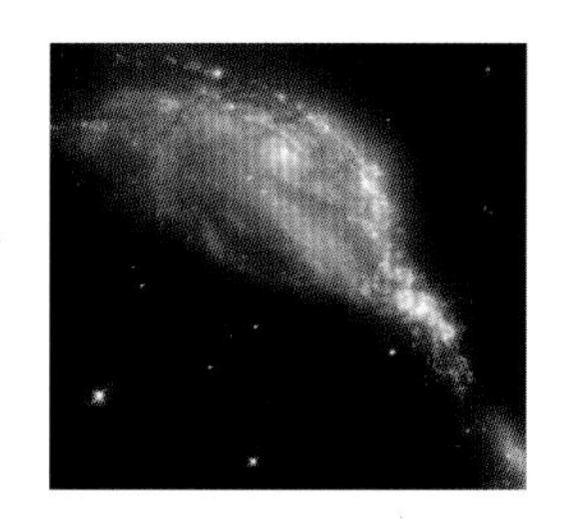

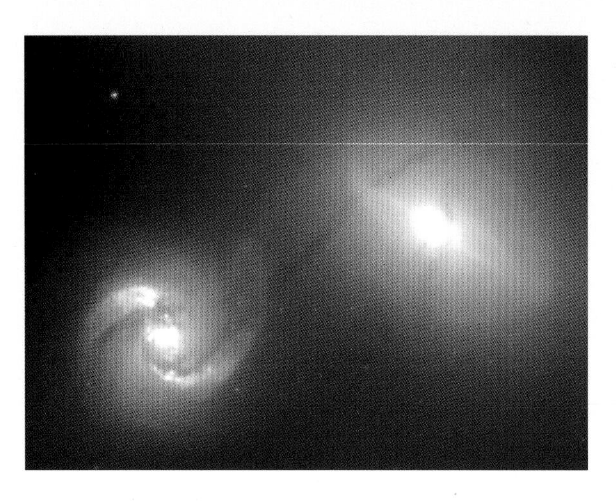

HE1013-2136
This quasar caught by ESO's VLT shows a long tidal tail pointing to the near galaxy.

Matter exchange
Observed by the Hubble Telescope, the two galaxies in collision NGC1410 and NG1409 appear bound together by a cord of matter longer than 20,000 l/ys.

Cirnicus galaxy
The Hubble Telescope has taken the picture of the galaxy which belongs to the group of Seyfert galaxies with a tight center that might contain a black hole. It emits jets of gas at an unimaginable speed.

in groups or clusters that are more or less densely populated. Even the Galaxy belongs to a cluster: the Local Group together with about several tens of galaxies among which the Magellan Cloud, Andromeda and its satellites, (N GC2 05, N GC221, M33), and many others, occupying a volume of about 10 Mpc^3 (10^4 kpc^3). It is a modest cluster, if you think that some of them include several thousands of galaxies. On the other hand, clusters tend to clump and form super masses with dimensions in the order of 100 Mpc^3. The super masses originated a very long time ago: after the big bang, the primordial elements (hydrogen and helium) supposedly banded together in huge clouds with a mass a million times greater than the one present in the Milky Way. Here, around the gravitational inequalities that resulted spontaneously, numerous galactic nuclei have produced: the globular clusters orbit still today around bigger galaxies. Some observations suggest that, in the further galaxies, of which we can therefore observe remote moments of activity, there are multiple nuclei besides an intense activity of stars formation with great mass. This scenario is compatible with the assumption that, originally, the greater stellar systems originated from the collision of several groups of stars (the clusters): the big galaxies known today would have grown by incorporating nearer stellar groups.

GALACTIC RADIATIONS

The radio and X spectroscopic observations have shown that galaxies' nuclei often emit huge amount of energy, so much that sometimes it is incompatible with the anomalous existence of the galaxy itself. Despite the fact that the mechanisms that lead to the production of this energy are not

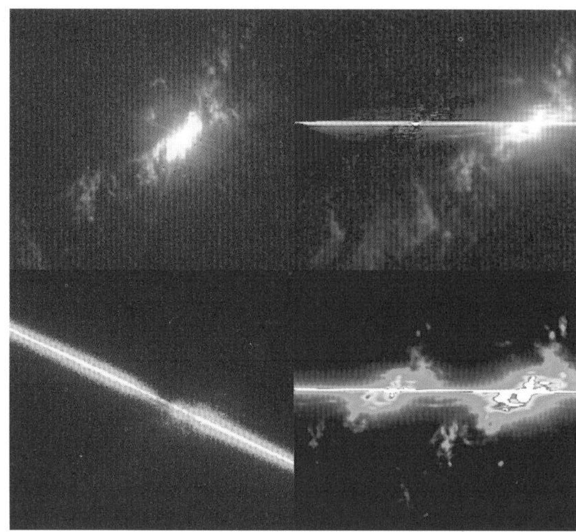

Galactic black hole
Images in the visible (on top) and in the ultraviolet (below to the left) contrasting an image of the STIS, the instrument mounted on the Hubble that is able to visualize and measure the speed of gas accelerated by the double cone of radiation coming out of a super massive black hole located at the center of Seyfert Galaxy NGC4151.

Quasar
On the left: the image captured by Hubble of the quasar PG1115 +080, at a distance of about 8 x 10^9 l/ys in the constellation of Leo, is distorted by a gravitational lens due to the presence of an elliptical galaxy located at 3 x 10^9 l/y away from Earth .
On the right: corretta the corrected and reconstructed image of the same object shows at infrared a ring that is almost complete. It is the galaxy containing the active nucleus. Similar observations are useful in measuring the dimensions of the universe and its speed of expansion.

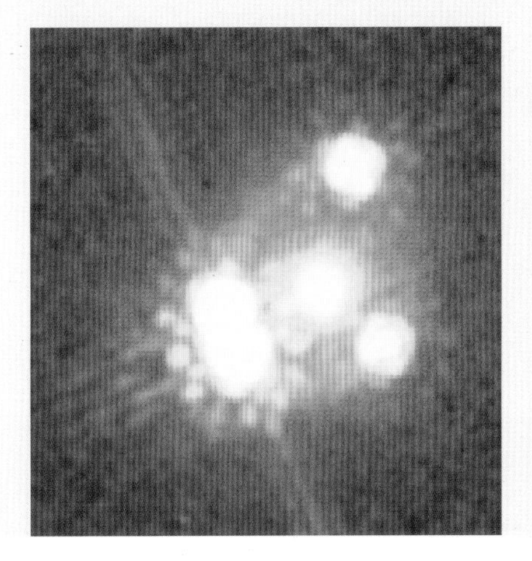

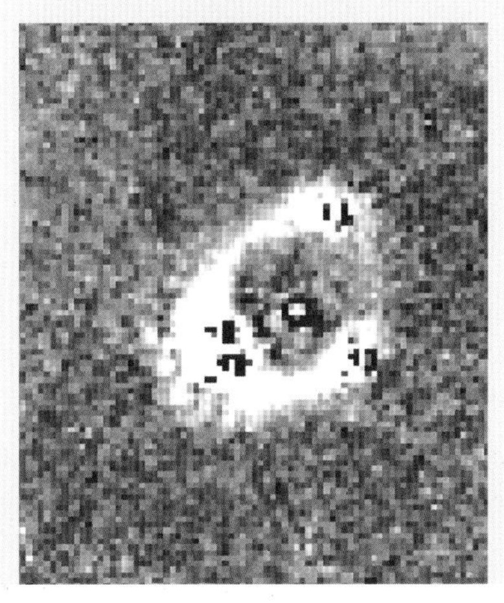

known yet, the theory that a black hole could be located at the center of these galaxies could explain, at least partially, some of the data. The incoming radiation from the galactic nuclei, for example, is for the most part a "non-thermal" type: it is the so called Synchrotron Radiation, a kind of radiation produced when electrons move at great speed in a direction that is not parallel to the lines of forces of a magnetic field where they are submerged, and that has an intensity which increases as the wave length decreases. Among all objects with active nuclei we distinguish: Seyfert Galaxies, with spectra full of very enlarged lines of intense emission; Markarian Galaxies, characterized by an intense ultraviolet emission and divided into normals- with an excess of circumscribed radiation at the nucleus - and diffused - that emit radiation from all over the galaxy: Compact galaxies, with a very high luminosity; Radio Galaxies, elliptical galaxies that emit, in the radius even greater than 10^7, more than other galaxies in zones located in two large lobes (even 5 Mpc of diameter); the Quasars, phenomena that are still hard to interpret, and, according to some, are the most energetic known phenomenon, characterized by a speed of removal compatible only with huge distances, while according to others these phenomena are less impressive and closer to us. Very small objects by nature (< 0.1 pc^3), they emit 100 times more energy than that emitted by a normal galaxy. Finally, the BL Lacert Objects, originally thought to be variable stars, are objects with intermediate characteristics between Seyfert galaxies and the Quasars: with a luminosity that is variable, they have a spectrum completely absent of absorption or emission lines.

Double nucleus
In these two images from Hubble Telescope we can observe the NGC4486B and an enlargement of its central area which shows clearly two nuclei.

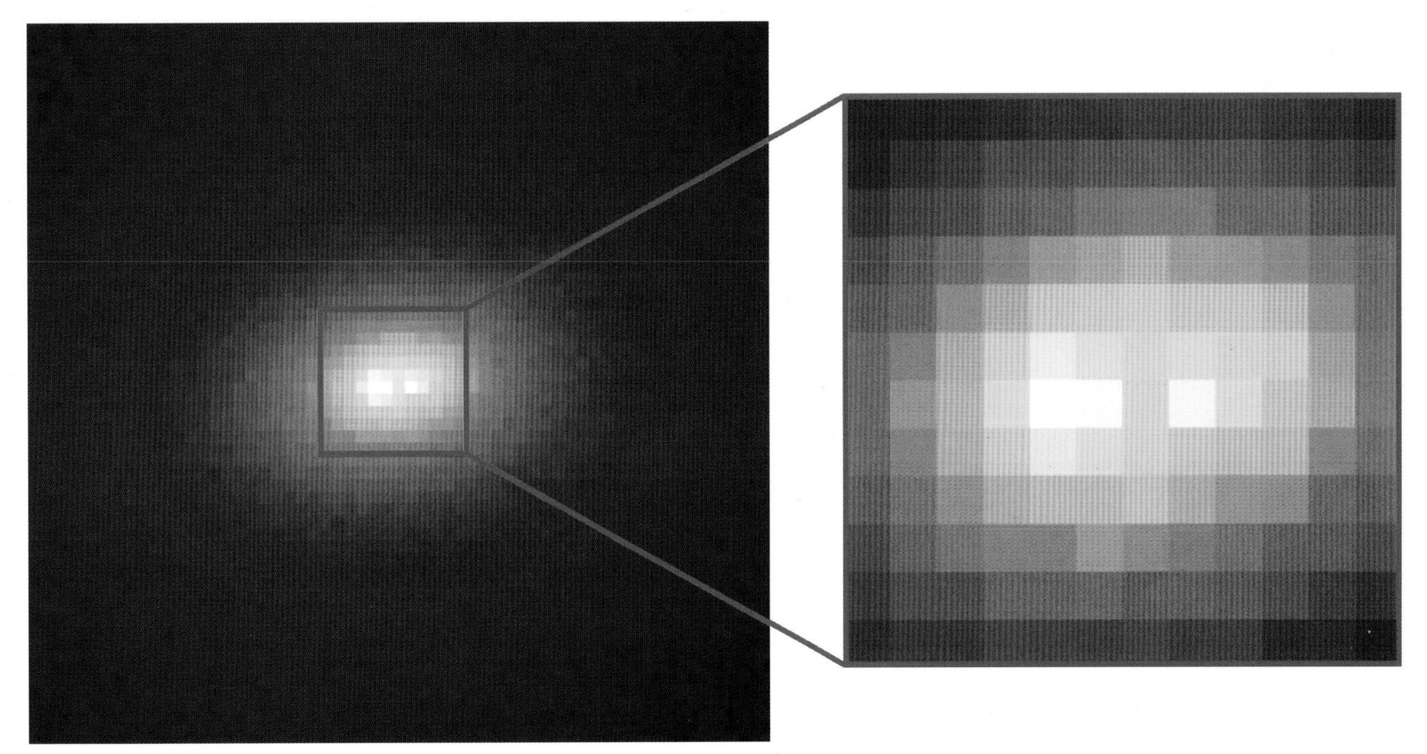

HOW LONG AGO DID THE UNIVERSE FORM? HOW BIG IS IT? HOW WILL IT END? WE ARE BACK TO THE QUESTIONS POSED BY THE ANCIENT PHILOSOPHERS. WITH ONE DIFFERENCE: THE ANSWERS THAT WE CAN PROVIDE TODAY ARE MORE "CONCRETE" THAN THE ONES GIVEN 4,000 YEARS AGO. TODAY WE CAN BASE OURSELVES ON OBSERVATION, MEASUREMENTS, CALCULATIONS, COMPUTER SIMULATIONS, AND WE CAN HOPE THAT THESE ARE NOT ALL ABOUT FANTASIES.

FROM THE BIG BANG TO THE FUTURE

AT THE EDGE OF THE UNIVERSE
The Hubble's powerful instruments allow us to observe further galactic objects and to establish their speed. According to the latest data, the far galaxies are less fast than originally expected: will the universe ever center back to a singularity?

Researchers' opinions about the beginning of the universe are relatively in agreement. The universe supposedly originated from a Singularity, that is a point in which the universal mass was concentrated at infinite density and temperatures in such conditions that every law under which our knowledge of the universe and every concept about space and time is based on had no meaning. The original cause of the birth of the universe, however, remains a mere philosophical , mathematical, or religious speculation. Then we had the big bang, the Great Explosion, theorized by Friedmann and Gamow in 1940, even though we can't really talk about "explosion" in the common sense of the term. Indeed, everything stays within the limits of a complex mathematical theory that distinguishes a succession of very rapid initial events:

- **between 10^{-43} and 10^{-35} seconds after the big bang** : an unknown force produces an inflation, that is an expansion of the universe at greater than light speed. From a small ball with less than a millimeter of diameter, in less than a moment the universe reached dimensions that we cannot observe today even with the most powerful telescopes. Space is born;
- **between 10^{-35} and 10^{-12} seconds after the big bang**: once the inflation was exhausted, the force that produced it splits between gravitational force and unified force (nuclear and electromagnetic): the elementary particles produced (electrons, quark, gluons, neutrinos) occupy a space at 10^{27} K. Einstein 'laws come to be valid, the universe continues to extend and get cold;
- **10^{-11} seconds after the big bang**: the temperature has dropped to 10^{15} K: electromagnetism starts acting along with rather fundamental physical forces that were already existing (gravitational, strong nuclear, weak nuclear) creating more complex nuclear particles;
- **10^{-6} seconds after the big bang**: quark gather in small groups of three forming protons and neutrons. The antimatter and the matter collide annihilating each other. For some unknown reason there is more matter than antimatter: matter, therefore, composes everything that is left off. The temperature drops to 10^{9} K;
- **10^{2} seconds after the big bang**: neutrons and protons form hydrogen, helium and lithium nuclei. The universe temperature keeps dropping so rapidly that no heavier atoms form. From this moment on, all events happen at immense lapses of time:
- **3 - 10^{5} years after the big bang**: the universe has a temperature of 3 x 10^{5} K and light, which was not able to come across due to the high density of electrons and photons, starts spreading. The electrons, indeed, can bind to the atomic nuclei and the photons are free to constitute the first electromagnetic signal in the universe. Space has

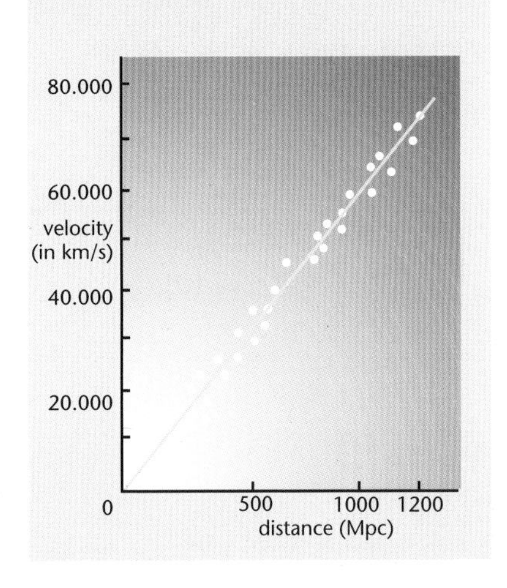

SPEED AND DISTANCE
As the Earth distance increases, the galaxies seem to increase their speed of removal (Hubble's law), but new observations give contrasting data.

OPEN UNIVERSE
All we need to do is change the value of mass at hand and the future changes: once a maximum dimension has been reached, it concentrates back into a singular point, just as it happened during the big bang.

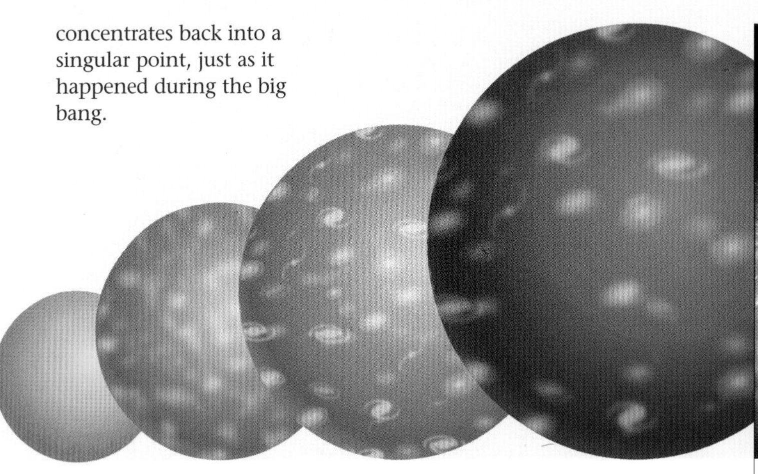

become transparent to radiation, and it is this radiation that we observe today: it is the Cosmic Radiation Background discovered by Penzias and Wilson in1965 and characterized by a spectral distribution typical of a black body ➢206 with an absolute temperature of 2.7 K;

- **2-3 x 10^9 years after the big bang**: at the center of huge gas clouds the first stars start lighting up generating new chemical elements and forming galaxies;
- **0.5 x 10^{10} years after the big bang**: the first solar system forms;
- **1.5-2 x 10^{10} years after the big bang**: man develops the big-bang theory;

And what about the future? The question that we try so hard to give an answer to, based on the observations, is: if galaxies are getting further and further away from one another, as we observed, will they continue to do so forever, or not? In other words, is the universe bound to expand until it has energy and die out in darkness, or does it have a mass sufficient enough for it to concentrate back into a singular point and then start all over? It all depends on the mass of the universe. It is not easy to understand whether the mass is sufficient to develop the universal gravity necessary to slow down galaxies and bring them back. The values brought by the different scenarios are, indeed, very close. It is, therefore, necessary to observe the galaxies that are more far, to the limit of the visibility, and to understand if their speed diminishes. A difficult task indeed.

SUPERMASS
Captured by the Hubble Telescope, the cluster of galaxies A b e ll 2 2 1 8 is located nearly 2 x 10^9 l/ys away from Earth in the direction of the constellation of the Dragon. It is extremely difficult to precisely calculate the mass of the innumerous galaxies that populate the universe and of the dark matter that they contain. And between one cosmological theory and another, the difference in mass is minimum.

YOU DON'T HAVE TO BE A SUPERNOVA TO START OUT IN ASTRONOMY: HOWEVER, EXPERIENCE, EXCHANGING IDEAS AND OPINIONS AND SUGGESTIONS MADE BY THOSE WHO HAVE WORKED IN THIS FIELD OR HAVE A PASSION FOR IT ARE ESSENTIAL. YOU ALSO NEED A DESIRE TO READ AROUND THE SUBJECT, MAKE NOTES AND TO SEARCH FOR NEW INFORMATION AND TO CONSIDER WHAT YOU THINK YOU KNOW WELL. THE STUDY OF THE SKIES, THE MOTIONS OF STARS AND BODIES, THE STUDY OF THAT HAS BROUGHT TO OUR KNOWLEDGE ARE ONLY ONE PART OF AN INFINITE JOURNEY AND A DEPARTURE POINT FOR NEW DISCOVERIES. IN THIS SECTION OUR AIM WAS TO GIVE THIS "RIGHT" TO ANYONE WHO WANTED TO REACH HIS OWN HORIZONS, A LITTLE HELP TO FOLLOW YOUR OWN CHOSEN PATH.

SUGGESTIONS

LIBRARIES AND BOOKSHOPS

The best books on astronomy are to be found in public libraries and those belonging to observatories and university institutions and these will inform you of the latest thinking and the opinions of expert astronomers.
In some libraries, you can find or order, in addition to books, large selections of sky charts (including luminous ones) as well as simple laminated cardboard plastic astrolabes used to determine the position of the visible stars.
You can also by astrolabes that take into account the latitude of the observer, but this is almost irrelevant to those who are only looking at constellations.

PLANETARIUMS

Planetariums are places with a hemispherical ceiling on which is projected an image of the firmament. The special apparatus used (called the planetarium) is located in the centre of the room and reproduces the apparent motion of the heavens with a surprisingly realistic effect. The constellations are shown, as well as the features of the most important visible celestial objects.
Planetariums also host lectures, courses and conferences of an astronomical nature. There are more than 50 in Italy alone. The Friends of the Planetarium Association, based in Brescia, supplies information on planetariums close to where you live.

ASTRONOMERS ASSOCIATIONS

Astronomers often meet in groups to exchange information on computed observations, techniques and instruments used, as well as updating their knowledge and drafting research programmes. Almost all cities have an astronomers' club or association. Activities can include observations using the club's telescopes or in collaboration with research institutes, training activities for new astronomers, cultural activities and lectures followed by debates, conferences, exhibitions of astronomy materials for schools and the public and guided visits to instruments and observatories. Subscription is normally open to anyone interested.
This association between fellow astronomers allows clubs to acquire or hire astronomical instruments that, otherwise, would be too expensive for individuals. Different associations have set up small, but well-equipped observatories that allow them to carry out observation programmes that have some scientific value and they often collaborate with public observatories.
All astronomer associations and clubs are linked nationally to the Italian Astronomers' Union based at the Department of Astronomy at Padua University that can supply any information or can put individuals into contact with astronomers' clubs in his or her area.

POPULAR BIBLIOGRAPHY

FOR MORE INFORMATION – BOOKS IN ITALIAN

Those marked with a dot are suitable for beginners

• AA.VV., *Astronomia pratica*, De Agostini, Novara 1998
AA. VV., *Guida pratica all'astronomia*, Sirio, Milano 1995.
• Asimov I., *C'è vita su altri pianeti?*, Scienza, Trieste 1992.
• Asimov I., *Com'è nato l'universo?*, Scienza, Trieste 1994.
• Asimov I., *Pulsar, quasar e buchi neri*, Scienza, Trieste 1994.
• Asimov I., *Supernovae*, Rizzoli, Milano 1990.
Barrow J.D., *Origini dell'universo*, Sansoni, Firenze 1995.
• Bellini U., a cura, *Stelle e pianeti*, Giunti, Firenze 1998.
Bergia S., *Dal cosmo immutabile all'universo in evoluzione*, Boringhieri, Torino 1995.
Bianucci P. e Ferreri W., *Atlante dell'universo*, Utet, Torino 1997.
Bianucci P., *Il Sole*, Giunti, Firenze 1992.
Bianucci P., *La Luna*, Giunti, Firenze 1990, 1999.
Bianucci P., *La Terra*, Giunti, Firenze 1990.
• Bianucci P., *Stella per stella*, Giunti, Firenze 1997.
Bourge P. e Lacroux J., *Il manuale pratico di astronomia*, Zanichelli, Bologna 1987
Braccesi A., Caprara G., Hack M., *Alla scoperta del sistema solare*, Mondadori, Milano 1993.
Briggs G. e Taylor F., *Atlante Cambridge dei pianeti*, Zanichelli, Bologna 1989.
• Brown S., *Come montare un telescopio*, Cosmo-Media, Como 1985.
• Brown S., *Fotografare con il telescopio*, Cosmo-Media, Como 1982.
Bussoletti E. e Melchiorri F., *Astronomia infrarossa: una nuova rappresentazione del cosmo*, Mondadori, Milano 1983.
• Cavedon M., *L'ABC dell'astronomia*,

Mondadori, Milano 1999.
De Angelis U., *A due passi da noi – Esplorazioni spaziali di Terra e dintorni*, Bibliopolis, Napoli 1992.
Falorni M. e Tanga P., *Osservare i pianeti*, Media Press, Milano 1994.
Ferreri W., *Fotografia astronomica*, Il Castello, Milano 1994.
Ferris T., *Galassie*, Fabbri, Milano 1990.
• Foresta Martin F., *Laboratorio di astronomia*, Edizioni Dedalo, Bari 1988.
Giacconi R. Tucker W., *Universo in raggi X*, Mondadori, Milano 1985.
• Gribbin J., *Astronomia e cosmologia*, Garzanti, Milano 1998.
Guest J.E. e Greeley R., *La geologia della Luna*, Newton Compton, Roma 1979.
Hack M., *Atlante di astronomia*, Sperling & Kupfer, Milano 1992.
Hack M., *L'universo alle soglie del Duemila*, Rizzoli, Milano 1980.
Hack M., *La galassia e le sue popolazioni*, Scienza, Trieste 1994.
Hawking S.W., *Dal big bang ai buchi neri*, Rizzoli, Milano 1988.
Henbest N e Marten M., *La nuova astronomia*, Hoepli, Milano 1996.
Herrmann J., *Atlante di astronomia*, Sperling & Kupfer, Milano 1992.
Hoyle F., *L'universo intelligente*, Mondadori, Milano 1984.
Jastrow R., *Incontro con una stella*, Mondadori, Milano 1990.
Krauss L., *Il cuore oscuro dell'universo*, Mondadori, Milano 1994.
• Levy, D.H., *Il cielo*, De Agostini, Novara 1996
Luminet J.-P., *I buchi neri*, Marco Nardi, Firenze 1992.
Meeus J., *Astronomia con il computer*, Hoepli, Milano 1990.
Menzel D.H. e Pasachoff J.M., *Stelle e pianeti: guida all'osservazione a occhio nudo e con il telescopio*, Zanichelli, Bologna 1990.
• Ranzini G., *Atlante dell'universo*, De Agostini, Novara 2000.
Reeves H., *L'evoluzione cosmica*, Rizzoli-BUR, Milano 1993.
• Ridpath I., *Mitologia delle costellazioni*, Muzzio, Padova 1994.
• Rigutti M., *Atlante del cielo*, Giunti, Firenze 1997.
Rigutti M., *Astronomia*, Giunti, Firenze 1994.
• Rigutti M., *Cento miliardi di stelle*, Giunti Firenze, 1995.
Rigutti M., *Comete, meteoriti e stelle cadenti*, Giunti, Firenze 1997.
• Rigutti M., *La vita nell'universo*, Rizzoli, Milano 1981.
• Rigutti M., *Storia dell'astronomia occidentale*, Giunti, Firenze 1999.
Smoluchowski A., *Il sistema solare*, Zanichelli, Bologna 1989.
• Verdet, J.-P., *Il cielo: caos e armonia del mondo*, Electa/Gallimard, Milano 1993.
Wald R.M., *Teoria del big bang e buchi neri*, Bollati Boringhieri, Torino 1980.
Weinberg S., *I primi tre minuti*, Mondandori, Milano 1977.
Wheeler J.A., *Gravità e spazio-tempo*, Zanichelli, Bologna 1994.

JOURNALS – IN ITALIAN

Coelum, Focus, Galileo, La macchina del tempo, L'Astronomia, Le Scienze, Le Scienze e il loro insegnamento, Newton, Nuovo Orione, Quark

MONOGRAPHS, CD-ROMS AND VIDEOS

Le Scienze

FURTHER INFORMATION: OTHER LANGUAGES

AA.VV, *Lunar Sourcebook*, Cambridge University Press, Cambridge 1991.
Acker A., *Formes et couleurs dans l'Univers*, Masson, Paris 1987.
Bourge P., Dragesco J. e Dargery Y., *La photographie astronomique d'amateur*, P. Montel, Paris 1979.
Cadogan P., *The Moon, our sister planet*, Cambridge University Press, Cambridge 1981.
Clark S., *Stars & Atoms*, Andromeda Oxford Ldt., Oxford 1994.
Cohen N., *Gravity's lens*, John Wiley and Sons, Chichester 1988.
Collins M., *Liftoff*, Grove Press, New York 1988.
Couderc P., *Les éclipses*, Presses Universitaires de France, Paris 1971.
Dragesco J., *High Resolution Astrophotography*, Cambridge University Press, Cambridge 1995.
Gowstad J., *Astronomy: The Cosmic Perspective*, John Wiley and Sons, Chichester 1990.
Hack M. e Struve O., *Stellar spectroscopy*, Del Bianco, Udine 1969.
Hartmann W.K., *Astronomy: the Cosmic Journey*, Wadsworth Pub. Co., Belmont, (Cal.) 1982.
Kaufmann, W.J.III, *Discovering the Universe*, Freeman & Co., New York 1990.
Kitchin C.R., *Journey to the Ends of the Universe*, Adam Hilger, Bristol 1990.
Kitchin C.R., *Stars, Nebula and the Interstellar Medium*, Adam Hilger, Bristol 1987.
Kuiper G., *Photographic Lunar Atlas*, University of Chicago Press, Chicago 1960.
Levinson e Taylor, *Moon Rocks and Minerals*, Pergamon Press, New York 1971.
Link F., *La Lune*, Presses Universitaire de France, Paris 1981.
Marau, S.P. (editor), *The Astronomy and Astrophysics Encyclopedia*, van Nostrand Reinhold, New York 1992.
Osterbrock D.G., *Stars & Galaxies: Citizens of the Universe*, Freeman, Oxford 1990.
Price F.W., *The Moon Observer's Handbook*, Cambridge University Press, Cambridge 1988.
Ross Taylor S., *Lunar Science: a post-Apollo View*, Pergamon Press, New York 1975.
Viscardy G., *Atlas-guide photographique de la Lune*, Masson, Paris 1986.

JOURNALS IN OTHER LANGUAGES

Astronomy, Astronomy Now, La Reserche, Sky & Telescope, Nature, Scientific American

FURTHER INFORMATION – THE INTERNET

- *Useful links*
www.ster.kuleuven.ac.be
www.tng.iac.es/links/links.html
www.algonet.se/~sirius/astro/links.html

- *International space and astronomy research institutes*
www.nasa.gov
www.estec.esa.nl
www.esa.int
www.eso.org
www.iac.es/eno
www.iau.org
www.hubblesite.org
http://oposite.stsci.edu/pubinfo/
www.stsci.edu/resources

- *Information and didactic*
www.telescope.org/rti
www.telescope.org/rti/nuffield.html
http://science.nasa.gov
www.algonet.se
http://hou.lbl.gov/research
http://solar-center.stanford.edu/resources.html
www.eso.org/outreach/info-events
http://micro.magnet.su.edu/primer/java/scienceopticsu/powersoft10
www.fourmilab.ch/yoursky/cities
www.hubblesite.org/news_ and _views/pr.cgi/2002.02
http://astroclub.net/saturne/antares_on_line

- *Astronomy societies*
www.uai.it
http://astrolink.mclink.it/assoc.htm
http://w3c.ct.astro.it/cnaa
www.astrofili.org

- *Popular and scientific journals on sale to the public*
www.lescenze.it
www.skypub.com/sktel/skytel.html
www.astronomy.com/home.asp
www.astronomynow.com
www.coelum.com
www.orione.it
www.galileonet.it
http://astroemagazine.astrofili.org
www.filemazio.net

- *Bioastronomy*
http://bioastronomy.uws.edu.au
www.seti.org/education
www.seti.org
www.nai.arc,nasa.gov
www.links2gonet/topic/SETI

- *Sun*
www.solar-center.stanford.edu
http://solar-center.stanford.edu/folklore.hyml
http://mesola.obspm.fr
http://sohowww.nascom.nasa.gov
www.solarviews.com/upgrade.html

- *Solar system*
http://seds.lpl.arizona.edu/nineplanets/nineplanets
www.planetary.org
www.solarviews.com/upgrade.html

GLOSSARY

A

Absorption: phenomenon through which a body attracts a fraction of the energy that falls on it.

Absorption band: field of frequency (or wavelength) in which energy is absorbed by a body or a group of particles. Each chemical element has a spectrum with characteristic absorption bands.

Acceleration: vector size that measures the variation in velocity compared with time (positive: increase; negative: decrease in velocity). Depends on the system of reference chosen. Of gravity; indicated by g. the acceleration to which are subject free bodies in fall near to terrestrial surfaces; its standard value (9.8062 m/s2) is measured at 45° latitude. It varies with the distance from the centre of the Earth; it is used to measure the shape of our planet indirectly.

Albedo: fraction of light reflected by a body.

Altitude: vertical distance between a conventional reference level (e.g. sea level) and the point in question. It is positive if the point is above the reference level, negative if below.

Amino acid: organic molecule essential to life containing an amino group (NH2) and an acid group (COOH). 21 main ones are known and these make up all living matter.

Anomalistic: the mean time interval that occurs between two passages of the sun at perihelion. Comprises 365.259 mean solar days.

Antimatter: matter comprising chemical anti-elements.

Apparent diameter: diameter that celestial bodies seem to have when seen from the Earth. Because of the distance it is less than the real diameter.

Apse: points at the extremity of the greater axis of an elliptical orbit. The line of the apse coincides with the greater axis of the orbit. The line perpendicular to the line of the terrestrial apses at the mid point and the plane of the ecliptic touches the celestial sphere at the two poles of the ecliptic.

Aphelion: point in an orbit furthest from the sun. The opposite, i.e. the point of the orbit closest to the sun, is called the perihelion.

Asteroid belt: region of the solar system between the orbits of Mars and Jupiter in which most of the asteroids orbit the sun.

Astronomical Unit (A.U.): unit used to measure length in astronomy, especially planetary distances. It originally represented the mean distance between the Earth and the sun during one revolution. The standard value is now 1 AU = 149,597,870 km = 1,496*108km = 499,005 light seconds.

Ascension line: angular distance measured clockwise (to the East) along the celestial equator from point g. With the declination constitutes one of the two co-ordinates in the mobile equatorial system. Is measured in hours –h-, minutes – min- and seconds –s-, with 1h = 15°.

Atomic model: model of the structure of an atom which gives a coherent explanation of phenomena observed and allows experimental verification.

Atom: from the Greek atmos = indivisible: was used in Antiquity for the smallest particle of matter deprived of structure. The modern atom, however, has a corpuscular and electromagnetic structure. It comprises a massive central nucleus (divided into corpuscular "sub-nuclears" or elementary particles) with a positive electrical charge, around which move electrons with negative electrical charges. In a stable complete atom (neutron) the total electrical charge is nil.

Axis of rotation: line around which each point of a body in rotation traces a circumference. The exceptions are the points of the body through which passes the axis of rotation itself.

Azimuth: angular distance measured clockwise (to the West) on the circle of the horizon from the South cardinal point. Usually measured in an angle and, together with the height, constitutes one of the two height-azimuth co-ordinates.

B

Bar; measurement of pressure.

Basalt: basic, dark, fine-grained igneous rock, with no quartz, containing a percentage of SiO2 between 45 and 50%. It is lava that constitutes around 90% of volcanic rocks. Characterised by the rock on the ocean bed and in lunar seas.

Batteries: single-cell organisms with no nucleus and protected by an external wall found in all terrestrial areas, even the most extreme. It is thought that the first living organisms in the history of the Earth were similar to modern archeobatteries and that there are probably similar forms of life on other planets or minor celestial bodies.

Big bang: the initial event in the history of the universe according to the most popular theories today the huge quantity of energy concentrated in one single point (singularity) "exploded" immediately beginning to be transformed, according to Einstein's Law of expanded matter.

Biomass: or organic matter: generic term that includes all organic matter. Contains solar energy in the form of chemical bonds.

Bi-polar: with two poles.

Black body: theoretical body that absorbs all radiation that hits it. Used to find the link between wavelength of irradiated energy and absolute temperature

Black hole: final phase of evolution of stars with very large mass or entire galaxies that, according to the theory, is reduced to objects with densities that produce gravitational forces able to "imprison" almost all of its radiation.

Brilliance: phenomenon of solar activity that, close to the surface of the star, randomly frees large quantities of energy in a very short time.

By the Earth: the ration between the light flow emitted by the surface of our planet and that arriving from the sun. It varies according to altitude and the type of ground. Water reflects 10% of sunlight, vegetation between 15 and 20%, sea ice between 50% and 60%, fresh snow 80%-90% and at its maximum in the Antarctic.

C

Catalyst: atom, molecule or substance that takes part in a reaction but remains unchanged. It can reduce reaction time, reduce the activation energy that stops the start of a reaction or modify the reagents in order to enable a reaction to take place.

Chemical element: primary component of matter. A type of atom corresponds to each chemical element (with its family of isotopes) with specific characteristics and, vice versa. There are 105 known elements – 92 or which are natural and 13 artificial.

Civil: based on solar year, it is the calendar year and comprises a whole number of mean solar days. To avoid slippage compared to the solar year, after 3 successive 365 day years, there is a leap year of 366 days. To "recover" slippage compared with the solar year produced by minor motion (for example precession) the year at the end of a century is a leap year if it is divisible by 4. 1700, 1800 and 1900 were not leap years, but 1600 and 2000 were.

Continental drift: advanced theory used several times in the study of geology, but developed by Alfred Wegener in 1912. was validated during the Sixties by Hess who researched the expansion of the ocean floor and expanded in McKenzie and Parker's theory of plate tectonics.

Cycle: phenomenon characterised by one or two variables that, after a certain period of time, return to their initial values.

Collapse of a star: implosion of a star towards its centre of gravity with destruction of its structure.

Compound: matter constituted by molecules formed by two or more atoms by various chemical elements.

Conjunction: position in which two stars are found when they have the same longitude as seen from the Earth. The Moon, for example, is in conjunction with the sun when it is a new moon.

Convection: phenomenon that takes place in fluids (gas or liquids) when the temperature rises. The transfer of heat in the fluid happens by means of convective currents.

Coronagraph: optical instrument in which a circular disc hides the solar disc in order to study the corona and chromosphere, which are visible from the Earth only during a solar eclipse. Used when there is little diffused light: on high mountains or on board satellites.

Current: mass of a fluid (liquid or gas) in continual motion (flow) mainly in one direction in terms of the reference system used.

Convective: current that occurs in the mass of a fluid when the temperature rises. Warm currents are formed by fluids at a higher temperature and lower density compared with the basic fluid. which is colder and denser. The final result is a transmission of energy. Currents exist until all the relevant mass is at the same temperature and the thermal rise is cancelled out.

Cromlech: megalithic construction built of large stones arranged in a circle.

Crust: outer solid layer of a planet.

D

Day:

Sidereal: by convention, this is the time interval that separates two consecutive passages of point g or of any star over the observer's meridian. Since the dimensions of the terrestrial orbit are negligible in terms of stellar distances (point g and stars can be considered to be an infinite distance from the Earth)) the sidereal day always has the same length: 24 sidereal hours – equal to 23h56m4s of mean solar time. The sidereal hour is measured from the moment point g is located over the meridian of the observer (zero hour).

Mean solar or civil: this is the time interval with a constant length of 24h3m56s that separates the two successive passages of the Sun over the observer's meridian. The mean Sun moves over the celestial equator with a uniform circular motion at a velocity equal to the mean of the true Sun and crosses the equator in one year.

True solar: this is the time interval slightly longer than 24 hours in length and variable during the course of a year that separates two consecutive passages of the true Sun over the observer's meridian. Its length varies over the course of the year because of the fact that, whilst the Earth is rotating on its own axis, it is also rotating around the sun at a non-uniform speed.

Dating: determination of the age OF rocks, fossils and other objects. One of the most popular methods used is that based on the quantification of proportions present in a sample of isotopes of a single radioactive element whose half-life is known.

Degree centigrade or Celsius: the unit of measurement of temperature on the thermometrical scale most often used (the centigrade scale) in which by convention at atmospheric pressure ice forms at 0 degrees C and water boils at 100 degrees C.

Decadence: spontaneous process of radioactive emission that takes place in nuclei of "heavy" atoms that exist in nature, through which they are transformed into atoms of different chemical elements.

Declination: angular distance of an object from the celestial equator. It is positive in the northern hemisphere and negative in the southern. It is measured in degrees and is one of the co-ordinates used in the equatorial system (fixed and mobile).

Deferential: on the Ptolemaic system this was a circle along which moves the centre of an epicycle. Synonymous with eccentric.

Density: ratio between mass of a substance and unit of volume – expressed in kg/m3 or g/cm3.

Determinism: cultural belief that any phenomenon is mechanically and necessarily caused by another that precedes it and that it is therefore possible to predict the behaviour of any given object or group of objects based on their initial condition and strict physical laws.

Diffraction: phenomenon due to the wave effect of light, for which propagation does not follow set patterns based on geometric optical rules. This also happens with sound and is due to interference.

Discontinuity: area in space where there is a rapid change in some physical or chemical parameters.

Dispersion: phenomenon in which the ray of polychromatic light separates into its various wavelengths forming a light spectrum. This happens with all types of electromagnetic radiation and is a consequence of refraction.

Dwarf:

White: last phase in the evolution of a star with a mass similar to that of the Sun. it comprises only what remains of the nucleus at a very high density (tens of thousands of times more than that of water).

Brown: star that is formed from the gravitational collapse of gas clouds but with insufficient mass to trigger nuclear reactions (lower than 1.8% of the solar mass). Radiates weak light and gravitational energy converts into heat.

Black: remains of a white dwarf in which the processes of nuclear fusion have been arrested.

E

Earthquake or seism: movement of racks due to a fracture in the crust. Comprises a series of shocks (seismic shocks) that move across the planetary body radiating from the place where the fracture took place (hypocentre). Each shock releases kinetic energy that is measured on the Richter scale. The shocks move through the rocks at different speeds and in different ways and their force is at its maximum in the area immediately above the epicentre.

Eccentricity: ratio between the distances of points of a cone from a focus and corresponding source. This constant ratio is >1 in hyperboles, = 1 in parabola and<1 in ellipses. E = o in circumferences. It is one of the basic elements used in the calculation of the orbits of celestial bodies. For planets and asteroids that have quasi-circular orbits, the values are low, with comets they are close to the unit.

Effect:

Doppler: change of frequency ion a signal produce by the relative motion of the source and the receiver.

Catapult: gravitational interference used by space vehicles to increase their velocity.

Greenhouse: gradual warming of the planet that, on the Earth and Venus, has lead to the increase of carbon dioxide in the atmosphere that prevents infrared radiation from dispersing into space.

Synergic: effect due to joint action with two or more causes, where the effect of the first cause facilitates the realisation of the second. The compound effects are greater than the effects of each cause taken individually.

Electrical charge: origin of an electrical field – a body that exercises an attraction or repulsion on other electrically charged bodies. There are positive and negative electrical charges. Charges with the same sign repulse, opposites attract. A body that has no electrical charge or where the electrical charge is negative is said to be electrically neutral.

Electrical: field where the effects of an electric force are felt. It is a bi-polar field: the poles – positive at the North and negative at the South – are characterised by the direction of the negative electrical charges that migrate from North to South.

Electromagnetic: field where the effects of an electrical force and a magnetic force are felt.

Electromagnetic Radiation: flow or form of propagation of electromagnetic energy that sometimes behaves like an electromagnetic wave (propagation of an electrical and magnetic field at the speed of light, equal to $3x10^8$ m/s in a vacuum), like a photon flow (each with energy h-v where h=6.6 $x10^{27}$) s and V = frequency).

Infrared: electromagnetic radiation with a lesser frequency and greater wavelength than visible light. It is emitted spontaneously by bodies at temperatures above 0K and most emissions take place when the temperature is lower than several thousands Kelvin. Its wavelength is between 1mm and 1mm.

Ultraviolet: electromagnetic radiation with wavelength between 400 and 4 nm. It is invisible to our eyes and has a high energy content.

X: together with g rays constitutes a very high frequency electromagnetic radiation (very short wavelength).

Elementary particle: dot-shaped particle with no structure. These are usually 6 quarks and 6 leptons.

Elliptic: plane traced by the Earth's orbit around the sun. this term is sometimes used to signify the trajectory of the Earth around the sun or the projection of the plane on the celestial sphere, i.e. the circular trajectory taken by the Sun in one year.

Electron: sub-atomic particle discovered in 1897 by J. Thomson, with an elementary negative electrical charge (all negative electrical charges are entire multiples of that of the electron) and indivisible, negligible elemental mass, by means of the atomic nucleus and equal to around 2 thousandths of the mass of a proton. To "tear" the electron from the attraction of a positively charged atomic nucleus and to create an ion, the atom needs to be supplied with energy.

Elipse: closed curved line described as the intersection of the surface of a circular cone with an inclined plane compared with the vertical axis of the cone. It is characterised by two vertices joined by a major axis. On the major axis, at a point equidistant from the vertices can be found the centre of the ellipse, through which passes, perpendicular to the major axis, the minor axis. Two points exist called the ellipse foci, located on the major axis such as, for every point on the curve, the sum of the distances between the same point and the two foci is constant.

Electrically neutral: body with positive and negative charges that cancel each other out.

Energy balance sheet: calculation of the total energy involved in the function of a certain system.

Energy flow: amount of energy that a surface unit uses in a unit of time (e.g. 1 m2) perpendicular to its direction of propagation.

Endogen: of internal origin.

Energy: capacity of a body to complete a work owned by virtue of its chemical and physical state. Measured in joules.

Chemical: energy held by a body because of its chemical structure, i.e. that which bonds the atoms in molecules. When these bonds are broken part of the energy can be transferred to another chemical bond or can be released in the form of heat.

Kinetic: energy held by a body to keep it in motion.

Degraded: energy that can no longer be used to perform a task. Equal to "low temperature" heat.

Nuclear bond: energy that keeps each atomic nucleus together. It is equal to the energy needed to be supplied to an atom to tear away the sub-atomic particles from the bonds.

Electromagnetic: energy issued into space in non-mechanical form (radiant energy) and that takes its name from the electrical and magnetic phenomena that determine it. A flow of electromagnetic energy is also called electromagnetic radiation. When radiation enters an object in order to absorb energy, it fulfils a task. It provokes the movement of electrical charges that allow it to be used.

Light: electromagnetic energy linked to light radiation.

Magnetic: energy that derives from the action of a magnetic force, linked to a magnetic field. It is measured in terms of the task that must be completed to create the magnetic field that it characterises or in terms of the work that is achieved at the moment when the field is destroyed.

Nuclear: energy released by reactions that involve the nucleus of the atom.

Radiation: energy dispersed in the form of electromagnetic radiation.

Solar: energy produced by nuclear reactions that take place in the Sun, irradiated into space (electromagnetic energy).

Calorific; energy contained in a body by virtue of the motion of agitation of the molecule that constitute it and that increases with the rise in temperature.

Thermal radiation: synonymous with infrared radiation.

Epicycle: according to the theories of antiquity, this is the circle on which each planet moves in a circular motion and that in its turn traces a concentric or eccentric path when seen from the Earth.

Equator:

Celestial: circumference produced by the intersection of the equatorial terrestrial plane with the celestial sphere.

Solar: maximum circumference of the sun, perpendicular to the axis of rotation.

Terrestrial: maximum circumference perpendicular to the axis of rotation and the reference parallel and in the equatorial plane.

Equinox: from the Latin aequa – equal and nox=night indicating two points in the Earth's orbit where the elliptic intersects with the celestial equator. When this is at these points (21st March and 21st September) the Earth is illuminated according to the perpendicular to its own axis of rotation; this determines the equal length of the day and night that both last 12 hours.

Erosion: wearing of rock due to the action of eroding agents (water, wind, temperature changes, ice etc.)

Exogen: of external origin.

F

Field: area in space in which the effects of a force are felt.

Fluid: state in which liquid or gaseous matter exists. According to the internal viscosity (i.e. the internal friction) its behaviour may follow the laws of fluid dynamics.

Force: physical property based on the dynamic and static defined as a physical agent capable of altering the state of rest or motion of a body, or able to produce deformation. It is characterised by intensity, a flow and point of application. The second principle of dynamics states that it is linked to the body's mass m which is applied and to the acceleration a which is determined by formula $F = m * a$. in many instances, a force is not directly applied to a body, but is seen as a remote action (field).

Electrical: force exercised by any field by virtue of its own electrical charge.

Electromagnetic: force developed by an electrical charge in motion. Force used by any electrical field that is found in an electromagnetic field.

Gravitational: force exercised by any body by virtue of its own mass: or force used by any mass that is found in a gravitational field.

Magnetic: force exercised by a magnet or by any magnetic body. Or force to which any electrical charge or magnetic body is subjected that is found in a magnetic field.

Nuclear: force that binds the particles that makes up atomic nuclei (protons and neutrons) at distances below 10-15m.

Gravity: close to the Earth and other celestial bodies, this is the force with which objects are attracted to the core of the body. The force itself results from the attraction and centrifugal forces, produced by the rotation of the body on its own axis.

Forces

Of friction: forces that exist at the contact point between two bodies and which oppose the relative motion of one against the other.

Electromagnetic: forces that act between electrically charged bodies, transported by photons: they also bind electrons to the nucleus.

Gravitational: a consequence of the mass, also acting at astronomical distances. It is thought that these are transported by gravitons.

Fossil: this term originally was only used to indicate the subterranean origin of a body or the combustible materials from some rock layers. The concept was then extended to a body that had undergone a process of fossilisation or, generically, something that has reached us over along period of time (fossil radiation, living fossils).

Friction: group of dissipative forces that hinder the relative motion of the two surfaces in contact. Friction causes motion to take place using a large amount of energy.

Frequency: physical property that indicates the number of repetitions of a periodic phenomenon in the unit of time (generally a second). With a wave, it is the number of wave crests that pass through a pre-determined point in a second. It is equal to the ratio between velocity of wave propagation and its length.

Of electromagnetic radiation: unit that quantifies the number of oscillations that the magnetic and electrical fields present at a point in space invested by radiation have in the unit of time. It is measured in hertz (Hz); visible light, for example, has a frequency of approx 1015 oscillations per second = 1015 Hz.

G

Galaxy: common name that indicates an enormous concentration of gas, dust and billions and billions of evolving stars linked by gravitational forces.

Gauss (G): unit of measurement for magnetic indication. 1 G is the magnetic induction of 1 maxwell on a surface of 1 cm2. 1 G is equal to 9.99*105 V s/m2.

Geomagnetic or magnetic field on Earth. Magnetic field originating in the Earth: similar to a bi-polar magnetic field> it is probable that it has changed many times (inversion of terrestrial magnetic field): if it had stayed unchanged there would be two magnetic poles in inverse positions to those of today.

Gnomon: elongated object whose shadow projected onto a horizontal or vertical plane indicates the direction and height of the sun. The ancients used it to determine latitude and the passage of time.

Gradient: variation of a physical property in space, according in a given direction.

Gravitational: area of space where the effects of gravitational pull are felt.**Geoids**: irregular form – approximately spherical, typified by the Earth.

Granite: acid igneous rock, large grained, primarily formed from quartz, alkaline feldspar and mica. It is a crystalline rock formed from a magma below the surface of the crust.

Gravimetry: geological analysis that, by means of the quantification of differences in the measurement of gravity, results in differences of mass. Suince it indicates the presence of masses with different densities, it is used to have indications about rocky bodies that is found within the crust.

Gravity (or rather gravitational force): force that acts on any object with mass that is to be found close to another object with a mass. According to the law of Newton's law of gravity, it is proportional to the product of the masses (m*M) of the bodies under consideration and inversely proportional to the square of their distance ®. In the formula F=G*(m*M)/r, where G is the universal gravitation constant measured by Cavandish in 1798 and equal to 6.67*10-11 Nm2kg2.

H

Half-sky: point at which the celestial equator intersects with the meridian at the place from the part of the south cardinal point. Used to measure the hourly angle of an observed body.

Helical dawn: the rising of a celestial body just before the sun. for a certain horizon, a given star rises every day around 4 mins earlier compared to the following day and rises at a given hour once a year. Therefore, every day different stars rise a little before the Sun. the people of Antiquity and North American Indians observed the helical dawn of very bright stars or specific constellations to correct their calendars to match the real length of the year. The Egyptians used the helical dawn of Sirius to calculate their calendar.

Hertz (Hz): S.I. unit of measurement of frequency. 1 Hz is the frequency of a phenomenon with a period of 1 second: 1Hz = 1s-1.

Helium: chemical element characterised by a nucleus with 2 protons and 1 or 2 neutrons. After hydrogen, it is the simplest element, with a small mass. It is the most abundant element in the Universe.

Hemisphere: Sun's sphere of influence, extending beyond the orbit of Pluto.

Heat: form of energy belonging to a body connected to the motion of agitating molecules that constitute it. The hotter a body, the quicker and more randomly its molecules move. Under optimal conditions, their motion can give rise to a very visible mechanical work. It also takes place in a vacuum under the form of electromagnetic radiation.

Height: angular measurement of location in which a celestial body above the horizon of the observer. Used with the azimuth to calculate the height-azimuth co-ordinates.

High plane: more or less extensive and flat region characterised by a mean altitude higher than 200m. Usually delimited by escarpments and with deep valleys.

Horizon: maximum extent of the celestial sphere at right angles to the vertical of the place for the observer.

Hour angle: angular distance measured along the celestial equator form Halfway Point to the celestial meridian passing through the relevant body. With the declination, it constitutes one of the two co-ordinates in the fixed equatorial system. It is measured in hours – h-, minutes – min-and seconds –s- or in degrees. Positive towards the west, negative towards the east.

Hubble: theoretical distance, according to the laws of Hubble, to which the galaxies must go reciprocally to the speed of light , equal to around 20 billion light years (-6*106 kpc). This distance corresponds to the distance that the light can have travelled from the nig bang to today: the universe, therefore, would have an age in years equal to the same value of this distance. Any object beyond that distance would not be visible.

Hydrogen: the simplest chemical element and the one with the lowest mass. Its nucleus is made up of one proton.

I

IAU: International Astronomy Union: the official international association covering all astronomers and space research.

Ionised gas: gas made up of electrically charged particles.

Interplanetary (or interstellar) matter: low density matter made up of gas and dust.

Infrared: part of electromagnetic radiation emitted by each body that has a temperature above 0Kand with a total wavelength between approx 1 mm and 1mm.

Interference: phenomenon characteristic of wave motion, superimposition of waves that originate in coherent sources, i.e. that emit waves of equal frequency and whose phase difference is uniform. If the waves are concordant with the phase, through interference a wave is created that has amplitude equal to the sum of the amplitudes of the two waves (constructive interference). If the two waves have a phase difference equal to an unequal number of semi-wavelengths, the amplitude of the resultant wave is nil (destructive interference). In intermediate cases can be seen intermediate situations.

Ions: term of Greek derivation (ion comes from ienai – to go, referring to the movement of ions towards an electric pole) indicating and atom or molecule with a positive or negative charge compared with their normal electrically neutral system configuration. The charge is due to the loss or acquisition of one or more charged electrons or atomic groups. The process that leads to the formation of one or more ions through the loss of electrons or the snapping of molecular bonds is called ionisation: it supplies energy to the system.

Isotope: the atoms of each chemical element are characterised by a precise number of protons: the nucleus also contains a variable number of neutrons. There are therefore atoms with equal nuclear charge (equal number of protons) and so equal chemical characteristics but different weight (different numbers of neutrons) and so different physical characteristics. This is the reason we don't speak of different elements but of isotopes of a single chemical element. Each isotope is characterised by:

Its atomic number: indicated by Z and representing the number of protons (i.e. with a positive charge) present in the nucleus of an atom of an element. characteristic of each chemical element (in a neutral atom it is equal to the number of electrons): 0

Its mass number: indicated by A and representing the total number of protons (Z) and neutrons (N) contained in the atomic nucleus, characteristic of each isotope of each chemical element. The formula used is A=Z+N. Each isotope then is identified by the symbol of the original element preceded by the atomic number and the mass number. For example, Carbon - C – is present in nature in different proportions in three isotopes. 98.9% is made up of isotope 126C; 1.1% isotope 136C and 0.1% in the unstable isotope 146C. This method of writing the symbol of the element tells us that the carbon element is distinguished by an atomic number Z=6 9i.e. a nucleus with 6 protons) and can be found in one of three isotopes that have nuclei with N=6, N=7 and N=8 neutrons respectively.

J

Joule (J): the unit of measurement of work, energy and quantity of heat.

K

Kelvin or absolute degree (K); the unit of measurement of temperature on the absolute thermometrical scale unlike other thermometrical scales, the Kelvin is defined using thermodynamics without reference to specific physical phenomena. Temperatures expressed in Kelvin are linked or those in Celsius using the equation T=t+273.

Kilo- (k): prefix that indicates one thousand times the unit. For example, 1km = 1,000m.

kPs (kilopascal): multiple of the Pascal (Pa), S.I. unit of measurement of pressure. 1kPa = 1,000 Pa = 1,000 Nm2 = 9,869*103 atm = 102 bar.

Kpc (kiloparsec): unit of measurement of length used in astronomy; 1 kpc = 1000 pc = 3.08*1016 km = 3.2558*103 light years.

KREEP or norite: lunar rock formed from Potassium (K), Rare Earth (REE) and phosphorus (P) with rich traces of uranium and thorium.

Kulper Body: solar system body presumably of the same type as a comet's nucleus with an orbit that lies between Neptune and Pluto. Some scientists believe that Pluto is the largest Kulper Body.

L

Laser: system that generates and amplifies radiation with a specific optical frequency.

Light year: unit of measurement of length used in astronomy, corresponding to the distance travelled in one year by light at a constant speed of 300,000 km/s and equal to approx. 9.46*1012 km. one light year corresponds to 0.3066 parsecs. Smaller sub-multiples also exist: light-year, light-minute and light-second.

Lines (of flow, of field, of force...): imaginary lines that in conventional representations of fields, link points in space characterised by the same value of size considered; they are more numerous (directly proportional) where the variation is more rapid and vice versa.

Light: radiant energy visible to the naked eye, between 380 and 700 nm of wavelength.

Luminosity: intense brightness of a star defined as total energy irradiated per unit of time. It is proportional to the surface of the star and its effective temperature.

Solar: total quantity of radiant energy that the Sun emits during a unit of time.

M

Magnetic storm: disturbance of the terrestrial magnetosphere due to an increase in solar activity (caused, for example, by brilliance emitting into the solar wind large quantities of charged particles); the intensity of the magnetic field on the Earth's surface can be altered fro short periods and radio transmissions affected.

Magnetic: where the effects of magnetic force are felt. A bi-polar field – the North and South, positive and negative, poles are characterised by the direction of the lines of force that come from the South pole and enter the North pole. In a magnetic field, there are zones through which force lines do not pass – these are the neutral zones where the magnetic force has no effect. Many celestial bodies are characterised by an internal magnetic field in the surrounding space (magnetosphere).

Magnetic: line around which the lines of force developed by a magnetic field are arranged symmetrically.

Magnetosphere: area in space relatively near to any celestial body equipped with a magnetic field, within which can be perceived the magnetic effects of the body itself. The magnetospheres of planets contain "bundles" of electrically charged particles originating in the solar wind "trapped" by lines from the terrestrial magnetic field (van Allen bundles).

Magnitude or size: luminosity of heavenly bodies. The magnitude is also called the size but the two should not be confused. Since when we talk about the size of a star, we are not referring to its dimensions of its mass, but to its luminosity.

Apparent: luminosity of a star as it appears to an observer. The reference system used from magnitude 5 to a star with luminosity 100 times greater compared with a star of magnitude 1, according to a scale that follows historical classification.

Maser: acronym for Molecular Amplification by Stimulated Emission of Radiation = amplification of microwaves by means of stimulated emission of radiation used in particular for microwaves used to receive very weak symbols. This instrument uses the fact that when a molecule or an atom is found in a suitable energy state, the passage through its vicinity of an electromagnetic wave with a certain wavelength may induce it to emit radiation in the form of other electromagnetic radiation with the same wavelength that reinforces the passage wave triggering a cascade of phenomena that result in an increase in the intensity of the original impulse. In some clouds of interstellar matter excited by the radiation of neighbouring stars is verified the same phenomenon that determines the formation of an intense bundle of radiation with well defined wavelengths.

Mass: one f the most difficult concepts in Physics to define in acceptable and unequivocal terms. In classical Physics it is defined as the quantity of matter constituting a body. Afterwards, the accent has been placed on the property of the body when velocity is changed (mass = inertia, resistance to change). Einstein stated that the mass was another aspect of energy – a type of "condensed" energy; according to Einstein's Second Law mass in linked to energy using the equation E=mc2 where c is the speed of light in a vacuum (constant). The mass, therefore, is also a state of energy; when mass disappears it creates energy and vice versa.

Atomic: ratio between the mass of an atom and that of a hydrogen atom taken as a unit of reference; it is slightly less than the total sum of the mass of protons and neutrons that form the nucleus; this defect of mass is equal to the energy required for the formation of nuclear bonds.

Mega- (M): prefix that indicates one hundred times the unit. For example; 1 MW (megawatt) = 1 million watts.

Menhir: large and tall rough-hewn stone, approximately square in shape inserted into the ground and with known religious, funereal and astronomical functions

Meridian: maximum perpendicular circumference at the equator and passing through the poles. With the equator it defines geographical and astronomical co-ordinates.

Methane: the simplest of the hydrocarbons, with molecules formed by 1 atom of carbon and 4 of hydrogen.

Meteor: luminous phenomenon produced in the Earth's atmosphere by a meteor that, when falling to Earth, super-heats.

Meteorite: celestial body of varying dimensions and variable constitution (metallic and rocky) that falls to the Earth.

Micro (m): prefix that indicates the millionth part of a unit. For example, 1 ms (microsecond) = 10-6 s.

Microwave: electromagnetic radiation with wavelength less than the metre and frequency more than 300Hz.

Milli- (m): prefix that indicates the thousandth part of a unit. For example 1mm = 1 thousandth of a metre.

Mineral: natural crystalline substance with a defined chemical composition and a regular structure.

Molecule: union of two or more atoms linked by electromagnetic forces (or van der Waals). According to the number of atoms that constitute it can have minimum dimensions (as the hydrogen molecule) or microscopic proportions (like the bacterial chromosome formed by a single DNA molecule.

Inorganic: molecule without a skeleton of carbon atoms.

Organic: molecule belonging to the carbon family, with a skeleton of carbon atoms.

Unstable: molecule that, because of its bonds, can spontaneously split.

Moon: the name of Earth's only natural satellite and, as a common noun, any satellite of a planet. The name derives from leuk - an old Indo-European root that has influenced many terms such as leucos = white, pure, clear in Greek, luceo = to illuminate in Latin, lux = light in Latin and lumen = light in Latin.

Motion:

Direct: movement around own axis or around a planet characteristic of most planets and satellites, observed at the elliptic with the head towards the celestial North Pole, appears anticlockwise.

Galactic: rotary movement of Galaxy around its own axis.

Retrograde: motion around own axis or around a planet in the opposite direction to the most usual; characteristic of some planets and satellites.

Turbulent: irregular motion in which the particles do not follow a linear path but follow trajectories with no continuity or regular timeframe, at variable velocities at all points and with strong transversal components compared to the main direction. Turbulent motion determines a strong dissipation of energy.

Mya: acronym for million years ago, commonly used to indicate geological periods.

N

Nano-(n): prefix that indicates a millionth part of a unit. E.g. 1nm = 10-9m.

Nebula: more or less condensed interstellar matter. Being made of gas and dust, condensing and then giving birth to thousands of stars.

Planetary: mass of vary rarefied cosmic dust formed from matter a star with a mass similar to that of our Sun during its final phase.

Neutrino: particle with no electrical charge and extremely small mass that, together with antineutrinos, elements and quarks form the family of leptons, retained the only really elementary subatomic particles (i.e. invisible).

Neutrons: atomic particle with no electrical charge, with a mass equal to that of a proton into which it can be transformed through radioactive decadence b. It is formed from subnuclear particles called quarks.

Nuclear or atomic fusion: opposite process to nuclear fission in which an atomic nucleus is produced with a higher mass from the union of two nuclei with a lower mass. To be able to unite the original nuclei must counteract the repulsion forces that are due to their identical positive charge and can reach high speed. This only happens in very high temperatures. In fusion, the nuclei produced usually have a different mass from that of the original nuclei and the energy difference is to be found in the form of nuclear bonds and energy that is liberated (emission of rapid neutrons, electromagnetic radiation). If the mass increases, it is because the fusion needs to be supplied with external energy.

O

Opposition: position in which two stars are to be found when, as seen from the Earth they are 180 degrees apart in longitude. The Moon, for example, is in opposition to the sun when it is at Full Moon.

Orbit; trajectory followed in space by a celestial body around another. Usually an ellipse. The shape and dimensions are determined by the eccentricity and length of the greater semi-axis. The orbit's orientation is due to the inclination of the orbital plane compared to a plane of reference, whilst the position of the body along the orbit is defined by its anomaly, i.e. the angle, measured in the direction of movement, between the perihelion, the body around which it moves and the body itself. The theoretical plane around which a celestial body orbits is called the orbital plane.

P

Parallax: angle under which the Earth's radius (or the Earth's diameter), or the Earth's orbit according to the measurements that must be made) is seen from a celestial body. It is calculated using trigonometry measuring the height of a celestial object at two sides of the selected length.

Parsec (pc): unit of measurement of length very popular in astronomy: 1pc is the distance at which 1 UA subtends an arc of 1": 1pc = 3.08*1013km = 3.2558 light years.

Pascal (Pa):S.I. Unit of measurement of pressure. 1 Pa = 1 NM2 =9.869*10-6 atm = 10-5 bar.

Photo dissociation: break in chemical bonds of a molecule by means of electromagnetic energy of light (from the Greek photos).

Photons: elementary particles deprived of mass and defined as indivisible quantities of electromagnetic energy (amount of light or energy). Photon energy is proportional to the frequency of radiation (v) in the equation E = h*v in which h is Planck's constant. The photon moves at the speed of light with a wavelength g inversely proportional to its frequency (and therefore its energy).

Photosphere: visible "surface" of the Sun from which comes the light that we observe. It corresponds to an area around 300 km thick in which gases that make up the Sun are transparent to solar radiation.

Photosynthesis: production of chemical bonds and a molecule by means of light.

Planets: "cold" celestial bodies of largish dimensions (mass of the order of 1019-1027 kg) even if a lot smaller than the stars, approximately spherical , in which there are no spontaneous nuclear fusion reactions. In the outer solar system are Jupiter, Saturn, Uranus, Neptune, and Pluto. Jupiter, Saturn, Uranus and Neptune are gaseous and young. Apart from Pluto, they are composed of a high percentage of volatile components and have masses a lot greater than the inner planets and are accompanied by a large number of satellites and rings.

Inner solar system: either rocky or terrestrial, they are Mercury, Venus, Earth and Mars and have relatively small masses. They are predominantly composed of minerals of iron, nickel silicon and magnesium.

Plane:

Equatorial: theoretical plane cut be the Equator

Orbital: theoretical place cut by an orbit.

Planetoids: matter condensed by primordial nebula that, according to a currently accepted theory, made the planets of the solar system by "concretion".

Plasma: gas composed of highly ionised matter characteristic of a large part of the stellar structure and of the interstellar and interplanetary medium. In Physics, it is considered as the fourth state of matter, after solids, liquids and gases.

Polar cap: on the surface of the Earth and Mars, areas near to geographical poles covered in ice (water and carbon monoxide).

Polar dawn: visible diffuse luminosity visible in the sky at high terrestrial latitudes and in the upper atmospheres of planets with large magnetic fields. They are light phenomena caused by the interaction between high energy electrically charged particles originating in the sun (solar wind) "captured" by the magnetic field and extremely rarefied gases (plasma) of the upper atmosphere, according to a process similar to that which causes the fluorescence in neon tubes.

Pole: each of the tow points at which an axis intersects a sphere.

Celestial: point at which the axis of terrestrial rotation meats the celestial sphere.

Of the ellipse: point at which the axis perpendicular to the ellipse at its centre intersects with the celestial sphere.

Galactic: point at which the axis perpendicular to the galactic equator meets the celestial sphere.

Geographic; point at which the axis of terrestrial rotation meets the Earth's surface.

Magnetic: point at which the force lines of a magnetic field converge.

Positron: subnuclear particle identical to an electron but with a positive electrical charge (= anti-electron). It is part of subnuclear particles of antimatter.

Proton: nuclear particle with positive elementary electrical charge (equal in intensity to that of an electron), with a mass equal to 2,000 times that of an electron. Formed from subnuclear particles called quarks.

Q

Quark: elementary particles that comprise protons and neutrons whose existence is proved indirectly. They are thought to be elementary particles of matter together with leptons and are thought to have no structure. To make sense of observations, they are divided into 6 types (up, down, strange, charm, top, and bottom) each respectively formed by 3 possible charges of colour 9green, red, blue). The fact that these are coloured makes them sensitive to the forces of colour. Quark anti-particles also exist which are subdivided into anti-types and anti-colours.

R

Radioactive emissions: three types are known, b rays constituted by electrons, g rays constituted by energy-rich electromagnetic radiation.

Radial direction: direction of rays diverging from one point (origin or source). If the source is spherical in shape the direction of each ray is always perpendicular to the tangential plane at the ray's exit point.

Radio-telescope: instrument that shows radiation with wavelengths in radio bands emitted by celestial objects.

Rays: corpuscular emissions and/or energy emissions produced by radioactive decadence or nuclear reactions.

a: comprise helium nucleus (2 protons and 2 neutrons),

b: comprise one electron.

g: comprise high-energy photons.

X: comprise photons with a large amount of energy, but still less than g rays.

Reaction: in chemistry, this term indicates any variation or transformation of the chemical state of one or more substances. Each chemical reaction is accompanied by production of energy (exothermic reaction) or by an absorption of energy (endothermic).

Chain: successive series of reactions usually exothermic that, once triggered, develops independently of the preceding reaction.

Red shift: the shift of the lines present in the spectrum of stars and galaxies towards a greater wavelength. Similarly, these objects can show a "blue shift2 if they come closer. It is caused by the Doppler Effect. There is a lot of discussion regarding the red shift of distant galaxies, about whether it is due to the Doppler effect of some other origin.

Refraction: deviation in the direction of propagation of waves (seismic, electromagnetic or light, sound) from their path through matter with different densities.

Reflecting power: capacity to reflect radiation. It is calculated by measuring the fraction of radiation that arrives that a given surface reflects out.

Regolith: derives from the terms regular and lithos (= rock) and is the name given by astronauts to the lunar surface.

Revolution: movement of rotation of a celestial body or a system of bodies around a centre of gravity.

Rocks: minerals aggregates.

Rotation:

Differential: rotation characterising a system in which the different parts move at different speeds. Typical of fluid bodies (Jupiter, Sun) or similar (galaxies).

Sidereal (terrestrial): movement of the Earth around its own axis observed by taking a reference from a celestial body fixed on the celestial sphere.

S

Selene: name given to the Moon derived from sela (= light, splendour, lamp, flame in Greek).

Size of magnitude: luminosity of celestial bodies.

Sidereal; the time interval taken by the Earth to return to the same point in the orbit in terms of the fixed stars. Comprises 365.256 mean solar days. The difference of 20 minutes during the solar year is due to the precession of the equinoxes.

Solar constant: energy flow emitted by Sun and measured above terrestrial atmosphere at a distance from the Sun equal to the mean Sun-Earth distance. Equivalent to around 1.400 watt/m2.

Solar: cycle of solar activity 11 years in length in the course of which the intensity of solar phenomena varies from a maximum to a minimum to return to a new maximum.

Solar activity: complex of variable processes and phenomena in the visible part of the sun.

Solar or tropical: the time interval taken by the Earth to see the Sun newly projected at point g. Comprises 365.242 mean solar days.

Solstice: from the Latin Sol = Sun, literally means static sun, the name given to the day when the Earth is in one of two points of the ellipse that are furthest from the celestial equator. The name derives from the fact that at this time, the variation in the maximum height of the sun above the horizon is minimal from one day to another, the lowest for the whole year.

Spectroscopic analysis: Study of radiation emitted or absorbed by substances, carried out using the spectrum index. Chemical-physical characteristics can therefore be calculated from the emitting or absorbing object (e.g. the type, chemical composition or the physical state of stellar objects).

Spectrum: range of coloured "bundles" and dark lines into which is divided the Sun's visible luminous radiation or that from any other celestial object through a prism. In reality, they are part of a spectrum that also includes invisible "bundles" – gamma rays to microwaves.

Gas absorption: group of black lines that are seen on a continuous spectrum of white light that a gas has passed through. Each substance is capable of absorbing the radiation that it admits: the absorption spectrum is the "negative2 of the emission spectrum. Absorption spectra can also include frequencies that are beyond the visible.

Prismatic dispersion or prismatic spectrum: luminous figure that is obtained by collecting on a screen the light emerging from a prism hit by a bundle of light to be analysed. Phenomenon closely linked to diffraction.

Gas emission: group of luminous and coloured lines on a dark background showing the breakdown of radiation emitted when passed through a prism. Emission spectra can be obtained for frequencies outside the visible.

Spectrophotometer: instrument able to compare the light intensity of various sources with different wavelengths.

Spectrograph, spectroscope: instruments able to produce observe and record spectra.

Spontaneous process: process that happens without outside energetic help.

Sun mean(theoretical or fictional): theoretical artifice – a Sun that moves on the celestial equator at speeds equal to the mean of the true Sun. this takes a year to circle the equator and is used to calculate the length of the mean solar, or civil, day .

Supernova: final stage in the evolution of a star that has a mass equal to various solar masses and that do not collapse gradually.

T

Temperature: the scale that measures the mean kinetic energy of a molecule, atoms, particles...that makes up any given body.

Kinetic: derived from calculations of the kinetic energy of a molecule and not from an experimental measurement carried out on the substance under examination. It is used for rarefied gases whose molecules have large amounts of energy (plasma).

Terminator: line of separation between the illuminated part of a solid or liquid celestial body and the part in the shade – i.e. the line separating night from day.

Tectonics: word derived from the Greek tectaino = to build) indicating the branch of Geology that deals with structure and deformation of the Earth's crust, such as those that result from the effects of ice and the distribution of certain rocks around the globe. It also investigates the causes that are at the origin of observations studying the processes that determine the basic structural characteristics of the Earth's crust.

Tide: periodic change in the level of the atmosphere, seas or oceans, continents and crust of a celestial body due to the gravitational pull of another nearby celestial body.

V

Velocity: ratio between distance travelled and the time take.

Angular: ratio between radiant angle from a point along a circumference or a line on a circle rotating in a given direction, around a given axis and the time taken in the rotation.

Escape velocity: velocity that a body must have to escape from the gravitational pull of a celestial body.

Volcanic erosion: eruption, often fluid, of fluids through a break in the crust. It is produced by gas pressure produced deep below the surface.

W

Watt (W): it isa unit of measure of power. It is estimated measuring the amount of energy freed or absorbed in the time unit : 1 W = 1 J/s

Wavelength: distance between two crests or two successive dips of a wave of radiation: distance between two successive crests of an electromagnetic wave: in space the wavelength l of electromagnetic radiation is linked to the frequency (v) in the relation l x v = c, where c is the speed of light (3.108 m/s).

Z

Zenital astronomy: study of the stars based on a system of height-azimuth co-ordinates, typical of the Ancients, who lived in areas close to the equator.

POWERS OF TEN

10^3 = 1,000
10^4 = 10,000
10^5 = 100,000
10^6 = 1,000,000
10^7 = 10,000,000
10^8 = 100,000,000
10^9 = 1,000,000,000

10^{-2} = 0.01
10^{-3} = 0.001
10^{-4} = 0.0001
10^{-5} = 0.00001
10^{-6} = 0.000001

ANALYTICAL INDEX

The pages in which the topic is dealt with in greatest detail are in bold, while proper names of celestial objects,; instruments and titles of books are in italics